The Food Web of a Tropical Rain Forest

MW00466348

QH
109
.P6
F66
1996

The Food Web of a Tropical Rain Forest

Edited by

Douglas P. Reagan and Robert B. Waide

WITHDRAWN

The University of Chicago Press / *Chicago and London*

GOSHEN COLLEGE LIBRARY
GOSHEN, INDIANA

DOUGLAS P. REAGAN is national manager of natural resource assessment practice at Woodward-Clyde Consultants in Denver.

ROBERT B. WAIDE is director of the Terrestrial Ecology Division at the University of Puerto Rico.

The University of Chicago Press, Chicago 60637
The University of Chicago Press, Ltd., London
© 1996 by The University of Chicago
All rights reserved. Published 1996
Printed in the United States of America

05 04 03 02 01 00 99 98 97 96 5 4 3 2 1

ISBN (cloth): 0-226-70599-4
ISBN (paper): 0-226-70600-1

Copyright is not claimed for chapter 3, "Microorganisms," by D. Jean Lodge.

Library of Congress Cataloging-in-Publication Data

The food web of a tropical rain forest / edited by Douglas P. Reagan and Robert B. Waide.
p. cm.
Includes bibliographical references (p.) and index.
ISBN 0-226-70599-4—ISBN 0-226-70600-1 (pbk.)
1. Food chains (Ecology)—Puerto Rico—Luquillo Experimental Forest. 2. Rain forest ecology—Puerto Rico—Luquillo Experimental Forest. I. Reagan, Douglas P. II. Waide, Robert Bruce.
QH109.P6F66 1996
574.5′2642′097295—dc20 95-50299
CIP

∞ The paper used in this publication meets the minimum requirements of the American National Standard for Information Sciences—Permanence of Paper for Printed Library Materials, ANSI Z39.48-1984.

Contents

Preface

COMPLEXITY in ecological communities varies from the most simple ecosystems persisting under severe conditions that preclude high species richness to the diverse assemblages that characterize tropical forests. Ecologists have used food webs as a device to depict the relationships among species in a community. The development of a food web for a community organizes many individual observations of natural history into a consistent framework and provides a common point of view for comparing systems of varying complexity. One of the underlying hypotheses of such an exercise is that the principles of organization inherent in these communities remain constant as the scale of complexity changes.

Investigations concerning the properties of food webs have concentrated on the simplest natural systems, which lend themselves more readily to definition. Theoretical studies often focus on even simpler artificial systems, the better to develop mathematical rigor. Aquatic ecosystems offer many advantages over terrestrial systems (e.g., the boundaries of the community are easily defined), and most of the published studies concerning food webs are from streams, rivers, lakes, or oceans. Terrestrial systems, especially complex, species-rich tropical forests, are daunting to those studying food webs, with good reason. Determination of interactions among plants and animals in a community where the number of vertebrate species is in the hundreds, and the number of invertebrate species is orders of magnitude greater, is an intimidating task. For this reason, only one highly aggregated tropical forest web is described in the literature.

Species-rich communities, however, may hold the key to understanding the organization of food webs. Emphasis on simple or highly aggregated food webs already has been shown to lead to conclusions that are not applicable to more complex webs (Polis 1991a; Martinez 1991). Moreover, because complex tropical communities harbor most of the world's animal diversity, questions relating to the maintenance of biodiversity, the preservation of rare species, and the importance of redundancy in ecosystems may best be addressed by studying these communities.

Our goal is to summarize the natural history and trophic dynamics of a relatively simple tropical rain forest community. The community consists of the plants and animals inhabiting a 40 ha area of forest around the El Verde

Field Station in the Luquillo Experimental Forest of Puerto Rico. Our understanding is based on three decades (1963 to 1993) of investigations conducted or coordinated by the biologists in the Terrestrial Ecology Division of the University of Puerto Rico (formerly the Center for Energy and Environment Research) and by many visiting scientists who have worked at El Verde. We construct a comprehensive food web documenting the relationships among species in this community as a means of organizing the information we have collected. Lay-people, students, academics, resource managers, professional scientists, and others interested in the natural history of tropical forests should find points of interest in this book. In addition, ecologists specializing in the study of trophic dynamics are provided with a detailed food web from a biome underrepresented in the available data base and with our interpretations of the importance of this web.

The general background and environmental setting of the study is described in the Introduction. Chapters 2 through 13 address different components of the community. Each chapter begins with a diagram illustrating the location of the subject organisms in the overall food web and then describes the diversity and origin of these organisms. Subsequent sections of each chapter describe abundance and biomass (including temporal and spatial variation), diet and principal predators, and consumption rates. Each chapter maintains consistency in terminology and style. In expressing densities we have used the units and spatial scales appropriate for the organisms under consideration; for ease of comparison, we present these data in the same units in appendix 14.B. A glossary and list of references cited are also provided.

Chapter 2 describes the vegetation of the forest, emphasizing the trophically important components such as wood, flowers, and fruit. Chapter 3 discusses the microorganisms that are important food for some animals, but are also important in transferring energy and cycling nutrients through the ecosystem.

Chapters 4 through 12 address the other components of the terrestrial community, classified along taxonomic lines but subdivided according to their importance or categorized into appropriate ecological subdivisions. Thus, anoline lizards are in a chapter separate from other reptiles, and invertebrates are subdivided into arboreal invertebrates, litter arthropods, etc. While taxonomic detail is provided for the sake of completeness, it is not essential to understanding the role of each group in the food web. Appendixes to the chapters contain lists of plants and invertebrates as well as detailed methods.

Chapter 13 describes the food web of aquatic organisms in forest streams. The types and strength of trophic interactions between the aquatic and terrestrial food webs of the forest are discussed.

Chapter 14 synthesizes information from chapters 2 through 13 and presents conclusions regarding food web organization based on analyses of the combined data. The complete food web is presented as a matrix suitable for subsequent mathematical development (See appendix 14.A).

As a compilation of observations by a generation of biologists, this book represents not only the natural history of tabonuco forest at El Verde, but also a contribution to further studies of food webs in the tropics and elsewhere. Some of the observations chronicled in the following chapters reinforce ideas drawn from a wide range of studies; others are unique and contribute to our emerging view of the dynamics of species-rich communities. The challenge of describing, analyzing, and understanding these tropical communities provides a worthy goal for the future.

Food web studies at El Verde began as part of a comprehensive investigation of cycling and transport processes in a rain forest and were a continuation of research sponsored by the Office of Health and Environment Research of the U.S. Department of Energy beginning in 1963. More recently (1988 to the present), the National Science Foundation has supported work at El Verde through its Long-Term Ecological Research (LTER) program. The University of Puerto Rico has provided both in-kind and direct support for work at El Verde. A special acknowledgment is due the Oak Ridge Associated Universities, which have provided travel support for visiting scientists and students.

The number of people who have contributed to this book is very large, perhaps exceeding the number of links in the food web. We thank all of those reviewers who have helped us improve earlier drafts of the manuscript. We give special thanks to Stuart Pimm for his insightful comments, which greatly improved the final form of the book. Gerardo Camilo helped edit the final version and was especially helpful in designing diagrams and preparing the final photographic material. Eva Cortes, Nilda Sosa, Maria Villamil, and Ana Correa typed parts of the manuscript. Albert Muñiz and Pedro Sotelo drafted figures, and Rosser Garrison produced line drawings of some of the species in the web. We also thank all of our field assistants, students, and colleagues, who have provided us much of the information on natural history in this book. Chief among the latter, we are pleased to acknowledge Don Alejo Estrada Pinto, who managed and was the principal naturalist at the El Verde Field Station from 1970 to 1994.

1

The Rain Forest Setting

Robert B. Waide and Douglas P. Reagan

Panoramic view of tabonuco (*Dacryodes excelsa*) forest at the El Verde Field Station (elevation 350 m) in the Luquillo Experimental Forest, Puerto Rico. The study area is in the subtropical wet forest life zone in the Holdridge system of classification (Ewel and Whitmore 1973). (Photograph by D. Reagan)

A Chain of Jungle Life

This is the story of Opalina
Who lived in the Tad,
Who became the Frog,
Who was eaten by Fish,
Who nourished the Snake,
Who was caught by the Owl,
But fed the Vulture,
Who was shot by Me,
Who wrote this Tale,
Which the Editor took,
And published it Here,
To be read by You,
The last in The Chain,
Of Life in the tropical Jungle.

I OFFER a living chain of ten links—the first a tiny delicate being, one hundred to the inch, deep in the jungle, with the strangest home in the world—my last, you the present reader of these lines. Between, there befell certain things, of which I attempt falteringly to write. To know and think them is very worth while, to have discovered them is sheer joy, but to write them is impertinence, so exciting and unreal are they in reality, and so tame and humdrum are any combinations of our twenty-six letters.—William Beebe, *Jungle Days*

DESTRUCTION of tropical rain forests and the consequent loss of biodiversity they represent are among the most urgent ecological concerns facing humankind as we move into the twenty-first century. Our knowledge of the taxa comprising the animal communities of these ecosystems is incomplete, and we know little of the principles governing their organization. Such knowledge is needed in order to create appropriate preserves or to manage tropical rain forests on a sustainable basis. Because a food web is a map of feeding interactions within a community, it provides a framework for understanding how that community is organized. Moreover, the number of species, the connections between species, the lengths of food chains, and other web parameters can be quantified, which permits comparisons with other communities, even quite dissimilar kinds of communities. Constructing a food web requires detailed knowledge of the natural history of the organisms making up the community as well as the implementation of a variety of approaches to determine who eats whom.

This book uses the construct of a food web to integrate information on the biology of the organisms of a tropical forest community. Our interest in the natural history of this community is foremost; detailed descriptions of entire tropical communities are rare. Most treatments focus on single or multiple groups of organisms with little effort to establish relationships among those groups. However, the presentation of natural history information in the format of a food web allows us to examine a level of biological organization beyond the population or species. Key ecological questions (e.g., the relationship between productivity, biodiversity, and predation; the relative importance of top-down and bottom-up control of community structure; and the functional importance of redundancy in ecosystems) require coordinated studies of whole communities.

This book does not attempt to address all of the issues pertinent to understanding the structure of food webs. Much of our collective insight regarding trophic dynamics comes from comparative studies of many webs and mathematical models of hypothetical communities. These two approaches are quite different than the one we take and better suited to examine many of the unresolved questions about food webs. Our contribution comes from the first detailed description of the web of a tropical forest, an ecosystem that is underrepresented in the cumulative data base on food webs, and the evaluation of how the characteristics of our web correspond to those of other webs.

A TROPICAL FOREST COMMUNITY

The forest community near the El Verde Field Station in northeastern Puerto Rico has characteristics common to many insular communities in the tropics. Although it does not have the tremendous species richness that characterizes mainland tropical forests, it does have the abiotic characteristics of tropical forests (long growing season, relatively little seasonality, high rainfall and temperature) as well as many of the major biotic components of those forests (fig. 1.1). The food web derived from the animal community at El Verde is in many ways intermediate between mainland temperate and tropical webs, which provides a means for comparing webs of differing species richness. Hence, our studies at El Verde are a first step toward inclusion of species-rich tropical food webs in the data base that we use to develop ecological theory.

THE BIOGEOGRAPHICAL CONTEXT OF THE FOOD WEB AT EL VERDE

The assemblage of species that occurs in rain forest in Puerto Rico is largely determined by two biogeographical patterns or gradients in biodiversity. The well-known increase in species richness with decreasing latitude (Pianka 1966) results in Puerto Rico having a much greater number of species of some taxa than do temperate forests. For example, the number of plant species in the Luquillo Experimental Forest is much higher than comparably sized temperate areas (Lawrence, this volume).

Other groups, such as birds and ants, have about the same number of species in Puerto Rico as in temperate forests, even though these same taxa show increases in species number in low latitude mainland tropics. For these groups, the trend toward reduced species richness on smaller, more isolated islands (MacArthur and Wilson 1967) counteracts the latitudinal increase. The isolation gradient acts both through the absence of entire families on oceanic islands (e.g., among birds, manakins, antbirds, furnariids, trogons; Waide, this volume) and the reduction of the number of species per family with increasing isolation. Thus, the most remote islands often have only one or two representatives of each family of birds present, and proportional reductions occur for other animal groups (fig. 1.2).

Finally, some groups have many fewer species (e.g., Coleoptera and Lepidoptera; Garrison and Willig, this volume), are nearly absent (terrestrial mammals; Willig and Gannon, this volume), or do not occur (e.g., Plecoptera; Covich and McDowell, this volume) on tropical islands compared to both temperate *and* tropical mainland habitats. In this case, the reduction because of isolation overshadows the latitudinal influence. The variable response of different groups to these two gradients determines the particular subset of mainland taxa that are characteristic of the food webs of islands. Smaller, more isolated islands (e.g., St. Martin) have even more restricted subsets of the communities of large islands (e.g., Puerto Rico) and hence have simpler food webs. On these smaller islands, the lack of refugia from catastrophic disturbances such as hurricanes may also play an important role in determining species richness.

Figure 1.1. Understory of tabonuco forest at El Verde. The sierra palm (*Prestoea montana*) and the tree fern (*Cyathea arborea*) are common elements of the understory.

Figure 1.2. The Puerto Rican Tody (*Todus mexicanus*). Todies are the only resident insectivorous birds inhabiting the understory of tabonuco forest. They are endemic to the Greater Antilles, with one species each on Puerto Rico, Cuba, and Jamaica, and two on Hispaniola.

THE ECOLOGICAL CONTEXT OF THE FOOD WEB AT EL VERDE

El Verde Field Station and its associated research areas are located at elevations between 300 and 500 m on the northwest slope of the Luquillo Mountains in northeast Puerto Rico (fig. 1.3). The field station is situated within the Luquillo Experimental Forest (LEF), whose boundaries are the same as the Caribbean National Forest. The environment at El Verde is described in detail by Odum and Pigeon (1970). The best source of information on other areas of the LEF is Brown et al. (1983).

The general topography of the area is rough, with deeply dissected drainages and steep northeast- and southwest-facing slopes. The beds of the larger streams consist of exposed boulders, as does much (up to 80% in some places) of the forest floor (fig. 1.4). The forest is generally cool and extremely humid, with a closed canopy at about 20 m. Scattered emergent trees and understory shrubs complete the three recognizable layers of vegetation. A few tree species have well-developed buttresses. Bromeliads and lianas are conspicuous elements of the vegetation.

Climate

Mean annual rainfall at El Verde Field Station is 346 cm (McDowell and Estrada Pinto 1988) and is distributed evenly over the year except for a somewhat drier period from January to April (fig. 1.5). Precipitation is greater than evapotranspiration during all months. The predictable seasonal differences in rainfall at El Verde are disrupted by passing weather systems that dictate short-term cloudiness and rainfall frequency throughout the Caribbean (Odum et al. 1970b). Under normal conditions, winds from the north of east produce rainy weather, whereas winds from the south of east are associated with warmer, drier weather. Weather systems that produce thunderstorms over mountain peaks are most common in the summer but possible at any time of the year.

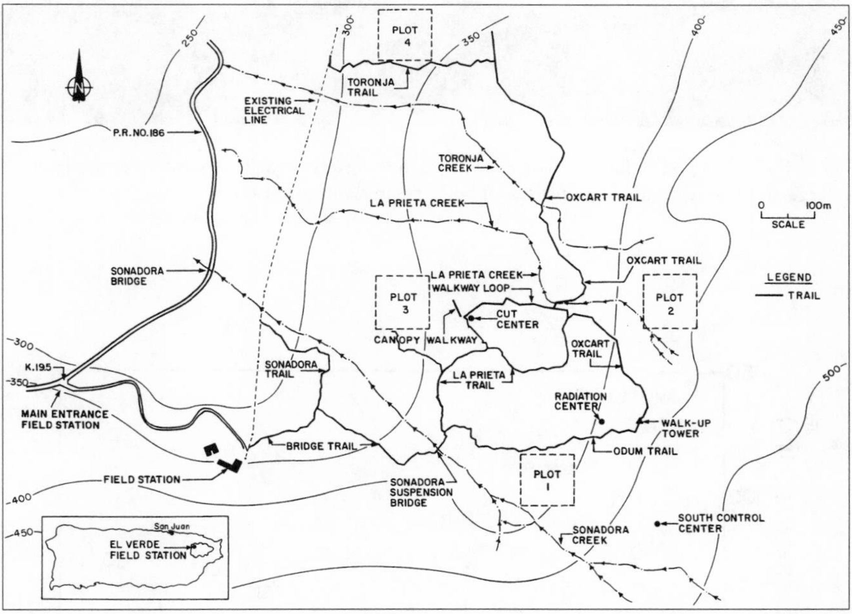

Figure 1.3. The field station at El Verde showing the location of research areas. Contour lines from 250 to 500 m are also shown. Research at El Verde during the mid-1960s focused on the effects of irradiation on forest structure and function (Odum and Pigeon 1970). This initial study concentrated on three principal areas at El Verde: an experimental site exposed to radiation (Radiation Center), a control site of similar size (South Control Center), and a small area where trees were girdled and understory cut to mimic the effect of radiation (Cut Center). The location of these experimental sites as well as areas of more recent research are included in the figure for the purpose of orientation.

Figure 1.4. The Quebrada Sonadora runs through the study area. The presence of moderate- to large-sized boulders is characteristic of both streams and forest floor.

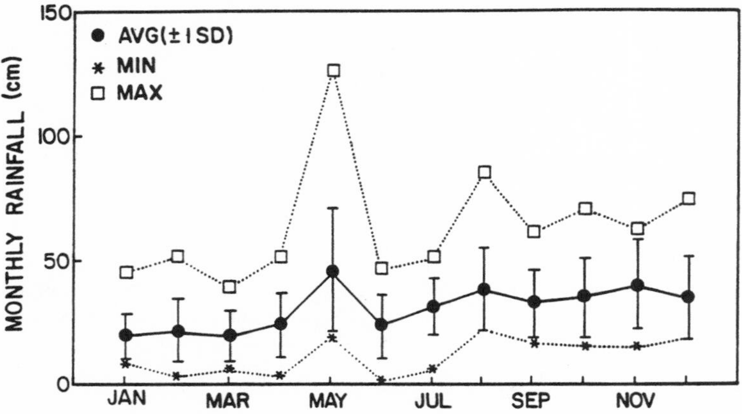

Figure 1.5. Average monthly precipitation (cm). Data from 1964–1967 and 1974–1986 at El Verde Field Station.

Using insolation measured above the forest canopy, Odum et al. (1970b) developed a five-stage classification of cloud regime at El Verde. The January to March period has 15% clear days, whereas during April to December only 5% of the days are clear. About 5% of days are continuously cloudy, regardless of season.

Relative humidity remains above 95% for most of the night, both at ground level and above the canopy. During the late morning and early afternoon, relative humidity can decrease to 89% at ground level and to 81% above the canopy, especially during drier months (Odum et al. 1970b). Absolute humidity changes little during the day and is almost always higher at ground level than above the canopy. Mean air temperature above the canopy is 22.6°C from April to December and 2°C less from January through March (Odum et al. 1970b). Diurnal temperature ranges are small, averaging 4°C above the canopy and 3°C at ground level. Prevailing direction of wind at El Verde is from the southeast, across and occasionally downslope from the main mountain mass (Odum et al. 1970b). Wind velocity averages 4.10 kph above the canopy and 1.19 kph within the canopy.

Geology and Soils

The area near El Verde is underlain by thick bedded volcanic sandstones and calcareous mudstones. The volcanic rocks have an andesite to basaltic-andesite composition and originated from a volcanic complex standing at or near sea level (Seiders 1971). Alternation of volcanico-clastic deposition and marine sedimentation produced a net accumulation of over 7500 m of sediment (Seiders 1971) containing fossils of Upper Cretaceous (Cenomanian) age (Odum 1970a).

The soils at El Verde and throughout the rest of the LEF are highly variable. The most common soil types are upland Ultisols belonging to the Los Guineos series, which are principally well-drained clays and silty clay loams (Edmisten 1970a). These soils are deep, highly weathered, leached, and acidic (Beinroth 1971; Fox 1982).

Vegetation

Five life zones occur in the LEF (subtropical moist forest, subtropical wet forest, subtropical rain forest, lower montane wet forest, and lower montane rain forest; Ewel and Whitmore 1973). The LEF also comprises four major vegetation types. Life zones and forest types are not congruent. Below 600 m, the tabonuco forest community takes its name from the dominant tree, *Dacryodes excelsa,* which is best developed on protected, well-drained ridges. At El Verde, the tabonuco forest occurs in the subtropical wet forest life zone.

Above the average cloud condensation level (600 m), palo colorado (*Cyrilla racemiflora*) replaces tabonuco, except in areas of steep slope and poorly drained soils, where the palm *Prestoea montana* occurs in nearly pure stands. The dwarf forest occupies ridge lines and comprises dense stands of short, small-diameter trees and shrubs that are almost continually exposed to wind-driven clouds. Palm and dwarf forests have relatively few plant species compared to tabonuco and colorado forests.

El Verde lies within the subtropical wet forest life zone and the tabonuco forest type. The tabonuco forest, covering 42% of the LEF's 11,231 ha (Brown et al. 1983), has been the subject of the most extensive ecological studies of any of the four forest types. The most detailed of these was the Rain Forest Project of the U.S. Atomic Energy Commission, which took place from 1963 to 1968 (Odum and Pigeon 1970). Long-term studies of forest growth have recorded 168 tree species in the tabonuco zone (Wadsworth 1951). In addition, the tabonuco forest has been the subject of studies of tree growth (Crow and Weaver 1977) and community composition (Crow and Grigal 1979); the extensive data base has served in the development of simulation models of forest regeneration (Doyle 1980, 1981). A survey of vegetation at El Verde was conducted in four 1 ha permanent plots established in 1980 (Zou et al. 1995; fig. 1.3). In this survey, the eight most important tree species in each plot constituted 72 to 83% of the basal area and 64 to 80% of the density. The most common species (*P. montana, D. excelsa,* and *Sloanea berteriana*) were similar in three of the plots and similar to earlier surveys. Subdominants varied, probably as a result of local differences in slope, exposure, drainage, or history of disturbance. The fourth plot differed from the others by being dominated by the early successional species *Casearia arborea*. A detailed description of forest structure at El Verde is given in Odum and Pigeon (1970) and Lawrence (this volume).

Disturbance

Hurricanes

The repeated occurrence of tropical storms forms part of the backdrop against which food web structure must be evaluated. The LEF is subjected to the direct impact of a hurricane on average once every sixty years (Scatena and Larsen 1991). Less severe storms affect the forest at more frequent intervals. The cumulative effect of this disturbance regime maintains the forest in a perpetual state of secondary succession, with mature and younger stands of trees mixed together in a mosaic whose pattern is determined by the path and intensity of recent hurricanes and the topography of the area (Waide and Lugo 1992). The effect of this kind of drastic disturbance on the distribution of animal populations is known only in general terms. Hurricanes create environmental heterogeneity and redistribute animal populations along new gradients of forest structure and abiotic conditions. Reshuffling of consumer populations takes place in both horizontal and vertical dimensions. Species groups such as anoline lizards and birds that are stratified within the vertical structure of the forest are profoundly affected when the forest is subjected to high winds that remove leaves and branches and create piles of debris at the ground surface. In addition, changed conditions may cause loss or gain of species. Hurricane Hugo in 1989 provided the opportunity to study the effects of hurricanes on community structure at El Verde (Walker et al. 1991; fig. 1.6). While studies are continuing at El Verde, initial results show that population density, diet, and microhabitat use by animals are drastically altered by hurricanes (Walker et al. 1991).

Landslides and Treefall Gaps

The most frequent disturbances in tabonuco forest are landslides and treefall gaps. Both kinds of disturbance generally affect relatively small areas, although the area of landslides ranges from 20 to 4800 m^2 (Scatena and Larsen 1991). The conditions produced by these two disturbance types are quite similar at canopy level, but quite different at ground level. Treefalls add structure and biomass to the understory, providing both shelter and food for decomposer species and their predators. In contrast, landslides remove structure from the understory, leaving behind bare soil and patches of surviving vegetation. Hence, landslides usually have a negative effect on understory animals whereas treefalls may have a positive effect, at least on some taxa such as leptodactylid frogs (L. Woolbright pers. comm., 1991).

THE EL VERDE FOOD WEB

A simplified version of the food web (fig. 1.7) provides a framework for visualizing the relationships among the major plant and animal groups discussed in this book. The complexity of the full web (app. 14.A) prevents us from attempting a graphic representation; even a relatively simple Caribbean web is too involved to comprehend (Goldwasser and Roughgarden 1993a). In each of the following chapters, we will focus on one of the components of this simplified web. The subjects of some of the chapters (e.g., termites) have very specific roles in the web; other groups (e.g., birds) are important at different places in the web and therefore are shown more than once. Some of the groups are shown straddling lines between trophic levels, reflecting omnivorous (*sensu* Pimm 1982) diets. All groups eventually feed back into decomposer organisms.

The most obvious property of the animal community at El Verde is the absence of large mammalian herbivores and predators. Not as obvious from the diagram but equally important is the relatively low faunal richness of the

Figure 1.6. The effect of Hurricane Hugo at El Verde. Canopy trees on north- and northeast-facing slopes were stripped of leaves, but leeward slopes were spared to some degree. The hurricane resulted in the formation of a mosaic of open and closed-canopy sites within the study area.

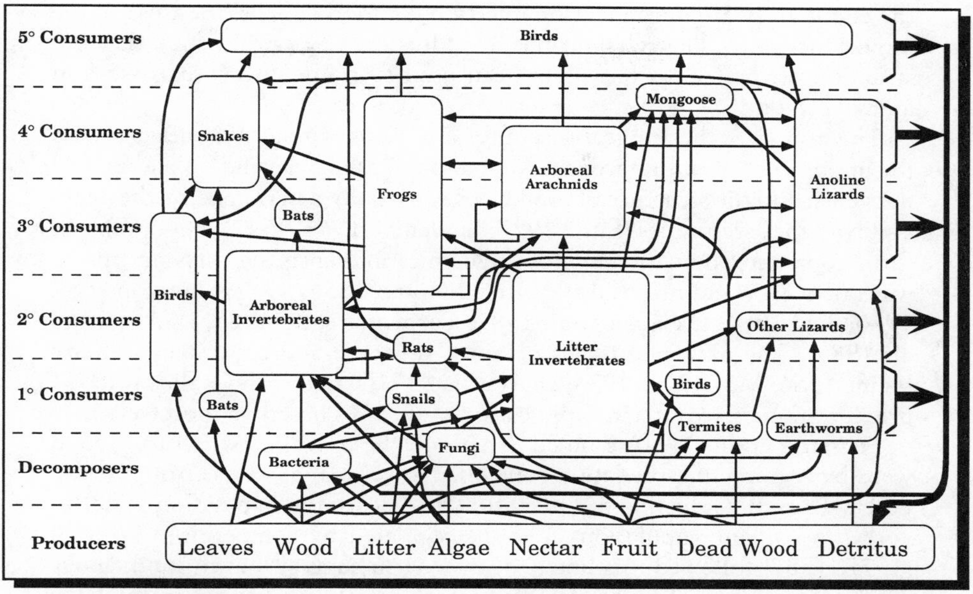

Figure 1.7. An aggregated food web of the El Verde community presenting the groups discussed in this book.

community when compared to tropical continental sites. The dominant predators in the web are frogs and lizards, whose population densities at El Verde are among the highest ever recorded. The compartmentalized nature of the community is indicated by the separation of decomposer and consumer pathways at low trophic levels. All of these properties affect the structure of the food web and contribute to some of the differences we see between our web and webs from other areas.

THE STUDY OF FEEDING RELATIONSHIPS IN ANIMAL COMMUNITIES

Many of the important topics of contemporary ecology are inextricably related to the relationship between prey and predator. Understanding of population dynamics, energy flow, nutrient cycling, primary productivity, ecosystem stability, and biodiversity all depend to a great degree on our knowledge of feeding relationships. In his 1927 book on animal ecology, Charles Elton stated: "Feeding is such a universal and commonplace business that we are inclined to forget its importance. The primary driving force of all animals is the necessity of finding the right kind of food and enough of it. Food is the burning question in animal society, and the whole structure and activities of the community are dependent on questions of food supply." Because of the importance of feeding relationships in addressing ecological issues, the construction of food webs has often been used to frame conceptual issues in animal ecology.

The history of the applications of food webs to ecological issues confirms the importance of this approach. Animal populations depend on energy and nutrients, and the structure of food webs is partially determined by the availability of these resources (Elton 1927; Lindeman 1941, 1942). Food webs, in turn, comprise populations whose dynamics influence the rates and pathways of energy and materials flow (e.g., Odum 1956, 1957; Hairston et al. 1960; Teal 1962; Oksanen et al. 1981; Carpenter et al. 1985; DeAngelis et al. 1989). The entwined issues of biotic diversity and ecosystem stability (Paine 1966; May 1972, 1973; DeAngelis 1975; Lawton and Rallison 1979) provide another area for the application of theories based on food webs.

Hairston et al. (1960) kindled debate on the processes structuring food webs by arguing that predation controlled lower trophic levels from the top down. Thus, if competing top predators held down the population of herbivores, then plant populations must be regulated by competition. Fretwell (1977, 1987) followed this line of reasoning to suggest that in food chains with odd numbers of trophic levels grazers are always limited from above and the system should accumulate plant biomass. In systems with even numbers of trophic levels, herbivores should prevent the accumulation of green plants. Menge and Sutherland (1976) proposed an alternative model, that predation and competition were complementary, predation dominating at lower trophic levels and competition dominating at higher trophic levels. Field experiments relevant to these points of view were reviewed by Connell (1983) and Schoener (1983, 1985). Trophic cascade models (Carpenter et al. 1985) incorporate the idea of top-down control of plant standing crops.

The alternative notion, that consumers are food-limited regardless of their trophic level (bottom-up), has intuitive appeal. As Hunter and Price (1992) point out, the removal of plants from a system destroys the system while the removal of consumers only modifies it. Predictions from Fretwell's model (1977, 1987) indicate that primary productivity determines the number of trophic levels, which in turn determine plant standing biomass (Power 1992).

A more recent synthesis suggests that both top-down and bottom-up forces are important in structuring communities (Hunter and Price 1992), a view which finds accumulating support (Power 1984; Mittelbach 1988; Liebold 1989; McQueen et al. 1989; Arditi and Ginsberg 1989). Hunter and Price (1992) further suggest that the inclusion of environmental heterogeneity and the behavior of individual species is necessary in models of trophic dynamics.

By the early 1980s, the first major theory concerning the structure of food webs emerged (Pimm 1982). This theory, known as the species dynamic interactions model, predicts many of the observed patterns of food web structure and some of the demographic attributes of species in food webs. A major drawback of the theory concerned the many underlying assumptions and *a posteriori* knowledge required to make specific predictions.

Pimm (1979, 1980, 1982, 1991), Critchlow and Stearns (1982), May (1983b), Cohen et al. (1990), and others have deduced some of the general properties of food webs through comparative studies. These properties include a small number of trophic levels (usually three or four; Pimm and Kitching 1987), an almost constant ratio (1:1) of predator to prey species (Cohen 1989), and a constant ratio (1:2) between the number of species and the number of observed links (Warren and Lawton 1987). The ratio of observed to potential number of links is termed connectance, and represents a measure of system complexity (Pimm 1982). Food web analyses suggest that omnivory (i.e., feeding at two or more different trophic levels; Pimm and Lawton 1977) and loops (i.e., either "A eats B eats A," or "A eats B eats C eats A") tend to destabilize system dynamics and consequently should be rare in nature (Pimm 1982).

Acceptance of these properties as general features of most food webs has been slow because of the difficulty of defining interactions among components of real food webs. For example, it is customary to consolidate poorly known taxa into a single trophic box (e.g., phytoplankton), while at the same time treating better-known taxa with different life stages (e.g., caterpillars and moths, tadpoles and frogs) as separate trophic species (Cohen 1989). These conventions distort the actual relationships, oversimplifying some and exaggerating others. The lack of consistency in expressing even the most simple systems in quantitative terms has been a notable impediment to progress in understanding food web organization (Closs 1991; Cohen et al. 1993).

The trade-off between taxonomic consistency and resolution is difficult to resolve. We present two webs, one with maximum resolution (app. 14.A) and the other with less resolution but more consistency in aggregating related taxa. Certain attributes of the species involved have complicated the development of these webs. Many animal species at El Verde could not be classified into discrete or easily identified trophic levels. For example, some large invertebrates routinely preyed upon vertebrates (fig. 1.8). Species interactions were far more complex than had been envisioned previously, and some patterns (e.g., looping) were more common than we expected. The pattern and rates of consumption of some consumers (e.g., sucking insects that feed in the forest canopy) have not been quantified.

Despite these problems, we have analyzed the structural characteristics of our two webs, and we present the conclusions from these analyses in detail in chapter 14. In the El Verde web, cross predation (food loops) is a widespread phenomenon, contrary to published predictions. The web is dominated by small ectotherms rather than large endotherms as in continental

Figure 1.8. A centipede of the genus *Scolopendra* feeding on a living *Bufo marinus*. This interaction reflects the importance of large invertebrates as predators at El Verde. (Photograph by D. Reagan)

sites, which may contribute to the extreme length of food chains at El Verde. Connectance within the web is an inverse function of taxonomic resolution. The proportion of top predators is much less than the "scale-invariant" values reported in the literature. The food web at El Verde becomes more compartmentalized (*sensu* Pimm 1982) as prey taxa are more finely resolved.

We begin to develop the arguments leading to these conclusions with a description of the food resources available at the bottom of the food chain. After describing how primary production is packaged into different forest components, we follow the movement of this energy through increasing trophic levels using our aggregated web as a guide. We thus build the web from the bottom up, adding groups of consumers sequentially until we have a complete picture of the food web. The emergent characteristics of the web are then defined and compared to other webs.

2

Plants: The Food Base

William T. Lawrence, Jr.

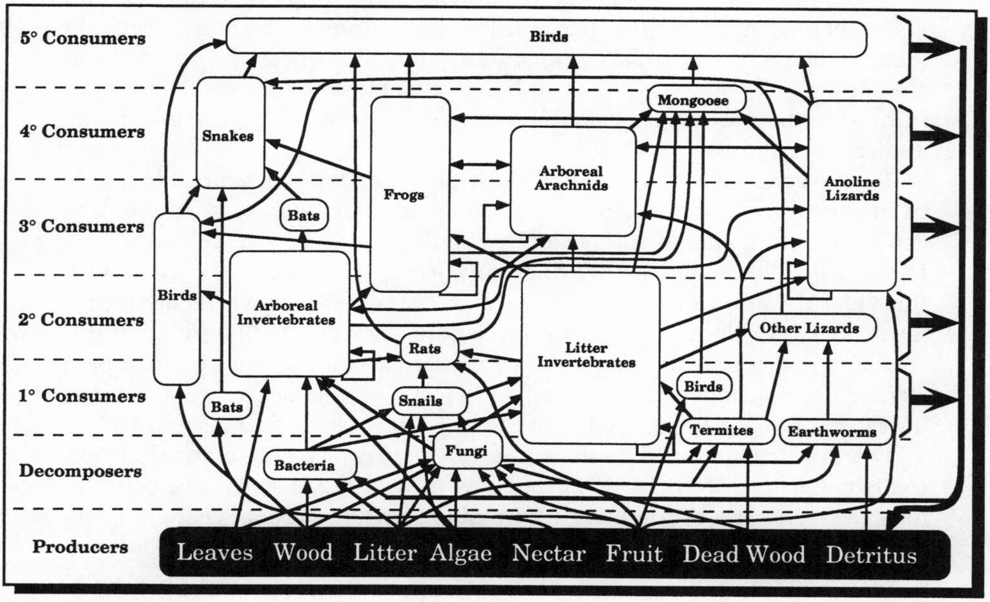

THE ENERGETIC BASIS OF THE FOOD WEB

PLANTS are the primary producers in the ecosystem, fixing solar energy and converting it into the carbon-based compounds and vegetative structures upon which all other organisms in the food chain eventually depend. Decomposers and primary consumers are directly dependent on plants, and even higher order consumers may use plant foliage, fruit, and flowers to some degree. Plants determine the availability of food to consumers by their energy allocation pattern. The distribution of this energy between leaves, flowers, fruit, nectar, roots, and wood influences the pattern and strength of links between plants and their consumers. Plants also affect consumers' feeding behavior by concentrating soil nutrients in accessible tissues and vascular fluids.

The existence of a complex food web thus depends on a predictable supply of consumable plant parts, and variation in the availability of such resources can have strong effects on consumer populations and food web structure. Large-scale disturbances have the capacity to reorganize the food web through their effect on producers. At El Verde, the dominant disturbance is hurricanes (Waide and Lugo 1992), whose periodic disruption of forest structure and productivity provokes extreme variation in the resource base available to consumers (Walker et al. 1991). Given the recurrence interval of hurricanes passing within 10 km of the LEF (sixty years; Scatena and Larsen 1991), the structure and composition of the vegetation at El Verde is extremely dynamic on the scale of the longest-lived organisms in the community, the trees. Succession after a hurricane provides gradual changes in resource availability, to which consumer populations must adapt in order to survive.

The most visible type of consumption in most tropical forests is herbivory on live leaves, although this is only a small part of the overall consumption of plant material (see food web diagram above). Other live plant parts such as flowers, fruit, and seeds are harvested by primary consumers, but the majority of all these resources reach the forest floor uneaten. By far the greatest amount of plant material is consumed by detritivores in the litter layer. This can be inferred since little or no litter accumulates at El Verde even though over 800 g m^{-2} of litter reach the forest floor every year (Zou et al. 1995). This consumption of dead plant material can take place within the canopy on trapped vegetative debris or unshed plant parts, in the litter layer on the forest floor, or on sloughed root material in the soil. This latter class of detritus exists in tremendous quantities as fine roots are constantly being turned over in the forest floor litter and upper soil horizons (Odum 1970c; Parrotta and Lodge 1991).

The activities of sap- and nectar-feeding organisms are another important but often overlooked component of the food web. Consumption of plant fluids takes place both above and below ground by insects and fungi feeding on the vascular systems of leaves, stems, and roots. The amount of consumption is very difficult to quantify or even estimate. Some carbon is even leaked from roots as an exudate, although it is not clear in most cases whether this is a natural process or caused by pathogens.

Once consumed, plant materials continue to pass through the food web until all that remains are intractable, low-nutrient content, long-lived carbon molecules (i.e., lignins and condensed tannins) that eventually end up in the pool of soil organic matter (SOM) with extremely long turnover rates (Schlesinger 1977; Parton et al. 1989; Sanford et al. 1991). Some plant substrates may only be suitable for ingestion after "pretreatment" by a consumer group. Lodge (this volume) cites an example of obligate precolonization of leaves by fungi and other microorganisms before consumption by herbivores.

The major goal of this chapter is to examine the ways in which plant diversity and disturbance act to determine the availability of resources to consumers over time. The categories of consumable plant parts that we use in our food web (see app. 14.A) reflect basic similarities in plant morphology among forested ecosystems. The differences that distinguish diverse communities and their food webs relate to the variation in the properties of plant parts relevant to consumers (e.g., size, shape, and secondary compounds). This variation is strongly affected by differences in diversity and disturbance patterns among sites. In areas of high plant diversity, a wider range of resources is available to consumers than in areas of low diversity. However, large-scale disturbance (e.g., hurricane, fire, drought) can subject the consumer community to short- to long-term periods in which resource abundance and availability is drastically restricted. These periods following disturbance may have disproportionate effects on the structure of the food web.

I will examine the spatial and temporal distribution of major plant components used by consumers and decomposers at El Verde and compare my results with other tropical sites. The relative quantities of each of these components will also be discussed so that the resource base of each level of the food web can be better understood. Some of the links between producers and consumers include leaves to arboreal invertebrate herbivores and epiphytic fungi (Pfeiffer, chap. 5, this volume; Lodge, this volume); flowers and fruit to birds, bats, and arboreal invertebrates (Garrison and Willig, this volume; Waide, this volume); live roots to litter invertebrates (Pfeiffer, chap. 5, this volume); litter to bacteria, fungi, and litter invertebrates; dead wood to termites (McMahan, this volume); and detritus to earthworms and litter invertebrates (see food web diagram above).

DIVERSITY OF THE FLORA

Tropical ecosystems are among the most diverse in the world (Wilson 1988), but within the tropics plant species richness varies widely among different life zones. The El Verde flora comprises 468 species of vascular plants made up of seventy-eight species of ferns and fern allies and 390 species of flowering plants (Taylor 1994). Seventy-eight dicotyledonous families contain 269 species, while 119 monocot species are found in sixteen families. The families with the most species include the Orchidaceae (twenty-eight), Rubiaceae (twenty-six), Poaceae (twenty-one), Asteraceae (seventeen), Cyperaceae (seventeen), Fabaceae (sixteen), Euphorbiaceae (fourteen), and Melastomaceae (fourteen). Although many species from El Verde are endemic to Puerto Rico or the West Indies, others have wide distributions in Central or South America (Taylor 1994). Flowering plants at El Verde are divided into woody (170 species), herbaceous (172 species), and vine or liana (forty-eight species) life forms. Further details, including a complete list of plant species, can be found in Taylor (1994) and appendix 2.

The flora of El Verde represents only a small part of the plant diversity found in Puerto Rico. A census of a 16 ha plot near the El Verde Field Station (Zimmerman et al. 1994) found eighty-eight species among trees greater than 10 cm in diameter. Of these, twenty-six species made up 91.4% of all stems. Tabonuco forest within the LEF has 153 species of tree, whereas the LEF as a whole has 225 tree species, forty-seven of which are introduced (Little and Woodbury 1976). Little et al. (1974) reported 750 tree species from Puerto Rico, of which 203 are naturalized.

The flora of tropical islands is often described as depauperate compared to tropical mainland sites, but comparisons among sites controlled for life zone, elevation, plot size, and differences in species/area relationships are rare. Attempts to control for these factors suggest that climate and landscape heterogeneity contribute as much or more to differences in species richness as insularity per se (Holdridge et al. 1971; Lugo and Brown 1981c; Gentry 1982; Rice and Westoby 1983; Lugo 1987). As water availability and temperature increase among tropical forests, so does tree species richness (Lugo and Brown 1981c; Gentry 1982). As Reagan et al. (this volume) discuss, communities with greater species richness potentially have more complex food webs with greater numbers of links between consumers and producers.

Larger-scale biogeographic issues are also important in assessing species richness in tropical forests. One-hectare plots in upper Amazonia and the Chocó (155 to 283 tree species ha^{-1}) have similar species richness to a site in Manaus (1979 tree species ha^{-1}) that is drier and on poorer soil (Gentry 1990). All three of these sites have more species than La Selva (eighty-eight to 118 tree species in plots ranging from 2 to 4 ha) or Barro Colorado Island (seventy-six to 116 tree species ha^{-1}) in Central America, from which Gentry (1990) concludes that Central American forests are less rich floristically than those of South America. One-hectare plots at El Verde have from forty-three to fifty-four tree species (Waide unpublished).

DISTURBANCE EFFECTS

Common physical disturbances at El Verde include branch fall, tree fall, and landslides (Larsen and Torres-Sánchez 1990). Branch and tree falls cause small gaps and little perceptible change in the forest other than canopy openings, whereas landslides can create large areas of bare mineral soil (4500 m^2 at one site near El Verde; Fetcher et al. unpublished). Previously, human-mediated disturbance was common in the forests of Puerto Rico and included clearing for subsistence agriculture and selective logging of valuable timber species. Such disturbance is now rare, and the area covered by forest in Puerto Rico is increasing.

Hurricanes are the principal agents of large-scale disturbance in the LEF and have significant impact. The effect of hurricanes on resource availability is particularly germane to this discussion since Hurricane Hugo made landfall in Puerto Rico on 18 September 1989, heavily damaging the forest at El Verde. Scatena and Larsen (1991) report the recurrence intervals for the physical characteristics of Hurricane Hugo: similar rainfall occurs on average every five years, stream discharge every ten to thirty-one years, and wind velocity every fifty years. Hugo was thus a relatively dry hurricane whose principal effect was the result of high winds. I examine below the impact of this hurricane on the resources supporting the food web at El Verde.

PLANT BIOMASS AND PRODUCTIVITY

In an ecosystem as dynamic as tabonuco forest, consumer organisms are exposed to large temporal and spatial fluctuations in resource availability. The importance of fluctuations in resources available to herbivores and detritivores in the food web cannot be assessed without accurate estimates of plant biomass and productivity. Live plant biomass and litter standing stocks have been estimated many times by both direct harvest and regression techniques at El Verde and other places in the LEF. Because some of these estimates are from mature forest plots and others from sites recently affected by Hurricane

Hugo, the degree of resource fluctuation can be evaluated for many of the consumable plant parts underpinning the food web.

Aboveground Biomass

Pre-hurricane

In 1963, Ovington and Olson (1970) estimated total biomass of leaves, branches, boles, and entire understory palms at three plots (Radiation Center, South Control Center, and North Cut Center; see Introduction) at El Verde (table 2.1). Their estimates were based on allometric regressions for total biomass of three major tree life forms plus palms based on harvest and analysis of over 100 individual trees. The total aboveground biomass reported by Ovington and Olson (1970), as well as the separate compartments, corresponds well with similar measurements made at another tabonuco forest site (Bisley) in the LEF in 1989 (table 2.1). Assuming a net woody biomass accumulation characteristic of tabonuco forest (250 g m^{-2} yr^{-1}; Weaver and

Table 2.1. Biomass distribution at El Verde

Compartment	1	2	3	4	5	6
Aboveground						
Leaves	922	551	1,024	939	700	60
Branches	4,508	2,737	3,924	3,677*	2,810	1,290
Boles	19,162	11,162	16,173	15,331*	18,647	9,698
Palm	87	322	59	181*	360	250
Subtotal	24,679	14,772	21,180	20,128	22,517	11,298
Litter	685	511	—	598*	—	—
Belowground						
Fine roots (<0.2 cm)	—	—	—	—	206	175
Small roots (<0.5 cm)	—	—	—	750	220	—
Large roots (>0.5 cm)	7,813	5,447	5,807	6,480*	7,240	—
SOM (0–2.5 cm)	—	—	—	5,960	—	—
SOM (2.5–15.2 cm)	—	—	—	7,540	—	—
SOM (15.2–30.5 cm)	—	—	—	1,130	—	—
Subtotal	—	—	—	21,860	—	—
Total	33,175	20,750	26,987	42,586	30,252	—

Sources: Data in (1) and (3) are from Ovington and Olson (1970), as predicted by regression equations based on a harvest of 100 trees. Litter data are from Wiegert (1970a), and fine root data are from Odum (1970c). Data in (5) are from Scatena et al. (1993).

Notes: Column headings are as follows: (1) Radiation Center, (2) South Control Center, (3) Cut Center, (4) means of multiple measures from points in and around sites 1, 2, and 3, (5) pre-hurricane data from Bisley, another site in tabonuco forest, and (6) post-hurricane data from Bisley.

Units are g dry weight m^{-2}. SOM, soil organic matter.

* Weighted average of (1) and (3).

Murphy 1990), total aboveground biomass at El Verde extrapolated to 1989 would range from 21,272 to 31,179 g m^{-2}. (All references to mass, unless otherwise specified, are on a dry weight basis.) Aboveground biomass at Bisley falls at the lower end of that range. All of these values are in the mid-range of estimates from other tropical montane forests reported by Scatena et al. (1993).

Total biomass varies significantly among the three El Verde sites, the Radiation Center having 17% greater mass than the Cut Center and 67% greater than the South Control Center. The relatively high biomass of palms in the South Control Center may point to the reason for the lesser total biomass. Palms are found in areas of high soil moisture where other species do not grow well. Because of their growth form, individual palms contribute less to woody biomass than other trees. Total aboveground biomass is three times as high on ridges in tabonuco forest than in riparian valleys where palms predominate (Scatena et al. 1993).

The distribution of biomass among leaves, branches, and roots within each of the three El Verde plots is fairly similar. Leaves vary from 3.7 to 4.8% of total aboveground biomass, palms from 0.3 to 2.2%, branches from 18.3 to 18.5%, and boles from 75.6 to 77.6%. A large proportion of the total biomass at El Verde is found in wood (over 15,000 g m^{-2}), placing this site at the upper end of the range of forest volumes (6,100 to 17,600 g m^{-2}) found in a wide survey of tropical forests (Brown and Lugo 1984).

Total leaf biomass and leaf area index are important ecosystem parameters both because they reflect the forest's capacity to capture energy for production and because they represent one of the major resources available to herbivores. Other leaf parameters (e.g., specific weight, nutrient and defensive compound content) are also important since they relate to the palatability of foliage to chewing, scraping, and cutting insects. Live leaf biomass estimates for El Verde (table 2.1) are highly variable but broadly overlap the range of values for other lowland and montane evergreen tropical forests (Medina and Klinge 1983; Scatena et al. 1993).

Leaf area index (LAI) values at El Verde are in the mid-range for tropical forests as compiled by Jordan (1985). A number of sources (summarized in Odum 1970c) present results from El Verde obtained with methods ranging from direct measurement of harvested leaves to plumbline intercept studies and estimates based on the spectral quality of light under the canopy. Maximum values of LAI were 6.2 to 6.6 from spectral and direct harvest, respectively. Slightly lower LAI values were recorded in plumbline studies (5.2 to 5.6).

Leaf distribution through the canopy can have an important effect on consumer populations, both in its spatial and temporal aspects (Janzen 1983a). This distribution depends not only on tree species and growth form, but also on the surrounding vegetation, competition, recent disturbance (tree falls, limb breakage, lightning), and the available resources (light, nutrients, moisture, space). Odum et al. (1963) sampled leaf distribution at El Verde by harvesting and weighing all vegetation in 5 m × 5 m prisms from the forest floor to the top of the canopy (fig. 2.1). These data show a relatively uniform distribution of leaf biomass from the forest floor to canopy top without the stratification that has been observed in other tropical forests (Richards 1952; Walter 1971; Medina and Klinge 1983). More recently, Brokaw and Grear (1991) constructed vegetation height profiles at El Verde (fig. 2.2). Differences in the distributions obtained in these two studies result in part from the variable sampling interval used by Brokaw and Grear (1991). Although the forest structure at El Verde is fairly open at ground level, there is nearly 40 g m^{-2} of foliage available in the lower 2.5 m of the canopy which would be accessible to herbivores that shelter in litter when not feeding. Garrison

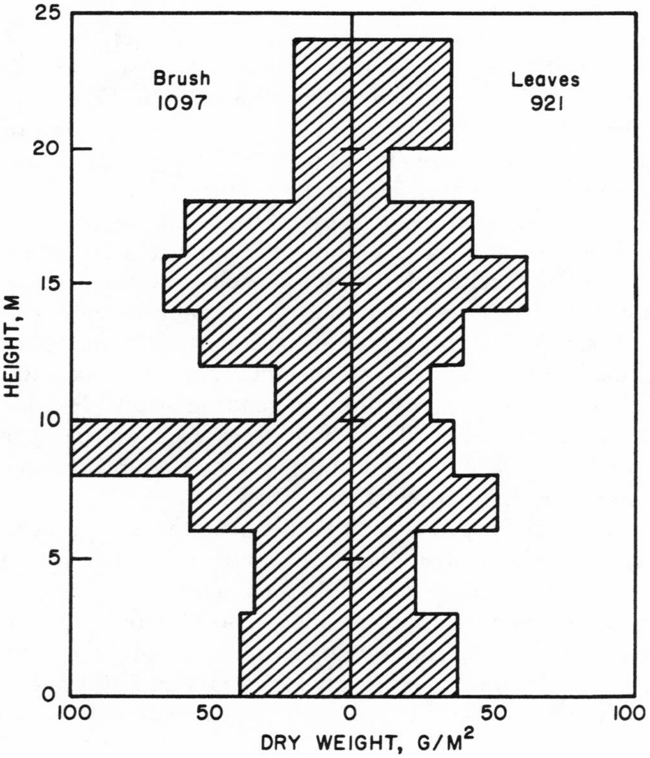

Figure 2.1. Vertical distribution of branches (brush) and leaves from 5 m × 5 m rectangular sample prisms through the vegetation canopy at El Verde. Average of ten prisms cut by Odum (1970c).

and Willig (this volume) found higher arboreal invertebrate populations in the lower 2 m of the canopy than in the upper 17 m.

Fruits produced in tabonuco forest are either consumed or fall to the forest floor. Although the amount of annual consumption has not been measured, fruit fall has been surveyed in several studies. Fruit fall varies considerably from stand to stand within tabonuco forest (Lugo and Frangi 1993). Zou et al. (1995) found 207, 24.2, 16.4, and 9.6 kg ha^{-1} of fruitfall in 1981 in four 1 ha plots at El Verde. The mean annual fruit fall for tabonuco forest was estimated at 332 kg ha^{-1} yr^{-1} based a number of studies, but large variation exists between years of high and low production (Lugo and Frangi 1993). Marked annual variation in fruit production might cause shifts in consumer diets and resulting changes in food web structure.

The fruiting pattern of sierra palm (*Prestoea montana*) is well known in the LEF, in part because it is an important element in the diet of the endangered Puerto Rican parrot (*Amazona vittata*). Sierra palm is the most abundant tree at El Verde and contributes about 10% of the total annual fruit fall

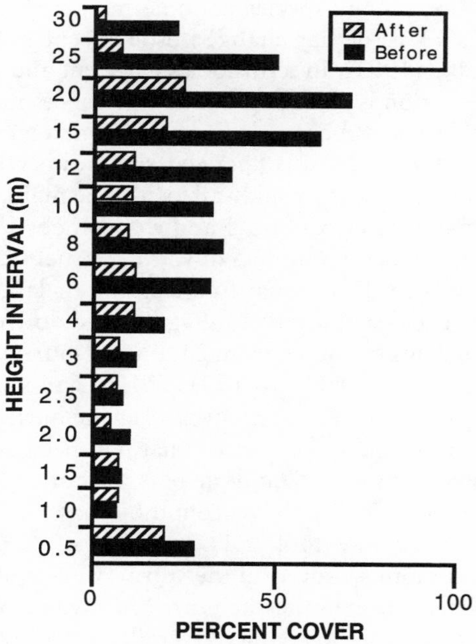

Figure 2.2. Vegetation height profiles in tabonuco forest plots at El Verde before and after Hurricane Hugo. The horizontal axis shows the number of points with cover in each height category as a proportion of the total number of points sampled. The vertical axis is graduated and shows the upper limit of each height interval.

(Lugo and Frangi 1993). Fruit fall in sierra palms declined by an order of magnitude between 1980 and 1981 in a floodplain forest studied by Frangi and Lugo (1985). The observed strong annual variation in sierra palm fruit has important implications for the diet of frugivores, and caused Lugo and Frangi (1993) to reduce the minimum estimate of forest carrying capacity for the parrot from 51,000 to 3,000 individuals.

Post-hurricane

Although first impressions of the El Verde site after Hurricane Hugo suggested that complete defoliation and massive mortality of canopy trees were widespread, subsequent measurements found less severe damage. Some sites seemed undisturbed, while others on more northerly exposures and ridges were extensively damaged (Walker 1991). Although more than 50% of all trees were defoliated (Walker 1991) and 25% suffered damage to the main stem, mortality estimates near El Verde ranged from 7.4% at 54 weeks post-hurricane (Walker 1991) to 9% (Zimmerman et al. 1994). Delayed mortality raised the estimate to 13.3% after 171 weeks (Walker 1995). Because of heavy sprouting among many species (Zimmerman et al. 1994), forest species composition at El Verde has changed little except in the most severely damaged sites (Walker 1991). In terms of areal extent and dominance, however, canopy composition is changing due to differences in branch regrowth and relative growth rates of successional species (Walker 1991).

The structure of the forest was changed significantly by the hurricane, with canopy architecture most strongly affected. The major effects of canopy alteration were a massive influx of litter and woody debris to the forest floor (Lodge et al. 1991), a concomitant loss of vertical structure (fig. 2.2), and an increase in light levels to those equal to a clearing or large gap (Fernandez and Fetcher 1991). The distribution of foliage shifted downward, with maximal mean canopy heights changing from 21.1 m pre-hurricane to 9.3 m after Hugo (fig. 2.2; Brokaw and Grear 1991). Most fruits and flowers were stripped from the trees, and several months after the hurricane the principal sources of fruits were successional species that had been established in gaps existing before the hurricane (J. Wunderle pers. comm.).

Estimates of the reduction in aboveground biomass because of the hurricane are available from Bisley (table 2.1). Live leaves suffered the most damage: only 9% of leaf biomass survived the storm. Canopy fruits and flowers were nearly completely absent after the hurricane (Waide pers. observation), but no quantitative data are available. Overall, aboveground biomass was reduced to 50% of its pre-hurricane value. Since Bisley was more directly in the path of hurricane winds than El Verde, damage was both greater and more uniform. Given the low rainfall and rapid passage of Hurricane Hugo, more extreme damage is possible and has probably occurred in the past from storms of higher intensity.

Table 2.2. Components of litter fall at El Verde

Component	1	2	3
Leaves	485.0	490.5	—
Fruit, seed	51.3	30.2	—
Flowers	—	14.6	—
Branches	—	156.1	412.5
Logs/Boles	—	—	197.1
Miscellaneous[a]	—	181.2	—
Total	536.3	872.6	—

Sources: (1) Wiegert (1970a), (2) Zou et al. (1995), and (3) Odum (1970c).

Note: All values in g dry weight m^{-2}.

[a]Due to rapid decomposition, and significant litter processing in the canopy (Lodge, this volume), some litter is difficult to classify.

Litter

Pre-hurricane

Several studies provide annual means of litter standing biomass at El Verde. Wiegert (1970a) assessed standing crop at two sites (table 2.1) and found the average litter biomass to be 598 g m^{-2}. Pfeiffer (chap. 5, this volume) reports a mean litter standing crop of 588 g m^{-2} measured over the course of a year in association with invertebrate collections. Zou et al. (unpublished) compared standing stocks of litter in mature forest and secondary forest on rehabilitated agricultural land in both wet and dry seasons. Values ranged from 346.3 to 453.1 g m^{-2}; the mature forest had significantly more litter only during the wet season. These three studies provide a range of background values against which the effects of the hurricane can be assessed.

Litter fall is a key measure of production and energy and nutrient input to the detritivore food chains in any forest. Estimates of annual litter fall at El Verde range from 536.3 g m^{-2} (leaves, fruit, and seeds only; Weigert 1970a) to 872.6 g m^{-2} (Zou et al. 1995), with the difference being due nearly entirely to the inclusion of very fine material and branches in only the latter study (table 2.2).

Wood is an important component of litter fall because it provides food and shelter for specialized groups of detritivores (e.g., termites; McMahan, this volume) both on the ground as fallen branches and boles and while still standing as dead branches and standing dead trees. Odum (1970c) found that brush fall was nearly as great as small litter fall at El Verde (table 2.2). Log fall delivers less biomass to the forest floor, but provides a long-lasting substrate for a succession of fungal and insectivorous decomposers (Lodge, this volume). The decomposition of wood on the ground may take seventy to 100 years for smaller size classes (<30 cm diameter) or 100 to 200 years for boles >30 cm diameter (Brown and Lugo 1986). Much faster rates are expected for twigs and small stem wood, which is readily colonized by fungi (Lodge, this volume).

Post-hurricane

Hurricane Hugo more than doubled the amount of litter standing crop (including fine wood and detritus) at both El Verde and Bisley (Lodge et al. 1991). Leaf fall for one day alone exceeded the highest previously recorded value for leaf standing crop at El Verde (615 g m^{-2}, Lodge et al. 1991, versus 598 g m^{-2}, Wiegert 1970a). In addition, 928 g m^{-2} of litter was also found suspended in the understory up to 3 m from the ground at El Verde (Lodge et al. 1991). Since shed leaves were for the most part actively growing rather than senescent, their nutrient contents were higher than the normally senescent leaf material in litter. Litter with high nutrient content should provide excellent substrate for large detritivore populations (Pfeiffer, chap. 5, this volume).

Increased litter fall had direct effects on forest populations. Seedlings on the forest floor were killed by the debris (You and Petty 1991), but sapling populations increased due to enhanced light availability on the forest floor. A thick, structurally complex litter layer was strongly correlated to 4-fold increases in frog populations due to increases in retreat sites and reduced predators (Woolbright 1991). *Anolis* lizards were reduced in number, with both structure and microclimate involved in population reductions (Reagan 1991).

Rates of litterfall were greatly reduced after the hurricane (Zimmerman et al. 1994). Six months after the hurricane, litter fall was about 40% of its pre-hurricane value, but recovered to about 80% within three years. LAI was zero at many points after the hurricane but recovered rapidly. Within four years after Hurricane Hugo, LAI at El Verde had reached 3.96 (Zimmerman et al. 1994).

Belowground Biomass

A wide range of consumable material is available below ground, including living and dead roots as well as litter of a variety of sizes, nutrient concentrations, carbon qualities, and states of decomposition (see Pfeiffer, chap. 5, this volume; Lodge, this volume). There is also a pool of very fine textured detritus, or soil organic matter, which is consumed by earthworms and other soil organisms.

Pre-hurricane

The amount of belowground biomass at El Verde is equivalent to the aboveground biomass (table 2.1). Estimates of living belowground biomass (7,230 g m^{-2}; table 2.1) fall in the mid- to upper range compared to figures for other tropical sites reported in the literature (range: 1,120 to 13,200 g m^{-2}; Jordan 1985). Most of the standing stocks at El Verde occur in large roots and soil organic matter, but rapid turnover in fine roots may make them more important in the food web than their biomass suggests. Fine roots are extended, sloughed off, and regrown continuously, and associated hyphal filaments of soil fungi attached to the roots in mycorrhizal associations are constantly being turned over (see Lodge, this volume).

Root Biomass. The root densities at El Verde are heavily concentrated in the top 30 cm of the soil profile, with very little penetration below 70 cm. The densities and distribution of roots vary widely over short distances. Root distributions are only weakly associated with soil development and slope, except on level areas with deep organic layers, in which roots tend to be more abundant. The localization of roots near the surface is correlated with observed nutrient interception at or near the surface of the forest floor and poor drainage. The high root biomass concentration in the upper layers of soil maximizes the probability of nutrient interception and absorption by roots rather than their loss to deeper, inaccessible layers of the soil profile (Stark and Spratt 1977; Jordan 1985). Such patterns of root distribution are common in tropical areas with their high precipitation and tight nutrient cycling (Odum 1970b; Jordan 1985).

Soil Organic Matter. Organic matter in soil is the product of decomposition of litter and dead root material of an extremely wide range of particle sizes and nutrient qualities. As litter passes through links of the decomposer food chain, it is fractionated and becomes more and more refractory as labile carbon compounds and nutrients are removed, leaving only poorly utilized compounds such as lignins (Schlesinger 1977; Sanford et al. 1991). At El Verde the soil organic matter (SOM) has been quantified to a depth 30.5 cm (Odum 1970c; table 2.1). The greatest concentration of SOM is found in the upper 2.5 cm of soil. Soil organic matter accounts for 34% of the total biomass at El Verde (table 2.1).

Post-hurricane

The effect of Hurricane Hugo on belowground biomass is known with accuracy only for fine roots. At El Verde, live fine root biomass declined to zero within three months of the hurricane and fluctuated widely thereafter (Parrotta and Lodge 1991). Standing stocks of dead roots (mean, 423 g m^{-2}) remained fairly constant for at least thirteen months following the hurricane. No significant reduction in fine roots was found at Bisley (table 2.1). Measurements of small and large roots were not made after the hurricane, but it is reasonable to assume that their mortality was proportional to the loss of aboveground biomass.

Simulations of the effect of hurricanes on SOM (Sanford et al. 1991) show that the quantity of carbon in the soil at El Verde is maintained at high levels because of disturbance. The quality of soil organic matter (i.e., its carbon content) increases following massive litter inputs, potentially increasing phosphorous mineralization and ecosystem fertility. The model suggests that overall forest productivity is enhanced by repeated disturbance, with a burst of growth expected after each disturbance. According to model results, production, soil organic matter, and mineralization may remain at high levels long after the disturbance (Sanford et al. 1991).

Conclusions

Hurricane Hugo resulted in a short-term scarcity of many of the resources required by primary consumers. Live leaves, fruit, flowers, and fine roots were drastically reduced for periods of a few to many months after the hurricane. Much of the standing leaf biomass was shunted from primary consumer to decomposer pathways in the food web within twenty-four hours. The large influx of carbon to the forest floor stimulated microbial growth, which temporarily sequestered soil nutrients (Zimmerman et al. 1995) and reduced aboveground productivity. The variability in the standing crops of live leaves, wood, fruit, and litter resulting from disturbance at El Verde spans the range of values found in most tropical forests. Given the relatively small changes in animal community composition after the hurricane, consumer populations must be adapted to respond to these large changes in resource availability, despite the fact that they occur infrequently.

High-diversity systems have the potential for high connectance within the food web. However, many tropical systems have highly coevolved plant-herbivore relationships that restrict the number of possible interconnections (Janzen 1983a,b). Large-scale disturbance works counter to these finely adjusted relationships by periodically breaking the usual spatial and temporal patterns of resources and consumers. Flexibility in diet or foraging behavior are key characteristics of animal populations in frequently disturbed ecosystems.

LEAF, FLOWER, AND FRUIT PHENOLOGY

The distribution and availability of leaf, flower, and fruit biomass both in time and space is of critical importance to consumer populations, especially if they are at all selective in their feeding and consequently dependent on any particular suite of species or leaf age classes. Janzen (1983a) points out that the strong coevolution of herbivores with a host plant and plant secondary defensive chemicals makes host switching difficult (see also Dirzo 1987). He hypothesizes that half the herbivore species in the tropics have but a single host plant, and that no more than 10% of herbivores can feed on even 10%

of plant species present (Janzen 1983b). This high degree of specificity, coupled with the preference for leaves early in their development (Coley 1983), suggests that knowledge of the phenology of host species is necessary to understand the dynamics of herbivore populations.

Islands have been described to have more phytophagous insects than mainlands (Tanaka and Tanaka 1982). Garrison and Willig (this volume) report most of the herbivores at El Verde to be polyphagous, with the exception of larval lepidopterans which are more than 50% monophagous. Eleven of the fifteen species of lepidopteran larvae whose populations erupted after Hurricane Hugo were found to feed on a single host species (Torres 1992).

Plant phenology is closely related to climate and in particular the pattern of rainfall. The El Verde site has an even temperature regime and lacks a pronounced dry season. Despite the absence of a drought period, leaf fall is highly seasonal (fig. 2.3; Odum 1970c; Ovington and Olson 1970; Wiegert 1970a; Zou et al. 1995), as is leaf flush (Odum 1970c). New leaves are generally the preferred diet of chewing insects due to their higher nutrient content (per area) and lower fiber content (Coley 1983, 1987a,b; Cooke et al. 1984; Dirzo 1984, 1987). Odum (1970c) analyzed photographs from aircraft and canopy towers for the presence of newly flushed leaves in the canopy. The greatest apparent new leaf activity occurred from June to Au-

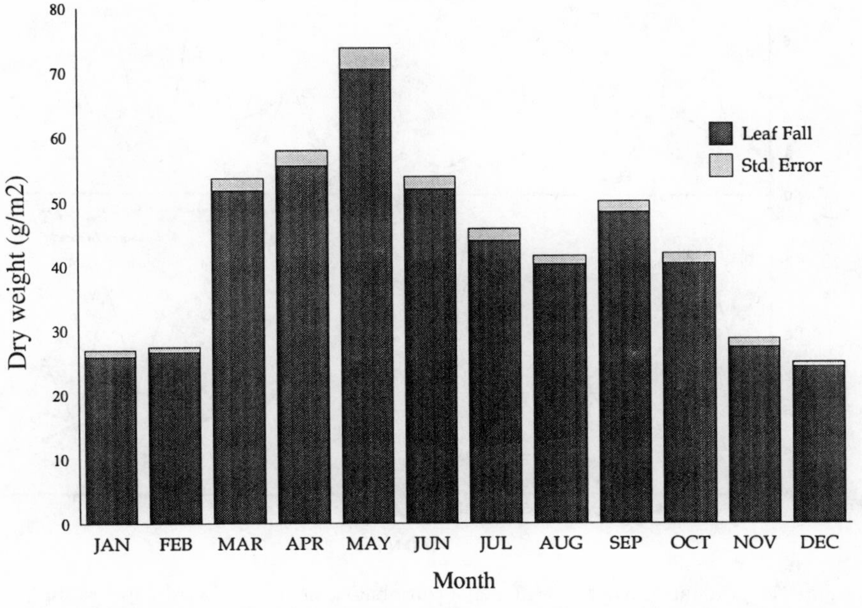

Figure 2.3. Mean monthly leaf fall estimated from several studies done at El Verde. Data from Zou et al. (1995) covering six-year sampling periods from 1966, 1970 to 1973, and 1981.

gust, when 21% of canopy trees had new growth. Values for the rest of the year varied between 6 and 9%. This pattern is consistent with data from 1991, which show an increase from a mean LAI of 3.25 determined from April to May to 4.10 in July to August (V. Quiñones pers. comm.).

Numerically dominant species have a large effect on the pattern of leaf fall and leaf flush. At the peak of its leaf fall, *Dacryodes excelsa* contributes more than 50% of total leaf fall (fig. 2.4). During nonpeak periods, *D. excelsa* still contributes as much to total leaf fall as less abundant species. The species composition of leaf material reaching the forest floor is highly seasonal (fig. 2.4), which means that detritivores are faced with seasonal changes in the abundance of particular food plants. Diaz (unpublished) found no difference in invertebrate aggregations from litter bags containing leaves from different plant species. This result suggests that detritivorous invertebrates are not particularly selective in their diet, but this question deserves further study.

Studies of the phenology of flowering, fruiting, and fruitfall at El Verde (Bannister 1970; Estrada Pinto 1970; Odum and Pigeon 1970; Devoe 1990; Lugo and Frangi 1993; Zou et al. 1995) show that fruit fall is steady

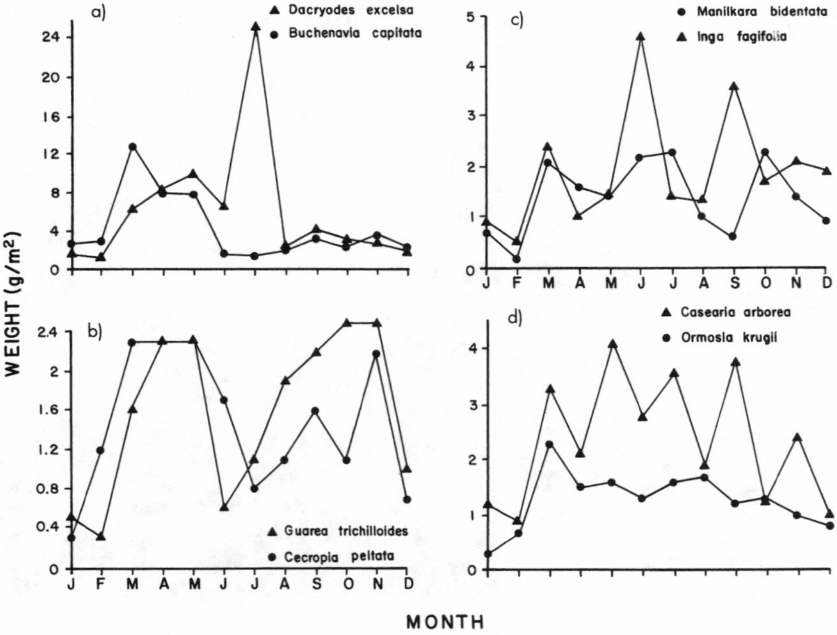

Figure 2.4. Monthly leaf fall for eight major contributing species at El Verde. Species are grouped by seasonality of yearly leaf litter fall as single (*a*), double (*b*), or triple (*c*), peaked or continuous (*d*). Data from Zou et al. (1995) for 1981.

throughout the year except for two months (May to June) at the beginning of the wet season (fig. 2.5). Flower fall is greatest from July to December. These general patterns are a composite of species patterns that show high interspecific variability in the season, place, and quantity of fruit and flower production. Consumers that depend on the fruit or nectar of particular plant species will be most strongly influenced by the pattern of those species rather than the general pattern of the whole plant community. Because the relationships between individual consumer and plant species are still being worked out for El Verde, species-specific effects are in general not included in the food web.

Almost every pattern of flowering has been observed for common species at El Verde (Estrada Pinto 1970). The most common phenological pattern was flowering between June and September with fruit drop from September to December. A few understory species fruited and flowered year around, and several species had short flowering seasons but long-term fruit drop. Such phenological complexity has been observed in many tropical areas (Walter 1971). Disturbance, however can strongly influence this observed pattern. You and Petty (1991) found no production whatsoever of flower, fruit, or seed in *Manilkara bidentata* nine months after Hurricane Hugo.

This temporal and spatial patchiness of supply makes specialization on the

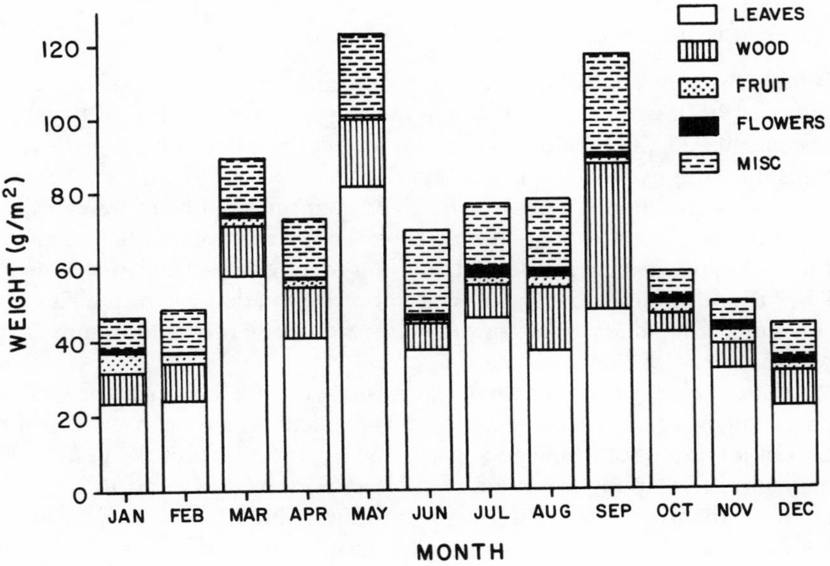

Figure 2.5. Seasonal distribution of the components of litter fall (leaves, wood, fruit, flowers, and miscellaneous) during 1981 at El Verde. The miscellaneous category reflects fine material whose origin could not be determined. From Zou et al. (1995).

fruit of any one species particularly risky. As an example, the Puerto Rican parrot (*Amazona vittata*) feeds heavily on sierra palm fruit (Snyder et al. 1987). Peak production of palm fruits has been reported in February, from May to August, and from December to February depending on the year and location within the LEF (Lugo and Frangi 1993). If other plant species are equally variable, diet specialization may not be a viable option for many frugivores in the LEF.

CONTROLS ON PRIMARY CONSUMPTION

Given the seemingly tremendous quantity of resources available to herbivores in the tropics, why isn't there more evidence of large scale loss of foliage? There are at least two answers to this question. The first is that herbivore populations may be kept in check by their predators (Hairston et al. 1960) and the second is that herbivory is, in fact, high, but we fail to recognize its true magnitude. This latter point of view is supported by Janzen (1983b) who points out that herbivory costs are those of the lost tissue plus energetic investments in antiherbivore defenses. In fact, primary consumers are suppressed both by their predators and parasites, who reduce their populations directly, and by their food plants, which invest energy to reduce palatability or digestibility of their tissues.

Rates of Consumption

Whether or not control of herbivores is primarily top-down or bottom-up (Power 1992), what is the level of herbivory expected in tropical forest ecosystems like El Verde? How might that level change over the range of conditions imposed by frequent disturbance? I focus on folivory over other types of primary consumption because it can be quantified readily; frugivores, nectarivores, and sap-feeding insects leave little evidence of their activities. Moreover, plants may develop adaptations to encourage frugivory or nectarivory if useful services such as dispersal or pollination are provided by the consumer. The majority of consumers do not enjoy this kind of mutualistic relationship with their host plants.

While the detection of folivory in a forest setting is straightforward, accurate quantification is more difficult. The patchy nature of insect consumption of leaves in both time and space (Jordan 1985) makes the design of an appropriate sampling scheme extremely important. Losses to folivores can manifest themselves in many ways, including aborted leaves, meristematic losses, metabolic costs for repair of damaged leaves, and leaf area reductions not normally considered herbivore losses, such as those to leaf rolling insects. These kinds of losses must be included for an accurate estimate of herbivory (Aide 1993).

At El Verde, herbivory based on missing tissue from fallen leaves was measured for twenty-three months at the Radiation and South Control Centers (Odum and Ruíz-Reyes 1970). Consumption increased in both sites from initial values between 4 and 7% to final values between 8 and 11%. The mean for the entire sampling period was 7%. Seasonally low values were observed in both plots in March to May and November, but the record is too short to put much faith in this pattern. Seasonal and annual changes were extremely consistent between these two plots despite the fact that one was severely disturbed by irradiation.

Estimates of tropical herbivory vary widely and to a great extent due to the methodologies involved in the experiments. The 8 to 13% range of leaf consumption at El Verde is consistent with similar studies in New Guinea, Mexico (Dirzo 1987), Panama, and Costa Rica. Long-term studies of herbivory have shown 14.6% losses in a subtropical Australian rain forests, and 21% average losses at Barro Colorado in Panama (Lowman 1984, 1985a,b).

Costs of Herbivory

The true cost of herbivory to a plant must include not only the replacement of the missing parts, but also the metabolic costs of producing the secondary compounds devoted to leaf defense (Janzen 1983) that prevent or reduce herbivore injuries. Herbivory rates are often moderated by defensive characteristics which include physical barriers to herbivory such as pubescence, high fiber content, low water content, and leaf toughness; chemical barriers like poisonous or digestibility reducing compounds (nitrogen-based alkaloids and nonprotein amino acids, carbon-based tannins, turpenes, polyphenolics); low nutrient concentrations which increase predator handling time and may tax their digestive system; and symbiotic insect relationships. Coley (1983) found 70% of variation in grazing to be explained by leaf toughness and structure. Chemical defenses may not reduce the amount of leaf material removed by any individual insect, but may effect reductions in the total population by slowing growth, reducing reproduction, and increasing susceptibility to disease and predation (Price et al. 1980).

The timing of herbivory is also critical. Insects usually consume leaves between dusk and midnight, a time at which the leaves are still loaded with photosynthate prior to its conversion to structural components or translocation to other parts of the plant. Consumption of such "enriched" leaves is a direct foliar loss as well as a loss to potential production. Herbivory during the first two to four weeks of leaf presence is also costly as the plant has invested energy in structure, but photosynthesis has not yet begun to bring a positive energy flow to the plant (Janzen 1983b). Total costs of herbivory may never be completely evaluated as exclusion experiments eliminate loss of leaf area but can never quantify the allocation of energy to evolutionarily mediated defensive compounds rather than reproductive or growth uses (Janzen 1983a). The suite of defensive techniques found in a plant are potentially controlled by the environment and resource availability; slow-growing plants in low-resource situations invest more energy in defenses than fast-growing species (Coley et al. 1985).

Species Palatability

A species susceptibility to herbivory depends on many factors including leaf structure, leaf age, nutrient content, and secondary chemical defenses. Leaf specific area (LSA; cm^2 leaf area g^{-1} leaf weight) is a rough indicator of leaf toughness and the best predictor of vulnerability to herbivores. The denser the leaf (higher LSA), the more sclerophyllous, fibrous, and resistant to herbivores. There are strong trade-offs for a folivore trying to maximize nutrient uptake against a backdrop of leaf toughness, density, and potential secondary defensive compounds. On a strict nitrogen concentration basis, there are large differences in leaf nitrogen among species at El Verde (table 2.3). *Ormosia krugii* and *Buchenavia capitata* have the highest nitrogen contents among species sampled by Medina et al. (1981), with 20.2 and 16.7 mg N g^{-1}, respectively. However, these two species clearly demonstrate the paradox of consumption versus concentration. Converting the nutrient data to an area basis (table 2.3) shows that the herbivore must consume more than 2.5 times as much leaf area of *Buchenavia* as *Ormosia* to get the same amount of nitrogen. The controls of herbivore behavior are often more complex than these simple measures would predict. Garrison and Willig (this

Table 2.3 Nutrient content for selected tabonuco forest species

Species	P[a] mg g^{-1}[d]	N[b] mg g^{-1}[d]	P mg cm^{-2}[e]	N mg cm^{-2}[e]	LSA[c] $cm^2\ g^{-1}$
Buchenavia capitata	0.68	16.7	4.42	0.108	153.8
Matayba domingensis	0.52	11.2	9.61	0.207	54.1
Linociera domingensis	0.44	10.8	5.47	0.134	80.4
Manilkara bidentata	0.42	10.7	6.09	0.155	69.0
Dacryodes excelsa	0.53	10.4	7.98	0.156	66.4
Ormosia krugii	0.59	20.2	8.10	0.277	72.8
Schefflera morototoni	0.59	15.3	6.91	0.179	85.4
Average all species listed	0.54	15.3	6.50	0.163	83.1

Source: Nutrient contents calculated from data in Medina et al. (1981).

[a]Phosphorus. [b]Nitrogen. [c]Leaf specific area.

[d]Mass P or N per g leaf dry weight.

[e]Mass P or N per cm^2 leaf surface area.

volume) found walking sticks to be extremely general in their diets, optimizing a species mix for constant nutrient intake. In some cases the older rather than younger leaves of a shrub were preferred, seemingly contrary to research results reviewed above.

Effects of Hurricane Hugo on Herbivory

As described above, forest structure at El Verde changed from a late secondary forest to a mosaic of mature trees and gaps after Hurricane Hugo. Early successional species, such as *Cecropia schreberiana* and *Schefflera morototoni,* increased rapidly and provided a source of leaves with high nutrient content per unit area. In addition, refoliation created a large crop of young leaves of all species. A three-month drought and higher temperatures were part of the changed environmental conditions. Under drought conditions, nutrients and carbohydrates may be concentrated in the leaves, increasing their palatability. Increased air temperatures may increase insect growth rates, reducing the effectiveness of predators and pathogens in reducing their numbers. The effects of leaf secondary defensive compounds may also be reduced by the same air temperature-induced increase in metabolism (Mattson and Haack 1987). Thus, many of the effects of the hurricane were beneficial to herbivores.

The most intriguing response to these changed conditions occurred seven months after the passage of the hurricane. Large-scale outbreaks of fifteen species of lepidopteran larvae were recorded for the first time on in the LEF. Large-scale defoliations of outbreak species (Mattson and Addy 1975) are not a dominant feature of tropical forests, probably due to the diversity of species and distance between individuals, but total defoliation is sometimes observed on the level of individual trees (Odum and Ruíz-Reyes 1970; Janzen 1983a; Dirzo 1987).

Larval stages of moths from seven different families caused massive defoliation of many herbaceous plants that flourished in the altered conditions after the hurricane (Torres 1992). One species (*Spodoptera eridania;* Noctuidae) fed on fifty-six plant species from thirty-one families. Several abundant plant species were heavily defoliated (e.g., *Phytolaca rivinoides, Impatiens wallerana, Ipomea tiliacea, Cestrum macrophyllum*) and suffered outright mortality. Less severe defoliation occurred in areas of lighter hurricane damage, like El Verde. The outbreak of *S. eridania* terminated in May when most of the preferred host plants were consumed. Parasitism by two ichneumonid wasps increased from 18% to 27% at the end of the outbreak (Torres 1992). All of the other fourteen species of lepidopteran involved in outbreaks fed on one or a few host plants, principally early successional species. Their outbreaks ended by August 1990, eleven months after the hurricane. Outbreaks of lepidopterans were attributed by Torres (1992) to the flush of new leaves in the understory caused by increased sunlight reaching the forest floor.

Other insect species benefited from the hurricane. An increase in Diptera, principally fruit flies, in October 1989, was attributed to an increase in rotting fruit on the forest floor (Torres 1992). Bark beetles (Coleoptera: Scolytidae) and pin-hole borers (Coleoptera: Platypodidae) increased in November and December in response to the expanded supply of dead or damaged trees. Biting flies temporarily increased at El Verde after the hurricane (Waide pers. observation). Scale insects increased in abundance in coffee groves in Puerto Rico after Hurricane San Ciprian in 1932 (Torres 1992).

Direct negative effects were seen on some herbivore populations. These included 75% decreases in snail (four species) and walking stick (two species) populations measured nine months after Hugo (Willig and Camilo 1991; Garrison and Willig, this volume), which were attributed to changes in microclimate rather than lack of food. The populations of adults of these species was severely reduced and virtually no recruitment was found. This is believed to be due to intolerance of eggs and initial stadia to low humidity and high light levels. These changes in the populations of the most generalized of the phytophagous arboreal invertebrates (Tanaka and Tanaka 1982; Garrison and Willig, this volume) could have a profound effect on the food web.

Schowalter (1994) compared herbivory levels and insect abundance in canopy trees in intact forest and gaps seventeen months after the hurricane. There were no differences in the amount of leaf area missing from trees in intact forest and gaps. However, there was significantly more herbivory on the late successional trees *Manilkara bidentata* and *Sloanea berteriana,* which were numerous in the sample, than in the early successional *Casearia arborea* and *Cecropia schreberiana,* which were less numerous. Lepidoptera, detritivores, and predaceous beetles were more common in intact forest, whereas several species of sapsucking phytophages were more common in gaps. Since losses to sapsucking insects were not measured in this study, the actual amount of consumption (leaf area plus sap) may have been higher in gaps. However, tree species effects still seem to override disturbance effects with regard to herbivory. Invertebrate biomass and herbivory was positively related to tree species abundance in the forest with the exception of *Dacryodes excelsa. Dacryodes* has high content of aromatic terpenoids in the leaves, which probably discourages herbivores.

Very similar results were obtained by Odum and Ruíz-Reyes (1970) in the Radiation Center. Increased light penetration and the subsequent flush of new leaves provided conditions favorable to herbivores after irradiation. However, herbivory rates in the gap did not diverge from the forested control during the year following irradiation. Disturbance at the scale of a small gap apparently has little effect on the consumption of leaves by herbivores.

Conclusions

The large-scale nature of the hurricane disturbance caused forestwide reductions in some herbivores (e.g., snails, walking sticks) and prompted outbreaks of others. However, except for the burst of production and consumption in understory vegetation, folivory remained about the same in heavily disturbed and intact areas. The synchronous burst of new foliage in disturbed areas may not have given canopy insects time to respond before new leaves developed defensive compounds. Alternatively, early differences in consumption between intact and disturbed areas may have been missed. The greater number of sapsucking insects found in gaps underscores the unmeasured effect of this functional group on total consumption. The mosaic nature of the disturbance at El Verde may have prevented the development of more notable differences between disturbed and intact areas.

SUMMARY

The growth form and community structure of plants bears great importance for the structure of the food web at El Verde. Consumers are dependent upon plants for converting solar energy to accessible sources of carbon and for obtaining necessary nutrients and water. The distribution and nature of acquired energy stores among plant parts determines the form of connections within the food web, and the diversity of species in the community influences the number of possible links among consumers and food. Disturbance acts to reorganize the resource base available to consumers, and thus has strong potential effects on food web structure.

The specialization of consumers on living or dead plant parts directs the flow of energy through two principal paths in the food web. A large and diverse group of consumers feeds on living leaves, roots, and other plant tissues, but more than 10 times as much plant consumption occurs among the detritivores, whether their activities take place on the forest floor, in the soil, or on perched litter or dead wood in the canopy. Even more biomass is consumed by detritivores in the standing dead wood of boles and branches eaten by termites, wood-eating beetles, and microorganisms.

Consumer diversity has often been related to the diversity of plant resources. Plant species richness is high in the tropics compared to temperate zones, but varies widely among different tropical life zones. The flora at El Verde comprises 468 species of vascular plants, seventy-eight ferns, and 390 flowering plants. The number of tree species found at El Verde (88) represents a small fraction of the species in tabonuco forest zone (153), the LEF (225), and the island of Puerto Rico (750). El Verde has fewer tree species (43 to 54 ha^{-1}) than lowland neotropical forests, but this may result from elevation, climate, and landscape heterogeneity as well as insularity.

Hurricanes, the principal agents of large-scale disturbance at El Verde, have significant effects on resource distribution and availability. Mortality of trees at El Verde reached 13.3% 171 weeks after Hurricane Hugo, and aboveground biomass was severely reduced. Massive defoliation results in a synchronous flush of new leaves in many canopy species. Fruits and flowers were stripped from the trees and only recovered slowly. Damage to the canopy deposited large amounts of nutrient-rich litter on the forest floor and resulted in large changes in canopy architecture. Light levels and temperature increased at ground level after the hurricane, and soil moisture decreased because of a drought. Early successional plant species invaded heavily damaged areas.

The hurricane resulted in a short-term scarcity of many resources required by primary consumers. Leaves, fruit, flowers, and fine roots were drastically reduced, and standing leaf biomass was shunted from primary consumer to decomposer pathways. Differences in the standing crops of live leaves, wood, fruit, and litter before and after the hurricane approach the range of values found in tropical forests worldwide. Consumer populations must be adapted to respond to these large changes in resource availability in disturbance-prone ecosystems.

Temporal heterogeneity in resources was evident even before the hurricane. Both leaf fall and leaf flush are highly seasonal in mature tabonuco forest, with the pattern determined largely by a few numerically dominant species. Leaf fall is greater between March and October, and leaf flush from June to August. The quantity of fruit fall is steady throughout the year, but the major fruiting and flowering species change with season. Most species flower between June and September and drop fruit from September to December.

Primary consumers are suppressed by their predators and by chemical defenses developed by their food plants. Herbivory at El Verde ranged between 4 and 11%, with a mean of 7%. This falls within the range of herbivory estimates from many other tropical sites. Although many of the effects of disturbance should be beneficial to herbivores, two different studies failed to show differences in the degree of herbivory between gaps and intact forest. Insect outbreaks occurred after the hurricane as a result of a burst of plant growth in the understory. The mosaic pattern of disturbance at El Verde may have masked differences between disturbed and intact areas.

Consumption of dead plant parts is performed principally by microorganisms and specialized insects such as termites. These groups play a very important role in conditioning litter and dead wood for consumption by other organisms. They also are key prey for higher consumers in the food web. By mediating the release of nutrients from litter, they can also control the productivity of living plants to some degree. These groups thus influence the development of food web structure by controlling access to the resources stored in different plant compartments.

Appendix 2. Plants at El Verde

Species	Family
Adiantum latifolium Lam.	Pteridaceae (P)[a]
Adiantum petiolatum Desv.	Pteridaceae (P)
Adiantum pulverulentum L.	Pteridaceae (P)
Adiantum tetraphyllum Humb. & Bonpl.	Pteridaceae (P)
Aechmea lingulata (L.) Baker	Bromeliaceae
Aeschynomene americana L.	Fabaceae
Alchornea latifolia Sw.	Euphorbiaceae
Alchorneopsis floribunda (Benth.) Muell. Arg.	Euphorbiaceae
Alocasia cucullata (Lour.) Don	Araceae
Alpinia purpurata Vieill. ex K. Schum.	Zingiberaceae
Alsophilus portoricensis (Spreng. ex Kuhn) Conant	Cyatheaceae (P)
Andira inermis (W. Wright) DC.	Fabaceae
Andropogon bicornis L.	Poaceae
Andropogon leucostachys Kunth	Poaceae
Anthurium crenatum (L.) Kunth	Araceae
Anthurium dominicense Schott	Araceae
Anthurium scandens (Aubl.) Engler	Araceae
Antirhea obtusifolia Urb.	Rubiaceae
Apteria aphylla (Nutt.) Small	Burmanniaceae
Arachniodes chaerophylloides (Poir.) Proctor	Dryopteridaceae (P)
Ardisia glauciflora Urb.	Myrsinaceae
Arthridostylidium sarmentosum Pilg.	Poaceae
Artocarpus altilis (Parkinson) Fosberg	Moraceae
Asclepias curassavica L.	Asclepiadaceae
Asplenium auriculatum Sw.	Aspleniaceae (P)
Asplenium cuneatum Lam.	Aspleniaceae (P)
Asplenium serratum L.	Aspleniaceae (P)
Axonopus compressus (Sw.) P. Beauv.	Poaceae
Bacopa monnieri (L.) Pennell	Scrophulariaceae
Begonia decandra Pav. ex DC.	Begoniaceae
Begonia nelumbiifolia Schltdl. & Cham.	Begoniaceae
Beilschmiedia pendula (Sw.) Hemsl.	Lauraceae
Bidens alba (L.) DC. var. *radiata* (Sch. Bip.) Melchert	Asteraceae
Blechnum occidentale L.	Blechnaceae (P)
Blechum pyramidatum (Lam.) Urb.	Acanthaceae
Bolbitis nicotianifolia (Sw.) Alston	Dryopteridaceae (P)
Brunfelsia portoricensis Krug & Urb.	Solanaceae
Buchenavia tetraphylla (Aubl.) R. A. Howard	Combretaceae
Byrsonima spicata (Cav.) HBK	Malpighiaceae
Brysonima wadsworthii Little	Malpighiaceae
Calophyllum calaba L.	Clusiaceae

This appendix was prepared by Charlotte M. Taylor.

Calycogonium squamulosum Cogn.	Melastromaceae
Calyptranthes krugii Kiaersk.	Myrtaceae
Campylocentrum micranthum (Lindl.) Rolfe	Orchidaceae
Caperonia palustris (L.) St.-Hil.	Euphorbiaceae
Casearia arborea (Rich.) Urb.	Flacourtiaceae
Casearia decandra Jacq.	Flacourtiaceae
Casearia sylvestris Sw.	Flacourtiaceae
Cassipourea guianensis Aubl.	Rhizophoraceae
Cayaponia americana (Lam.) Cogn.	Cucurbitaceae
Cecropia schreberiana Miq.	Moraceae
Ceiba pentandra L.	Bombacaceae
Cestrum macrophyllum Vent.	Solanaceae
Chamasyce hyssopifolia (L.) Small	Euphorbiaceae
Chamasyce prostrata Aiton	Euphorbiaceae
Chionanthus domingensis Lam.	Oleaceae
Chione venosa (Sw.) Urb.	Rubiaceae
Chloris barbata Sw.	Poaceae
Chrysophyllum argenteum Jacq.	Sapotaceae
Cinnamomum elongatum (Nees) Kosterm.	Lauraceae
Cissampelos pareira L.	Menispermaceae
Cissus obovata Vahl	Vitaceae
Cissus verticillata (L.) Nicolson & Jarvis	Vitaceae
Citharexylum caudatum L.	Verbenaceae
Citus paradisi Macfad.	Rutaceae
Clibadium erosum (Sw.) DC.	Asteraceae
Clidemia strigillosa (Sw.) DC.	Melastomaceae
Clusia clusioides (Griseb.) D'Arcy	Clusiaceae
Clusia gundlachii Stahl	Clusiaceae
Clusia rosea Jacq.	Clusiaceae
Cnemidaria horrida (L.) C. Presl	Cyatheaceae (P)
Coccoloba diversifolia Jacq.	Polygonaceae
Coccoloba pyrifolia Desf.	Polygonaceae
Cochleanthes flabelliformis (Sw.) R. Schult. & Garay	Orchidaceae
Cocos nucifera L.	Arecaceae
Codiaeum variegatum (L.) Blume	Euphorbiaceae
Coffea arabica L.	Rubiaceae
Colocasia esculenta (L.) Schott	Araceae
Columnea tulae Urb.	Gesneriaceae
Commelina diffusa Burm.f.	Commelinaceae
Commelinopsis persicariifolia (DC.) Pichon	Commelinaceae
Comocladia glabra (Schult.) Spreng.	Anacardiaceae
Cordia alliodora (Ruiz & Pav.) Oken	Boraginaceae
Cordia borinquensis Urb.	Boraginaceae
Cordia polycephala (Lam.) I. M. Johnst.	Boraginaceae
Cordia sulcata DC.	Boraginaceae
Cordyline fruticosa (L.) A. Chev.	Liliaceae
Cranichis muscosa Sw.	Orchidaceae

Croton poecilanthus Urb.	Euphorbiaceae
Ctenitis subincisa (Willd.) Ching	Dryopteridaceae (P)
Cucurligo cf. *latifolia* Dryand.	Hypoxidaceae
Cyathea arborea (L.) J. E. Sm.	Cyatheaceae (P)
Cyathea borinquena (Maxon) Domin	Cyatheaceae (P)
Cuphea strigulosa Kunth	Lythraceae
Cyclopogon cranichoides (Griseb.) Schltr.	Orchidaceae
Cyclopogon elatus (Sw.) Schltr.	Orchidaceae
Cyperus distans L.f.	Cyperaceae
Cyperus odoratus L.	Cyperaceae
Cyperus rotundatus L.	Cyperaceae
Cyperus surinamensis Rottb.	Cyperaceae
Cyperus tenuifolius (Steud.) Dandy	Cyperaceae
Cyrilla racemiflora L.	Cyrillaceae
Dacryodes excelsa Vahl	Burseraceae
Danaea elliptica J. E. Sm.	Marattiaceae (P)
Danaea nodosa (L.) J. E. Sm.	Marattiaceae (P)
Daphnopsis philippiana Krug & Urb.	Thymelaeaceae
Dendropanax arboreus (L.) Decne. & Planch.	Araliaceae
Dennstaedtia bipinnata (Cav.) Maxon	Dennstaedticeae (P)
Desmodium adscendens (Sw.) DC.	Fabaceae
Desmodium axillare (Sw.) DC.	Fabaceae
Desmodium barbatum (L.) Benth.	Fabaceae
Desmodium wydlerianum Urb.	Fabaceae
Dieffenbachia seguine (Jacq.) Schott	Araceae
Digitaria ciliaris (Retz) Koeler	Poaceae
Digitaria sanguinalis (L.) Scop.	Poaceae
Diodia sarmentosa Sw.	Rubiaceae
Dioscorea alata L.	Dioscoreaceae
Dioscorea pilosiuscula Bertero	Dioscoreaceae
Dioscorea polygonoides Humb. & Bonpl. ex Willd.	Dioscoreaceae
Ditta myricoides Griseb.	Euphorbiaceae
Drypetes glauca Vahl	Euphorbiaceae
Eclipta prostrata (L.) L.	Asteraceae
Elaphoglossum apodum (Kaulf.) Schott ex J. Sm.	Dryopteridaceae (P)
Elaphoglossum crinitum (L.) H. Christ.	Dryopteridaceae (P)
Elaphoglossum herminieri (Bory & Fée) T. Moore	Dryopteridaceae (P)
Elaphoglossum maxonii Underw. & Morton	Dryopteridaceae (P)
Elaphoglossum petiolatum (Sw.) Urb. var. *dussii* (Underw. & Maxon) Proctor	Dryopteridaceae (P)
Elaphoglossum rigidum (Aubl.) Urb.	Dryopteridaceae (P)
Elaphoglossum simplex (Sw.) Schott ex J. Sm.	Dryopteridaceae (P)
Eleocharis elegans (HBK) Urb.	Cyperaceae
Eleocharis flavescens (Poir.) Urb.	Cyperaceae
Elephantopus mollis Kunth	Asteraceae
Eleusine indica (L.) Gaertn.	Poaceae
Elleanthus cordidactylus Ackerman	Orchidaceae

Eltroplectris calcarata (Sw.) Garay & H. R. Sweet	Orchidaceae
Emilia fosbergii Nicolson	Asteraceae
Emilia sonchifolia (L.) DC.	Asteraceae
Encyclia cochleata (L.) Dressler	Orchidaceae
Encyclia pygmaea (Hook.) Dressler	Orchidaceae
Epidendrum boricuarum Hágstater & L. Sánchez	Orchidaceae
Epidendrum nocturnum Jacq.	Orchidaceae
Epidendrum ramosum Jacq.	Orchidaceae
Epidendrum secundum Jacq.	Orchidaceae
Epipremnum pinnatum (L.) Engl. cv. *aureum* (Lindl. & André) Nicolson	Araceae
Erechtites valerianifolia (Spreng.) DC.	Asteraceae
Erigeron belloides DC.	Asteraceae
Eryngium foetidum L.	Apiaceae
Erythrodes hirtella (sw.) Fawc. & Rendle	Orchidaceae
Erythrodes plataginea (L.) Fawc. & Rendle	Orchidaceae
Eugenia domingensis Berg	Myrtaceae
Eugenia stahlii (Kiaersk.) Krug & Urb.	Myrtaceae
Eupatorium microstemon Cass.	Asteraceae
Euphorbia heterophylla L.	Euphorbiaceae
Euphorbia pulcherrima Willd. ex Klotzsch	Euphorbiaceae
Fadyenia hookeri (Sweet) Maxon	Dryopteridaceae (P)
Faramea occidentalis (L.) A. Rich.	Rubiaceae
Ficus americana Aubl.	Moraceae
Ficus crassinervia Desf.	Moraceae
Fimbristylis dichotoma (L.) Vent.	Cyperaceae
Forsteronia portoricensis Woodson	Apocynaceae
Fuirena umbellata Rottb.	Cyperaceae
Garcinia portoricensis (Urb.) Alain	Clusiaceae
Genipa americana L.	Rubiaceae
Gleichenia bifida (Willd.) Spreng.	Gleicheniaceae (P)
Gonzalagunia spicata (Lam.) M. Gómez	Rubiaceae
Grammitis flabelliformis (Poir.) Morton	Polypodiaceae (P)
Grammitis seminuda (Willd.) Willd.	Polypodiaceae (P)
Grammitis serrulata (Sw.) Sw.	Polypodiaceae (P)
Graptophyllum pictum (L.) Griff	Acanthaceae
Guarea glabra Vahl	Meliaceae
Guarea guidonia (L.) Sleumer	Meliaceae
Guatteria caribaea Urb.	Annonaceae
Guazuma ulmifolia Lam.	Sterculiaceae
Guettarda valenzuelana A. Rich.	Rubiaceae
Guzmania berteroniana (Roem. & Schult.) Mez	Bromeliaceae
Guzmania lingulata (L.) Mez	Bromeliaceae
Guzmania monostachya (L.) Rusby	Bromeliaceae
Gymnosiphon niveus (Griseb.) Urb.	Burmanniaceae
Haenianthus salicifolius Griseb.	Oleaceae
Hamelia axillaris Sw.	Rubiaceae

Hedychium coronarium Koenig	Zingiberaceae
Hedyosmum arborescens Sw.	Chloranthaceae
Heliconia caribaea Lam.	Heliconiaceae
Hemidictyum marginatum (L.) C. Presl	Dryopteridaceae (P)
Hemidiodia ocymifolia (Willd. ex Roem. & Schult.) K. Schum.	Rubiaceae
Henriettea fascicularis (Sw.) M. Gómez	Melastomaceae
Heteropterys laurifolia (L.) A. Juss.	Malpighiaceae
Heterotrichum cymosum (J. C. Wendl. ex Spreng.) Urb.	Melastomaceae
Hibiscus pernambucensis Arruda	Malvaceae
Hibiscus rosa-sinensis L.	Malvaceae
Hillia parasitica Jacq.	Rubiaceae
Hippocratea volubilis L.	Hippocrateaceae
Hirtella rugosa Pers.	Chrysobalanaceae
Homalium racemosum Jacq.	Flacourtiaceae
Hymenophyllum polyanthos (Sw.) Sw.	Hymenophyllaceae (P)
Hypolepis repens (L.) C. Presl	Dennstaedtiaceae (P)
Hypoxis decumbens L.	Hypoxidaceae
Ichnanthus pallens (Sw.) Munro	Poaceae
Ilex sideroxyloides (Sw.) Griseb.	Aquifoliaceae
Impatiens balsamina L.	Balsaminaceae
Inga laurina (Sw.) Willd. ex L.	Fabaceae
Inga vera Willd. ex L.	Fabaceae
Ipomoea repanda Jacq.	Convolvulaceae
Ipomoea cf. *setifera* Poir.	Convolvulaceae
Ipomoea tiliacea (Willd.) Choisy	Convolvulaceae
Ixora ferrea (Jacq.) Benth.	Rubiaceae
Jacquiniella globosa (Jacq.) Schltdl.	Orchidaceae
Jacquiniella teretifolia (Sw.) Britton & P. Wilson	Orchidaceae
Justicia martinsoniana R. A. Howard	Acanthaceae
Laetia procera (Poepp. & Endl.) Eichler	Flacourtiaceae
Lantana camara L.	Verbenaceae
Laplacea portoricensis (Krug & Urb.) Dyer	Theaceae
Lasiacis divaricata (L.) Hitchc.	Poaceae
Lasianthus lanceolatus (Griseb.) M. Gómez	Rubiaceae
Leochilus portoricensis M. W. Chase	Orchidaceae
Lepanthes rubripetala Stimson	Orchidaceae
Lepanthes rupestris Stimson	Orchidaceae
Lepanthes selenitepala Rchb.f.	Orchidaceae
Lepanthes veleziana Stimson	Orchidaceae
Lepanthes woodburyana Stimson	Orchidaceae
Lindernia crustacea (L.) F. Muell.	Scrophulariaceae
Lindsaea lancea (L.) Bedd.	Dennstaedtiaceae (P)
Lindsaea quadrangularis Raddi subsp. *antillensis* Kramer	Dennstaedtiaceae (P)
Liparis nervosa (Thunb.) Lindl.	Orchidaceae
Lisianthus laxiflorus Urb.	Gentianaceae

GOSHEN COLLEGE LIBRARY
GOSHEN, INDIANA

Lobelia cliffortiana L.	Lobeliaceae
Lomariopsis amydrophlebia (Sloss. ex Maxon) Holttum	Dryopteridaceae (P)
Lonchocarpus latifolius (Willd.) DC.	Fabaceae
Ludwigia octovalvis (Jacq.) P. H. Raven	Onagraceae
Lunania ekmannii Urb.	Flacourtiaceae
Lycopodium aqualupianum Spring.	Lycopodiaceae (P)
Lycopodium cernuum L.	Lycopodiaceae (P)
Lycopodium dichotomum Jacq.	Lycopodiaceae (P)
Lycopodium funiforme Cham. ex Spring.	Lycopodiaceae (P)
Lycopodium linifolium L.	Lycopodiaceae (P)
Lycopodium reflexum Lam.	Lycopodiaceae (P)
Lycopodium taxifolium Sw.	Lycopodiaceae (P)
Lycopodium verticillatum L.f.	Lycopodiaceae (P)
Macfadyena unguis-cati (L.) A. Gentry	Bignoniaceae
Magnolia portoricensis Bello	Magnoliaceae
Magnolia splendens Urb.	Magnoliaceae
Malaxis massonii (Ridl.) Kuntze	Orchidaceae
Malpighia furcata Ker Gawl.	Malpighiaceae
Mammea americana L.	Clusiaceae
Mangifera indica L.	Anacardiaceae
Manilkara bidentata (A. DC.) A. Chev.	Sapotaceae
Marcgravia rectiflora Triana & Planch.	Marcgraviaceae
Margaritaria nobilis L.f.	Euphorbiaceae
Matayba domingensis (DC.) Radlk.	Sapindaceae
Matelea variifolia (Schltdl.) Woodson	Asclepiadaceae
Maxillaria acutifolia Lindl.	Orchidaceae
Maxillaria coccinea (Jacq.) L. O. Williams	Orchidaceae
Maytenus elongata (Urb.) Britton	Celastraceae
Mecardonia procumbens (Mill.) Small	Scrophulariaceae
Mecranium amygdalinum (Desr.) C. Wright	Melastomaceae
Meliosma herbertii Rolfe	Sabiaceae
Miconia impetiolaris (Sw.) D. Don	Melastomaceae
Miconia laevigata (L.) DC.	Melastomaceae
Miconia mirabilis (Aubl.) L. O. Williams	Melastomaceae
Miconia prasina (Sw.) DC.	Melastomaceae
Miconia racemosa (Aubl.) DC.	Melastomaceae
Miconia sintenisii Cogn.	Melastomaceae
Miconia tetrandra (Sw.) D. Don	Melastomaceae
Micropholis garciniifolia Pierre	Sapotaceae
Micropholis guayanensis (A. DC.) Pierre	Sapotaceae
Mikania cordifolia (L.f.) Willd.	Asteraceae
Mikania fragilis Urb.	Asteraceae
Mimosa ceratonia L.	Fabaceae
Mimosa pudica L.	Fabaceae
Musa sp.	Musaceae
Myrcia deflexa (Poir.) DC.	Myrtaceae
Myrcia fallax (A. Rich.) DC.	Myrtaceae

Myrcia leptoclada DC.	Myrtaceae
Myrcia splendens (Sw.) DC.	Myrtaceae
Myrsine coriacea (Sw.) R. Br.	Myrsinaceae
Nectandra turbacensis (Nees) Mez	Lauraceae
Neorudolphia volubilis (Willd.) Britton	Fabaceae
Nephrolepis biserrata (Sw.) Schott	Davalliaceae (P)
Nephrolepis exaltata (L.) Schott	Davalliaceae (P)
Nephrolepis rivularis (Vahl) Mett. ex Krug	Davalliaceae (P)
Nepsera aquatica (Aubl.) Naudin	Melastomaceae
Neurolaena lobata (L.) Cass.	Asteraceae
Nicolaia elatior (Jack) Horan.	Zingiberaceae
Ochroma pyramidale (Cav. ex Lam.) Urb.	Bombacaceae
Ocotea floribunda (Sw.) Mez	Lauraceae
Ocotea leucoxylon (Sw.) Laness.	Lauraceae
Ocotea moschata (Meisn.) Mez	Lauraceae
Ocotea portoricensis Mez	Lauraceae
Ocotea spathulata Mez	Lauraceae
Odontonema cuspidatum (Nees) Kuntze	Acanthaceae
Odontosoria aculeata (L.) J. Sm.	Dennstaedtiaceae (P)
Odontosoria scandens (Desv.) C. Chr.	Dennstaedtiaceae (P)
Oldenlandia lancifolia (Schumach.) DC.	Rubiaceae
Oleandra articulata (Sw.) C. Presl	Dryopteridaceae (P)
Olyra latifolia L.	Poaceae
Oplismenus hirtellus (L.) P. Beauv.	Poaceae
Ormosia krugii Urb.	Fabaceae
Oxalis barrelieri L.	Oxalidaceae
Oxalis violacea L.	Oxalidaceae
Oxalis corniculata L.	Oxalidaceae
Oxandra laurifolia (Sw.) A. Rich.	Annonaceae
Palicourea croceoides Ham.	Rubiaceae
Panicum laxum Sw.	Poaceae
Paspalum conjugatum Berg.	Poaceae
Paspalum millegrana Schrad.	Poaceae
Paspalum pleostachym Doell.	Poaceae
Pasplaum plicatulum Michx.	Poaceae
Passiflora sexflora Juss.	Passifloraceae
Paullinia pinnata L.	Sapindaceae
Pavonia fruticosa (Miller) Fawc. & Rendle	Malvaceae
Pelexia adnata (Sw.) Spreng.	Orchidaceae
Peltapteris peltata (Sw.) Morton	Dryopteridaceae (P)
Pennisetum setosum (Sw.) Rich.	Poaceae
Peperomia alata Ruiz & Pav.	Piperaceae
Peperomia emarginella (Sw.) C. DC.	Piperaceae
Peperomia pellucida (L.) Kunth	Piperaceae
Peperomia rotundifolia (L.) Kunth	Piperaceae
Peperomia sintenisii C. DC.	Piperaceae
Pharus latifolius L.	Poaceae

Philodendron angustatum Schott	Araceae
Philodendron giganteum Schott	Araceae
Philodendron lingulatum (L.) C. Koch	Araceae
Philodendron scandens C. Koch	Araceae
Phoradendron hexastichum (DC.) Griseb.	Loranthaceae
Phoradendron piperoides (Kunth) Trel.	Loranthaceae
Phyllanthus niruri L.	Euphorbiaceae
Physalis angulata L.	Solanaceae
Phytolacca rivinoides Kunth & Bouché	Phytolaccaceae
Pilea inaequalis (Juss. ex Poir.) Wedd.	Urticaceae
Pilea krugii Urb.	Urticaceae
Pilea microphylla (L.) Liebm.	Urticaceae
Pilea nummulariifolia (Sw.) Wedd.	Urticaceae
Pilea semidentata (Juss.) Wedd.	Urticaceae
Pinzona coriacea Mart. & Zucc.	Dilleniaceae
Piper aduncum L.	Piperaceae
Piper blattarum Spreng.	Piperaceae
Piper glabrescens (Miq.) C. DC.	Piperaceae
Piper hispidum Sw.	Piperaceae
Piper jacquemontianum Kunth	Piperaceae
Piper umbellatum L.	Piperaceae
Piptocarpha tetrantha Urb.	Asteraceae
Pisonia subcordata Sw.	Nyctaginaceae
Pitcairnea angustifolia Aiton	Bromeliaceae
Pityrogramma calomelanos (L.) Link	Pteridaceae (P)
Plantago major L.	Plantaginaceae
Plectranthus scutellariodes (L.) R. Br.	Lamiaceae
Pleurothallis pruinosa Lindl.	Orchidaceae
Pleurothallis ruscifolia (Jacq.) R. Br.	Orchidaceae
Pluchea symphitifolia (Mill.) Gillis	Asteraceae
Polybotrya cervina (L.) Kaulf.	Dryopteridaceae (P)
Polygala paniculata L.	Polygalaceae
Polypodium astrolepis Liebm.	Polypodiaceae (P)
Polypodium aureum L.	Polypodiaceae (P)
Polypodium crassifolium L.	Polypodiaceae (P)
Polypodium dissimile L.	Polypodiaceae (P)
Polypodium latum (T. Moore) T. Moore ex Sodiro	Polypodiaceae (P)
Polypodium loriceum L.	Polypodiaceae (P)
Polypodium lycopodioides L.	Polypodiaceae (P)
Polypodium pectinatum L.	Polypodiaceae (P)
Polypodium phyllitidis L.	Polypodiaceae (P)
Polypodium piloselloides L.	Polypodiaceae (P)
Polyscias guilfoylei (W. Bull.) L. H. Bailey	Araliaceae
Polystachya concreta (Jacq.) Garay & H. R. Sweet	Orchidaceae
Polystachya foliosa (Hook.) Rchb.f.	Orchidaceae
Polytaenium feei (W. Schaffn. ex Fée) Maxon	Vittariaceae (P)
Prescottia oligantha (Sw.) Lindl.	Orchidaceae

Prescottia stachyodes (Sw.) Lindl.	Orchidaceae
Prestoea montana (Graham) Nicholson	Arecaceae
Pseudolmedia spuria (Sw.) Griseb.	Moraceae
Psidium guajava L.	Myrtaceae
Psychotria berteriana DC.	Rubiaceae
Psychotria brachiata Sw.	Rubiaceae
Psychotria deflexa DC.	Rubiaceae
Psychotria grandis Sw.	Rubiaceae
Psychotria guadalupensis (DC.) R. A. Howard	Rubiaceae
Psychotria maleolens Urb.	Rubiaceae
Psychotria uliginosa Sw.	Rubiaceae
Pterocarpus officinalis Jacq.	Fabaceae
Quararibaea turbinata (Sw.) Poir.	Bombaceae
Rajania cordata L.	Dioscoreaceae
Ravenia urbanii Engl.	Rutaceae
Renealmia jamaicensis (Gaertn.) Horan.	Zingiberaceae
Renealmia occidentalis (Sw.) Sweet	Zingiberaceae
Rhipsalis baccifera (J. S. Mill.) Stearn	Cactaceae
Rhynchospora nervosa subsp. *ciliata* (Vahl) T. Koyama	Cyperaceae
Rhynchospora radicans (Schltdl. & Cham.) H. Pfeiff.	Cyperaceae
Rolandra fruticosa (L.) Kuntze	Asteraceae
Rondeletia portoricensis Krug & Urb.	Rubiaceae
Rourea surinamensis Miq.	Connaraceae
Roystonea borinquena O. F. Cook	Arecaceae
Ruellia coccinea (L.) Vahl	Acanthaceae
Samyda spinulosa Vent.	Flacourtiaceae
Sapium laurocerasus Desf.	Euphorbiaceae
Sauvagesia erecta L.	Ochnaceae
Scaphyglottis modesta (Rchb.f.) Schldl.	Orchidaceae
Schefflera morototoni (Aubl.) Maguire *et al.*	Araliaceae
Schlegelia brachyantha Griseb.	Scrophulariaceae
Schradera vahlii Steyerm.	Rubiaceae
Scleria canescens Boeck.	Cyperaceae
Scleria cubensis Boeck.	Cyperaceae
Scleria microcarpa Nees ex Kunth	Cyperaceae
Scleria pterota C. Presl	Cyperaceae
Scleria secans (L.) Urb.	Cyperaceae
Securidaca virgata Sw.	Polygalaceae
Selaginella krugii Hieron.	Selaginellaceae (P)
Selaginella willdenovii (Desv.) Baker	Selaginellaceae (P)
Sesbania sericea (Willd.) Link	Fabaceae
Setaria parviflora (Poir.) Kerguélen	Poaceae
Sida rhombifolia L.	Malvaceae
Sida stipularis Cav.	Malvaceae
Simaruba cf. *glauca* DC.	Simarubaceae
Simaruba tulae Urb.	Simarubaceae
Sloanea berteriana Choisy	Elaeocarpaceae

Smilax domingensis Willd.	Smilacaceae
Smilax havanensis Jacq.	Smilacaceae
Solanum americanum Mill. var. *nodiflorum* (Jacq.) Edm.	Solanaceae
Solanum rugosum Dunal	Solanaceae
Solanum torvum Sw.	Solanaceae
Spathodea campanulata P. Beauv.	Bignoniaceae
Spermacoce assurgens Ruiz & Pav.	Rubiaceae
Spermacoce prostrata Aubl.	Rubiaceae
Spigelia anthelmia L.	Loganiaceae
Sporobolus indicus L.	Poaceae
Swietenia macrophylla King	Meliaceae
Symplocos martinicensis Jacq.	Symplocaceae
Synedrella nodiflora (L.) Gaertn.	Asteraceae
Syzygium jambos (L.) Alston	Myrtaceae
Tabebuia heterophylla (DC.) Britton	Bignoniaceae
Tabernaemontana citrifolia L.	Apocynaceae
Tectaria plantaginea (Jacq.) Maxon	Dryopteridaceae (P)
Tectaria trifoliata (L.) Cav.	Dryopteridaceae (P)
Teliostachya alopecuroidea (Vahl) Nees	Acanthaceae
Ternstroemia luquillensis Krug & Urb.	Theaceae
Tetragastris balsamifera (Sw.) Kuntze	Burseraceae
Tetrazygia urbanii Cogn.	Melastomaceae
Theobroma cacao L.	Sterculiaceae
Thelypteris angustifolia (Willd.) Proctor	Thelypteridaceae (P)
Thelypteris deltoidea (Sw.) Proctor	Thelypteridaceae (P)
Thelypteris opposita (Vahl) Ching	Thelypteridaceae (P)
Thelypteris patens (Sw.) Small	Thelypteridaceae (P)
Thelypteris poiteana (Bory) Proctor	Thelypteridaceae (P)
Thelypteris reticulata (L.) Proctor	Thelypteridaceae (P)
Tillandsia utriculata L.	Bromeliaceae
Tournefortia hirsutissima L.	Boraginaceae
Tournefortia maculata Jacq.	Boraginaceae
Trema micrantha (L.) Blume	Ulmaceae
Trichantha ambigua (Urb.) Wiehler	Gesneriaceae
Trichilia pallida Sw.	Meliaceae
Trichomenes crispum L.	Hymenophyllaceae (P)
Trichomenes punctatum Poir.	Hymenophyllaceae (P)
Trichomenes rigidum Sw.	Hymenophyllaceae (P)
Trimezia steyermarkii R. Foster	Iridaceae
Triphora surinamensis (Lindl.) Britton	Orchidaceae
Tripogandra serrulata (Vahl) Handlos	Commelinaceae
Turpinia occidentalis (Sw.) Don	Staphyleaceae
Urena lobata L.	Malvaceae
Urera baccifera (L.) Gaudich.	Urticaceae
Vanilla poitaei Rchb.f.	Orchidaceae
Vigna juruana (Harms) Verdc.	Fabaceae
Vitex divaricata Sw.	Verbenaceae

Vitis tiliifolia Roem. & Schult.	Vitaceae
Vittaria lineata (L.) J. E. Sm.	Polypodiaceae (P)
Vriesia macrostachya (Bello) Mez	Bromeliaceae
Wallenia pendula (Urb.) Mez	Myrsinaceae
Wullschlaegelia aphylla (Sw.) Rchb.f.	Orchidaceae
Xylosma schwaneckeanum (Krug & Urb.) Urb.	Flacourtiaceae
Youngia japonica (L.) DC.	Asteraceae
Zanthozylum martinicense (Lam.) DC.	Rutaceae

[a]Pteridophyte. All others are Anthophyta or flowering plants.

3

Microorganisms

D. Jean Lodge

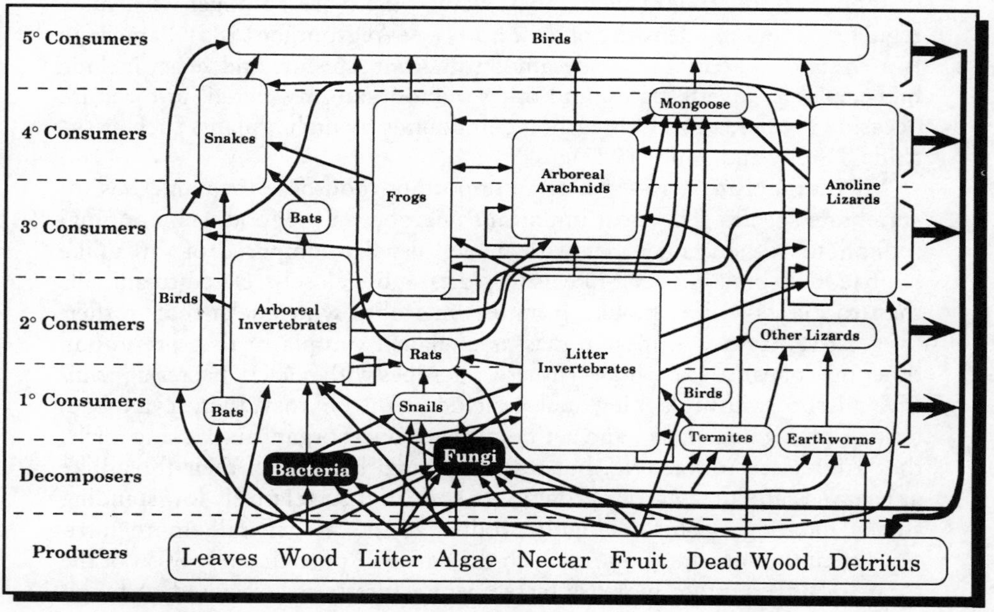

MICROORGANISMS are particularly important in making nutrients available to higher plants through the decomposition of organic matter; however, they also play other important roles in the food web. For instance, some bacteria and fungi are beneficial symbionts that provide mineral nutrients to their host in exchange for energy-containing compounds (e.g., mycorrhizal fungi, and nitrogen-fixing bacteria and actinomycetes), some live in the guts of invertebrates and are essential for digestion of food (e.g., microflora in termites) while others are pathogens that may regulate population densities of their hosts (see Augspurger 1983). Parasitism is a common nutritional mode among microorganisms, and hosts include higher plants, invertebrates, and other microorganisms. Fungi, in turn, are occasionally parasitized for carbon compounds by higher plants such as orchids (Harley and Smith 1983).

Microorganisms also serve as an important food base for numerous invertebrate species. The most important microbes in the food web are fungi (nonphotosynthetic microbes with nuclei, usually composed of threadlike hyphae, but sometimes cell-like as in yeasts; more closely related to animals than to plants), slime molds (primitive microbes with nuclei that in their "vegetative" state are mobile masses of naked cytoplasm that prey upon other microbes), algae (photosynthetic microbes with nuclei), bacteria (primitive microorganisms lacking nuclei, including the photosynthetic cyanobacteria or blue-green algae), and actinomycetes (microorganisms lacking nuclei that are related to bacteria, but resemble fungi because of their threadlike appearance and filamentous growth). In spite of their relatively low standing stocks, microorganisms are highly productive and are generally more nutritious than the substrates upon which they grow. Approximately 60% of the total soil invertebrate biomass in temperate forests is composed of fungivores that feed directly on fungal mycelia and fruiting bodies (McBrayer and Reichle 1971). Many invertebrates are specialized for feeding on microorganisms (e.g., fungivorous nematodes, plant hoppers and orabatid mites,

The Forest Products Laboratory is maintained in cooperation with the University of Wisconsin. The final revisions of this article were prepared by U.S. Government employees on official time, and it is therefore in the public domain and not subject to copyright.

certain staphalynid rove beetles, fungus-gardening ants, and fungus gnats; Pfeiffer, chap. 5, this volume; Garrison and Willig, this volume), whereas others are detritivores that depend on previous degradation and nutrient enrichment of their substrates by microorganisms (e.g., *Nasutitermes* termites; McMahan, this volume). Microorganisms play many roles in food webs, and not all of these relationships can be discussed in this chapter. Emphasis has therefore been placed on reviewing published and unpublished data on microbes that were collected at El Verde.

TAXONOMIC SURVEYS

Most taxonomic studies of microorganisms at El Verde are on fungi, but even this group is poorly known. Of the 400 species of higher basidiomycetes and ascomycetes we have identified at El Verde (an estimated 30% of the total number occurring), only 27% were previously reported in Puerto Rico (Stevenson 1975). Only 53% of the mushrooms are found in Pegler's (1983) agaric flora of the nearby Lesser Antilles, and 10 to 20% of the macrofungal species are undescribed (e.g., Lodge 1988; Singer and Lodge 1988; Lodge and Pegler 1990; Rogers et al. 1991; Laessøe and Lodge 1994). These data suggest that even for the most conspicuous group of fungi, Caribbean floras are too poorly known to speculate on such subjects as dispersal and island biogeography. However, a high incidence of homothalism (selfing) and secondary homothalism among agaric fungi at El Verde as compared to the mainland United States (R. H. Petersen pers. comm.) suggests that colonization of islands may be more successful among variants that do not need to encounter another individual of the opposite mating type in order to reproduce sexually.

DENSITY AND BIOMASS

Limitations to the Study of Microorganisms

Our knowledge of microorganisms at El Verde, as in most other places, is only fragmentary. Barriers to our understanding include the overwhelming diversity of microorganisms and the limitations of techniques used to detect them. Fortunately, estimates of microbial biomass and production can be made without reference to specific taxa. Such estimates provide no information about linkages among microorganisms, but are useful measures of potential availability of microorganisms as food for other organisms in the food web.

Unfortunately, all of the techniques used for quantifying microorganisms have limitations. For example, dilution-plate culture methods may give reasonable estimates of viable bacteria populations, but they cannot be used to determine fungal biomass or the relative abundances of all fungal species. Similarly, although we have reasonable estimates for total microbial respiration at El Verde, we cannot separate the relative contributions of fungi and bacteria to soil respiration using the popular selective inhibition technique (Anderson and Domsch 1975). This is in part because of the high incidence of streptomycin resistance in our bacterial populations (Holler 1966; Perry 1970). The high rates of resistance to antibiotics may indicate exposure to these or related substances in the environment. Antibiotic production is thought to be a mechanism by which one microorganism interferes with closely related species or strains competing for the same resources (Fredrickson and Stephanopoulos 1981).

Microbial Biomass in Wood

Microbial biomass (all values given are for dry weight) in wood has not been measured directly at El Verde, but production of fungal and bacterial biomass can be roughly estimated from dead wood standing stocks (table 3.1), wood decomposition rates (table 3.2), and microbial carbon assimilation efficiencies (table 3.2). If carbon losses are equated with inputs, annual production estimates in woody debris at El Verde range from 317 to 793 g m^{-2} yr^{-1} for fungi and from 15 to 35 g^{-1} m^{-2} yr^{-1} for bacterial plus actinomycetes (table 3.2). The production of fungal mycelium is likely to be greater in (wet) tropical than in temperate forests because the rate of organic matter

Table 3.1. Input of dead wood and loss through decomposition at El Verde

Annual input of logs	197 g m^{-2} y^{-1} [a]
Annual input of branches	412 g m^{-2} y^{-1} [a]
	156 g m^{-2} y^{-1} [b]
	47 g m^{-2} y^{-1} [c]
Annual weight loss in butt logs	10.9% y^{-1} [a]
Half-life for logs	6 y
Dead wood standing stocks	242,000 g m^{-2} [d]
Annual carbon loss from wood decomposition	13,189 g m^{-2} y^{-1} [e]
	122–305 g m^{-2} y^{-1} [f]
	64 g m^{-2} y^{-1} [g]

[a] From Odum 1970c (I-10). Branches were defined as greater than thumb sized.

[b] From Zou et al. 1995.

[c] From Ovington and Olson 1970 (H-65).

[d] From A. Lugo and V. Marinelly-Ortiz pers. com.

[e] Calculated from annual weight loss in butt logs, 50% carbon and dead wood standing stocks.

[f] Calculated assuming the rate of wood input equals decomposition.

[g] Calculated from Odum 1970 using an estimate of 0.35 g dry wood loss m^{-2} d^{-1} and 50% carbon by weight.

Table 3.2. Microbial production in wood at El Verde

Annual C loss from wood decomposition	122–305 g m^{-2} yr^{-1}[a]
Fraction of C respired by microbes	90%[b]
Annual C loss from wood through microbial respiration	110–275 g C m^{-2} y^{-1}
Fungal fraction of live microbial biomass in wood	92%[c]
Fungal C assimilation efficiency	60%[d]
Bacterial C assimilation efficiency	45%[d]
Fungal fraction of microbial respiration	90.4%[e]
Bacterial fraction of microbial respiration	9.6%[e]
C respired by fungi in wood	99–249 g C m^{-2} y^{-1}
C respired by bacteria in wood	11–26 g C m^{-2} y^{-1}
C flow into fungal production in wood	149–373 C m^{-2} y^{-1}
C flow into bacterial production in wood	9–21 g C m^{-2} y^{-1}
Production of dry fungal biomass in wood	317–793 g dry m^{-2} y^{-1}
Production of dry bacterial biomass in wood	15–35 g dry m^{-2} y^{-1}

[a]From table 3.1.

[b]From Reichle 1977.

[c]From Frankland 1982, Cumbria, United Kingdom.

[d]From Holland and Coleman 1987. Carbon assimilation efficiencies range from 40 to 70% in fungi, and from 20 to 51% in bacteria; the remaining carbon goes to respiration.

[e]Calculated from the fractions of live biomass constituted by fungi and bacteria, and their production efficiencies.

[f]Calculated from the microbial efficiencies above and the total microbial carbon from wood per year.

[g]Based on the percentage of carbon by dry weight for fungi in wood (Lodge 1987a), and the percentage of carbon by dry weight for bacteria (Van Veen and Paul 1979).

input is greater and because higher temperatures and wet conditions result in rapid growth and decay (Frankland 1982). Standing stocks of fungi and bacteria cannot be estimated for El Verde because turnover times for microorganisms in wood are unknown.

Microbial Biomass and Nutrient Stores in Fine Litter

Fungi are frequently so abundant in the litter layer at El Verde that their mycelial fans obscure leaf surfaces (fig. 3.1). A direct count method (Lodge and Ingham 1991; Lodge 1993) was used to estimate mean fungal biomass (5.2 mg g^{-1} litter, range 1.7 to 9.5 mg g^{-1} litter, Lodge 1993) from samples collected throughout 1985. Biomass estimates for fungi in litter at El Verde (mean 1.5 g m^{-2}, range 0.5 to 2.7 g m^{-2}, based on 282 g fine liter m^{-2}, Zou et al. 1995) were similar to estimates by Witkamp (1974) for yellow poplar forest in Tennessee (8 g m^{-2}), considering that forest floor litter mass was higher in the temperate forest than at El Verde. Fungal biomass changed rapidly in response to moisture. It tripled in response to rains during the first 2 weeks of March, then decreased by half as the litter dried in the following 2 weeks (Lodge and Ingham 1991; Lodge 1993).

Figure 3.1. Heavy growths of mushroom mycelia (such as this mycelium of *Collybia johnstonii*) on leaf litter contain a significant proportion of the phosphorus pool in the litter layer at El Verde. Mycelial fans and threadlike hyphal strands or cords are visible. The fungus uses these cords to colonize fresh litter and transport nutrients into its new food base.

Most fluctuation in fungal biomass probably occurs in the upper litter layer (zero to three- or four-month-old leaves at El Verde) because this layer experiences the greatest fluctuation in moisture (Hedger 1985). Peaks in litter fungal biomass were observed later in the year when rain followed short dry periods, but fungal biomass subsequently decreased even when wet weather persisted (Lodge 1993). A similar pattern was found for hyphal length on dialysis tubing placed in the litter layer at three-day intervals during the rainy season (Lodge unpublished). Decreases in fungal populations during rainy periods might be caused by grazing of fungi by invertebrates. Newell (1984) showed that selective grazing by a collembollan in spruce litter in the United Kingdom altered the vertical distribution of fungi in the litter layer. Litter invertebrate populations at El Verde fluctuate with litter moisture, but they increase slowly following long dry periods (Pfeiffer, chap. 5, this volume). Bacterial biomass in fine litter has not been studied at El Verde.

Nutrient stores in fungal biomass in fine litter (table 3.3) were calculated from an estimated fungal mass (Lodge 1993) and the nutrient concentrations determined in field-collected mycelia at El Verde (Lodge 1987a). According to the total nutrient store estimates for fine litter at El Verde (Edmisten 1970a; Odum 1970c) fungi contain 1.6% of the total nitrogen pool and

Table 3.3. Fungal nutrient stores at El Verde

	Nutrient content (g m^{-2})					
Substrate	N	P	K	Ca	Mg	Na
Litter	0.054–0.154	0.013–0.066	0.008–0.031	0.097–0.258	0.006–0.015	0.006–0.015
Soil[a]	0.595–15.130	0.170–3.400	0.090–2.040	0.680–16.070	0.085–1.450	—

Sources: Estimates of fungal nutrient stores were obtained using estimates of fungal biomass in Lodge (1993) and fungal nutrient concentrations in Lodge (1987a).

[a]Upper 10 cm; estimates based on a mean soil bulk density of 0.85 g cm^{-3}.

22.2% of the total phosphorus pool in litter (Lodge 1993). Fungal phosphorus concentrations are highly variable, however, and can increase by an order of magnitude in mycelia just prior to fruiting (Lodge 1987a). Therefore in litter mats dominated by vigorous prefruiting mushroom mycelia, up to 85% of the litter phosphorus can be accounted for by fungal biomass (Lodge 1993).

Soil Microbial Biomass

I estimated a mean of 207 g m^{-2} of fungal biomass (live and dead) in the upper 10 cm of soil not covered by boulders (daily means ranged from 14 to 333 g m^{-2}) using a direct-count method (Lodge 1993). Landscapewide fungal biomass in the upper 10 cm of soil may be lower because about 20% of the study area (Plot 3) was covered by boulders that prevented inputs of aboveground organic matter to the soil below. The total mean fungal biomass in litter plus the upper 10 cm of soil at El Verde (216 g m^{-2}) was greater than for temperate forest study sites in the United Kingdom and southeastern United States (Frankland 1982, United Kingdom; Ausmus and Witkamp 1973, 74 g m^{-2} for fungi + bacteria, southeastern United States) but only a fraction of the soil fungal biomass is alive (Frankland 1975, 1982, 53 to 61%).

Witkamp (1970) determined the densities of viable bacteria in the upper 2.5 cm of soil at El Verde by dilution plating, and these ranged from 30 to 531 × 10^{6} bacteria cm^{-3}. The higher densities were found in the irradiated area, but the increased densities were likely because of the change from dry to wet season, and possibly from greater nutrient inputs and more favorable microclimate (comparable samples were not taken in the undisturbed control area or the cut area; Cowley 1970a). Total (live and dead) bacterial biomass at El Verde was estimated to be 1,364 to 24,091 g m^{-2} in the upper 10 cm of soil, but this appears to be an overestimate (see assumptions in app. 3.C).

A second estimate of bacterial biomass for El Verde (381 to 421 g m^{-2}) was extrapolated from data on total soil microbial biomass for a nearby site with seventeen-yr-old native second growth forest and pine plantation (same elevation and exposure). This estimate for total microbial biomass less the estimated fungal biomass in soil at El Verde (207 g m^{-2} to 10 cm) gives an estimated 174 to 214 g m^{-2} to 10 cm depth for the remaining microbial biomass (mostly bacteria and actinomycetes). However, more recent data obtained by using a direct count method (X. Zou and E. Ingham pers. comm.) indicate that fungal biomass is greater than bacterial biomass in the upper 10 cm of tabonuco forest soil.

Perry (1970) isolated 23,500 to 42,000 actinomycete colonies g^{-1} substrate in the lower litter plus the upper 4 to 6 cm of undisturbed soil at El Verde, but the colony count decreased to 9,000 g^{-1} after irradiation. These numbers are difficult to convert to biomass for the same reasons that fungal biomass cannot be estimated from plate counts.

Soil Microbial Nutrient Stores

Microbial nutrient stores in the upper 10 cm of soil represent substantial portions of the total soil pool for some elements at El Verde (Lodge 1993; table 3.3). Mean fungal stores represent 0.8 to 20.0% of the acid extractable soil phosphorus, and 23.6% of the total soil calcium, but only 3.6% of the extractable soil potassium, 3.1% of the magnesium, and 0.5% of the nitrogen in the upper 10 cm of soil. Bacterial and actinomycete nutrient stores based on the second method of biomass estimation (nutrient concentrations from Van Veen and Paul 1979 and densities from Witkamp 1970) were 21.4 to 24.9 g N m^{-2} and 3.7 to 4.3 g P m^{-2} of soil to a depth of 10 cm. Bacterial nutrient stores at El Verde represent 6.1 to 7.1% of the total nitrogen and 21.8 to 25.3% of the total phosphorus in the upper 10 cm of soil.

EFFECTS OF DISTURBANCE ON FUNGI

Litter basidiomycetes with superficial mycelia are especially sensitive to fluctuations in moisture (Hedger 1985). For example, I found that mycelia of *Collybia johnstonii* (fig. 3.1) disappeared from all but three of its twenty regularly surveyed sites in the El Verde Research Area for two years following Hurricane Hugo. Mycelia reappeared in nine previous sites within five years post-hurricane, as well as in three new locations. In another study, Asbury and Lodge (unpublished) placed rain canopies over the forest floor and found that eight mushroom species stopped colonizing new leaf litter under dry conditions. A study of abundance and frequency of basidiomycete fruiting bodies on fine litter at El Verde (O. O. Molina-Gomez, D. J. Lodge, J. Zimmerman, and S. Cantrell unpublished) showed that repeated fertilization of forest plots dramatically reduced litter basidiomycete reproduction and diversity, while a one-time removal of litter following Hurricane Hugo only moderately reduced fungal fruiting body abundance and diversity as compared to control plots. Superficial mycelia disappeared from the fertilized plots, suggesting salt stress.

Various studies of fungal communities were conducted in the 1960s in an area irradiated by a gamma source (Radiation Center), an area where trees were cut and the logs and slash were left in place (Cut Center), and a matched area of undisturbed forest (South Control Center; see fig. 1.3). Cowley (1970a) examined the effect of radiation on populations of microfungi in litter and soil at El Verde. A high degree of similarity was noted between populations of microfungi on like litter species in the control and radiation areas before irradiation. Irradiation reduced the similarity in microfungal communities between the disturbed and the control sites. Diversity of microfungi in soil was higher in undisturbed forest than in either the cut or the irradiated area (Holler and Cowley 1970). Bray-Curtis ordination showed that microfungal populations in soil were similar in all areas before treatment, but the similarities declined after irradiation and cutting. The soil populations in the undisturbed and radiation sites were slightly more similar to each other than either was to populations in the cut site.

Cowley (1970a) studied changes in frequencies of macrofungal fruiting bodies following irradiation and cutting. Most of the differences in fruiting were attributable to differences in moisture caused by differences in canopy closure, and to availability of organic substrata (Cowley 1970b). *Schizophyllum commune* was only found in open areas and is known to tolerate drying, whereas seven species of basidiomycetes and ascomycetes were only found in shaded areas.

Production of fruiting bodies is often a reproductive response of the mycelium to environmental stimuli or "cues" (Hedger 1985), rather than a reflection of changes in mycelial abundance (Hering 1982). For example, David Janos (pers. comm.) notes that fruiting in *Schizophyllum* species in Latin America is stimulated by burning. Therefore, the absence of fruiting bodies of a given species does not necessarily mean that the mycelium is absent, and the importance of a species should not be judged by the abundance of fruiting bodies. I have found many wood-rotting basidiomycetes that rarely fruit at El Verde (e.g., *Hydropus citrinus, Gerronema rhyssophyllus, Lentinula boryana, Xerulina asprata,* and *Mycena cuspidatapilosa*), but which are abundant and widespread when they fruit. Infrequent fruiting in tropical basidiomycetes has also been noted by Singer and Araujo in Brazil (1979) and Hedger in Ecuador (1985). This pattern is most common in species with stable microenvironments (e.g., soil, large wood, and humus) and massive fruitings (Hedger 1985). Fruiting in these species may depend on the buildup of sufficient mycelial biomass and is sometimes seasonal (Hedger 1985).

For example, among the terrestrial species at El Verde, *Hygrocybe melleofusca* has been found only during late August and early September, whereas *H. batistae* has been found only from late February through April. Some fungi fruit every second or third year at El Verde, including most terrestrial and humicolous species in the Entolomataceae, many humicolous *Lepiota* species, and some *Collybia* species on wood.

MICROORGANISMS IN THE FOOD WEB

Wood Decomposers

Although fungi are usually the primary agents of wood decomposition (Swift 1977; Rayner and Todd 1979) and the fruiting bodies of macrofungi are conspicuous on wood at El Verde, there has been relatively little taxonomic work on fungal wood decomposers in Puerto Rico. Stevenson (1975) summarized all previous collections of fungi from Puerto Rico, but few of those are from the Luquillo Mountains. United States Department of Agriculture-Forest Service surveys of forest pests in Puerto Rico were conducted in 1973, 1976, 1981, and 1984 and included fungi that infect trees, but emphasis was placed on introduced tree species and plantation forestry. From 1984 to 1995, I conducted the most extensive taxonomic survey of macrofungi growing on wood at El Verde, but more emphasis was placed on agarics (especially in the Tricholomataceae) and members of the Xylariaceae than on the equally important members of the Polyporaceae, Ganodermatacea, Corticiaceae, and Thelephoraceae. These species and previous fungal records are listed by substrate size in appendix 3.A as follows: logs, >10 cm diameter; branches, >1 to <10 cm diameter; and twigs, <1 cm diameter.

Fine Litter

Fine litter is primarily composed of dead leaves, but flowers and soft fruits are also included. The only extensive taxonomic survey of macrofungi growing on leaf litter at El Verde was conducted by the author (app. 3.A). A few litter fungi were previously reported from the Luquillo Mountains, according to a bibliographic survey of the fungi of Puerto Rico (Stevenson 1975). Holler (1966) identified some of the microfungi growing on leaf litter at El Verde (app. 3.A). Cowley (1970b) examined the effect of leaf litter species on microfungal communities. Although fungal species composition was largely the same among different litter species, populations of microfungi differed among litter species due to changes in dominance among fungi. Padgett (1976) found a low diversity of aquatic hyphomycetous fungi in six leaf litter species submerged in a stream at El Verde (app. 3.A). Diversity may have been low because of fungal specificity for leaf species and the low number of leaf types examined. In contrast, a foam spore trap placed in the same stream collected a high diversity of fungal spores.

Humus

Some of the fungal species that occur in the lower litter layer also occur on soil with high concentrations of organic matter or on well-rotted wood. These species (e.g., *Leucoagaricus spp., most Lepiota* spp., and some members of the Entolomataceae at El Verde) are more accurately called humicolous. Hedger (1985) has demonstrated in Ecuador that humicolous species of *Leptiota* require substrates which have been partially decomposed by other fungi. Humicolous species have not been listed separately in appendix 3.A because data are insufficient to classify many species, but fungi occurring on very rotten wood and those species occurring on more than one substrate are likely to be humicolous.

Soil

Holler (1966) was able to identify some species of microfungi isolated from soil at El Verde (app. 3.A). The majority of microfungi isolated from soil were not found in either roots or surface litter (fifty-three of seventy-seven morphospecies, Holler and Cowley 1970). Cowley (1970b) surveyed macrofungi at El Verde and found that 20 to 30% of all species found and 5 to 15% of the fruiting bodies occurred on soil. Unfortunately, Cowley (1970b) identified only a few species of macrofungi. I conducted a survey of terrestrial macrofungi at El Verde from 1983 to 1992 (app. 3.A). The most conspicuous, brilliantly colored, and abundant soil mushrooms are in the Hygrophoraceae (Lodge and Pegler 1990). The Hygrophoraceae is common in wet forests above 250 m elevation in the Lesser Antilles (see Pegler 1983), and Costa Rica (pers. observation). Many members of the Entolomataceae are also found on mineral soil at El Verde.

Predators and Parasites of Other Organisms

Alexopoulos (1970) conducted a brief survey of slime molds (myxomycetes) at El Verde (app. 3.A). Myxomycetes get their nutrition by digesting other microorganisms. Hagelstein (1932) published a revision of the myxomycetes in Puerto Rico and the Virgin Islands that includes his own records and those of Seaver and Chardon (1926). Although none of the species listed by Hagelstein (1932) were collected from the Luquillo Mountains, many are widespread species that probably occur in El Verde and have therefore been included in appendix 3.A. Some hyperparasitic fungi (parasites on other fungi) such as *Hypomyces* and *Nectria* from El Verde have been identified by G. J. Samuels, C. T. Rogerson, and B. Spooner (app. 3.A), but many of the species remain undescribed. The mycelium of a wood decomposer fungus at El Verde (*Hyphodontia* sp., Corticiaceae, H. H. Burdsall, Jr. pers. comm.) was found to have spines for trapping nematodes. Various other species of wood-decomposing fungi (e.g., *Hohenbuehelia, Pluteus,* and *Pleurotus* spp.) are likely to be similarly equipped for killing and digesting nematodes to obtain their nitrogen.

Epiphyllic and Endophytic Microorganisms

Cowley (1970b) examined vertical stratification and host specificity of microfungal communities of leaves of two tree species at El Verde. Samples of leaves were collected at heights of 3 to 6 m, 10 m, and 15 to 20 m above ground from two tree species, *Manilkara bidentata* and *Dacryodes excelsa* (methods in app. 3.C). The occurrence of only twenty-two of the 101 fungal species on both hosts suggested some specificity, but most of the fungi are rare. Of the forty-nine more frequently isolated fungal species, however, only twenty-one were isolated from both tree species. Spores lying on leaf surfaces may have contributed to the greater number of fungal isolates that were recovered from the lower levels of the canopies (Cowley 1970c). Comparisons of the most frequent fungal isolates in the uppermost canopies, where spore contamination was less, showed that five species were only on *D. excelsa,* eight were only on *M. bidentata,* and just three were found on both tree species. Some of these fungi are epiphyllic organisms that grow on leaf surfaces and absorb nutrients leached from their host, some are probably pathogenic parasites, some are asymptomatic and possibly beneficial endophytes, and some are facultative parasites that colonize senescent tissue and then play a role in early decomposition (e.g., some *Phoma* and *Pestalotia* spp.).

Edmisten (1970b) identified epiphyllous algae as species of *Nostoc, Scytonema, Anabeana,* and *Calothrix.* I have found several species of sooty molds growing on excrement of mealy bugs (Pseudococcidae or Ericoccidae) on canopy leaves. Other fungi grow in or on leaves at El Verde, including *Mycena citricolor* (a plant pathogen) on *Psychotria berteriana,* and two apparently undescribed mushrooms (*Crinipellis* sp. and *Marasmius* sp.) on diseased petioles of *Schefflera morototoni* that were collected in the canopy (Laessøe and Lodge 1994). Nine species of xylariaceous ascomycetes (e.g., *Xylaria cubensis,* and *X.* aff. *arbuscula*) were cultured from healthy petioles of *S. morototoni* collected from the canopy and are probably nonpathogenic endophytes (Laessøe and Lodge 1994). Most of the surface-disinfested 1 cm long petiole segments tested contained fungi in the Xylariaceae. Species of *Corticium* and *Marasmiellus,* and *Marasmius crinis-equi* grow on the surfaces of living shrubs and saplings at El Verde and trap organic debris from the canopy using aerial networks of rootlike structures called rhizomorphs (fig. 3.2). Hedger (1985, 1990) has described a similar community of aerial agarics from a seasonal forest in Ecuador.

Fine Roots

Associations with Mycorrhizal Fungi and Nitrogen-Fixing Bacteria

Edmisten (1970a) found active nitrogen-fixing nodules on roots of four leguminous tree species at El Verde. Plants associated with nitrogen-fixing bacteria were *Andira inermis, Inga fagifolia, I. vera,* and *Neorudolphia volubilis.* Edmisten (1970a) also mentioned the abundance of active nodules on an understory herb (*Desmodium* sp.) at El Verde. Taxonomy of mycorrhizal fungi was studied at El Verde by M. K. Kellam and D. J. Lodge (app. 3.A). Spores of eight species of *Acaulospora, Gigaspora,* and *Glomus* that form vesicular-arbuscular (VA) endomycorrhizae were identified from roots and soil (two other species could not be identified, Kellam and Lodge unpublished). In addition, I isolated a mycorrhizal *Rhizoctonia*-like fungus from

Figure 3.2. Rhizomorphs (rootlike structures) of the horse-hair mushroom (*Marasmius crinis-equi*) form litter-trapping nets in the understory at El Verde. This litter decomposer fungus avoids competition from other fungi that are confined to the forest floor.

rhizomes of an achlorophyllus plant (*Apteria aphylla* [Nutt.] Burm.). I have collected several species of *Russula* and *Lactarius* (Basidiomycotina, Agaricales, Russulaceae) under *Coccoloba swartzii* at El Verde that are apparently forming peritrophic mycorrhizae with this tree species.

Mycorrhizae

Most higher plants form symbiotic relationships with mycorrhizal fungi (Harley and Smith 1983) which facilitate nutrient uptake (reviews by Bowen 1980; Tinker 1980; Harley and Smith 1983) and enhance water relations (Safir et al. 1972; Reid 1979; Sieverding 1984). There are several different types of mycorrhizae (Harley and Smith 1983) but almost all plants form vesicular-arbuscular endomycorrhizae (VAM) with primitive water mold fungi in the Endogonaceae (Gerdemann 1968). Most low-elevation wet tropical forests are dominated by plants which form VAM (Janos 1983), although some successional communities (Singer and Morello 1960; Janos 1980a) and areas with seasonal dry periods or very nutrient-poor soils are dominated by ectomycorrhizal fungi (e.g., the white sand region of the Amazon, some leguminous forests in Africa and dipterocarp forests in Asia). Ectomycorrhizae are usually formed with basidiomycetous fungi, but some are formed with higher ascomycetes.

Formation of VA-endomycorrhizae versus ectomycorrhizae may be ecologically significant, because VAM fungi do not appear to be capable of litter decomposition whereas some ectomycorrhizal fungi apparently have the capacity to break down organic matter (Trojanowski et al. 1984; Haselwandter et al. 1990; Read et al. 1989). This is sometimes called direct nutrient cycling (e.g, Went and Stark 1968). Although there is ample evidence that mycorrhizal fungi remove nutrients directly from leaf litter, there is controversy over the importance of ectomycorrhizal fungi in litter decomposition. Ectomycorrhizal forests frequently have greater accumulations of litter than forests dominated by VA-endomycorrhizal plants because of slower decomposition rates. It was therefore hypothesized that ectomycorrhizal fungi might inhibit litter decomposition by competing for nitrogen and phosphorus with strict decomposers (Gadgil and Gadgil 1971). Poor litter quality, however, may contribute to slow rates of decomposition in forests dominated by ectomycorrhizal species. Ectomycorrhizal plant communities typically develop on nutrient-poor soils, and these plant species often have low nutrient concentrations in their foliage.

Regardless of the roles of mycorrhizal fungi in litter decomposition, ectomycorrhizal fungi may provide their hosts with a competitive advantage over VAM hosts under extreme conditions of nutrient or water stress. Ectomycorrhizal fungi may also promote regeneration of their hosts through transfers of carbohydrates from adult trees to conspecific seedlings growing in the shade beneath them (Newman 1988). In addition, there is some host-specificity among ectomycorrhizal fungi but less specificity in VAM fungi, and this has implications for competition among different plant species interconnected by VAM fungi (Janos 1983; Newman 1988). Interspecific transfer of ^{32}P via VAM fungi has been demonstrated by Chiarello et al. (1982).

Mycorrhizal fungi contribute a significant proportion of root weight and might be expected to increase respiratory demand for carbohydrate by more than 10 to 50% of that used by host tissue (Bevege et al. 1975; Harley and Smith 1983). Experiments have shown three to eight times as much carbon is translocated to mycorrhizae as to nonmycorrhizal roots on a relative weight basis (Shiroya et al. 1962; Nelson 1964; Bevege et al. 1975; Harley and Smith 1983). Some of the additional translocated carbon supports extensive hyphal systems of the mycorrhizal fungi in the soil. This may explain why total carbon loss measured by belowground respiration exceeds the sum of organic matter inputs and root respiration in many ecosystems (Harley and Smith 1983) including El Verde (W. T. Lawrence pers. comm.). Estimates of mycorrhizal fungal carbon costs typically range from 10 to 25% of gross primary production, which would amount to 1,197 to 2,993 g C m^{-2} yr^{-1} at El Verde, given an estimated gross primary production of 32.8 g m^{-2} d^{-1} (Odum 1970b).

Edmisten (1970c) examined thirty-two tree species for associations with mycorrhizal fungi in the El Verde rain forest, but his percentages of ectomycorrhizal (22%) and nonmycorrhizal (28%) species were unusually high when compared to most other low elevation wet tropical sites (Janse 1896; Johnston 1949; Redhead 1968, 1980; Thomazini 1974; Janos 1980b; St. John 1980b; St. John and Uhl 1983). Some of Edmisten's (1970c) records were divergent from the usual condition in these families or species (e.g., ectomycorrhizae on *Cecropia schreberiana* and on a palm, *Prestoea montana*).

I resurveyed mycorrhizal relationships of woody plants at El Verde from 1983 to 1987 (Lodge 1987b; methods in app. 3.C), including twenty-one of the thirty-two species examined by Edmisten (1970c). One of the forty-eight species examined was nonmycorrhizal and forty-seven had VA mycorrhizae, which is similar to proportions in other low-elevation wet tropical forests (Lodge 1987b; app. 3.B). All of Edmisten's reports of ectomycorrhizae are erroneous. Edmisten (1970a) may have mistaken luxuriant growths of extramatrical hyphae of VAM fungi, pathogenic fungi, nonpathogenic endophytes, or primary root cap hairs for ectomycorrhizal fungi (app. 3.B). Three species which Edmiston did not examine, however, (*Pisonia subcordata, Coccoloba swartzii,* and *C. pyrifolia*) had 50 to 82% of their fine root length covered by thick mantles of basidiomycetous hyphae with clamp connections, but the ecological significance of these peritrophic associations is unknown (app. 3.C).

Biogeography and Mycorrhizal Associations

Native ectomycorrhizal and peritrophic mycorrhizal plant species are apparently rare in the Puerto Rican flora. Only several species of *Coccoloba* (*C. uvifera,* Kreisel 1971; Pegler 1983), *Pisonia* (pers. observation), probably *Neea buxifolia,* and possibly some leguminous tree species of the dry forests are associated with basidiomycetes. Even though ectomycorrhizal *Pinus occidentalis* is native to the Dominican Republic on the neighboring island of Hispaniola, pine has always been absent from Puerto Rico.

Most ectomycorrhizal fungi are spread long distances by airborne spores, whereas VA-mycorrhizal fungi have very large spores which are normally soilborne. Although one might deduce that wind-blown ectomycorrhizal fungal spores could more easily colonize islands than soilborne VAM spores, VAM fungi are the dominant symbionts in island vegetation. VAM spores have been recovered from the feet of birds and wasps (McIlveen and Cole 1976) which may help explain how VAM fungi reach islands. Initial introductions of pine to Puerto Rico met with failure because of the lack of infection by ectomycorrhizal fungi (see history summarized by Hacskaylo and Vozzo 1967; see app. 3.C).

Miscellaneous Fungi Growing in Roots

Various nonmycorrhizal fungi grow in live plant roots, and these primary consumers may extract significant quantities of carbon from their host plant. Microfungi cultured from fine root mats at El Verde by Holler (1966) are listed in appendix 3.A. Except for some *Rhizoctonia* species, these fungi are generally nonpathogenic. Four of the forty-seven woody plant species I examined for mycorrhizae had additional nondestructive intracellular infections by septate fungi (app. 3.B). It is not known if these fungi are pathogenic, neutral, or beneficial. Similar infections have been found in Puerto Rico by Paul Bayman (Xylariaceae in orchid roots, pers. comm.), in Brazil by Thomazini (1974) and Sharron Rose (pers. comm.), in Australia by Sally Smith (*Pyronophora* sp., pers. comm.), and in Ecuador by John Hedger (*Diheterospora* sp.; pers. comm.).

LINKAGES INVOLVING MICROORGANISMS IN THE FOOD WEB

Fungi as Food for Invertebrates

Fungi may be an especially important food base for some invertebrates because fungal nutrient concentrations (Lodge 1987a) as well as fungal biomass production are relatively high. Nitrogen, calcium, and potassium are 2.4 to 2.8 times more concentrated, and phosphorus is thirteen times more concentrated in fungal mycelium growing in leaf litter than in freshly fallen

Table 3.4. Nutrient concentrations in fungi and fungal components at El Verde. Ratios of nutrient concentrations in fungal mycelium: fresh substrate concentrations are also given.

Fungus/Fungal component	C	N	P	Ca	Mg	Na	K
Vegetative mycelium in wood[a]							
Concentration mg g^{-1}	467.4	38.6	3.02	44.0	3.60	3.54	2.105
Standard deviation	±36.7	±21.7	±2.21	±35.1	±4.32	±4.74	±2.14
Ratio of nutrient[a] concentrations in fungus: fresh substrate	7.1	12.5	12.2	1.6	ND	1.0	ND
Leathery fungal fruiting bodies on wood, mg g^{-1}							
Phanerochaete sp. *nova-c. flava*	552.3	31.9	2.19	ND	0.75	1.15	2.56
Polyporus sp.	521.5	33.4	5.15	ND	0.86	1.06	12.12
Polyporaceae, white effused-reflexed	458.6	43.9	2.13	43.42	1.69	1.56	5.11
Thelephora sp.	536.2	43.9	2.61	2.52	1.13	1.47	5.50
Fleshy fungal fruiting bodies on wood, mg g^{-1}							
Auricularia fuscosuccinea	506.0	10.5	ND	ND	ND	ND	ND
Favolus braziliensis	493.0	70.6	5.78	ND	0.95	0.64	13.78
Psathyrella sp.	485.9	79.4	12.6	0.43	1.50	1.58	2.56
Fleshy fungal fruiting bodies on leaf litter, mg g^{-1}							
Collybia johnstoni	523.9	90.0	11.19	2.29	1.66	0.94	±21.48
Collybia sp. Sect. *Subfumosa*	475.0	74.5	9.63	ND	0.94	2.48	21.67

[a]Data are from four samples of three species of lignicolous fungi, *Phanerochaete* sp. -*Coniphora flava* (misidentified as *Phlebia chrysochreas*), *Cyathus pallidus*, and an effused-reflexed polypore (Lodge 1987a).

leaves at El Verde (Lodge 1987a). Wood-rotting fungi generally maintain even higher concentrations of mineral nutrients relative to the concentration in their substrate than fungi growing on leaf litter (Stark 1972; Frankland 1982; Lodge 1987a). Mean elemental concentrations of vegetative mycelia of 4 basidiomycete species growing in wood at El Verde are given in table 3.4. Nitrogen, phosphorus, and calcium were 7.1 to 12.5 times more concentrated in mycelium than in the corresponding fresh wood substrate (Lodge 1987b; table 3.4). Certain bark beetles are known to take advantage of the ability of wood-decaying fungi to concentrate nitrogen by consuming mycelium and fungal-infested wood (Levi and Cowling 1969), but this has not been studied at El Verde. Sodium is an important nutrient for animals and is also concentrated by fungi (Stark 1972; table 3.4) even though it is not an essential element for fungal growth (Foster 1949). Calcium concentrations vary greatly among species, and most calcium may be in the form of calcium oxalate crystals in those species with high calcium concentrations (Frankland et al. 1978; Cromack et al. 1979; Lodge 1987a). Fungi may serve as the primary source of calcium and sodium for forest floor invertebrates because fungi are much richer in these elements than plant materials (Cromack et al. 1977). Earthworms and collembolans have gut microflora that are able to break down fungal calcium oxalate (Cromack et al. 1977), but it is not known whether or not snails can use fungal calcium for shell deposition.

Fruiting bodies may be especially attractive to consumers because those on wood have higher concentrations of phosphorus and potassium, and those on leaf litter have higher concentrations of nitrogen and potassium than their corresponding vegetative mycelia (Frankland 1982; table 3.4). In general, the fruiting bodies of "wood rotters" are gregarious (figs. 3.3 and 3.4) or are larger (fig. 3.5) and collectively have more mass than the fruiting bodies of species growing on fine litter or soil in the wet tropics (see Hedger 1985). The size and number of fruiting bodies produced by higher fungi is partly dependent on the size of their food resource. Because fine litter turnover is relatively fast at El Verde, and because litter inputs and decomposition rates are fairly uniform when compared to seasonal forests, the litter layer is rarely deep enough to support abundant fungal fruiting. Woody substrates, however, provide substantial food resources that are capable of supporting massive fungal fruitings. For example, the leathery *Laetiporus percisinus* specimen shown in figure 3.5 grew from a 50 cm diameter *Inga vera* tree and weighed 14 kg fresh, while another specimen from El Yunque weighed 21 kg.

Epiphyllous and Endophytic Microorganisms in the Food Web

Epiphyllous and endophytic microorganisms frequently initiate the process of leaf decomposition before abscission (Odum 1970c), and pathogenic endophytes can regulate host populations (Augspurger 1983; Gilbert in press). At El Verde, a pathogenic mushroom (*Mycena citricolor*) has killed some of the *Psychotria berteriana* shrubs that recruited in after Hurricane Hugo, and a pathogenic ascomycete (*Rosellinia bunodes*) has killed most of the *Hybiscus* shrubs planted next to the driveway of the El Verde Field Station (pers. observation). Asymptomatic endophytic fungi may benefit their hosts by producing compounds that are toxic to herbivores (Clay 1988), but such relationships have not been studied in tropical forests. Epiphyllous algae may be important in providing nitrogen to plants through their leaves. Harrelson (1969) estimated that 6.15 g N m^{-2} yr^{-1} was fixed by the epiphyllic flora at El Verde, and some of the nitrogen fixed by epiphylls is available to the plants on which they reside (Bentley and Carpenter 1980).

Microorganisms growing on living leaves and bark (fungi, algae, bacteria, and lichens) frequently support large populations of arboreal orabatid mites (Carroll 1980; Pfeiffer, chap. 5, this volume) and Psocoptera (Garrison and Willig, this volume) and may also be a major source of food for snails at El Verde. Snail feeding trails are common on understory leaves and are identifiable by the radula-scraped clean areas which stand out against the remaining heavy epiphyll cover. Heatwole and Heatwole (1978) studied feeding behavior of the large *Caracolus caracola* snail at El Verde, but their conclusion that algae and plant material were the primary foods were based on fecal analyses (Garrison and Willig, this volume). Fungi and bacteria are usually more digestible than filamentous algae, and their importance is often

Figure 3.3. A large *Caracolus* snail with one of its preferred food items, fruiting bodies of *Favolus brasiliensis*. Feeding damage by the snail can be seen on the margins of several fruiting bodies. Wood-decomposer fungi are not generally limited by the availability of carbon and often fruit in large troops.

Figure 3.4. Signs of snail feeding damage by *Caracola* to fruiting bodies of *Favolus brasiliensis*. Snail feces can be seen near the center of the photograph.

Figure 3.5. Some wood-decomposer fungi produce massive fruiting bodies as they are not limited by the supply of carbon. The author is holding a fruiting body of an ephemeral polypore fungus (*Laetiporus percisinus*) that weighed 14 kg fresh and was growing from the base of an *Inga vera* tree. (Photograph by W. Pfeiffer)

underestimated in fecal analyses. I studied epiphyll composition on the surface of *Dracena* sp. leaves adjacent to fresh snail feeding trails and found that fungi outnumbered algae in the material ingested by *Caracolus* snails (77% fungi and 23% algae in 147 observations along transect lines).

Consumers and Parasites of Fungi and Bacteria at El Verde

Homoptera (probably in the families Derbidae and Achilidae) and Collembola are frequently found feeding on mushrooms (especially fungi in the Entolomataceae and Tricholomataceae). Ants are almost invariably found on mushrooms at El Verde. Fungus ants, *Cyphomyrmex ramosus,* may be opportunistically consuming fungi outside their fungal gardens, whereas *Solonopsis* ants may be foraging for invertebrates. Rove beetle larvae and adults feed on fruiting bodies of *Gymnopilus pampeanus* (Speg.) Sing., *Pleurotus* spp., and *Favolus brasiliensis* (Fr.) Fr. Most rove beetles are predatory, but species of *Eumicrota, Phanerota,* and *Brachychara* are wholly fungivorous (Ashe 1984). Fly larvae have been found feeding in *Lepiota* spp., *Pluteus* sp., and *Gymnopilus pampeanus* fruiting bodies. Mycetophilid fly larvae, orabatid mites, and certain nematodes feed on mycelia while snails graze on both mycelia and basidiomycete fruiting bodies (especially *Favolus brasiliensis* [figs. 3.3 and 3.4] and *Marasmius leoninus,* but they have also been seen eating *Pleurotus* spp., *Coprinus disseminatus, Auricularia fuscosuccineus, Daedalea elegans,* and *Collybia* spp.). Positive correlations have been found between populations of fungivorous nematodes and fungal biomass in Swedish soils (Bååth et al. 1978), and a collembolan has been found to control vertical distributions of basidiomycetes in litter in the United Kingdom through selective grazing (Newell 1984).

Some wood-decomposing fungi in the genera *Hyphodontia, Pluteus, Pleurotus,* and *Melanotus* create a food loop by preying on fungivorous nematodes. Nematode traps have been observed on mycelia of *Hyphodontia* and *Pluteus* spp. at El Verde, but the one species of *Melanotus* studied (*M. eccentricus* [Murr.] Sing.; Petersen 1992) did not produce nematode traps in culture. However, nematode presence is often needed to induce trap production by the fungi. Some nematodes and amoebae are known to prey upon and regulate populations of bacteria in soil (Bååth et al. 1978; Coleman et al. 1983) but they have not been studied at El Verde. Numerous invertebrates (especially beetles) prey on slime molds (Blackwell et al. 1982; Wheeler and Blackwell 1984), and predation may partly account for the rarity of fruiting bodies at El Verde (Alexopoulos 1970). Slime molds obtain their nutrition by digesting other microorganisms, and their plasmodia are common during the rainy season at El Verde. In 1988 and 1992, migrating plasmodia of *Physarum* destroyed most of the spore-producing tissue of shelf fungi (*Daedalea elegans* and *Favolus brasiliensis,* pers. observation), but slime molds more commonly prey upon inconspicuous microorganisms.

Fungal hyperparasites have been observed on mushroom fruiting bodies and VAM fungal spores at El Verde but they have not been identified. Fungal hyperparasites of shelf fungi (Polyporaceae) and dead man's fingers (Xylariaceae) were identified by G. Samuels (app. 3.A). The placement of such species in the food web is difficult. Other fungal hyperparasites, however, are clearly secondary consumers. For example, there are many reports for Puerto Rico of fungi that are hyperparasites on fungi that are obligate parasites of leaves (black mildews, in the Meliolaceae; Stevenson 1975). Furthermore, several fungal parasites of invertebrates at El Verde are tertiary consumers, including a *Bouvardia* sp. on a wasp, and a common fungal parasite of *Leucauge* spiders.

Some higher plants at El Verde are parasitic and obtain most or all of their carbon from their mycorrhizal fungi. Such plants include orchid seedlings, an achlorophyllous orchid (*Wullschlegelia aphylla*), and an achlorphyllous member of the Bermanniaceae (*Apteria aphylla*). The mycorrhizal fungi involved obtain their carbohydrates either through decomposition of organic matter or from another higher plant (Harley and Smith 1983). A reversal in carbohydrate flow (i.e., with carbohydrates moving from orchid to fungus) has not been demonstrated in species which are photosynthetic as adults (Harley and Smith 1983), and it is uncertain whether the mycorrhizal fungi obtain any benefits from the association.

MICROBIAL PROCESSES: RATES AND FLOWS

Production of microbial biomass can be estimated from litter decomposition rates, rates of carbon dioxide loss from litter, or from equating losses through decomposition with organic matter inputs (assuming steady state). Although microorganisms are responsible for most of the respiration in organic matter (Reichle 1977), invertebrates are known to accelerate microbial respiration (van der Drift and Jansen 1977; Addison and Parkinson 1978), organic matter decomposition (Lam and Dudgeon 1985), and mineralization (Bardate and Prentki 1974; Anderson et al. 1981; Woods et al. 1982). Almost all of the organic matter passes through the microbial system (Coleman et al. 1983), and carbon may pass through microbes several times before it is respired as carbon dioxide because dead microorganisms are readily recycled. Consequently, microbial biomass production may appear to exceed organic matter inputs if it is calculated by summing fungal productivity rather than the carbon dioxide that is respired. Similarly, nitrogen and phosphorus are known to cycle through successive generations of microbes eight to ten times per year (Coleman et al. 1983).

Wood

Wood Decomposition Rates

The best published data for rates of wood decomposition at El Verde are from a five-year study of butt logs of *Dacryodes excelsa, Schefflera morototoni, Drypetes glauca,* and *Duggena hirsuta* (Odum 1970). The mean loss of dry weight observed in that study was 10.9% per year (SD = 5.0), which corresponds to a mean half-life of six years (Odum 1970c). Much of the rapid weight loss was probably from sapwood, however, and long-term decomposition rates for the remaining heartwood is probably much slower based on surveys in Central America (A. Lugo and S. Brown pers. comm.). Furthermore, the estimated release of carbon as carbon dioxide (13,189 g m^{-2} y^{-1}; table 3.1), based on this rate of decomposition and the large amount of wood standing stocks, is unreasonably high. In addition, decomposition rates varied among species, and the mean rate may not be representative of the whole forest. Although rates of carbon dioxide evolution from logs have also been measured at El Verde (Odum 1970c), these data are difficult to interpret because they are expressed on an areal rather than on a wood dry weight basis. Approximate wood decomposition rates were calculated for El Verde by equating log and branch fall inputs (244 to 609 g m^{-2} y^{-1}; table 3.1) with losses, but it is unlikely that the standing stocks of dead wood are in steady-state equilibrium (Odum 1970c) because of selective cutting in some parts of El Verde during the 1940s and occasional high inputs during hurricanes and tropical storms (Walker 1991; Scatena et al. 1993).

Concentration and Transport of Nutrients by Wood-Decomposing Fungi

Wood-decomposer fungi are able to capture almost 100% of the nitrogen and 90% of the phosphorus in their substrate (Swift 1977; Frankland 1982). These fungi can maintain high nitrogen concentrations relative to wood apparently by digesting their dead parts in addition to extracting, concentrating, and transporting nitrogen from other food bases (Jennings 1982; Watkinson 1984). Watkinson (1984) found that *Serpula lacrimans* removed more than 75% of the nitrogen from one food source and transported it over an inert surface to a new food base. This phenomenon is probably important at El Verde for redistributing nutrients from fallen logs (see Holland and Coleman 1987), as judged by the extensive hyphal cord systems of several fungi (e.g., *Mutinus* and *Phanerochaete* spp.). Removal of woody debris from forest plots at El Verde immediately after Hurricane Hugo was found to stimulate tree productivity, possibly by reducing competition from wood-decomposer fungi for limiting soil nutrients (Lodge et al. 1994; Zimmerman et al. 1995). Migrating mycelia of a polypore (*Schizopora flavipora*) may also redistribute nutrients at El Verde. *Schizopora* mycelia frequently cover all

Figure 3.6. The author is holding twig and leaf litter from the forest floor that have been bound together by a migrating mycelium of a basidiomycetous decomposer fungus, *Schizopora flavipora*.

nearby organic and inorganic substrata after exhausting the resources in a log (fig. 3.6). One mycelium covered circa 0.75 m^2 of boulder and an equal area of adjacent forest floor to a depth of 2 to 3 mm. Therefore, it is not surprising that we found extremely high fungal biomass at El Verde in litter and soil near dead wood (Lodge 1993).

Fine Litter

Rates of Fine Litter Decomposition

The most representative estimates of the rates of leaf litter decomposition at El Verde is circa 60.5% and 70% of initial weight lost after 224 and 288 days in the field, respectively (Zou et al. 1995). The Zou et al. study used freshly fallen leaves (within twenty-four hours of abscission) in the same proportion as was found in litter baskets. Leaf decomposition rates did not vary significantly with season even though the litterbags had different species compositions at different starting dates because the timing of peak litter fall varied among species. Other studies of litter decomposition are discussed in appendix 3.C. The turnover time for all fine litter (i.e., leaves, fruits, and flowers) on the forest floor may be faster than for the leaf litter component alone (table 3.5; see discussion in appendix 3.C).

Short-term instantaneous rates of carbon dioxide evolution from fine litter were measured at El Verde in 1984 using a modified Licor photosynthesis meter (W. T. Lawrence unpublished). These measurements yielded a total of 586 g C evolved m^{-2} y^{-1} based on a fine-litter standing stock of 281 g C m^{-2} (Zou et al. 1995). This value is near the rates calculated by assuming that

Table 3.5. Input, storage, decomposition rate, and turnover time for nonwoody litter at El Verde

Annual nonwoody litterfall inputs	
Annual leaf litter input[a]	515.1 g m^{-2} y^{-1}
Annual leaf litter input[b]	490.5 g m^{-2} y^{-1}
Annual flower input[b]	14.6 g m^{-2} y^{-1}
Annual fruit input[b]	30.2 g m^{-2} y^{-1}
Annual miscellaneous nonwoody input[b]	181.2 g m^{-2} y^{-1}
Total input of nonwoody litter per annum[b]	741.4 g m^{-2} y^{-1}
Storage of nonwoody litter[b]	281.6 g m^{-2} y^{-1} ±28.2 s.d.
Turnover time for nonwoody litter (based on litter input divided by storage)	0.39 y ±0.05
Rate of leaf litter decomposition[c]	0.73 g g-initial^{-1} y^{-1}
Turnover time for leaf litter (based on rates of leaf decomposition)	1.25 y

[a]Data are from 1966 (Wiegert 1970a)

[b]Data are from 1970–73 and 1980–81 (Zou et al. 1995)

[c]Data are from Zou et al. (unpublished)

annual carbon inputs of fine-litter equal annual carbon losses through respiration (482 g C m^{-2} n^{-1}, Odum 1970c, to 741 g C m^{-2} y^{-1}, Zou et al. 1995; table 3.5). Odum et al. (1970d) also measured carbon dioxide evolution using a foil flow method, but these rates are probably inflated because of the forced air movement (1,401 g C m^{-2} y^{-1}).

A substantial amount of leaf litter may be exported from steep slopes into streams at El Verde (Lodge and Asbury 1988), where it decomposes faster than in the terrestrial subweb. Padgett (1976) found that leaf discs of five species (*Buchenavia capitata, Sloanea berteriana, Cordia borinquensis, Dacryodes excelsa,* and *Manilkara bidentata*) required only 6 to 9.4 months for total decomposition in a stream at El Verde (mean 7.3 months). Most of the weight loss was from biological decomposition rather than physical degradation. Biological decomposition was primarily caused by microbially produced enzymes, although grazing by detritivores may have been important in two species (*B. capitata* and *C. borinquensis*).

Microbial Production in Fine Litter

Microbial production estimates for fine litter are roughly 492 to 790 g C m^{-2} y^{-1} (see app. 3.C). Summation of biomass production over the year can equal or exceed organic matter input (741 g fine litter input m^{-2} y^{-1}, Zou et al. 1995) because dead microorganisms and saprovores are recycled by other microorganisms and invertebrates (see Fermor and Wood 1981; Atkey and Wood 1983; Seastedt 1984; Hedger 1985). Hedger (1985) found in Ecuador that *Lepiota* species in the F2 layer require litter that has been partially decomposed by other fungi in the F1 layer (especially *Marasmius* spp.) and suggested that dead fungal biomass of F1 fungi may comprise a significant portion of the substrate used by the F2 fungi. Stratification is also apparent in litter fungal communities at El Verde. Carbon may, therefore, cycle through fungi several times before it is respired as carbon dioxide. Some carbon is undoubtedly contributed to the soil organic matter, but the proportion of substrate carbon which is eventually respired by organisms in the litter layer and the proportion which enters the soil system is unknown.

SUMMARY

Microorganisms are important in the food web at El Verde as primary, secondary, and tertiary consumers and as a food base for other consumers. Microorganisms include algae, lichens, fungi, slime molds, actinomycetes, and bacteria. The trophic levels that these microbes occupy are diverse, ranging from secondary and tertiary consumers, such as the slime molds which prey on other microorganisms, to primary producers, such as algae and lichens. Although algae and cyanobacteria (blue-green algae) contribute only slightly to primary production, they are an important food for snails. Furthermore, cyanobacteria and some related bacteria play a crucial role in nutrient cycling by fixing nitrogen from the atmosphere.

Fungi are the best-known group taxonomically, with over 400 species of macrofungi (higher basidiomycetes and ascomycetes) and fifty-one species of microfungi identified so far. However, the number of known macro- and microfungi are thought to represent only about 25 to 30% of the total actually present at El Verde. As in higher plant communities in tropical forests, the majority of fungal species are infrequent to rare, and dominance is low. Fungi interact with other organisms in a variety of roles, such as decomposers, parasites, predators, neutral or beneficial endophytes, and as victims of parasitism, hyperparasitism, and predation. Parasitic and predaceous fungi probably contribute to regulation of population densities of their hosts, which include animals, plants, and other microorganisms.

Fungi that are nonmycorrhizal endophytes and parasites of plants drain a significant but unknown quantity of carbon from their hosts. In addition, 15 to 25% (1,197 to 2,993 g C m^{-2} y^{-1}) of gross primary production is thought to be allocated to mycorrhizal fungal symbionts in exchange for increased uptake of mineral nutrients. A few higher plants such as orchids, however, obtain all or some of their carbon from their mycorrhizal fungus and give little or nothing in return.

As a group, fungi are more abundant than other microorganisms in the forest floor, with mean standing stocks of 1.5 and 207 g m^{-2} in the fine litter and upper 10 cm of soil, respectively. Fungal biomass in the litter layer is positively correlated with moisture, showing sharp increases and declines with daily variation in the frequency of rainfall; fungal biomass in the soil changes more slowly, showing seasonal variation. Annual production of microbes in woody debris at El Verde before Hurricane Hugo was estimated to be 317 to 793 g m^{-2} y^{-1} for fungi, and 15 to 35 g m^{-2} y^{-1} for bacteria and actinomycetes, and was apparently much higher after the hurricane. Total microbial biomass production for the fine litter layer was estimated at 492 to 790 g C m^{-2} y^{-1}.

The role of microbes in litter and wood decomposition is especially important for many invertebrates (Pfeiffer, chap. 5, this volume) including termites (e.g., *Nasutitermes;* McMahan, this volume) because they generally require partially decomposed substrates that are enriched with microbes or the decomposer microorganisms themselves as their food base. Furthermore, the high rates of microbial productivity and the ability of microorganisms to concentrate mineral nutrients gives them a far more important role in the food web than their relatively small standing stocks of biomass carbon might suggest (Pfeiffer, chap. 5, this volume; Garrison and Willig, this volume).

ACKNOWLEDGMENTS

The manuscript benefited from comments by J. C. Frankland, J. Hedger, D. P. Janos, R. B. Waide, D. P. Reagan, and several anonymous reviewers. Much appreciated assistance in identifying and describing the fungal species that are listed in the appendix checklist (3.A) was provided by T. Laessøe, J. D. Rogers, G. Bills, J. Polishook, B. Sutton, G. Samuels, R. Singer, O. K. Miller, D. Desjardin, R. Halling, D. N. Pegler, R. H. Petersen, D. Pfister, L. Ryvarden, B. Spooner, R. Hanlin, S. Cantrell, H. H. Burdsall, K. Nakasone, S. Huhndorf, R. A. MaasGeesteranus, M. Barr Bigelow, E. Setliff, G. Muller, R. Gilbertson, and C. Overbo. Financial support for the taxonomic work was provided by The Field Museum of Natural History, U.S. Dept. of Energy OHER grant DE-AC05-76RO and the ORAU visiting scientist program, the University of Puerto Rico, and the Center for Forest Mycology Research, Forest Products Laboratory, U.S.D.A. Forest Service. The ecological work was supported by the U.S. D.O.E. OHER grant (op. cit.), the University of Puerto Rico, and an NSF LTER grant BSR-881192 to the University of Puerto Rico and the International Institute of Tropical Forestry, U.S.D.A. Forest Service for the Luquillo Experimental Forest Long-Term Ecological Research Program.

Appendix 3.A Partial Checklist of Fungi at El Verde

	Substrate								
				Leaves					
Taxon	Log	Branch	Twig	Terrest.	Stream	Roots	Soil	Microbes	Insects
EUMYCOTA									
Ascomycotina:									
Bitunicatea (=Loculoascomycetes)									
Dothidiales									
Dacampiaceae									
Chaetomastia sp.[a]		+	+						
Immothia hypoxylon[a]								+	
Lophiostomataceae									
Herpotrichia cf. *macrotricha*[a]				+					
Lophiostoma sp.[a]		+	+	+					
Massarina sp.[a]			+						
Hysteriaceae									
Glonium sp.[a]	+	+	+						
Hysterographium sp.[a]			+						
Phaeosphaeriaceae									
Kalmusia sp.[a]		+	+						
Pleomassariaceae									
Splanchnonema sp.[a]			+						
Tubeufiaceae									
Tubeufia cf. *pezizula*[a]		+							
Tubeufia roraimensis[a]		+							
Tubeufia. 5–6 other spp.[a]	+	+		+					
Melanommatales									
Didymosphaeriaceae									
Didymosphaeria sp.[a]			+						
Xylobotryum andinum Pat.[a]	+								
Massariaceae									
Massaria sp.[a]			+	+					
Melanommataceae									
Byssosphaeria jamaicana[a]		+							
Byssosphaeria schiedermayeriana[a]			+	+					
Mycopepon smithii[a]	+								
Xenolophium appalanatum[a]	+	+	+						
Platystomaceae									
Astrosphaeriella rhytidosporium[a]						+			
Astrophaeriella sp. nov.[a]				+					
Astrosphaeriella, 4 other spp.[a]	+	+		+					
Pseudotrichia, 2 spp.[a]		+							
Pseudotrichia, cf.[a]				+					
Trematosphaeria sp. (hyaline)[a]	+								
Phacidiales									
Hypodermataceae/Stictidaceae									
Coccomyces clusiae (Lev.) Sacc.[b]				+					
Coccomyces sp.[b]				+					

Taxon	Substrate								
				Leaves					
	Log	Branch	Twig	Terrest.	Stream	Roots	Soil	Microbes	Insects
Pleosporales									
Botryosphaeriaceae									
Botryosphaeria, 2 spp.				+					
Dimeriaceae									
Nematostoma, cf.[b]	+	+							
Pleosporaceae									
Pleospora sp.[b]				+					
Pyrenophora sp.[b]						+			
Unitunicatae									
Glaziellales									
Glaziellaceae									
Glaziella vesiculosa Berk.[b]							+		
Discomycetes									
Heliotiales									
Heliotiaceae									
Cordierites guyanensis Mont.[a]		+							
Dactylospora stygia, var. *tenuispora* (Dennis) Hafellner[a]		+							
Sorokina sp. nov. Spooner, Lodge & Laessøe, ined.[a]	+	+							
Geoglossaceae									
Trichoglossum hirsutum (Pers.) Boud.[a] (on soil and termite carton)							+		
Orbiliaceae									
Orbilia andina Pat. (on Xylariaceae)[a]	+								+
Pezizales									
Pezizaceae									
Galactinia luteorosella La Gal[a]				+					
Sarcoscyphaceae									
Cookeina sulcipes (Berk.) Kuntze[b]		+							
Cookeina tricholoma (Mont.) Kuntze[b]		+							
Nanoscypha tetraspora (Seaver) Denison[b]		+	+						
Plectania rhytidea (Berk.) Nannf. & Korf[a]		+							
Phillipsia domingensis (Berk.) Berk.[b]	+	+							
Sarcosoma aff. *orientale* Pat.[a]		+							
Hemiascomycetes									
Saccharomycetales									
Cryptococcaceae									
Hanseniaspora sp.[c]				+			+		
Rhodoturulaceae									
Rhodotorula sp.[c]				+			+		
Pyrenomycetes									
Calosphaeriales									
Calosphaeriaceae									
Jattaea, cf.[a]				+					
Clavicipitales									
Clavicipitaceae									
Barya aurantiaca Henn. (on *Xylaria*)[a]								+	
Cordyceps sp. nov. (probably on Coleoptera)[a]									+

Taxon	Substrate								
				Leaves					
	Log	Branch	Twig	Terrest.	Stream	Roots	Soil	Microbes	Insects
Daiporthales			+						
Diaporthaceae									
Gnomoniaceae									
Cryphonectria cubensis (Brunner) Hodges[a]	+								
Clypeoporthe, cf.[a]		+	+	+					
Diaporthe sp.[a]				+					
Diaporthopsis, cf.[a]	+								
Linospora sp.			+	+					
Plagiostoma/Plagiophiale sp.[a]				+					
Melanconidaceae, undet. genus[a]		+							
Schizoparme botrytidis Samuels, M. E. Barr & Lowen[a]	+								
Wuestnia, cf.			+						
Valsaceae									
Valsa sp.		+	+						
Diatrypales									
Diatrypaceae									
Diatrype spp.		+							
Eutypa spp.		+	+						
Hypocreales									
Hypocreaceae									
Calonectria, cf.			+						
Genus nov.[a]	+								
Hypocrea brevipes (Mont.) Sacc.[a]	+								
Hypocrea pallida Ellis & Everh.[a]	+							+	
Hypocrea poronoidea A. Möller[a]	+								
Hypocrea c.f. *gelatinosa* (Tobe) Fr.[b]				+					
Hypocrea lutea[a]	+								
Hypomyces rosellus (Alb. & Schw. ex. Fr.)[b] L.-R. Tulasne (on *Rigidoporus microporus*)								+	
Hypomyces chrysostomus Berk & Br.[a]								+	
Hypomyces lanceolatus Rogerson & Samuels, sp. nov. (on *Rigidoporus microporus*)[a]								+	
Hypomyces sp. nov. (on *Polyporus tenuicolus*)[a]								+	
Hypomyces subiculosus (Berk. & Curt.) Hohnel[a]								+	
Nectria flavolanata (*Actinostilbe* anamorph)[b]				+					
Nectria flocculenta Hohnel[a]		+							
Nectria jatrophae (Möller) Wollenweber[a]	+								
Nectria joca Samuels, sp. nov. (on *Biscogniauxia*)[a]								+	
Nectria jungneri Henn.[a]		+							
Nectria pseudotrichia[a]		+							
Nectria rugulosa[b]				+					
Nectria aff. *veuillotiana* Sacc. (on *Anthurium*)[a]				+					
Nectria vilior Starb. (on *Kretzschmaria* spp.)[a]								+	
Podostroma sp. nov. 1[a]	+								
Podostroma sp. nov. 2[a]	+	+							
Sarawakus sordidus (Doi) Sammuels & Rossman[a]	+								
Sphaerostilbe sp. (on live palm leaves)[b]				+					
Thyronectria sp.[b]		+							

Taxon	Substrate								
				Leaves					
	Log	Branch	Twig	Terrest.	Stream	Roots	Soil	Microbes	Insects
Ostropales									
Odontotremataceae or Stictidiaceae[a]		+	+						
Phacidiales									
Tryblidiaceae									
Rhytidhysterium rufulum (Spreng. ex Fr.) Petr.[b]	+	+							
Phyllachorales									
Phyllachoraceae									
Phyllachora cecropiae (Rehm.) v. Arx[b]				+					
Sordiariales									
Chaetomiaceae									
Chaetomium spherale Chilvers[i]				E					
Neurosporaceae									
Lasiosphaeriaceae, 2–3 spp.	+	+	+						
Cercophora sp.		+	+	+					
Chaetosphaeria spp.	+	+	+						
Lasiosphaeria, 6–8 spp.[b]	+	+	+						
Lasiosphaeriella sp.[a]	+								
Melanochaeta sp.[a]	+								
Ophioceras sp.[a]	+								
Phaeotrichosphaeria, 2 spp.[a]	+		+						
Porosphaerella sp.[a]	+								
Nitschkiaceae									
Acanthonitschkea sp.	+	+							
Bertia moriformis var. *latispora*[a]	+	+							
Bertia moriformis var. *multiseptata*[a]		+							
Nitschkia sp.	+								
Trichosphaeriales									
Trichosphaeriaceae									
Fluviostroma wrightii[a]	+								
Trichosphaeria, 2–3 spp.		+	+	+					
Xylariales									
Amphisphaeriaceae (*senus lato:* M. E. Barr recently excluded all but the type genus)									
Dyrithium lividum[a]		+							
Griphosphaerioma, cf.[a]			+						
Seynesia cf. *Cerumpens*[a]				+					
Neohypodiscus cf. *cerebrinus* (Fée) J. D. Rogers, Y.-M. Ju & Laessøe (with small spores)[a]		+							
Neohypodiscus rickii (Lloyd) J. D. Rogers, Y.-M. Ju & Laessøe[a]	+								
Boliniaceae									
"*Apiocamarops luquilloensis*," sp. nov., Laessøe & Lodge, ined.[a]		+							
Camarops biporosa J. D. Rogers & Samuels[a]	+	+							
Camarops ustulinoides (P. Henn.) Nannf.[a]	+	+							

Taxon	Substrate: Log	Branch	Twig	Leaves: Terrest.	Leaves: Stream	Roots	Soil	Microbes	Insects
Clypeosphaeriaceae									
Apiospora, aff.[a]			+						
Discostroma, cf.[a]			+						
Discostroma sp.[a]			+						
Hyponectriaceae									
Glomerella cingulata (Stonem.) Spauld. & H. Schrenk[i]				E					
Leiosphaerella, cf.		+	+						
Linocarpon sp.		+	+	+					
Oxydothis sp.[a]			+	+					
Phomatospora, cf.[a]			+						
Physalospora sp.				+					
Magnaporthaceae									
Khuskia oryzae H. J. Hudson[i]				E					
Magnaporthe, 3 spp.[a]		+	+	+					
Pleurotremataceae									
Duradens sp.[a]		+							
Phomatospora sp.[ai]			+	E					
Saccardoella, 5 spp.[a]	+	+	+	+					
Xylariaceae									
aff. '*Anthostoma*' *adusta* sp. & gen. nov.?[a]	+								
Anthostomella cf *clypeata* f. rubi-ulmifolii[a]									
Anthostomella, 4 other spp.			+	+					
Asterocystis mirabilis Berk. & Broomea[a]				+					
Astrocystis sp.[a]			+						
Biscogniauxia nummularia v. *pseudopachyloma*[a]	+	+							
Biscogniauxia pro parte, see *Hypoxylon*									
Camillea fossulata (Mont.) Laessøe, J. D. Rogers & Whalley[a]		+							
Camillea hainsii (J. D. Rogers & K. P. Dumont) Laessøe, J. D. Rogers & Whalley[a]		+							
Camillea hyalospora (Pat.) Laessøe, J. D. Rogers & Lodge[a]		+							
Camillea cf. *labellum* Mont. & *C. venezuelensis*[b]	+								
Camillea obularia (Fr.) Laessøe, J. D. Rogers & Lodge [*C. broomeiana*][b]	+	+							
Camillea tinctor (Berk.) Laessøe, Rogers & Whalley[b]	+	+							
Camillea verruculospora J. D. Rogers, Laessøe & Lodge, sp. nov. (only on *Miconia tetranda*)[a]	+	+							
Daldinia bakeri Lloyd[a]		+							
Daldinia eschscholzii (Ehr. ex Fr.) Rehm.[b]	+								
Daldinia, see '*Hypoxylon*' *mulleri*									
Hypoxylon cf. *annuliforme* (not a synonym of *H. truncatum*, as J. H. Miller suggests)[a]		+	+						
'*Hypoxylon*' *cerebrinum* (Fée) Cke., see *NeophypodisHcus cerebrinum*									
'*Hypoxylon*' *citrinum* Shear [may not be syn. of		+							

	Substrate								
				Leaves					
Taxon	Log	Branch	Twig	Terrest.	Stream	Roots	Soil	Microbes	Insects
'*Penzigia*' *discolor* (Berk. & Br.) J. H. Miller; belongs in *Xylaria* Laessøe, ined.][b]		+							
Hypoxylon crocopeplum Berk. & Curt.[a]		+							
Hypoxylon aff. *culmorum* Cke. (spores small; parasitic on *Arundinella,* not *Bambusa*)[a]			+						
"*Hypoxylon cinnabarinum*" (P. Hen.n.) Y.-M. Ju & J. D. Rogers, comb. nov., ined.[a]		+							
'*Hypoxylon*' *cyclopicum* Speg.[b]	+	+							
Hypoxylon dieckmannii Theiss. (specimen 420, NYBG cited in J. H. Miller, not verified)[b]									
Hypoxylon aff. *dieckmanii,* sp. nov. (Lodge PR 660) Laessøe & Lodge, ined.[a]		+							
Hypoxylon erythrostroma J. H. Miller[a]		+	+						
'*Hypoxylon*' *hypohlaeum* (Berk. & Rav.) J. H. Miller (belongs in *Biscogniauxia*)[b]	+	+							
Hypoxylon investiens (Schw.) Curt.[b]	+	+	+						
Hypoxylon nitens [*Rosellinia nitens*][a]	+	+							
Hypoxylon notatum (Lodge PR 673)[a]	+	+							
'*Hypoxylon*' *mulleri* J. H. Miller (belongs in *Daldinia* according to T. Laessøe)[b]	+	+							
'*Hypoxylon nummularium* Bull. ex. Fr., var. *pseudopachyloma* (Speg.) J. H. Miller (belongs in *Biscogniauxia*)[b]	+	+	+						
'*Hypoxylon*' sp. nov. aff. *quisquiliarum* Mont.[a]		+	+						
Hypoxylon stygium (Lev.) Sacc.[b]		+	+						
'*Hypoxylon*' *subannulatum* P. Henn. (belongs in *Nemania,* Laessøe & Lodge ined.)[a]		+	+						
Hypoxylon thouarsianum (Lév.) Lloyd[b]		+							
Hypoxylon truncatum (Schw. ex Fr.) J. H. Miller [*H. annulatum* (Schw.) Mont. & *H. marginatum* (Schw.) Berk.][b] = *H. moriforme* Henn.		+	+						
'*Hypoxylon*' *serpens* var. *macrospora* (*Nemania*)	+								
'*Hypoxylon*' *uniapiculatum* (Penz. & Sacc.) J. H. Miller, var. *uniapiculatum* (belongs in *Biscogniauxia*)[b]		+							
'*Hypoxylon*' *uniapiculatum,* var. *macrosporum* J. D. Rogers (belongs in *Biscogniauxia*)[a]		+							
Kretzschmaria coenopus (Fr.) Sacc. [*K. clavus* (Fr.) Sacc.][b]	+	+							
'*Kretzschamaria*' *heliscus* (Mont.) Masse (belongs in *Xylaria,* Laessøe ined.)[a]		+							
Kretzschamaria micropus (Fr.) Sacc.[b]	+	+							
Kretzschamaria cf. *zonata* (Lév.) P. Martin[a]		+							
Kretzschamria sp. nova, Laessøe, ined.[a]	+								
'*Krestzschmaria*' *rugosa* Earle [*Penzigia cantarierense* (P. Henn.) J. H. Miller; belongs in *Xylaria;* spores smaller than type][b]	+								

Taxon	Substrate: Log	Branch	Twig	Leaves: Terrest.	Leaves: Stream	Roots	Soil	Microbes	Insects
Nemania bipapillata (= *H. subannulatum sensu* Miller but not syn. with this taxon)[b]	+								
Penzigia sp. nov. Laessøe & Lodge, ined.[a]		+				+			
'*Poronia*' *turbinata* (Ell. & Ev.) J. H. Miller [belongs in *Xylaria*][a]	+	+							
Phylacia bomba (Mont.) Pat.		+							
Phylacia globosa Lév.[a]	+	+				+			
Rosellinia aquila (Fr.) de Not[b]		+							
Rosellinia bunodes (Berk. & Br.) Sacc. [=*Rosellinia goliath* Speg.][b]		+				+			
Rosellinia necatrix complex[a]		+							
Rosellinia aff. *pepo* Pat.[a]		+	+						
'*Rosellinia*' *subiculata* (Schw.) Sacc.[b]		+	+						
'*Stilbohypoxylon*' *rehmii* ss Martin (*S. molleri* ss. J. D. Rogers, on fronds, *Prestoea montana*)[a]				+					
'*Ustulina deusta*' (Hoffm. ex Fr.) Petr. (belongs in *Kretzschmaria;* differs from type)[b]	+	+							
'*Ustulina*' cf. *zonata* (Lev.) Sacc. (Lodge PR 603)[a]	+								
Xylaria adscendens (Fr.) Fr.[b,j]		+		M					
Xylaria allantoidea (Berk.) Fr.[b,h]	+	+		E					
Xylaria apiculata Cke.[a,b]		+							
Xylaria arbuscula Sacc. (=*X. mellissii*?)[i]		+	+	E					
Xylaria aff. *arbuscula* (PR 414, on *Ficus*)[a]	+	+							
Xylaria areolata (Berk. & Curt.) J. H. Miller[a]	+	+							
Xylaria aristata (Mont.) Dennis (on Guttiferae)[b]				+					
Xylaria axifera (Mont.) Laessøe & Lodge (only on petioles of *Schefflera*)[b]				+					
Xylaria berteri (Mont.) Cke.[a]		+	+						
Xylaria boergesenii Lloyd[b]		+	+						
Xylaria caespitulosa Ces. (on wood & fruits)[a,j]		+		E					
Xylaria chordaeformis Lloyd[a]			+						
Xylaria coccophora[h]		+	+	E					
Xylaria cubensis (Mont.) Fr. (both typical & penzigiod forms)[b]	+	+							
Xylaria aff. *cubensis*[a]		+		E					
Xylaria curta Fr.[a]	+	+		E					
Xylaria cf. *delicatula* Starb. (*Clusia* leaves)[a]				+					
Xylaria feejeensis (Berk.) Fr.[a]	+	+							
Xylaria frustulosa (Berk. & Curt.) Dennis[a]		+							
"*Xylaria fuscopurpurea*" Laessoe, ined. (spec. nov. from Ecuador)[a]	+	+							
Xylaria fockei (Miq.) Dennis[b]	+								
Xylaria guareae Laessøe & Lodge (only on *Guarea guidonia*)[a]	+	+							
Xylaria globosa (Spreng. ex Fr.) Mont. [=*X. anisopleura* (Mont.) Fr.][b]	+	+							

	Substrate								
				Leaves					
Taxon	Log	Branch	Twig	Terrest.	Stream	Roots	Soil	Microbes	Insects
Xylaria hyperythra (Mont.) Fr.[a]		+							
Xylaria ianthino-velutina (Mont.) Fr. (at EV on ma-Fhogany fruit; elsewhere on legume pods)[b]									
Xylaria kegeliana (Lév.) Fr[a]		+	+	E					
Xylaria meliacearum Laessøe, in Laessøe & Lodge (on petioles of *Guarea guidonia* in PR)[a]				+					
Xylaria mellissii (Berk.) Cke. (= *X. arbuscula*?)[a,j]		+	+	E					
Xylaria microceras (Mont.) Fr. [= *Xylaria muscula* Lloyd][a]		+							
Xylaria multiplex (Kze. ex Fr.) Mont.[b]		+	+	E		+			
"*Xylaria neoberkleyi*" Laessoe & Lodge, ined.	+	+							
Xylaria nigrescens sensu Martin & Rogers[a]	+								
Xylaria obovata (Berk.) Fr.	+	+							
Xylaria palmicola Wint. (on fruit of palm, *Prestoea montana;* coll. are from El Yunque)[b]									
Xylaria poitei (Lév.) Fr.[b]	+								
Xylaria polymorpha Pers. ex Merat. complex[b]	+								
Xylaria scruposa (Fr.) Fr.[b]	+	+							
Xylaria sp. nov. aff. *X. claviceps* (PR 973 & 1099, on buried wood)[a]						+			
Xylaria sp. nov.? aff. *X. feejeensis* (PR 875A&B, on petiole of *Schefflera morototoni*)[a]				+					
Xylaria sp. nov., *aff. longiana* (on *Schefflera morototoni* petioles)[a]				+					
Xylaria telfarii (Berk.) Fr.[b]	+	+							
Xylaria tuberoides[a]	+	+				+			
Xylaria cf. *warburgii* P. Henn. (on fruit capsules of *Sloanea berteriana;* hairy)[a]									
Basidiomycotina									
Homobasidiomycetidae									
Gasteromycetes									
Lycoperdales									
Broomeiaceae									
Lycogalopsis solmsii Ed. Fischer[a]	+	+							
Geastraceae									
Geastrum saccatum (Fr.) Ed. Fischer[b]							+		
Geastrum schweinitzii (B. & C.) Zeller[b]	+	+							
Nidulariales									
Nidulariaceae									
Cyathus pallidus Berk. & Curt. (dark form)[b]		+							
Phallales									
Clathraceae									
Clathrus baumii (P. Henn) E. Fischer var. nov.?; vulva lilac[a]	+						+		
Pseudocolus sp. Lloyd[a]	+						+		
Laternea triscapa Turpin[b] (attached to wood via cords)[b]						+	+		

Taxon	Substrate								
				Leaves					
	Log	Branch	Twig	Terrest.	Stream	Roots	Soil	Microbes	Insects
Phallaceae									
'*Dictyophora*' *indusiata* (Vent. ex Pers.) Desv. (probably conn. to wood by cords; =*Phallus*?)[a]	?						+		
Mutinus bambusinus (Zoll.) Ed. Fischer[b]	+	+					+		
Sclerodermatales									
Pisolithaceae									
Pisolithus tinctorius (Pers.) Coker & Couch (mycorrhizal on introduced *Pinus caribaea*)[b]						M			
Sclerodermataceae									
Schleroderma cf. *geaster* (mycorrhizal on introduced *Pinus caribaea*)[a]						M			
Hymenomycetes									
Agaricales									
Agaricaceae									
Agaricus spp.[b]				+			+		
Cystoderma cf. *austrofallax* Sing.[a]		+							
Lepiota guatopoensis Dennis[a]	+			+			+		
Lepiota spp. (>30 other spp.)[a]	+	+	+	+			+		
Leucoagaricus sp.[a]				+			+		
Leucocoprinus birnbaumii (Corda) Sing.[a]	+								
Leucocoprinus fragilissimus (Rav.) Pat.[a]						+	+		
Coprinaceae									
Coprinus disseminatus (Pers. ex Fr.) Gray[a]	+	+							
Coprinus mexicanus Murr.[a]	+								
Psathyrella spp.[a]	+	+	+	+					
Cortinariaceae									
Gymnopilus pampeanus (Speg.) Sing.[a]	+								
Gymnopilus sp.[a]	+	+							
Gymnopilus chrysopellus (Berk. & Curt.) Murr.[a]	+								
Galerina sp.[a]	+								
Pyrrhoglossum lilaceipes Sing.[a]		+							
Pyrrhoglossum pyrrhus (Berk & Curt.) Sing[a]	+								
Crepidotaceae									
Crepidotus palmarum Sing. (on *Prestoea*)[a]				+					
Crepidotus variisporus Sing.[a]		+							
Crepidotus spp.[a]	+	+	+						
Entolomataceae									
"*Alboleptonia flaviphylla*" Baroni & Lodge, sp. nov., ined.[a]				+					
Aboleptonia hyalodepas (Berk. & Br.) Pegler[a]							+		
"*Alboleptonia subrosea*" Baroni & Lodge, sp. nova, ined.[a]				+					
"*Alboleptonia sulcata*" Baroni & Lodge, sp. nova, ined.							+		
Alboleptonia stylophora (Berk. & Br.) Pegler[a]							+		
Entoloma bakeri Dennis[a]							+		
Entoloma dragonosporum Sing.[a]	+								

Taxon	Substrate								
				Leaves					
	Log	Branch	Twig	Terrest.	Stream	Roots	Soil	Microbes	Insects
Entoloma lowyi Sing. ss. Horak[a]		+							
Entoloma quadratum (Berk. & Curt.) Horak [=*E. salmoneum* (Pk.) Sacc.][a]							+		
Inopilus glycosmus Pegler[a]							+		
Inopilus aff. *glycosmus*[a]							+		
Inopilus inocephallus (Romagn.) Pegler[a]							+		
Inopilus maculosus Pegler[a]							+		
'*Inopilus*' *entolomoides* Pegler (belongs in *Entofloma;* 2-spored, not 4-spored basidia)[a]							+		
Inopilus cf. *speciosus* (Romagn.) Pegler[a]				+			+		
Leptonia ceruleocapitata (Dennis) Pegler[a]							+		
Leptonia cf. *howellii* (Peck)[a]				+			+		
"*Leptonia lazulinellum*" (Sing.) Baron, comb. nov., ined.[a]		+							
"*Leptonia lowyi*" (Sing.) Baroni, comb. nov.[a]	+								
"*Leptonia paravelutina*" Baroni & Lodge, sp. nova, ined.[a]	+					?	+		
"*Nolanea bispora*" Baroni, ined[a]	+			+			+		
Nolanea cf. *pinna* (Romagn.) Dennis[a]				+			+		
Pouzaromyces mazzeri Court.[a]							+		
Hygrophoraceae									
'*Camarophyllus*" [=subgenus *Cuphophyllus* Donk of the genus *Hygrocybe* (Fr.) Kummer]									
'*Camarophyllus*' aff. *buccinulus* (Speg.) Pegler[a]							+		
'*Camarophyllus*' *cremeus* Murr.[a]							+		
'*Camarophyllus*' *pratensis* (Pers. ex Fr.) Kummer var. *pratensis*[a]							+		
'*Camarophyllus*' *umbrinus* (Dennis) Sing. ex Pegler var. *clarofulvus* Lodge & Pegler[a]							+		
Hygrocybe atrosquamosa Pegler [=neotropical *H. astatogala* Heim.; =*H. conicus* (Scop. ex. Fr.) Fr. var. *peridenyca* Sacc. *sensu* Dennis][a]							+		
Hygrocybe batistae Sing. (caespitose form)[a]							+		
Hygrocybe chloochlora Pegler[a]							+		
Hygrocybe firma (Berk. & Br.) Sing.[a]							+		
Hygrocybe aff. *firma,* sp. nov., ined.[a]							+		
Hygrocybe hypohaemacta (Corner) Pegler[a]							+		
Hygrocybe incolor Pegler[a]							+		
Hygrocybe melleofusca Lodge & Pegler [=neotropic. *H. cinerascens* (Berk. & Br.) Pegler *ss* Dennis 1970 and Hesler & Smith 1963][a]							+		
Hygrocybe nigrescens (Quél.) Kuhn., var. *brevispora* Dennis[a]							+		
Hygrocybe occidentalis (Dennis) Pegler, vars. *occidentalis* and *scarletina* Pegler & Fiard[a]							+		
Hygrocybe prieta Lodge and Plegler[a]							+		
Hygrocybe sp. nov., Sect. *Neohygrocybe*[a]							+		
Hygrocybe viridula Lodge & Pegler[a]							+		

Taxon	Substrate								
				Leaves					
	Log	Branch	Twig	Terrest.	Stream	Roots	Soil	Microbes	Insects
Paxillaceae									
Neopaxillus plumbeus Sing. & Lodge[a]							+		
Pluteaceae									
Pluteus cf. *aurantiarugosa*[a]	+								
Pluteus olygocystis Sing. var. *olygocystis*[a]	+								
Pluteus sp. aff. *P. aethalus* (Berk. & Curt.)[a] Sacc. & *P. tephrostictus* (Berk. & Curt.) Sacc.[a]	+								
Pluteus sp. aff. *P. albostipitatus* (Dennis) Sing. and *P. cubensis* (Murr.) Dennis[a]				+					
Pluteus cubensis (Murr.) Dennis[a]	+								
Volvariella sp.[a]	+								
Russulaceae									
Russula sp. (assoc. with *Coccoloba swartzii*)[a]						M			
Lactarius sp. (mycorrhizal with *C. swartzii*)[a]						M			
Strophariaceae									
Melanotus eccentricus (Murr.) Sing.[a]		+	+	+					
Melanotus spp.[a]		+	+						
Nematoloma subviridae (Berk. & Curt.) Smith[a]	+	+							
Psilocybe aff. *caerulescens* Murr.[a]							+		
Stropharia spp.[a]	+	+							
Tricholomataceae									
'*Amparoina*' *spinosissima* (Sing. in Sing. & Digilio) Sing. [*Mycena* Sect. *Schhariferae*][a]				+					
Amyloflagellula pseudoarachnoidea (Dennis) Sing.[a]		+		+					
'*Armillareilla*' *affinis* Sing. [*Armillaria*][a]	+	+							
Anthracophyllum lateritium (Berk. & Curt.) Sing.[a]		+							
Chaetocalathus liliputianus (Mont.) Sing.[a]		+							
Chaetocalathus niduliformis (Murr.) Sing., var. nova "*bispora*" R. H. Petersen[a]		+	+						
Collybia aurea (Beeli) Pegler[a]	+								
Collybia dichroa[a]		+							
Collybia johnstonii (Murr.) Dennis[b]				+					
Collybia pseudo-omphalodes Dennis[a]	+	+							
Collybia subpruinosa (Murr.) Dennis[a]	+	+	+	+					
Collybia aff. *fascicularis* Rick ex Sing. sp. nov?[a]				+					
Collybia sp. nova (orange)[a]				+					
Crinipellis sp. nova (parasite on *Schefflera*)[a]				+					
Crinipellis sp. nova, aff. *C. stupparia* (Berk. & Curt.) Pat.[a]			+	+					
Cymatella sp.[a]		+							
Dictyopanus pusillus (Pers. ex Lév.) Sing.[b]		+	+						
Favolaschia aurantiaca[a]			+						
Favolaschia varariotecta Sing.[a]			+	+					
Filoboletus gracilis (Klotzsch ex Berk.) Sing.[a]	+	+							
Gerronema cyathiforme (Berk. & Curt.) Singer	+								
Gloiocephala occidentalis Sing.[a]			+	+					
Gloeocephala, (3 other spp.)[a]				+					
Hohenbuhelia spp. (2)[a]	+	+							

Taxon	Substrate								
				Leaves					
	Log	Branch	Twig	Terrest.	Stream	Roots	Soil	Microbes	Insects
Hydropus albus Sing.[a]	+	+							
Hydropus citrinus Sing.[a]	+								
Hydropus mycenoides (Dennis) Sing., var. *mycenoides*[a]	+								
Lentinula boryana (Berk. Mont.) Pegler[a]	+	+							
Lepista subisabellina (Murr.) Pegler[a]				+					
Leptoglossum sp. Karsta[a]				+					
Marasmiellus albofuscus (Berk. & Curt.) Sing.[a]		+	+						
Marasmiellus coilobasis (Berk.) Sing.		+	+						
Marasmiellus defibulatus Sing.[a]			+	+					
'*Marasmiellus' nigripes,* see *Tetrapyrgos*									
Marasmiellus pilosus (Dennis) Sing.[a]		+	+						
Marasmiellus semiustus (Berk. & Curt.) Sing., var. *sabali* Sing. (on palm)[a]				+					
Marasmius atrorubens (Berk.) Berk. [=*M. portoricensis* Murr.][b]				+					
Marasmius bezerrae Sing.[a]					+	+			
Marasmius cohortalis Berk.[a]			+	+					
Marasmius eorotula Sing.[a]			+	+					
Marasmius aff. *crescentiae*[a]				+					
Marasmius crinis-equi[a]			+	+					
Marasmius fiardii (Sing.) *apud. Pegler*[a]		+							
Marasmius guyanensis Mont., var. *guyanensis* and var. *emarginatus*[a]			+	+					
Marasmius haematocephalus (Mont.) Fr., var. *haematocephalus*[a]			+	+					
Marasmius hinnuleus Berk. & Curt.[a]				+					
Marasmius leoninus Berk., var. *leoninus*[a]			+	+					
Marasmius niveus Mont.[a]				+					
Marasmius oleiger Sing.[a]	+			+					
Marasmius pallescens Murr.[b]			+	+					
Marasmius pseudoniveus Sing. var. *amylocystis* Sing.[a]			+						
Marasmius aff. *tetrachrous*[a]				+					
Marasmius thwaitesii Berk. & Br. (*M. echinosphaerus* Sing.)[a]			+	+					
Marasmius trinitatis Dennis, var. *trinitatis*[a]			+	+					
and var. *immarginatus*[a]			+	+					
Marasmius wilsonii Murr.[b]				+					
Marasmius sp. nova, Sect. *Marasmius* (=*Rotulae*)[a]			+	+					
'*Micromphale*' *brevipes* (Berkley & Rav.) Sing.[a]			+						
Mycena araujae Sing.[a]		+							
Mycena chloroxantha Sing.[a]			+	+					
Mycena citricolor (Berk & Curt.) Sacc.[b]				+					
Mycena cuspidatapilosa Lodge[a]		+	+	+					
Mycena delica Sing.[a]			+	+					
Mycena gelatinomarginata Lodge[a]			+						
Mycena griseoradiata Sing.[a]	+	+							
Mycena holoporphyra (Berk. & Curt.) Sing. (both 2- and 4-spored forms)	+	+	+	+					

Taxon	Substrate								
				Leaves					
	Log	Branch	Twig	Terrest.	Stream	Roots	Soil	Microbes	Insects
Mycena sp. aff. *holoporphyra* (lacking pleurocystidia, with amyloid spores)[a]				+					
Mycena aff. *lamprospora* Corner ex Horak[a]			+	+					
Mycena levis Sing.[a]				+					
Mycena melandeta Sing.[a]				+					
Mycena cf. *osmundicola* Lange[a]				+					
Mycena aff. *polyadelpha* (Lasch.) Kühner[a]		+							
Mycena pseudocrocata Dennis[a]			+						
"*Mycena roseovenosa*" sp. nov. Lodge ined.			+	+					
Mycena singeri Lodge (on epiphytic fern roots)[a]						+			
Mycena aff. *singeri*, sp. nov., ined.[a]		+							
Mycena sotae Sing.[a]				+					
Mycena tesselata (Mont.) Dennis (pink form)[a]				+					
Mycena trichocephala Sing.[a]				+					
"*Mycena vitellina*" Lodge ined. (sp. nov., Sect. *Carolinenses* Maas Geesteranus)[a]		+	+						
Nothopanus eugrammus (Mont.) Sing.[a]	+								
Oudmansiella canarii (Jungh) Hohen.[a]	+	+	+						
Resupinatus aff. *striatus*[a]	+								
Tetrapyrgos nigripes (Schwein.) [=*Marasmiellus nigripes* (Schwein.) Sing. =*Pterospora nigripes* (Schwein.) Horak][a]		+	+						
Tetrapyrgos sp. (lilac lamellae)[a]			+	+		+		+	
Tricholoma aff. *pachymeres* (Berk. & Br.) Sacc.[a]						+			
Trogia montagnei Fr. (prob.=*T. mellea* Corner)[a]			+						
Xerulina asprata (Berk.) Pegler [=*Gymnopus chrysopeplus* Murr. =*Cyptotrama asprata* (Berk.) Redhead & Gins.][b]		+				+			
Aphyllophorales									
Clavariaceae									
Clavaria aurantio-cinnabarina Schw. (on fruit capsule of *Sloanea berteriana*)[a]									
"*Clavulina puerto-ricensis*" R. H. Pet., sp. nova[a]							+		
Deflexula sp. Corner[a]						+			
"*Lentaria caribbeana*" R. H. Pet., sp. nova[a]			+	+					
Pistillaria spp.[a]			+	+					
Ramariopsis cf. *antillarum* (Pat.) R. H. Pet. (small basidiomes and spores; orange form)[a]							+		
Corticeaceae									
Aleurodiscus sp.		+							
Athelia spp.[a]	+	+	+						
Botryobasidium spp.	+	+							
Ceraceomyces sp.[a]		+	+						
Crustoderma sp.[a]		+							
Gloeocystidiellum aff. *heterogeneum* (Bourd. & Galz.) Donk[a]		+							
Gloeocystidiellum triste Hjorst. & Ryv.[a]	+	+							
Hypochnicium sp.[a]		+							

	Substrate								
				Leaves					
Taxon	Log	Branch	Twig	Terrest.	Stream	Roots	Soil	Microbes	Insects
Hyphoderma guttuliferum	+								
Hyphoderma puberum (Fr.) Wallr.		+							
Hyphodontia cf. *pallidula* (Bres.) J. Erikss.[a]		+							
Hyphodontia sp. (on bamboo)[a]		+							
Hyphodontia sp. (on palm frond)[a]				+					
Kneiffia wrightii Berk. & Curt.[b]	+	+							
Lindtneria sp. nov. M. Larsen, Lodge ined.				+					
Mycoacia sp.[a]	+	+							
Phanerochaete filamentosa (B & C.) Burds.[a]	+	+							
Phanerochaete roumeguerii[a]		+							
Phanerochaete sordida[a]	+								
Peniophora sp. (steel gray)	+								
Phanerochaete sp. nov., Nakasone, ined.[a]	+	+							
Phanerochaete cf. *chrysorhiza*[a]	+								
Phlebia chrysocreas (Berk. & Curt.) Burds.[a]	+	+							
Resinicium bicolor (Fr.) Parm.[b]		+							
Scopuloides rimosa (Cooke) Jülich [=*Phanerochaete* cf. *rimosus*][a]		+							
Schizopora flavipora (Cke.) Ryv.[a]	+	+	+						
Schizopora paradoxa (Fr.) Donk[a]	+	+	+						
Schizopora trichiliae (Van der Bye) Ryv.[a]	+	+							
Subulicystidium longisporum (Pat.) Parmasto[a]	+								
Trechispora spp.[a]	+	+		+					
Tubulicium capitatum (on palm frond)[a]				+					
Tubulicrinis sp.[a]		+							
Xenasma sp.[a]		+							
Xenasmatella sp.[a]		+							
Cyphellaceae									
Hemmingsomyces cf. *candidus* (Pers.) Kuntze[b]			+						
Hydnaceae									
Hericium cf. *racemosum*[a]	+								
Sarcodon, 2 spp., probably ectomycorrhizal[a]						M	(+)		
Hymenochaetaceae									
Hymenochaetae damaecornis Link:Lév.[b]		+	+						
Hymenochaetae unicolor Berk. & Cur.[b]	+	+							
Phellinus contiguus (Fr.) Pat.[a]		+				+			
Phellinus gilvus (Schw.) Pat.[a]	+	+							
Phellinus hoehnelii (Bres.) Ryv.[a]	+								
Fistulinaceae									
Fistulina hepatica Schaeffer ex Fr.[a]	+								
Lachnocladiaceae									
Scytinostroma sp.[a]			+	+					
Vararia tropica Welden[b]	+	+							
Vararia sp.		+							
Polyporaceae									
Amauroderma cf. *macrosorum* J. Furtado [=? *Ganoderma aurantiacum* Torrend. in Bres.][a]						+			
Antrodiella leibmannii (Fr.) Ryv.[b]	+	+							

Taxon	Substrate								
				Leaves					
	Log	Branch	Twig	Terrest.	Stream	Roots	Soil	Microbes	Insects
Ceriporia sp.[a]		+							
Coryolopsis cf *rigida* (Berk. & Mont.) Murr.[a]	+	+							
Cyclomyces iodinus (Mont.) Pat.[a]	+								
Earliella scabrosa (Pers.) Gilbn. & Ryv. [=*Trametes corrugata* (Pers.) Bres.][b]	+	+							
Favolus, see *Polyporus tenuiculus*									
Fuscocerrena portoricensis (Fr.) Ryv.[a]	+								
Hapalopilus albo-citrinus (Petch.) Ryv.[a]	+					+			
Hexagonia hydnoides (Fr.:Sw.) M. Fidalgo [=*Trametes hydnoides* (Sw. ex Fr.) Fr.][b]	+								
Hydnopolyporus fimbriatus (Fr.) Reid[b]		+				+			
Lentinus swartzii Berk. [=*Lentinus crinitis* (L.:Fr.) Fr. sensu Berk.][b]	+	+							
Lentinus tigrinus (Bull. ex. Fr.) Fr.[a]	+	+							
Lentinus strigosus (Schwein.) Fr.[b]	+	+							
Nigrofomes melanoporus (Mont.) Murr.[a]	+	+							
Oxyporus latemarginatus (Dur. & Mont.) Donk[a]	+	+							
Pleurotus flabellatus (Berk. & Br.) Sacc.[a] (misidentified as *P. ostreatus*,[c])	+	+	+						
Polyporus badius (Pers.: S. F. Gray) Schw.[b]		+							
Pleurotus fockei (Miguel) Sing. [=*Lentinus striatulus*, Lév.][a]		+							
Polyporus tenuiculus (Beauv.) Fr. [=*Favolus brasiliensis* (Fr.) Fr.][b]	+	+							
Polyporus tricholoma Mont.[a]	+	+							
Porodisculus pendulus (Schw.) Murr.[a]		+							
Porogramme albocincta (Cke. & Masse) Lowe[a]	+								
Pycnoporus sanguineous (L.:Fr.) Murr.[b]	+	+							
Laetiporus cf *persicinus* (Berk. & Curt.) Gilbn.[a]	+					+			
Laetiporus sulfureus (Bull.:Fr.) Murr.[b]	+								
Rigidoporus microporus (Fr.) overeem.[b]	+	+							
Rigidoporus lineatus (Pers.) Ryv.[a]	+	+							
Rigidoporus vinctus (Berk.) Ryv.[a]	+	+							
Tinctoporellus epimiltinus (Berk. & Br.) Ryv.[a]	+	+							
Trametes elegans (Spreng. ex Fr.) Fr.[b]	+								
Trametes membranacea (Sw.:Fr) Kreisel [=*Hydnum palmatum* Hook.][a]		+							
Trametes pavonia (Hook) Ryv.[a]	+								
Trametes membranacea (Sw. ex Fr.) Kreisel[a]	+								
Tyromyces hypolateritius (Berk.) Ryv.[a]	+	+							
Schizophyllaceae									
Schizophyllum commune Fr.:Fr.[b]		+	+	+					
Steccherinaceae									
Steccherinum creneo-album Hjorst.[a]		+							
Steccherinum sp.	+								
Stereaceae									
Mycobonia sp.[b]	+	+							
Sterium ostrea (Blume & Nees:Fr.) Fr.[b]	+	+							
Stereopsis radicans (Berk.) Reid[b]						+			

Taxon	Substrate								
				Leaves					
	Log	Branch	Twig	Terrest.	Stream	Roots	Soil	Microbes	Insects
Boletales									
Boletaceae, Suilloideae									
Rhizopogon rubescens complex (introduced; mycorrhizal with introduced *Pinus caribeae*)						M			
Suillus sp. (introduced; mycorrhizal with introduced *Pinus caribaea*)						M			
Heterobasidiomycetidae									
Auriculariales									
Auriculariaceae									
Auricularia delicata (Fr.) Henn[b]	+	+							
Auricularia mesenterica Pers.[a]	+								
Auricularia polytricha (Mont.) Sacc. [= *A. fuscosuccinea* (Mont.) Farl.][b]	+	+							
Dacryomycetales									
Dacryomycetaceae									
Dacrymyces falcata Brasfield[b]		+							
Ditiola cf. *radicata* Fr.[a]		+							
Tremellales									
Tremellaceae									
Bourdotia spp.	+			+					
Heterochaete cf. *andina* & *albida*		+	+						
Sebacina sp.			+						
Tremella fuciformis Berk.[b]	+								
Tremella sp.[b]	+								
Sterile mycelia:									
Rhizoctonia sp. (with orchids & Burmaniaceae)[b]						M?			
Deuteromycotina									
Hyphomycetes									
Moniliales									
Dematiaceae and Moniliaceae									
Actinospora megalospora[e]					+				
Acremonium cf. *falciforme*[h]				+					
Acremonium terricola series, 4 types[h]				+					
Acremonium, 8 other spp.[h]				+					
Acrodontium crateriforme[h]				+					
Alternaria sp.[h]				+					
Anguillospora gigantea[e]					+				
Anguillospora longissima[e]					+				
Anguillospora sp., cf.[h]				+					
Arthrinium sacchari M. B. Ellis[i]				E					
Aspergillus flavus[h]				+					
Aspergillus janus[h]				+					
Aspergillus japonicus Saito[c]				+		+	+		
Aspergillus niger van Tieghem[c]				+		+	+		
Aspergillus versicolor (Vuillemin) Tiraboschi[c]				+		+	+		
Aureobasidium, cf.[h]				+					
Beltrania rhombica[h]				+					
Beltraniella portoticensis[h]				+					
Botryosporium longibrachiatum[h]				+					

Taxon	Substrate								
				Leaves					
	Log	Branch	Twig	Terrest.	Stream	Roots	Soil	Microbes	Insects
Botryotrichum sp.[h]				+					
Botrytis terestris Jensen[c]				+					
Camposporium sp.[e]					+				
Campylospora chaetocladia[e]					+				
Cephalosporium acremonium Corda[c]						+			
Chaetopsina cf. *fulva*[h]				+					
Chaetospina, 3 other spp.[h]				+					
Chalara cylindrosperma[h]				+					
Chloridium lignicola[h]				+					
Chloridium phaeosporum var. *cubense*[h]				+					
Chloridium sp. (clavate conidiophores)[h]				+					
Circinotrichum maculiforme[h]				+					
Cladobotryum mycophilum[h]				+					
Cladobotryum sp.[h]				+					
Cladosporium cladosporioides[h]				+					
Clonostachys sp.[h]				+					
Clonostrachys sp. (lacking sporodochia)[h]				+					
Cryptophiale udagawae[h]				+					
Cryptophiale sp.[h]				+					
Cylindrocarpon janothele[h]				+					
Cylindrocarpon sp.[h]				+					
Cylindrosympodiella sp.[h]				+					
Cyphellophora cf. *taiwanensis*[h]				+					
Dactylaria, 3 spp.[h]				+					
Dactylella sp.[h]				+					
Dendrodochium, 5 spp.[h]				+					
Dicrandion fragile[h]				+					
Diplocladiella sp.[e]					+				
Eriocercophora balladynae[h]				+					
Exophiala jeanselmii[h]				+					
Exophiala, 3 other spp.[h]				+					
Flabellospora verticillata[e]					+				
Flabellospora sp.[e]					+				
Flagellospora curvula[e]					+				
Flagellospora penicillioides[e]					+				
Gliocladiopsis tenuis[h]				+					
Gliocladium roseum[h]				+					
Gliocladium viride[h]				+					
Gliocladium sp. (brown reverse)[h]				+					
Gonotrichium cf. *chlamedosporium*[h]				+					
Graphium penicilliodes[h]				+					
Haplographium sp.[h]				+					
Helicosporium sp.[e]				+	+				
Heteroconium sp.[h]				+					
Hormodendrum hordei Bruhne[c]				+			+		
Hormodendrum viride (Fresenius) Sacc.[c]							+		
Humicola sp.[c]						+			
Hyalodendron, cf.[h]				+					
Idriella lunata[h]				+					

	Substrate								
				Leaves					
Taxon	Log	Branch	Twig	Terrest.	Stream	Roots	Soil	Microbes	Insects
Idriella cf. *ramosa*[h]				+					
Idriella, 2 other spp.[h]				+					
Isthmolonispora minima[h]				+					
Kazulia vagans[h]				+					
Lauriomyces pulchra[h]				+					
Leptographium sp.[h] +									
Lindochium sp.[h]				+					
Lunulospora curvula[e]					+				
Menisporiopsis theobromae[h]				+					
Microdochium sp.[h]				+					
Monielliella sp., cf.[h]				+					
Mycoenterlobium platysporum[h]				+					
Myrothecium leucotrichum[h]				+					
Nigrospora oryzae[i]				E					
Nodulisporium, 5 spp.[h]				+					
Paecilomyces carneus[h]				+					
Paecilomyces marquandii Abbott[c,h]				+		+			
Paecilomyces, 2 other spp.[h]				+					
Papularia sp. cf.[h]				+					
Penicillium citrinum[h]				+					
Penicillium corylophilum Dierckx[c]				+		+	+		
Penicillium duponti Griffin & Maublanc, emend. Emerson[c]						+	+		
Penicillium fellutanum[h]				+					
Penicillium funiculosum Thom.[c]							+		
Penicillium glabrum (Wehmer) Westl.[i]				E					
Penicillium humuli van Beyma[c]						+			
Penicillium lilacinum Thom.[c]				+		+	+		
Penicillium melinii[h]				+					
Penicillium multicolor Grigoriena-Manoifova & Poradielova[c]				+		+	+		
Penicillium olsonii[h]				+					
Penicillium resedanum[h]				+					
Penicillium roqueforti Thom.[c]							+		
Penicillium sclerotiorum[h]				+					
Penicillium tardum Thom.[c]						+	+		
Penicillium variable Westling[c]							+		
Penicillium verruculosum Dierckx[c]						+			
Phaeoisaria clematidis[h]				+					
Phialamonium, 2 spp.[h]				+					
Phialophora clavispora[h]				+					
Phialophora, 8 other spp.[h]				+					
Polycephalomyces, cf.[h]				+					
Polyscylatum, 2 spp.[h]				+					
Pseudobillardia sojae[h]				+					
Pseudobotrytis terrestris[h]				+					
Pyramidospora casuarinae[e]					+				
Ramichloridium clauvilsporum[h]				+					
Ramichloridium cf. *schulzeri*[h]				+					

Taxon	Substrate: Log	Branch	Twig	Leaves: Terrest.	Leaves: Stream	Roots	Soil	Microbes	Insects
Ramichloridium, 2 other spp.[h]				+					
Ramularia, cf.[h]				+					
Rhinocladiella atrovirens[h]				+					
Rhinocladiella sp.[h]				+					
Sagenomella verticilliata[h]				+					
Sarcopodium coffeanum[h]				+					
Scolecobasidium constrictum[h]				+					
Scolecobasidium sp. (4-celled pyriform conidia)[h]				+					
Scolecobasidium sp.[h]				+					
Septomyrothecium uniseptata[h]				+					
Sesquicillium candelabrum[h]				+					
Sesquicillium microsporum[h]				+					
Septofusidium sp.[h]				+					
Speiropsis sp.[e]					+				
Sporothrix sp.[h]				+					
Stachybotryna sp.[h]				+					
Stilbella, cf.[h]				+					
Thozetella habensis[h]				+					
Thozetella sp.[h]				+					
Tolypocladium cf. *inflatum*[h]				+					
Torulomyces lagena[h]				+					
Tricellula inequalis[h]				+					
Trichocladium anomalum[e]					+				
Trichocladium gracile[e]					+				
Trichocladium splendens[e]					+				
Trichoderma hammatum[h]				+					
Trichoderma harzianum[h]				+					
Trichoderma koeningii Oudermans[c,i]				E			+		
Trichoderma lignorum (Tode) Harz.[c]				+					
Trinacrium gracile[h]				+					
Tripospermum sp.[e]					+				
Triposporium elegans[h]				+					
Triscelophorus monosporus[e]					+				
Tritiachrium sp.[h]				+					
Verticillium cf. *aranearum*[h]				+					
Verticillium catenulatum[h]				+					
Verticillium psalliotae[h]				+					
Verticillium sp., Sect. *albo-erecta*[h]				+					
Verticillium sp., Sect. *prostata*[h]				+					
Verticillium, 4 other spp.[h]				+					
Verticimonosporium ellipticum[h]				+					
Virgatospora echinofibrosa[h]				+					
Volutella mimima[h]				+					
Xenobotrytis sp.[h]				+					
Zalarion sp., cf.[h]				+					
Zygosporium echinosporum Bunting & Mason[i]				E					
Tubernulariaceae									
Fusarium avenaceum (Fr.) Sac.[i]				E					
Fusarium decemcellularae Brick[h,i]				+					

Taxon	Substrate								
				Leaves					
	Log	Branch	Twig	Terrest.	Stream	Roots	Soil	Microbes	Insects
Fusarium cf. *lateritium*[h]				+					
Fusarium solani (Martius) Sacc.[h,i]				+E					
Fusarium spp.	+	+	+	+		+	+	+	
Coelomycetes									
Melanconiales									
Colletotrichum sp. (setose, red conidioma)[h]				+					
Leptodiscella sp.[h]				+					
Mycoleptodiscus terrestris[h]				+					
Mycoleptodiscus, 2 spp.[h]				+					
Pestalotia sp. (=*Pestolozzia*[c])[b,e,]				+	+				
Pestolotiopsis maculans[h]				+					
Pestalotiopsis theae[h]				+					
Pestalotiopsis versicolor (Speg.) Steyaert[i]				E					
Sphaeropsidales									
Aposphaeria sp.[h]				+					
Botryodiplodia sp.[h]	+								
Ceuthospora, 2 spp.[h]				+					
Chaetophoma sp.[h]				+					
Chaetosticta, 2 spp.[h]				+					
Colletotrichum acutatum[h]				+					
Calletotorichm crasspies (Speg.) Arx[i]				E					
Colletotrichum gloeosporiodes[h,i]				+					
Coniothyrium fuckelii Sacc.[i]				E					
Coniothyrium sp.[h]				+					
Cytospora sp.[h]				+					
Hainesia cf.[h]				+					
Hendersonia sp.		+							
Hendersonia, cf.			+						
Libertella sp.[h]				+					
Microsphaeropsis, 4 spp.[h]				+					
Ophiostoma sp.[h]				+					
Phyllosticta sapotae Sacc.[i]				E					
Phoma, 2 spp.[h]				+					
Phomopsis manilkarae R. K. Rajak & A. A. K. Chatterjee[i]				E					
Sclerophoma sp.[h]				+					
Zygomycotina									
Zygomycetes									
Mucorales									
Mucoraceae									
Absidia butleri Lender[c]							+		
Mortieriella isabellina[h]				+					
Mortieriella ramanniana var. *angulispora*[h]				+					
Mucor hiemalis[h]				+					
Endogonaceae									
Acaulospora foveata Janos & Trappe[a]						M	+		
Acaulospora scrobiculata Trappe[a]						M	+		
Acaulospora tuberculata Janos & Trappe[a]						M	+		
Gigaspora margarita Becker & Hall[a]						M			

	Substrate								
				Leaves					
Taxon	Log	Branch	Twig	Terrest.	Stream	Roots	Soil	Microbes	Insects
Glomus aggregatum[a]						M	+		
Glomus rubiforme (Gerdemann & Trappe) Almeida & Schenk[a]						M	+		
Glomus taiwanense (Wu & Chen) Almeida & Schenck[a]						M	+		
Glomus cf. *tortuosum*[a]						M	+		
MYXOMYCOTA									
Myxomycotina									
Myxomycetes									
Exosporae									
Ceratiomyxaceae									
Ceratiomyxa fruticulosa (Mull.) Macbr.[f,g]	+							+	
Ceratiomyxa morchella Welden[f,g]		+						+	
Myxogastres									
Liceales									
Liceaceae									
Licea operculata (Wing.) Martin[g]								+	
Tubifera bombarda (Berk. & Br.) Martin[g]	+							+	
Tubifera microsperma[f]	+	+						+	
Cribrariaceae									
Cribraria violacea Rex[g]	+							+	
Reticulariaceae									
Lycogala epidendrum (L.) Fr.[f]								+	
Physarales									
Didymiaceae									
Diderma chondrioderma (DeBary & Rost.) G. Lister[g]								+	
Diderma hemisphaericum (Bull.) Hornem.[g]								+	
Didymium iridis (Ditmar) Fr.[g]								+	
Didymium nigripes (Link) Fr.[g]								+	
Physaraceae									
Physarella oblonga (Berk. & Curt.) Morgan[g]								+	
Physarum tenerum Rex[g]								+	
Trichiales									
Trichiaceae									
Arcyria cinera (Bull.) Pers.[g]								+	
Arcyria denudata (L.) Wettst.[f,g]								+	
Hemitrichia stipitata (Massee) Macbr.[g]								+	
Perichaena chrysosperma (Currey) Lister[g]								+	

Sources:

[a] Species identified in the recent survey by the author and collaborators (especially T. Laessøe, S. Huhndorf, J. D. Rogers, G. Bills, J. Polishook, R. H. Petersen, H. H. Bursdall, Jr., and B. Sutton) that are new records for Puerto Rico.

[b] Species found in the recent survey that were previously recorded from Puerto Rico (Stevenson 1975).

[c] Species previously identified from El Verde by Holler (1966).

[d] Species previously listed for El Verde by Cowley (1970b).

[e] Species previously identified from El Verde by Padgett (1976).

[f] Slime molds found at El Verde by the author and previously recorded from other forests in Puerto Rico by Hagelstein (1932).

[g] Slime molds found at El Verde by Alexopoulos (1970).

[h] Species cultured from leaf litter of *Manilkara bidentata* and *Guarea guidonia:* J. D. Polishook, G. F. Bills, and D. J. Lodge. Microfungi from decaying leaves of two rain-forest trees in Puerto Rico. In prep.

[i] Endophytic species cultured from healthy leaves of *Manilkara bidentata:* D. J. Lodge, P. J. Fisher, and B. C. Sutton. Endophytic fungi of *Manilkara bidentata* leaves in Puerto Rico. *Mycologia,* in review.

Notes: A few species in the Xylariaceae are from comparable forest life zones elsewhere on the island and are likely to occur at El Verde (Laessøe et al. in prep.). Names enclosed in double quotation marks are undescribed species that have not yet been published. Names enclosed in single quotation marks belong in another genus, which is indicated in parentheses when known, but the appropriate new combination has not been published. E, endophytic fungi recovered from healthy leaves; M, mycorrhizal fungi; (+), fruiting bodies, were found on this substrate, but they probably do not derive their carbon from this source, and are instead connected by rhizomorphs or cords to a host or some other substrate.

Appendix 3.B Infections in Woody Species at El Verde

Plant family	Plant species	No. trees	VAM	Non.	Other
Annonaceae	*Oxandra lanceolata* (Sw.) Baill.	1	+	−	−
Araliaceae	*Dendropanax arboreus* (L.) Decne. & Planch.	2	+	−	−
	Didymopanax morototoni (Aubl.) Decne & Planch.	3	+	−	−
Bignoniaceae	*Tabebuia heterophylla* (DC.) Britton	3	+	−	−
Bombacaceae	*Ochroma pyramidale* (Cav.) Urban	2	+	−	−
Boraginaceae	*Cordia sulcata* DC.	2	+	−	−
Bursuraceae	*Dacryodes excelsa* Vahl.	3	+	−	−
	Tetragastris balsamifera (Sw.) Kuntze	1	−	+	−
Combretaceae	*Buchnavia capitata* (Vahl.) Eichl.	2	+	−	−
Elaeocarpaceae	*Sloanea berteriana* Choisy	4	+	−	−
Euphorbiaceae	*Alchornea latifolia* Sw.	3	+	−	−
	Alchorneopsis portoricensis Urban	2	+	−	Pathogenic
	Croton poecilanthus Urban	2	+	−	−
	Drypetes glauca Vahl.	3	+	−	−
	Sapium laurocerasus Desf.	3	+	−	−
Flacourtiaceae	*Casearia arborea* (L. C. Rich.) Urban	3	+	−	−
	Casearia bicolor Urban	2	+	−	−
	Homalium racemosum Jacq.	3	+	−	−
Lauraceae	*Nectandra sintenisii* Mez.	1	+	−	−
	Ocotea floribunda (Sw.) Mez.	2	+	−	−
Fabaceae					
Lotidae	*Andira inermis* (W. Wright) H. B. K.	3	+	−	+?
	Ormosia krugii Urban	3	+	+	Pathogenic
Mimosoideae	*Inga fagifolia* = *I. laurina* (Sw.) Willid	2	+	−	+?
	Inga vera Willd.	3	+	−	+?
Malpighiaceae	*Byrosima coriacea* (Sw.) DC.	3	+	−	−
Melastomataceae	*Miconia tetranda* (Sw.) D. Don	1	+	−	−
	Miconia prasina (Sw.) DC.	1	+	−	−
Meliaceae	*Guarea guidonia* (= *G. trichiliodes* L)	3	+	−	−
	Guarea ramiflora Vent.	1	+	−	−
	Trichilia pallida Sw.	1	+	−	−
Moraceae	*Cecropia schreberiana* Miq.	4	+	−	−
	Ficus laevigata Vahl.	2	+	−	−
Myrtaceae	*Eugenia stahlii* Krug & Urban	4	+	−	−
	Myrcia splendens (Sw.) DC.	3	+	−	−
Nyctaginaceae	*Pisonia subcordata* Sw.	1	+	−	Peritrophic
Oleaceae	*Linociera domingensis* (Lam.) Knobl.	4	+	−	Pathogenic
Palmeae	*Prestoea montana* (Grah.) Nichols = *Euterpe globosa*	4	+	−	−

Plant family	Plant species	No. trees	VAM	Non.	Other
Polygonaceae	*Coccoloba pyrifolia* Desf.	1	+	–	Peritrophic
	Coccoloba diversifolia	3	+	–	Peritrophic
Rosaceae	*Hirtella rugosa* Pers.	3	+	–	–
Rubiaceae	*Genipa americana* L.	2	+	–	–
	Palicouria riparia Benth.	3	+	–	–
	Psychotria berteriana DC.	1	+	–	–
Rutaceae	*Zanthoxylum martinicense* (Lam.) DC.	1	+	–	+?
Sapotaceae	*Micropholis garciniaefolia* Pierre	1	+	–	–
	Manilkara bidentata (A. DC.) Chev.	3	+	–	–
Urticaceae	*Urera baccifera* (L.) Gaud.	3	–	+	–

Notes: VAM: vesicular-arbuscular endomycorrhizae; Non.: non-mycorrhizal species; Other: includes necrotic infections (pathogenic), non-necrotic intracellular—some Pyrenophora-like—infections (+?), and peritrophic mycorrhizal infections (peritrophic). There was no evidence of true ectomycorrhizal infection on any of the specimens examined, but peritrophic mycorrhizae are formed with the same genera of basidiomycetes as those typically forming ectomycorrhizae with other hosts.

Appendix 3.C Mycorrhizal Classification and Methods

I resurveyed mycorrhizal relationships of woody plants at El Verde from 1983 to 1987 (Lodge 1987b), including twenty-one of the thirty-two species examined by Edmisten (1970c). I attempted to collect fine roots from three individuals (at least one young and one mature) of forty-eight woody species, including all six species reported to have ectomycorrhizae by Edmisten. Special emphasis was placed on plants that are common at El Verde, and also on species in families and genera where ectomycorrhizal relationships have previously been reported (except Sapotaceae). Fine roots were collected from a total of 112 trees. For each of 112 trees, a random subsample containing at least 100 cm root length was drawn from each collection, cleared, stained with phenolic aniline blue (Kormanik and McGraw 1982), and examined in whole mounts. Cross- and longitudinal sections were examined as needed.

I verified that one of the four species (*Tetragastris balsamifera*) listed as nonmycorrhizal by Edmiston lacked mycorrhizae, but the other three species had VAM. Edmisten (1970c) only looked at cross-sections of five root tips from one individual of each tree species, which may explain how he missed seeing VAM in *Croton poecilanthus, Miconia tetranda,* and *Linociera domingensis.* Intensity of VAM infections can vary among roots of the same individual, and among individuals of the same species depending on age, rate of root growth, shading, nutrient status, fungal inoculum availability, infections by other fungi, and soil conditions. Edmisten's (1970a) observations of VAM in twelve species are probably all correct, because VAM infections were verified in the ten species which I reexamined (Lodge 1987b; app. 3.B).

The association of *Pisonia subcordata* and *Coccoloba* spp. with basidiomycetes differ from true ectomycorrhizae because Hartig net formation (penetration of the fungus between the host cortex cells) was usually absent in all three species, and hyphal pegs or transfer cells were present in the outer cortex cells (Rupert, Hammill, and Lodge unpublished). Such structures probably represent peritrophic mycorrhizae (see Ashford and Allaway 1982, 1985). Peritrophic mycorrhizae have previously been found on *Pisonia grandis* on islands in the Great Barrier Reef in Australia by Ashford and Allaway (1982), but it is not known if the fungi are beneficial to the plants. The fungal mantles are so thick and extensive, however, that they must influence plant nutrient uptake and carbon costs. The coastal species of *Coccoloba* (*C. uvifera* L.) is also associated with basidiomycetous fungi, but it forms typical ectomycorrhizae.

In 1955, members of the U.S.D.A. Forest Service inoculated pine seedlings in Puerto Rico with humus and soil taken from beneath pines in North Carolina. The success of the first inoculation led to more experiments in 1964, 1965, and 1966, in which seedlings were inoculated with either mixed natural inoculum or a pure culture of one of four ectomycorrhizal fungi from North America (Hacskaylo and Vozzo 1967); all treatments stimulated growth in the pines. In a later survey of the outplanted pines, however, none of the species introduced in pure culture were recovered. Most of the fungi that I now find fruiting under pine in Puerto Rico (*Pisolithus tinctorius, Scleroderma geaster,* and *Rhizopogon rubescens* complex) were probably introduced in the mixed natural inoculum. It is unknown but unlikely that the basidiomycetous fungi associated with *Pisonia* and *Coccoloba* species can form ectomycorrhizae with pine. It is therefore possible but not certain that the biogeographical problems of simultaneous dispersal of the host and its ectomycorrhizal fungi prevented the colonization of Puerto Rico by pines. Another possible explanation is that pines are apparently poor dispersers as suggested by their high rates of speciation in Mexico due to geographic isolation on different mountains. Alternatively, the absence of native *Pinus occidentalis* from Puerto Rico might be attributable to the absence of high-elevation habitat with environmental and edaphic conditions comparable to those in native pine stands of Hispaniola.

Assumptions and Methods Used in Estimating Microbial Biomass and Nutrient Stores

The calculations for fungal production in wood were made by assuming that ratios of live fungi to bacteria plus actinomycetes were the same as Frankland (1982) found in woody debris in Cumbria, United Kingdom (30.5 : 2.6 kg ha $^{-1}$), and that invertebrate respiration accounts for 10% or less of the total carbon dioxide evolution during decomposition (Berthet 1967; Reichle 1977; Persson et al. 1980; Petersen and Luxton 1982).

A modified Jones and Mollison (1948) direct count method (Lodge and Ingham 1991; Lodge 1993) was used to determine hyphal length per gram of litter, and these data together with hyphal diameter distributions (Bååth and Soderstrom 1979) were used to calculate fungal volume. Five random 113.7 cm ^{2} samples of leaf and small woody litter were collected at two-week intervals from a 1 ha plot (Plot 3, fig. 1.3) and preserved. Each litter sample was blended at high speed for 5 min, diluted to 400 ml, and a 20 ml aliquot of suspension was added to 30 ml hot agar. Fifty μl aliquots of agar suspension were transferred to glass slides using a micropipetter, and 22 mm round coverslips were added to create agar films of known area and volume. Estimations of hyphal diameter distributions and hyphal lengths were calculated from six agar films per litter sample using a line intersect technique (Giovanetti and Mosse 1980). The dry weight specific gravity of fungi collected from fine litter at El Verde (0.26 g dry cm $^{-3}$ fresh, Lodge 1987b) was used to convert fungal volumes into biomass.

Nutrient stores in fungal biomass in litter were calculated based on the estimated fungal biomass (5.2 mg fungus g $^{-1}$ litter, Lodge 1993; 1.5 g m $^{-2}$), assuming 282 g fine litter m $^{-2}$ according to Zou et al. (1995).

Microbial production estimates for the fine litter layer can be calculated from litter inputs (table 3.5) or rates of carbon dioxide evolution from litter. Assuming that invertebrate respiration represents 10% of the total litter respiration, then between 328 and 527 g C m $^{-2}$ is lost from litter per year through microbial respiration. Gross microbial biomass production can be estimated by using a 60% production efficiency estimate for rapidly growing mixed microorganisms (Payne 1970; McGill et al. 1973). Although some of the respired carbon dioxide reflects maintenance energy for dormant microorganisms, this probably represents less than 10% of the litter microbial respiration, because mean live microbial biomass in the fine litter layer probably does not exceed 4 g m $^{-2}$ at El Verde, and this generous estimate multiplied by the maintenance coefficient of 0.001 g of substrate C g $^{-1}$ live microbial C hr $^{-1}$ (Shields et al. 1973) gives only 35 g C m $^{-2}$ yr $^{-1}$ for maintenance. Furthermore, litter fungal populations at El Verde (Lodge 1993) and probably also bacterial populations

(Witkamp 1974) change rapidly with wetting and drying, so their populations may be in the logarithmic growth phase for much of the time. Based on the above assumptions, annual gross microbial production in the litter layer is roughly 492 to 790 g C m^{-2} y^{-1}.

For soil, I estimated a mean of 207 g m^{-2} of fungal biomass (live plus dead) in the upper 10 cm of soil not covered by boulders (daily means ranged from 14 to 333 g m^{-2}) using a direct-count method (Lodge 1993), and assuming the same dry weight density as in mycelium in logs (0.20 g m^{-3} fresh; Lodge 1987b). The fungal density estimate derived from log mycelia instead of litter mycelia was used for soil fungi because microorganisms have different dry weight densities depending on the moisture regime under which they have grown (Van Veen and Paul 1979; Newell and Statzell-Tallman 1982), and moisture content in soil at El Verde (Lodge and Pfeiffer unpublished data) and in logs (Harvey et al. 1978; Dowding 1985) is high with only moderate fluctuations as compared to fine litter. Assuming a bacteria cell diameter of 2.5 μm (Witkamp 1974) and a dry weight density of 0.8 g cm^{-3} fresh (Van Veen and Paul 1979), the living (viable) bacterial biomass in the upper 2.5 cm of soil at El Verde is 1.5 to 26.5 g m^{-2}. I assumed that bacterial densities are the same in the upper 2.5 and 10 cm of soil (probably an overestimate), and that only 0.44% of the total bacterial biomass in soil is living (Frankland 1982, United Kingdom). Total (live plus dead) bacterial biomass at El Verde was estimated to be 1,364 to 24,091 g m^{-2} in the upper 10 cm of soil. This appears to be an overestimate. Bacterial densities probably decrease with soil depth, and the ratio of live to total bacteria may be higher at El Verde than in the United Kingdom.

A second estimate of bacterial biomass for El Verde (381 to 421 g m^{-2}) was extrapolated from data on total toil microbial biomass for a nearby site 0.5 km away with seventeen-yr-old native second growth forest and pine plantation (same elevation and exposure). In addition to having different vegetation from El Verde, this site is located on a different soil type (characterized by magnesium concretions), so these data should be used with caution. Total microbial biomass carbon in November (rainy season) of 1987 was estimated by the fumigation-incubation technique, and was 217.4 and 293.4 g m^{-2} in the top 10 cm of soil for pine and secondary forest, respectively (Lodge 1993). Soil organic carbon was 1.7% under pine and 2.7% under secondary forest in the upper 15 cm at this site (Cuevas et al. 1991), whereas the Los Guineos silty clay loam at El Verde (chap. 1) typically has 3.85% soil organic carbon (Brown et al. 1983). If the ratio of soil microbial carbon to organic carbon is the same for El Verde as in the nearby pine and second growth forest (6.4% and 5.8%, respectively), then total microbial carbon in the upper 10 cm of soil at El Verde is 171.9 to 189.7 g m^{-2}. Assuming that 45.1% of the microbial biomass is carbon (mean percent carbon for vegetative fungal mycelia from litter and logs in table 3.4), then total microbial biomass in the upper 10 cm of soil at El Verde is circa 381 to 421 g m^{-2}.

Litter Decomposition Methods

Zou et al. (1995) used freshly fallen leaves (within twenty-four hours of abscission) in the same proportion as was found in litter baskets during May in two 1 ha plots (see Plots 3 and 4, fig. 1.3). Similar trials were initiated during March and October in

Plot 3. The litter was surface dried, enclosed in 1 mm mesh litterbags (10 g dry wt per bag), and set out in the same plots from which it originated and also in a common control plot (a small homogeneous area, Plot 5) soon after collection. Mass loss over 224 days was slightly but not significantly faster in leaf litter of the midsuccessional forest (72.7%, Plot 4) than in a lightly disturbed forest plot (67.4%, Plot 3). La Caro (1982) demonstrated that leaf decomposition rates were not significantly influenced by the tree species canopy under which they were placed at El Verde, but his estimated mean rate of decomposition was slower than that of Zou et al. (21.1 and 34.3% of initial weight lost during 224 days in primary and successional species, respectively). Weigert and Murphy's (1970) rates of litter decomposition for various species (32.3 to 81.0% after 224 days) were more similar to the Zou et al. estimate than the La Caro (1982) estimate. Wiegert and Murphy's (1970) methods involved drying and weighing the litterbags at intervals and returning them to the field.

The longer turnover time of fine litter calculated from the litterbag decomposition rate (Zou et al. 1994, 1.25 y) as opposed to storage divided by litterfall (0.36 yrs) in table 3.5 might be caused by (1) faster decomposition of litter components other than leaves, (2) slowing of decomposition by litterbags (St. John 1980a), (3) retention of fine fragments in the litterbags that would normally become incorporated in the upper soil layer and missed in litter storage estimates, or (4) losses of litter from storage because of the effects of steep topography on litter export (Lodge and Asbury 1988). A comparison of leaves enclosed in litterbags and unenclosed (tethered) showed no significant difference in decomposition after two months (i.e., before fragmentation losses would cause tethered leaves to disappear faster than leaves in bags; Lodge and Pfeiffer unpublished) so it is unlikely that litterbags slowed decomposition though they may have retained more fine fragments. Litter export from slopes was estimated as 628 g m^{-1} length of stream trace yr^{-1} at El Verde (Lodge and Asbury 1988), and may significantly reduce litter storage.

Fungal Communities in Leaves

Cowley (1970b) examined vertical stratification and host specificity of microfungal communities of leaves of 2 tree species at El Verde using the following methods. Six samples of leaves were collected at heights of 3 to 6 m, 10 m, and 15 to 20 m above ground from two trees, *Manilkara bidentata* and *Dacryodes excelsa*. The unwashed leaves were ground and plated on modified Martin's medium (Holler and Cowley 1970), and twenty-five isolates were randomly selected from the cultures obtained from each leaf sample. John Polishook, D. Jean Lodge, and G. Bills (in review) compared microfungi in decomposing leaves of *Manilkara bidentata* and *Guarea guidonia* that were mixed together on the forest floor at two different sites and found that about half of the fungal species grew in only one of the two leaf litter species. Lodge, P. J. Fisher, and B. Sutton (in review) found twenty-eight species of endophytic fungi in healthy leaves of *M. Bidentata,* of which twenty-one species were ubiquitous and were found in several trees. Ninety to 95% of the leaf blades and 100% of the petiole lengths were occupied by endophytic fungi, especially three species of *Xylaria.*

4

Termites

Elizabeth A. McMahan

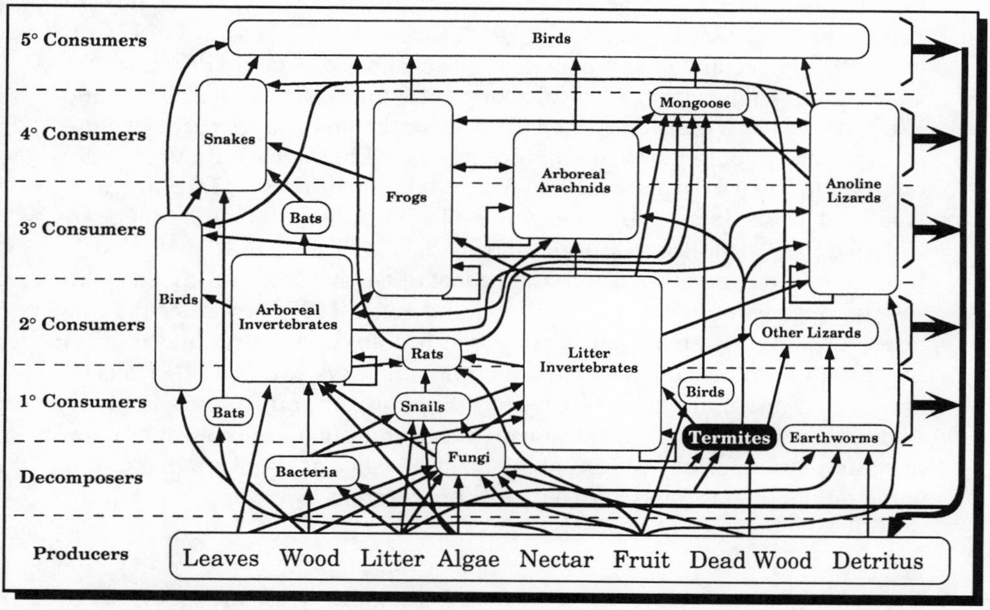

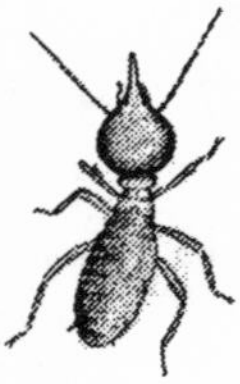

IT HAS long been assumed that termites play an important role in tropical ecosystems because of their feeding specialties and their enormous numbers. Until fairly recently, however, there have been almost no quantitative data available on such parameters as population densities, biomass, food consumption rates, and respiration rates that would permit a precise evaluation of termite impact. Such studies still are relatively few, mainly because of the great difficulties encountered in termite population sampling. Those published to date have concerned chiefly the termites of the tropical savannas of Africa (Josens 1973; Lepage 1974, 1984; Ohiagu 1979; Buxton 1981; Collins 1981a,b; Ferrar 1982; Wood et al. 1982) and the semiarid deserts of the United States (Haverty 1974; Johnson and Whitford 1975; Schaefer and Whitford 1981; Whitford et al. 1982). One of the first studies of the dynamics of termite populations in a rain forest was carried out by Wiegert (1970b) on *Nasutitermes costalis* in the El Verde tabonuco forest. More intensive studies in West Malaysia by Matsumoto (1976), Abe (1980), and others; by Collins (1979, 1980) in Sarawak; and by Salick et al. (1983) in Venezuela have followed. Summaries and assessments of the studies up to the mid-1970s are given in Brian (1978). Wood and Sands (1978) consider the role of termites in ecosystems under two main headings: (1) direct and indirect modification of the habitat by construction of nest systems and (2) effects on energy flow and nutrient cycling by consumption and transformation of cellulose. Figure 4.1 is their diagrammatic representation of the role of termites in ecosystems. Specific portions of the diagram (such as those relating to foraging on dung, grass, and herbs) do not refer to rain forest termites, but the overall view is certainly applicable.

TAXONOMIC AFFINITIES OF EL VERDE TERMITES

Termites (Order Isoptera) are ubiquitous in tropical and subtropical regions. There are over 2000 described species, and most are found within 35° of the equator.

Krishna (1969) classifies termites into six families: Mastotermitidae, Kalotermitidae, Hodotermitidae, Rhinotermitidae, Serritermitidae, and Termitidae. The first five possess, in the hindgut, symbiotic protozoans on which

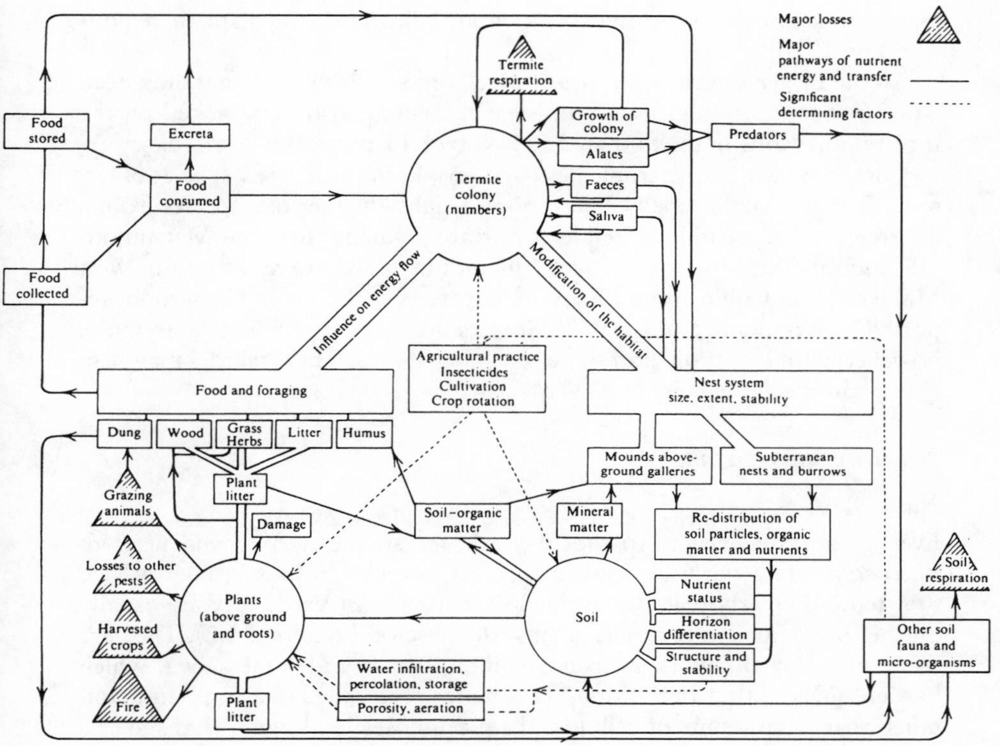

Figure 4.1. Diagrammatic representation of the role of termites in ecosystems, from Wood and Sands (1978)

the termites depend for cellulose digestion. The sixth, Termitidae, contains about three-fourths of all the termite species known, and these are the most advanced and diverse. They possess few symbiotic protozoa; their digestion is accomplished with the aid of bacteria.

Four species have been reported from the tabonuco forest: *Nasutitermes costalis* (Hölmgren), *N. nigriceps* (Haldeman), *Parvitermes discolor* (Banks), and *Glyptotermes pubescens* (Snyder). The four species represent about a third of the total number of termite species reported from Puerto Rico. The first three belong to family Termitidae, subfamily Nasutitermitinae. The Genus *Nasutitermes,* containing about 180 named species, probably arose in the Neotropical region but now occurs also in the Australian, Ethiopian, Malagasy, Oriental, and Papuan regions (Krishna 1970). The fourth species found at El Verde belongs to Family Kalotermitidae, the so-called "drywood termites" which build no discrete houses but usually nest in wood without soil connections. The genus *Glyptotermes* occurs in the Australian, Ethiopian, Madagascan, Neotropical, Oriental, Palearctic, and Papuan regions (Krishna 1970).

All the El Verde termites appear to seek out relatively firm, standing dead wood as food, although twigs and leaf litter may also be attacked on the forest floor. None of the four species is known to attack living wood.

Compared with other tropical ecosystems, El Verde has few species of termite. For example, Beebe (1925) reported eighty-five termite species from a quarter of a square mile of jungle at Kartabo, Guiana; Abe and Matsumoto (1979) found fifty-two species in a 1 ha plot in a lowland rain forest in West Malaysia; and Collins found forty-three species in 1.5 ha in Cameroon (reported in Wood and Sands 1978). Species numbers are known to decrease, however, with elevation, and to be fewer on islands. These factors may help to explain species paucity at El Verde.

Nasutitermes costalis

Nests

Even to a casual visitor to the El Verde forest, the dark, ovoid nests of *N. costalis* are a fairly obvious feature. At lower elevations in Puerto Rico they tend to be arboreal, but at the 350 m level of El Verde they are usually on the forest floor, often built against the base of a tree (fig. 4.2). They are constructed by the workers from semiliquid feces and some saliva, which harden quickly after deposition. This material is called "carton" and contains variable amounts of cellulose, hemicelluloses, and lignin (LaFage and Nutting 1978). The nests average about 45 cm in height and about 40 cm in diameter (range 12 to 80 cm and 12 to 55 cm, respectively).

Nasutitermes nests have been compared in structure to a rigid sponge. A thin surface skin of carton, full of pimples and fingerlike protrusions, covers an underlying mass of interconnecting chambers that house the colony. Near the lower center of the nest mass is located the enlarged royal chamber containing the reproductive pair. The carton immediately surrounding this chamber is much thicker and stronger than that nearer the periphery. The termites usually keep the nest surface intact, quickly repairing with anal cement any accidental damage such as that caused by falling branches or scampering lizards. At intervals, however, they remove portions of the skin in order to expand the nest. During these periods of spontaneous nest expansion the inhabitants are most vulnerable to predation. Jones (1979) has studied closely the nest expansion behavior of *N. costalis*.

McMahan and Blanton (1993) studied effects of Hurricane Hugo on *N. costalis* nest density and nest reconstruction at El Verde in the two years following the hurricane's destructive passage in 1989. No long-term effects, either deleterious or advantageous (e.g., from an increase in the supply of dead wood), were noted relative to number of nests and galleries constructed.

Figure 4.2. A large *N. costalis* nest (height approximately 60 cm) at the base of a tabonuco (*Dacryodes excelsa*) tree at El Verde

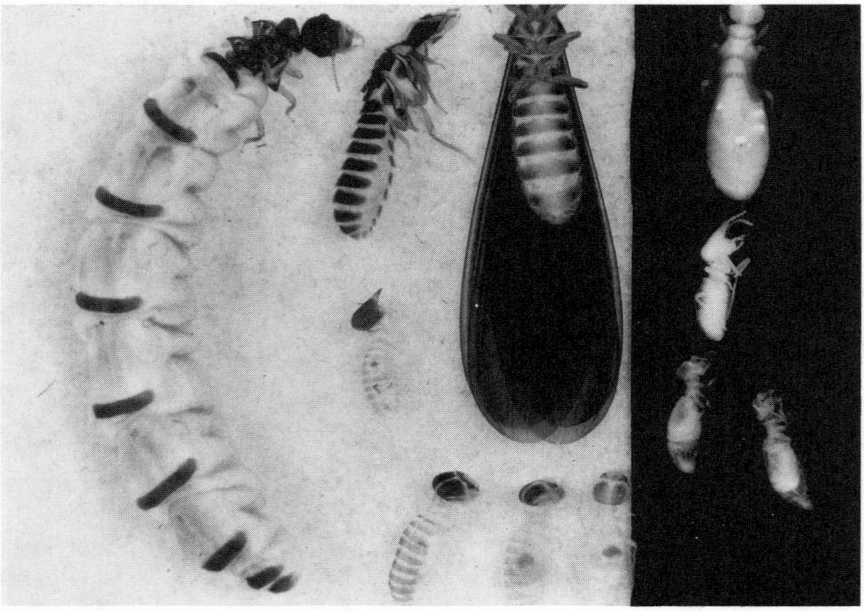

Figure 4.3. The castes of *N. costalis*. Light background (counterclockwise from upper right): female alate, male dealate (king), physogastric queen, 3 large worker instars, mature soldier (in center); dark background (top to bottom): reproductive nymph with wing pads, white soldier, 2 small worker instars.

Hugo appeared to have had an immediate effect, however, in causing a diminution in nest size through nest surface erosion. The post-Hugo nests were significantly smaller than those measured in pre-Hugo years and showed a significant, if temporary, acceleration in growth rate. This growth rate increase may have resulted from the smaller nests having an abnormally high nest population density for their size (the colony having been forced by the erosion to occupy a smaller volume) as well as a more favorable surface to volume relationship for nest gaseous exchange.

Reproductive Caste

In its caste organization, *N. costalis* is typical of the genus (fig. 4.3). The queen is physogastric (about 2.5 cm in length), her enlarged abdomen with its widely separated tergal plates giving her the appearance of a whitish grub. Her body is too large to pass through exit holes in the royal chamber, so she is confined therein, mating at intervals with the king and laying eggs throughout her lifetime. The king retains his original appearance as a dark dealate. The reproductives have functional compound eyes.

Queen and king longevity data are few for any species, and lacking for *N. costalis*. At El Verde one of the *N. costalis* nests that had been under investigation since 1966 was still active in 1984. Whether or not the original primary reproductives were present throughout that period is unknown.

Reproductive forms are derived from a line of individuals that develop wing pads which increase in size at every molt. Wing-padded individuals are called nymphs. In the spring they undergo a final molt, becoming dark-winged alates which are released from the colonies at El Verde in May and June in crepuscular swarmings. After losing their wings, surviving pairs creep into sheltered spots to begin new colonies.

Nasutitermes costalis alates are considerably larger than the largest workers (see fig. 4.3) and contain a higher fat accumulation. Wiegert and Coleman (1970) suggest that this ability of the sexuals to store large amounts of high-energy fat is a physiological adaptation for life in a crowded environment where energy reserves are important to the successful foundation of an incipient colony. Nevertheless, it is probable that only a few pairs, out of the thousands of alates that swarm, are successful.

In addition to the founding of incipient colonies by dealates, *N. costalis* may also form new colonies through sociotomy, the migration of a portion of the parent colony to a new location. Emerson (1929) reported such a *N. costalis* migration early one morning in Guiana. The royal couple was in the middle of the procession, which consisted of soldiers and workers and seven species of termitophiles. Although sociotomy has been reported for several termite species, little is known about this method of new colony foundation. What proportion of the colony migrates? Does the former nest continue to be active, with supplementary reproductives in place? During the nest monitoring at El Verde, when a formerly active nest was found to be abandoned, a new nest was almost always discovered in the near vicinity (usually within 5 m). It is not known if this represents total colony migration.

Worker Caste

The general work of the *N. costalis* colony is carried out by both a large and a small worker line, each developing through two dependent larval stages (fig. 4.4). The large worker line is female and consists of three stages of ascending age (LW1, LW2, LW3). Their average live weight is about 4.6 mg. The small worker line is male and consists of two age stages (SW1, SW2). Their average live weight is about 1.6 mg. Most SW1 individuals molt to become soldiers. Only about 25% become SW2 individuals. In the genus *Nasutitermes* pigmentation of the worker's head capsule usually increases with age, so the various worker instars in a given line can be identified. Neither workers nor soldiers have eyes.

The polymorphism of *N. costalis* workers is accompanied by polyethism (job allocation) based on size (sex) and age. The five worker types tend to

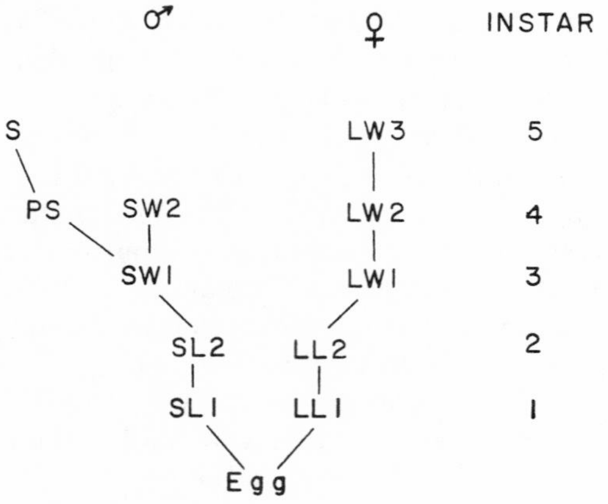

Figure 4.4. Diagram of the developmental pathway for *N. costalis* neuters. LL, large larva; SL, small larva; LW, large worker; SW, small worker; PS, presoldier; S, soldier.

participate at different frequencies in the various colony activities. So far, studies of polyethism have concentrated primarily on nest repair and foraging gallery initiation and construction. McMahan (1970a) found in field studies at El Verde that under conditions of continuous disturbance at surface breaches, LW3 workers did almost all of the nest repairs. When no disturbance had occurred at the repair site for half an hour or longer, however, or when spontaneous nest expansion was under way, the SW1 and SW2 workers rivalled the LW3 workers in importance as builders. LW2 and LW1 workers participated to a much lower degree on all occasions, but increased at a work site as the "hazard level" decreased.

As in the case of *N. lujae* (Pasteels 1965), LW3 workers of *N. costalis* probably also specialize in the early exploitation of food sources (although soldiers and not workers are apparently the initial scouts). Pasteels showed with *N. lujae* that this specialization is correlated with a greater degree of development of the sternal (trail-laying) gland. Traniello (1982) has concluded that the trail pheromone of *N. costalis* contains a short-lived excitatory component which stimulates recruitment, plus a long-lived component effective in orientation. Younger instars of both worker lines appear to be most involved with brood care and the feeding of the dependent castes.

Soldier Caste

Nasutitermes costalis soldiers, like those of other species of *Nasutitermes,* are derived from SW1 individuals and are therefore male (mean weight, 1.5 mg). The mature, dark-headed soldier stage is reached via a relatively brief and nonpigmented stage, sometimes called a "white soldier." This presoldier does not feed and has no defensive function. It molts to the mature soldier form, the nasute, which does not molt again. The nasute can spray a sticky and toxic secretion from its frontal gland, through a projecting snout or *nasus,* to distances up to several times its body length. Components of its secretion include α-terpinene, γ-terpinene, myrcene, camphene, terpinolene, β-pinene, limonen, and Δ^3-carene (see Deligne et al. 1981 for a summary). This defensive spray is fairly effective against small would-be predators, particularly ants. The soldiers align themselves around the margins of breaks in the nest surface and at the margins of foraging galleries under construction and eject their spray, usually with an accompanying lunge, when an intruder is detected. In spite of their lack of vision, *N. costalis* soldiers are effective in directing their spray. Traniello (1981) found that ants that had been heavily sprayed by *N. costalis* soldiers were immobilized within two minutes. On one occasion he observed that 67% of the 416 *Solenopsis* scouts entering a *N. costalis* foraging area were dead or disabled within three hours, whereas only eleven nasutes had been killed and retrieved by the ants.

So far, there has been no observation of *N. costalis* soldiers that were derived from LW1 (female) workers. In *N. exitiosus* in Australia, however, the female soldier is apparently a normal (though minor) component of the colony (McMahan 1974; Eisner et al. 1976), differing markedly from the usual (male) soldier in her larger size, nasus shape, and lack of aggressiveness.

In addition to their defensive role in the colony, *N. costalis* soldiers also regulate foraging activity in the initial stages (Traniello 1982). In laboratory tests they were observed to act as scouts, locating new food sources and then recruiting large numbers of additional soldiers from the home nest through the laying of trails from sternal gland secretions. Eventually the trails became concentrated enough to attract workers, which proceeded to build galleries over them. The soldier's sternal gland secretion served both to stimulate recruitment and as an arrestant for other soldiers.

Proportions of sterile caste members in *N. costalis* colonies at El Verde have been determined by Wiegert (1970b), who lumped large and small workers, and McMahan (unpublished data) through sampling termites extracted from entire nests. The results are given in table 4.1. Large workers predominate, SW2 individuals are few, and soldiers constitute a relatively large percentage of the nest population. Burns (see Odum 1983) has presented a model of the energy relationships between the various caste members (fig. 4.5).

Nasutitermes nigriceps

Nests

The largest termite species at El Verde in terms of body size is *N. nigriceps,* a species found throughout much of Central America. It builds very large,

Table 4.1. Caste and developmental stage proportions in *N. costalis* colonies at El Verde

	1		2		
Termite Type (Sex)	No.	% of Total	No.	% of Total	
Large Worker 3 (♀)	7501	72.9	2872	24.1	78.8
Large Worker 2 (♀)			2171	18.2	
Large Worker 1 (♀)			2257	18.9	
Small Worker 2 (♂)			547	4.6	
Small Worker 1 (♂)			1545	13.0	
Soldier, White (♂)	1803	17.5	214	1.8	13.1
Soldier, Mature (♂)			1345	11.3	
Larva 3 (♀)	980	9.5	640	5.4	8.1
Larva 2 (♂)			138	1.1	
Larva 1 (undeterm.)			188	1.6	

Sources: 1: Wiegert (1970b), four nests; 2: McMahan (unpublished), one large nest.

Note: All *N. costalis* colonies were totally harvested.

dark, ovoid arboreal nests; in Puerto Rico, at least, they are usually at elevations below 200 m (Martorell 1945).

Light (1933) noted that a single colony might have two or more nests connected with galleries and that individuals foraged over a considerable area. This species is definitely established at El Verde above the 400 m level, although only four nests had been discovered up to 1993. All nests appeared to be more than 75 cm in height and 60 cm in length and width, but because of their inaccessibility (on tree trunks at heights above 7 m) none had been measured precisely until 1990. In September 1989, Hurricane Hugo uprooted the tree on which Nest #2 (whose active galleries had been first noted in 1966) was attached. Its measurements were 95 × 85 × 90 cm, showing it to be almost spherical. It remained active at least through the summer of 1991, but was found to be abandoned in 1993. Nests #1 and #2 were discovered in 1970, Nest #3 in 1984, and #4 in 1990. Their foundation dates are unknown, but #1 and #2 probably predate 1966, and all except the fallen nest appeared to be active in 1993. Such nests probably contain several hundred thousand individuals each (Light 1933).

Castes

The caste system of *N. nigriceps* is approximately that of *N. costalis,* with one reproductive pair, large (7.0 mg, mean weight) and small (2.5 mg, mean weight) workers determined by sex, and nasutiform soldiers (2.4 mg, mean weight). The workers appear to be considerably more aggressive than those of *N. costalis*. Their large size also permits a better grip with the mandibles on human skin, and once attached do not readily release their grip. Attempts to brush them off leave a row of clinging *N. nigriceps* worker heads.

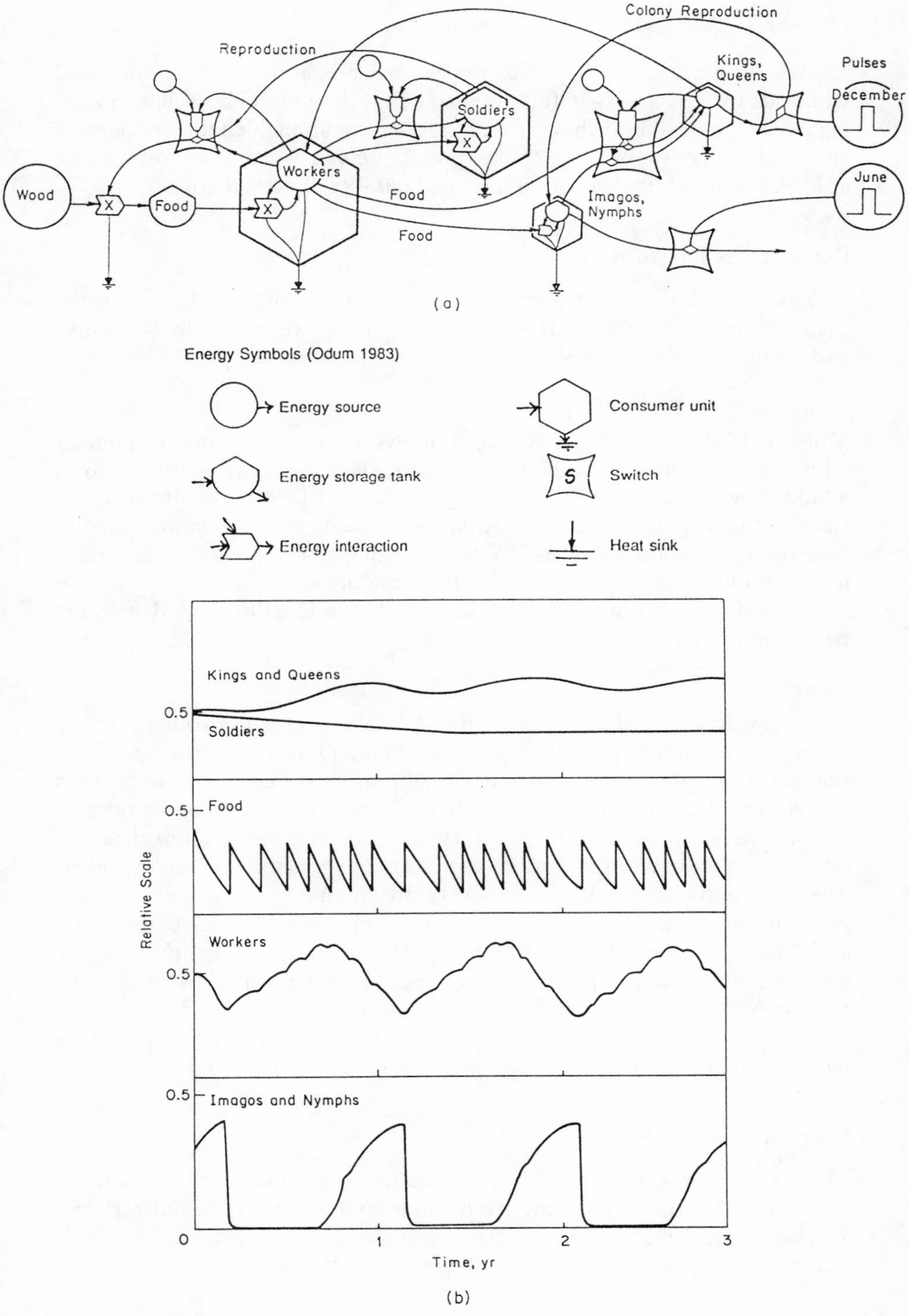

Figure 4.5. Model by L. A. Burns of the relationships of caste members of *N. costalis,* using Odum's energy flow diagrams. (*a*) Energy diagrams; (*b*) seasonal cycles resulting from simulation of termite model in *a* (from Odum 1983).

Like *N. costalis, N. nigriceps* appears to prefer relatively sound dead wood. Its tunnels are to be found on standing dead wood as well as on living trees with dead branches. The aggressiveness with which this species forages and the unusually large colony populations may give it an importance at El Verde out of proportion to the small number of nests there.

Parvitermes Discolor

Because it does not build discrete nests, this species is not obvious to the casual observer at El Verde. It is a physically smaller termite than *N. costalis* and is much smaller than *N. nigriceps.*

Nests

Wolcott (1948) states that *P. discolor* hollows out a rotten stump and lines it with a very dark brown building material, presumably anal cement, to form a hidden nest area. In 1970 I found a young colony living in chambers excavated within old frond scales at the base of a young *Prestoea montana* palm near the Radiation Center at El Verde (see chap. 1). Only a thin (approximately 4 mm) leaf scale separated it from the outside. No other *P. discolor* nest was discovered at El Verde, although foraging galleries and foraging parties were frequently encountered.

Castes

The incipient *P. discolor* colony on the *Prestoea* palm had a young primary queen, not obviously physogastric, which had laid hundreds of eggs. When she was removed with the colony to the laboratory she continued to lay eggs at the approximate rate of one per hour, although some came at twenty-minute intervals. *P. discolor* has two types of workers, both about the same size (2.0 mg mean weight), but one with greater pigmentation than the other. Their sex and developmental pathways are unknown. Both worker types participated in egg tending and brood care, in stomodeal feeding of the queen and other colony mates, and in foraging. The soldiers of *P. discolor* (0.8 mg mean weight) are nasutiform but much less aggressive than those of *N. costalis* or *N. nigriceps.* When their foraging tunnels are broken into or their foraging areas exposed, they tend to flee with the workers. Almost nothing has been published on the biology of *Parvitermes discolor.*

Glyptotermes Pubescens

This species has been reported from the Luquillo Mountains at elevations up to 760 m (Martorell 1945) and from Aibonito, Puerto Rico (Snyder 1925). Workers of a *Glyptotermes* species, presumably *G. pubescens,* were collected from a partly dead tree at El Verde in 1966, from the same tree in from several other El Verde sites in 1993. As one of the dry-wood termites, this species builds neither nests nor carton galleries but lives within the tunnels excavated in sound dead wood. No soil contact is made. Instead of producing anal cement for nest and gallery construction, *Glyptotermes* produces discrete, hard fecal pellets, each of which bears the imprint of the rectal muscles.

Like other kalotermitids, *Glyptotermes* has a nonphysogastric queen, and the work of the relatively small colony is carried out by immature members of the reproductive and soldier castes. Soldiers are of the mandibulate type. *G. pubescens* has not been the object of intensive search at El Verde, and because, like *Parvitermes,* it builds no visible nests its significance to the ecosystem may be underestimated.

POPULATION DENSITY AND BIOMASS

At El Verde, the *N. costalis* nests have been under surveillance since 1964 when studies of effects of gamma radiation on the rain forest ecosystem were begun (Odum and Pigeon 1970). Peter Murphy and Alejo Estrada began mapping the nests in 1964, and Wiegert (1970b) continued the studies. By 1966, twenty-two nests had been identified within two circular areas, each area with a radius of 80 m. One was the Radiation Center, which in 1965 was exposed to a 10,000-curie ^{137}Ce source for ninety-three days. The other was the South Control Center, which was similar in vegetation and topography but was not irradiated (see chap. 1). Only three nests were within 30 m of the cesium source and subject to its direct effect.

The twenty-two nests were monitored annually from 1966 to 1970, and again in 1979 (McMahan unpublished). Figure 4.6 shows the location of most of the active nests in 1966, and figure 4.7 graphs colony survival for these nests, using a log scale. Only one of the original nests still survived in 1979. The regression of number of nests surviving against time is well described by the following equation: $\log(Y + 1) = 1.32 - 0.079X$, where Y is number of nests surviving and X is number of years. The computed half-life is 3.81 years. The longest-lived *N. costalis* nest at El Verde survived for at least eighteen years, and another survived for at least thirteen. Neither was among the twenty-two original nests within the two centers. The former nest was just outside the 80 m radius circle of the Radiation Center and beneath a large overhanging rock that would have shielded it completely from the cesium source in any case.

The founding dates of the original twenty-two nests are unknown, but beginning in 1966 an annual search was made of the two centers for nests, plus an area of the forest extending some 50 m on either side of the trails that led from the two centers to the field station (a total of approximately 8.5 ha). Records of new nest foundings and old nest abandonments were kept to

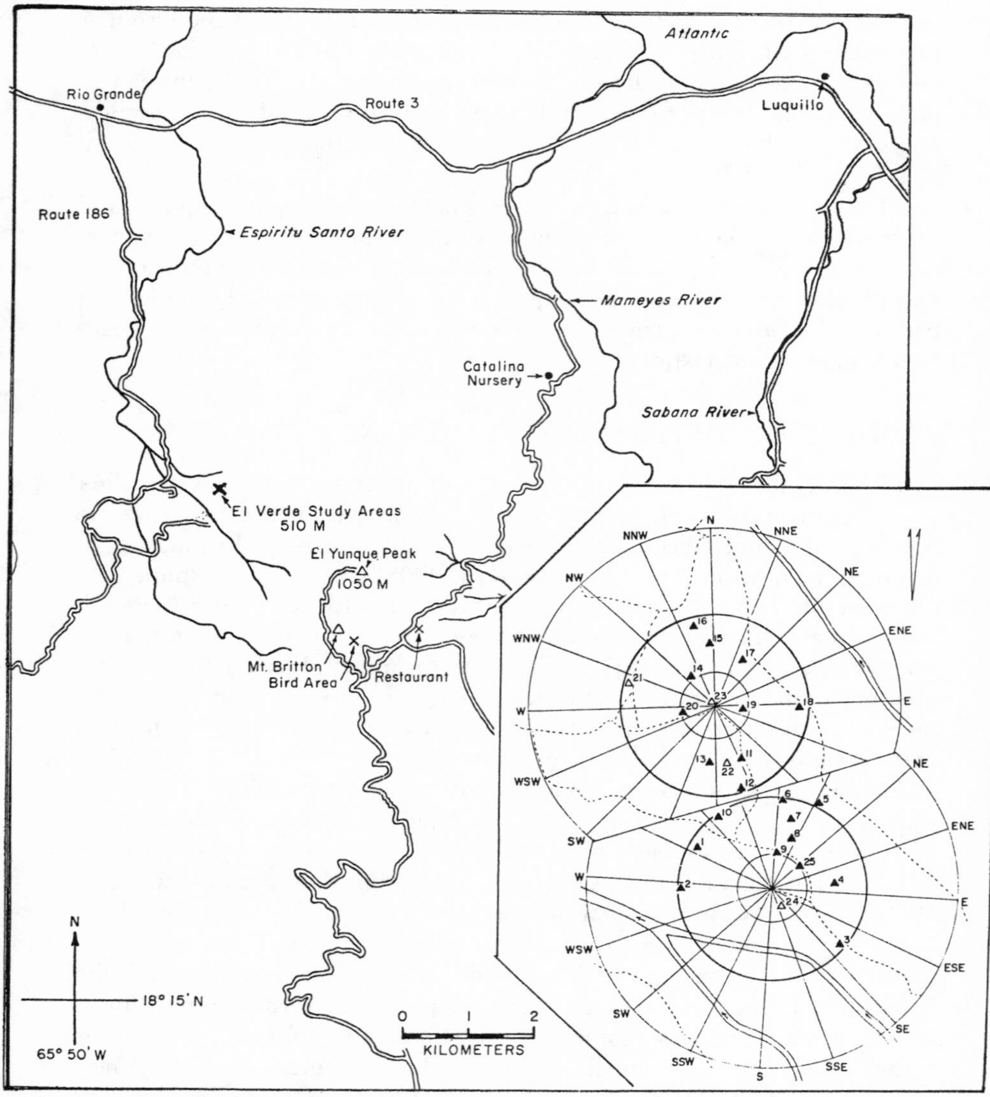

Figure 4.6. Map of *N. costalis* nests (superimposed on a map of the El Verde study area), in May 1966, within the 80 m radius in each of the two study centers. The upper circle is the Radiation Center; the lower is the South Control Center. Closed triangles = active nests; open triangles = abandoned nests (from Odum and Pigeon 1970 and Wiegert 1970b).

1979. Table 4.2 presents these data, including those for the original twenty-two nests, and indicates that rates of nest founding and nest demise are comparable. Approximately 20% of the nests became inactive each year. Lubin and Young (1977) reported that 16% of all nests (of all species studied) were abandoned per year on Barro Colorado, Panama, and Salik et al. (1983) observed a 11% turnover per year at Rio Negro in Venezuela.

A density of about 4.5 *N. costalis* nests ha^{-1} was estimated for the El Verde forest. This is a very small number compared with nest density of other termite species in other rain forests. For example, the values found by Matsumoto (1976) for several nest builders in the Pasoh forest of West Malaysia were as follows: *Macrotermes carbonarius,* 15 to 41 ha^{-1}; *Dicuspiditermes nemorosus,* 60 to 110 ha^{-1}; and *Homallotermes foraminifer,* 85 to 165 ha^{-1}. Maldague gives a comparable value of 875 ha^{-1} for *Cubitermes fungifaber* nests in a Congo rain forest (quoted by Matsumoto).

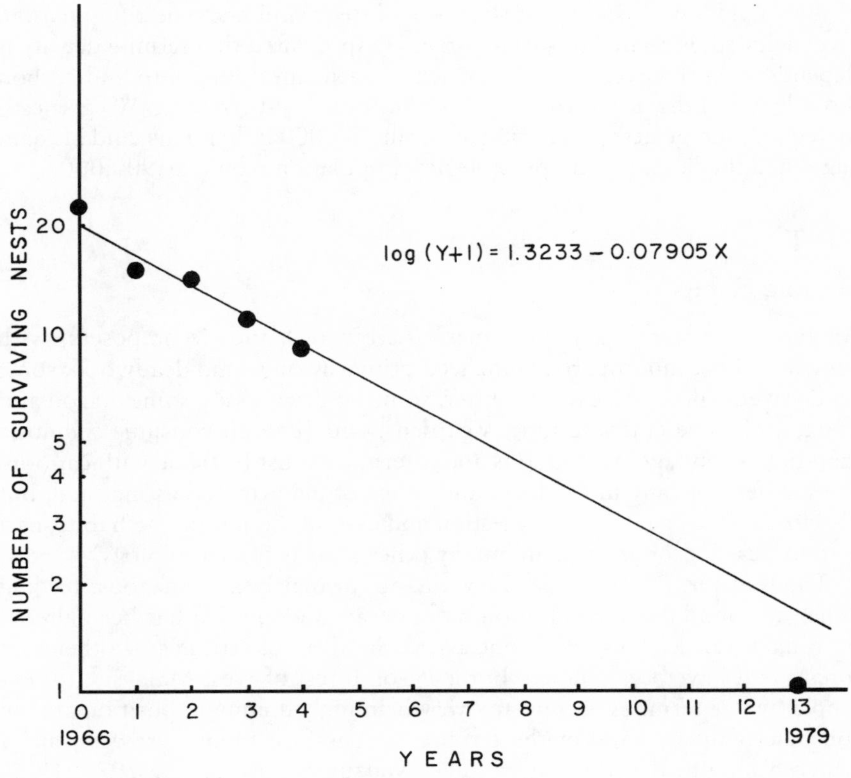

Figure 4.7. Plot of colony survival for 22 *N. costalis* nests at El Verde (mostly those of fig. 4.6) over a 13 y period

Table 4.2. *N. costalis* nest surveys at El Verde, 1966–1970 and 1979

Nests	1966	1967	1968	1969	1970	1979
Surviving from previous year of study		25	29	28	27	5
Inactive since previous year of study		10	3	5	9	25
New (or newly discovered since previous year of study)		7	4	8	3	36
Total active	35	32	33	36	30	41

In Wiegert's study of the *N. costalis* nests at El Verde, he measured base length and width and nest height not only of the nests of the study area but also of off-site nests and calculated a volume index for each. He then harvested the inhabitants of four off-site nests of different sizes in order to get an estimate of caste proportions and population densities. His data indicated that nest population is a function not of nest volume but of the 0.67 power of the volume. *N. costalis* nest shapes tend to be similar, so nest population, he concluded, is related to surface area. He speculated that termite density is dependent on the overall rate at which oxygen can diffuse into and carbon dioxide out of the nest through the porous outer nest covering. Wiegert estimated an average nest population at about 100,000 individuals, and his data suggested that a maximum possible nest population would be 900,000.

DIET

Feeding Habits

All termites are primary consumers (herbivores) and decomposers, with chewing-biting mouthparts. Some feed primarily on sound deadwood, some on decayed wood, some on leaf litter, some on dry grasses, some on soil and dung, and some cultivate fungus gardens. Although all consume cellulosic material as food, generalizations about termites must be made with caution. They differ not only in the form and range of plant materials ingested, but also in the efficiency of their digestion and assimilation, foraging habits, nest systems, caste proportions, and many other aspects of their biology.

The four termite species at El Verde are primarily xylophagous, feeding either on sound dead wood or on more decayed wood that has been altered by fungal attack. This is in contrast to the diets of termites in other rain forests that have been studied. In the Pasoh forest of West Malaysia, for example (where termites occupy an overwhelmingly dominant position among soil macrofauna), 65% of the termite biomass are fungus growers, all of which belong to the Macrotermitinae (Matsumoto 1976; Abe 1979, 1980, 1982; Abe and Matsumoto 1979). None of the El Verde species belong to this subfamily. At Gunung Mulu in Sarawak (Collins 1980), and probably at

Rio Negro, Venezuela (Salick et al. 1983), soil feeders are the dominant termite species. Abe (1982) has speculated that soil feeders and fungus growers are especially efficient litter consumers in a rain forest because they are such excellent go-betweens in combining decomposer microorganisms with organic matter. They transport bacteria-laden soils into wood, they use bacteria in their digestive tracts for litter breakdown, and (in the case of fungus growers) they use their feces to build fungus combs, where further decomposition of organic matter occurs.

The termite species at El Verde possess some of these capabilities, but because of their relatively low densities and their greater concentration on xylophagy, their overall contribution to plant decomposition appears to be considerably less than in the case of the termite species at Pasoh forest and Gunung Mulu.

In the course of their feeding on dead wood the El Verde termites probably ingest significant quantities of wood-decomposing fungi as well. To some extent the termites (especially *Parvitermes*) feed on leaf litter, twigs, and small branches on the forest floor. Pfeiffer (chap. 5, this volume) found many *Parvitermes* foragers in the litter layer at El Verde. Lodge (this volume) points out the importance of microbes to termites and other litter invertebrates through their initial decomposition of litter and wood, and Lawrence (this volume) summarizes the energy and nutrient input of litter fall and wood fall to the El Verde forest.

Termites may play a special role in nitrogen cycling. In some termite species (including *Nasutitermes*) the gut symbionts have been shown to have the physiological adaptation for fixing nitrogen (Benemann 1973; Breznak et al. 1973; Prestwich et al. 1980). Since nitrogen is in low supply in dead wood, which chiefly constitutes the diet of the El Verde termites, such a capability would be especially advantageous. *Nasutitermes* also appears to be able to select wood litter that is relatively rich in nitrogen (see Prestwich et al. 1980).

Foraging Galleries

Foraging occurs mostly at night (Wood 1978), but foraging parties can be found at any hour of the day. Tunnels or galleries made of carton lead out from under the *N. costalis* nests, through the litter to tree trunks, and up to dead branches or other food sources, sometimes to distances of 30 m or more. These covered shelter tubes are semicircular in cross section, range in width from 5 to 14 mm, and average about 5 mm in outside height. The floor of these foraging galleries is covered with fecal material impregnated with trail pheromone deposited by soldiers and workers (McMahan 1970b; Traniello 1982). Newly made tunnel walls are damp and friable, easily broken or washed away by rain. As they age and are reinforced with fresh carton, however, they become hard and resistant to crumbling, and they may

Table 4.3. Foraging galleries occupied by *N. costalis* and *P. discolor*

	Radiation Center				South Control Center			
		Active galleries (%)				Active galleries (%)		
Year	Trees with galleries	*N.c.*	*P.d.*	Total	Trees with galleries	*N.c.*	*P.d.*	Total
1966	90	3.3	8.9	12.2	92	23.9	27.2	51.1
1967	102	1.0	7.0	8.0	108	19.4	31.5	50.9
1968	93	16.1	7.5	23.6	136	16.9	29.4	46.3
1969	89	43.8	11.2	55.0	120	36.7	14.2	50.9
1970	113	49.5	23.0	72.5	123	31.7	22.0	53.7

Notes: Approximately 900 trees were examined in each center in June of each year. *N.c.* = *N. costalis, P.d.* = *P. discolor.*

last for years. They are most often found on the trunks of living trees, reaching heights up to 20 m, and always ending on dead branches. Galleries of *N. costalis, N. nigriceps,* and *P. discolor* are similar but differ mainly in dimensions (*P. discolor* < *N. costalis* < *N. nigriceps*). Almost never are active galleries of two species to be found on the same tree. Abandoned *N. costalis* galleries are sometimes appropriated by *P. discolor.*

Foraging galleries indicate food preferences and degree of termite infestation. Martorell (1941, 1945) listed 108 different Puerto Rican trees attacked by termites. From 1966 to 1970 (as part of a study of the effects of gamma radiation on the rain forest ecosystem [Odum and Pigeon 1970]), the densities of *N. costalis* and *P. discolor* galleries were monitored within a 30 m radius circle at each of the two El Verde sites called the Radiation Center and the South Control Center (see chap. 1; McMahan 1970b,c). The lower trunk of every tree and sapling of 2 cm diameter or greater (approximately 890 in the Radiation Center and 920 in the South Control Center) was examined for foraging galleries or gallery fragments. (Lawrence, this volume, lists all the trees and shrubs found at El Verde.) Galleries were found on approximately 10% of the trees in each circle and on over thirty different tree species, but over half (51%) were concentrated on four dominant trees in both centers: *Dacryodes excelsa* (16%), *Sloanea berteriana* (18%), *Drypetes glauca* (10%), and *Manilkara bidentata* (7%). Each gallery was examined for activity and for species identification. Table 4.3 shows that gallery occupancy in the Radiation Center was especially low for both species in 1966 and 1967, but increased thereafter, especially for *N. costalis.* This trend may reflect initial effects of radiation and a later increase in attractiveness of the area due to an augmentation of dead wood there, killed by the radiation. Fluctuations were not as great in the South Control Center. The data indicate that at any given time, approximately 5% of the trees at El Verde will have active termite foraging galleries.

Foraging efficiency depends partially on the speed with which workers can travel within these galleries, transporting within their digestive tracts the wood fibers scraped from the food source. Large workers and soldiers of *N. costalis* were tested for running speed in small "speedways" similar to galleries. Large workers average 1.43 cm sec^{-1} and soldiers averaged 1.01 cm sec^{-1} (McMahan 1970b). At this rate, workers would require nearly twelve minutes to bring food to the nest from a distance of 10 meters.

The foraging galleries of *N. nigriceps* are considerably broader than those of *N. costalis,* ranging from 15 mm to sheaths of plaster 5 cm and more in width, almost covering the surfaces of the dead wood being consumed (fig. 4.8). The gallery surfaces often tend to be very rough, with a clotted

Figure 4.8. Enlarged portion of a *N. nigriceps* gallery on a tree trunk

appearance and uneven contours, and the galleries usually cover a deep groove eaten into the underlying wood. No *N. costalis* nest was ever found nearer than about 50 m to one of the *N. nigriceps* nests. An area being foraged by *N. nigriceps* appeared to have no foragers from *N. costalis* or *P. discolor.*

Foraging galleries of *P. discolor* are usually smaller than those of *N. costalis* (averaging about 6 mm in width) and are more likely than those of *N. costalis* to be found on totally dead, upright tree trunks. The galleries often disappear abruptly into the dead wood.

The prominence of *P. discolor* at El Verde is indicated by the prevalence of its foraging galleries on trees and by the fact that, in a preliminary study of termite attack on small (<5 cm diameter) dead and recently fallen branches along the forest trails, one in six was found to be infested with foraging members of this species (McMahan unpublished). The comparable figure for *N. costalis* was one in thirty-six. *P. discolor* may be more important than *N. costalis* in the El Verde ecosystem.

Termites as Food for Other Organisms

A significant route of return of energy from termite colony to ecosystem is via predation. Cintron (1970) analyzed the stomach contents of ten each of three species of frogs at El Verde and found that termites composed about 10% in *Eleutherodactylus eneidae,* 3% in *E. richmondi,* and none in *E. portoricensis* (probably E. *coqui*). Stewart and Woolbright (this volume) found termites to be very rare in the diet of 173 *E. coqui* and three other species examined. The seasonal swarming of alates provide a feast for opportunistic feeders, such as *Bufo marinus* toads, *Anolis* lizards, the Puerto Rican tody and other birds, bats, possibly mongooses and rats, and a wide variety of arthropods. Thomas and Gaa Kessler (this volume) report that 52% of the stomach contents of a blind snake, *Typhlops richardi platycephalus,* consisted of *Parvitermes* prey. Arthropod predators on termites include spiders, scorpions, centipedes, beetles, flies, wasps, and especially ants. This predation results in the mortality of nearly 100% of the alates that swarm each year, and ants probably account for the largest proportion. Traniello (1981) suggests that ant predation has been the major selective pressure that has led not only to the evolution of the well-armed soldier caste and the relatively high proportion of soldiers in *Nasutitermes* but also to the development of the integrated systems of foraging and defense exhibited by *N. costalis.* In addition to alates, foraging neuters and those repairing nest breaches or expanding the nest are also preyed upon by ants, spiders, lizards, and birds. No vertebrates specialized for predation on termites, such as anteaters and pangolins, are present at El Verde, so the termite nests are relatively safe from gross destructive action except for falling branches.

As food for predators, the El Verde termites probably make their greatest contribution during the swarming season, when alates may provide a large part of the diet of arthropods, lizards, bats (as opportunistic feeders), and possibly frogs (but see Stewart and Woolbright, this volume). At other times, individuals of a colony are relatively well protected by the walls of their nest, the carton tunnels in which they forage, and the wood they eat.

CONSUMPTION RATES AND ENERGY FLOWS (POPULATION ENERGETICS)

Termites at El Verde, in their role as decomposers, consume dead branches, logs, and other dead wood (ingesting, as well, any wood-decomposing fungi present). They also feed on twigs and leaf litter. They constantly produce brood (which develop mainly into neuters but also into alates). They return nutrients and unrespired energy to the ecosystem through (1) the feces and saliva used in the construction of their carton nests and foraging galleries, and (2) through the termite tissue that is consumed by predators or is decomposed.

An accurate estimation of energy flow through the termites of the El Verde forest requires quantitative measurements of population densities, biomass, production of new individuals in the colony, colony respiration rates, rates of wood consumption, rates of assimilation and feces or carton production, and predation pressures. Only a few of these data have been collected for *N. costalis* and almost none for the other three species at El Verde.

Wiegert has made the only quantitative evaluation so far of the role of *N. costalis* in the total decomposer energy budget of the El Verde tabonuco forest. He estimated total numbers of the various termite castes and developmental stages (lumping large and small workers) from harvested nests, obtained dry weights for the various castes, measured oxygen consumption of individual termites and carbon dioxide production of entire nests, and determined calorific content per gram of each caste and stage. He also measured the calorific content and ash percentage of the carton material from nests in the Radiation and South Control centers (see tables 4.1, 4.4, and 4.5).

Biomass values for the El Verde termites are low. Wiegert estimated about 75 *N. costalis* termites m^{-2}, each averaging 3.17 mg in weight. This gives termite biomass figures of 0.238 g m^{-2} wet weight and 0.052 g m^{-2} dry weight (using 0.22 as the dry weight/wet weight proportion). McMahan estimated only about 45 *N. costalis* termites m^{-2} at El Verde. Her termite biomass figures are 0.15 g m^{-2} wet weight and 0.033 g m^{-2} dry (based on her count of 4.5 nests ha^{-1}, and an average nest population estimate of 100,000). Wiegert's figures will be used in the tables that follow.

From his data on biomass, calorific content, respiratory energy loss, and secondary production of *N. costalis,* Wiegert estimated a total energy flow

Table 4.4. Dry weight, biomass, and calorific content for each caste of *N. costalis*

Sample	Mean Dry Wt (mg)	Biomass ($g\ m^{-2}$) R	SC	Calorific Content ($kJ\ g^{-1}$)	Ash %
Workers	0.86	0.0585	0.0688	23.937	0.16
Soldiers	0.36	0.0043	0.0054	23.753	0.20
Alates	5.35	alates flown		28.371	0.97
Larvae	0.10	0.0007	0.0009	27.990	1.26
Eggs	0.02	0.0002	0.0003	27.035	0.71
Nest material				21.333	3.11
Total		0.0637	0.0754		

Source: Data adapted from Wiegert 1970b.

Note: R = Radiation Center; SC = South Control Center.

Table 4.5. Respiratory energy loss, secondary production, and energy flow of *N. costalis* at El Verde

	Center	Workers	Soldiers	Worker and Soldier nymphs	Winged repro-ductives	Eggs	Total
Standing crop (mean)	SC	1.645	0.130	0.025	0.699	0.008	2.507
	R	1.398	0.100	0.021	0.494	0.004	2.018
Annual respiration	SC	16.287	1.465	0.293	4.522	—	22.567
	R	13.859	1.172	0.251	3.768	—	19.050
Annual secondary production	SC	0.544	0.042	—	1.398	—	1.984
	R	0.461	0.033	—	0.988	—	1.482
Annual energy flow	SC	16.832	1.507	0.293	5.904	—	24.536
	R	14.320	1.214	0.251	4.773	—	20.558

Source: Data adapted from Wiegert 1970b.

Notes: All values are in $kJ\ m^{-2}$. R = Radiation Center; SC = South Control Center.

of 20.5 to 24.5 $kJ\ m^{-2}\ y^{-1}$ through *N. costalis* at El Verde (see table 4.5). This represents such a small fraction of the total energy input into the detritus-decomposer food chains there that Wiegert concluded that *N. costalis* is not quantitatively of much importance in the total decomposer energy budget of the Puerto Rican rain forest. His study, of course, did not include the other three termite species of the area, and together their impact probably exceeds that of *N. costalis*. Pfeiffer (chap. 5, this volume) found *Parvitermes discolor* occupying hollow twigs in the leaf litter and calculated from his data a mean annual density of 374 *Parvitermes* individuals m^{-2} for the El Verde site. *P. discolor* workers and soldiers weigh about half as much as those of *N. costalis*, its workers averaging 2.0 mg and its soldiers 1.6 mg. The proportions of workers and soldiers are not known for this species, but if they are similar to those of *N. costalis*, the *Parvitermes* biomass would be about 0.71 $g\ m^{-2}$, wet weight. Using similar assumptions about worker and soldier proportions and estimating 200,000 termites per nest, with 0.47 nests ha $^{-1}$, the biomass value for the large species, *N. nigriceps,* is 0.06 g m $^{-2}$, wet weight. Although there are no data available for *Glyptotermes pubescens* on which to base biomass estimates, its biomass is probably below that of *N. costalis.* (This non-nest-building species lives and feeds entirely within relatively sound wood.)

When the figures for the *N. costalis, P. discolor,* and *N. nigriceps* are added, the biomass value becomes 1.01 g m $^{-2}$ wet weight or about 0.22 g m $^{-2}$, dry weight. Even if *G. pubescens* data were available the biomass figure for the termites of El Verde would still represent a relatively low energy input. Salick et al. (1983) similarly found that less than 5% of the annual litter and treefall is cycled through the termite populations of the forests of the Rio Negro.

In the return of nutrients to the ecosystem, some of the termite saliva and feces may be used by microorganisms, but probably most of these materials are tied up within the carton nest and are not available to other organisms until the nest is abandoned and disintegrates. In the case of *N. costalis* a nest's half-life is about 3.8 years; for *N. nigriceps,* nest longevity is unknown but is certainly longer. *P. discolor* builds no discrete nests but uses carton to line nesting areas and builds extensive carton foraging galleries. Its colony longevity is also unknown.

El Verde nests have not been analyzed for organic content apart from Wiegert's (1970b) study of the caloric and ash content of *N. costalis* nests (table 4.4). He found the calorific content of *N. costalis* carton to be 21.333 kJ g $^{-1}$, which was less than that of the termites (23.9 kJ g $^{-1}$) but greater than that of leaf detritus from the El Verde rain forest (18.898 kJ g $^{-1}$). Carton analyses for other species at other localities have indicated a high percentage of lignin, low carbohydrate content, low pH, a high carbon to nitrogen ratio and increased concentration of certain nutrients.

Salick et al. (1983) compared the nutrient content of termitaria in forests along the Rio Negro in Venezuela with that of wood litter and surrounding soils (both laterite and podzol) and found a decreasing nutrient concentration gradient. They calculated that greater proportions of nitrogen (12.8%), phosphorus (8.5%), and potassium (7.3%) were cycled by the termitaria than would be represented in a random sample of organic matter.

In addition to nutrients, the carton of termitaria may contain concentrations of resins, phenols, and other materials that are resistant to termite digestion (Wood and Sands 1978). It has been suggested that these factors may aid in inhibiting bacterial growth and therefore help to slow nest decomposition.

On the death or migration of the termite colony, other organisms may temporarily invade the carton nest and hasten its disintegration. The carton may become completely interlaced with fine rootlets from adjacent vegetation (fig. 4.9). Salick et al. (1983) found that in the nutrient-deficient rain forests of the Rio Negro a decomposed nest forms a nutrient patch that may

Figure 4.9. An abandoned *N. costalis* nest interlaced with rootlets

represent a microsite for tree seedling establishment. Whether the *N. costalis* nests at El Verde serve a similar function needs further investigation.

The most extensive studies to date of the role of termites in rain forest ecosystems are those carried out in the Pasoh Forest Reserve of West Malaysia (Matsumoto 1976; Abe 1979, 1980, 1982; Abe and Matsumoto 1979) and at Gunung Mulu in Sarawak (Collins 1980). Table 4.6 compares salient features in the four study areas: Pasoh, Gunung Mulu, Rio Negro, and El Verde. All except El Verde are at elevations below 220 m, are part of much larger, continuous land masses, and are within 4° of the equator. The Asian sites are described as lowland dipterocarp forest, the Venezuelan sites as moist tropical rainforest. Table 4.7 compares feeding habits, density, and biomass data where available. It is obvious that termite species variety and

Table 4.6. Four termite study sites in tropical rain forests

	Pasoh (West Malaysia)	Gunung Mulu (Sarawak)	Rio Negro (Venezuela)	El Verde (Puerto Rico)
Latitude	3° N	4° N	4° N	18° N
Elevation (m)	130	130–220	120	300–500
Precipitation, annual (mm)	2000	5100	—	3500
Termite genera	10	23	13	3
Termite species	—	57	20	4
Litter consumption	55.5 $g\ m^{-2}\ yr^{-1}$	—	59 $g\ m^{-2}\ yr^{-1}$	—

Sources: Pasoh, Matsumoto (1976) and Abe (1979); Gunung Mulu, Collins (1980); Rio Negro, Salick et al. (1983); El Verde, Wiegert (1970b).

Table 4.7. Feeding habits, density, and biomass (wet weight) for termites at three study sites in tropical rain forests

	Pasoh		Gunung	El Verde	
Feeding Habits	density ($N\ m^{-2}$)[a]	biomass ($g\ m^{-2}$)	density ($N\ m^{-2}$)	density ($N\ m^{-2}$)	biomass ($g\ m^{-2}$)
Lichen feeders	50	0.14	+	0	0
Soil feeders	1505	2.52	1579	0	0
Fungus growers	960	6.12	13	0	0
Wood feeders	970	0.64	11	75[b]	0.24[b]
				374[c]	0.71[c]
				9.4[d]	0.06[d]
Total				458.4	1.01

Sources: See table 4.6.

Notes: +, present but not quantified.

[a] $N\ m^{-2}$ = individuals m^{-2}. [b] *N. costalis.* [c] *P. discolor.* [d] *N. nigriceps.*

population densities at El Verde are considerably less than those found in the other study areas.

What factors may be responsible for limiting the size of the termite populations at El Verde and their consumption rates, as compared with termite populations in the other tropical rain forests?

Soil feeders and fungus growers, whose biomass greatly exceeds that of wood feeders in the other rain forests, are totally lacking at El Verde. What prevented their establishment here? Perhaps they simply failed to colonize Puerto Rico. Wood feeders might be favored in island colonization if the likely route of entry was via nests on floating wood. The ease in transport of fungus spores on the bodies of fungus eaters would appear to eliminate difficulty of spore transfer as a barrier to colonization. Perhaps Puerto Rican lizards and/or ants are potentially more efficient as predators on soil dwellers, preventing their establishment. Perhaps the apparently low consumption estimates for El Verde termites as compared with the other rain forest sites are due to a shortage of suitable (relatively undecayed?) wood or of acceptable nest sites, though these explanations seem unlikely. Perhaps it is due to El Verde's temperature effects on metabolic rates because of the slightly greater elevation and distance from the equator. Perhaps our estimates of consumption rates for El Verde termites, based sometimes on assumptions and on relatively few data, are too low. Only continued studies can provide answers.

Salick et al. (1983) pointed out that the importance of termites to an ecosystem is not necessarily reflected in the magnitude of energy flow through the population (see also Kitchell et al. 1979). Some consumers can exert a direct regulatory influence through affecting the rates of ecosystem processes like nutrient cycling. Schaefer and Whitford (1981) suggest such a role for termites in the desert ecosystem they studied. There are considerable data showing that termites not only redistribute organic matter within an ecosystem by concentrating nutrients in their nests and galleries but also bring about chemical changes in food material through digestion (Wood and Sands 1978; Golley 1983b). They may also play a role in nitrogen cycling. Until more data on these accessory roles are available for the El Verde termites, however, their exclusively wood-feeding habits and their relative paucity in species and numbers lead to the general conclusion that their importance in this rainforest food web is considerably less than that of their relatives in other tropical rainforests.

SUMMARY

The chief role of termites in the El Verde food web is the consumption of dead wood and litter, a function vital to the recycling of organic matter through the decomposer community. Along with microorganisms, and other litter invertebrates, they are the principal agents of wood decomposition in tabonuco forest. In addition to consuming and transforming cellulose, termites facilitate the entry of other decomposers by creating channels in the dead wood. They also serve as food for a variety of predators (especially during the swarming season) and may play a role in nitrogen cycling.

Only four termite species have been found at El Verde, a much lower termite diversity than has been reported for other tropical ecosystems. None of the soil-eating or fungus-growing types so prevalent in other tropical rain forests are present at El Verde. The explanation for this absence is unknown but it may be due to factors related to colonization, the physical and chemical environment, or predation.

The four El Verde species all feed on standing dead wood, fallen logs, brush, and litter. Two of the species, *Nasutitermes costalis* and *N. nigriceps,* build carton nests of semiliquid feces and saliva. Their nests often serve as shelter for other organisms. One species, *Parvitermes discolor,* excavates stumps for its nests, and lines them with carton. These three species construct carton galleries from nest to feeding sites. The fourth species, *Glyptotermes pubescens,* is a drywood species. It builds neither nest nor galleries, but lives in chambers and tunnels within the sound wood on which it feeds.

Data for *N. costalis* at El Verde indicate a nest density of about 4.5 ha^{-1}, an average nest population of about 100,000 individuals, and a nest half-life of around 3.8 years. Biomass estimates are 0.238 g m^{-2} wet weight (0.052 g m^{-2} dry weight), with a total energy flow of 20.5 to 24.5 kJ m^{-2} y^{-1} through *N. costalis* at El Verde.

Active *N. costalis* foraging galleries were found on 5% of the trees at the El Verde study sites, but over half were concentrated on four dominant trees, *Dacryodes excelsa, Sloanea berteriana, Drypetes glauca,* and *Manilkara bidentata. N. costalis* foragers travel within their galleries at the rate of 1.43 cm sec^{-1}, bringing cellulosic material back to the nest inhabitants.

Only wood-eating termites are found at El Verde. Biomass of termites at El Verde is greater than that observed for wood-feeding termites of other rain forests, where wood-eating species lag considerably behind the soil feeders and fungus growers in species numbers and population densities.

The small number of termite species at El Verde, their relatively low densities, and the absence of species that feed primarily on soil and fungus gardens result in what appears to be a less dominant role for termites in the El Verde ecosystem than in many other tropical rain forests. It appears that more wood goes through free-living fungi and less through termites at El Verde. Nevertheless, the termites' contributions as consumers of dead wood, as food for many predators, and as possible links in the chain of nitrogen cycling make them an integral part of the El Verde food web.

5

Litter Invertebrates

William J. Pfeiffer

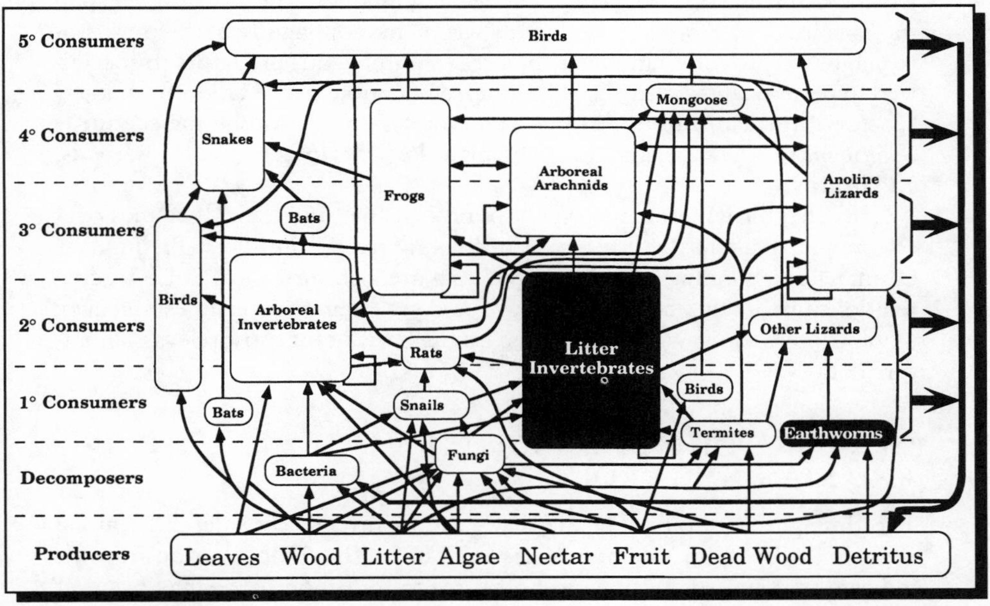

THIS broadly defined and diverse assemblage includes all of the major taxa of the phylum Arthropoda that feed in litter and soil: Chelicerata (e.g., mites, spiders, pseudoscorpions, scorpions, amblypigids, and schizomids), Myriapoda (millipedes and centipedes), Insecta (especially collembola, ants, flies, beetles, termites, and various Homoptera and Hemiptera), and Crustacea (isopods and copepods). Only one nonarthropod group is of functional significance in the litter/soil habitat at El Verde: the earthworms (Annelida: Oligochaeta). The feeding modes of the Arthropoda span the spectrum of trophic interactions, from monophagous herbivores, polyphagous saprovores, and microbivores to polphagous predators to specialized hyperparasites. Even vertebrates are susceptible to predation from a few species of the leaf litter arthropod community—predation by the centipede *Scolopendra alternans,* and the amblypigid *Phrynus longipes* on frogs, toads, and lizards.

This chapter will document arthropod densities and their dynamics over one annual cycle in the litter layer of the tabonuco rain forest. Preliminary estimates for standing stocks of some taxa are also furnished. Since feeding relationships in this community have not yet been rigorously established, trophic interactions will be advanced on the basis of the most relevant literature data.

DYNAMICS OF LITTER MOISTURE AND STANDING CROP

The timing and duration of the dry season can vary considerably among years in the Luquillo Mountains (see fig. 3 in Smith 1970b). Various records indicate that extended dry periods of three to four months, as may occur at other Neotropical sites (e.g., Barro Colorado in Levings and Windsor 1982; Rand and Rand 1982), are infrequent. From 1964 to 1979 the mean annual rainfall near the El Verde site was 346 cm at an elevation of 500 m (Brown et al. 1983).

During the 1984–85 study 310 cm of rainfall were recorded at the El Verde Field Station. A dry period extended from 22 February to 1 May (fig. 5.1), during which the mean daily precipitation was 0.15 cm · day $^{-1}$.

Correspondingly, litter moisture declined from 211% (100 × g H_2O g^{-1}) on 21 February to a minimum of 17% by 17 April and moisture in the upper 5 cm of soil fell from 88% to a minimum of 45% by 1 May (fig. 5.1). Peaks for litter moisture of approximately 310% occurred in mid-February and early November. These minimum and maximum values for litter moisture are similar to those reported from Barro Colorado Island by Levings and

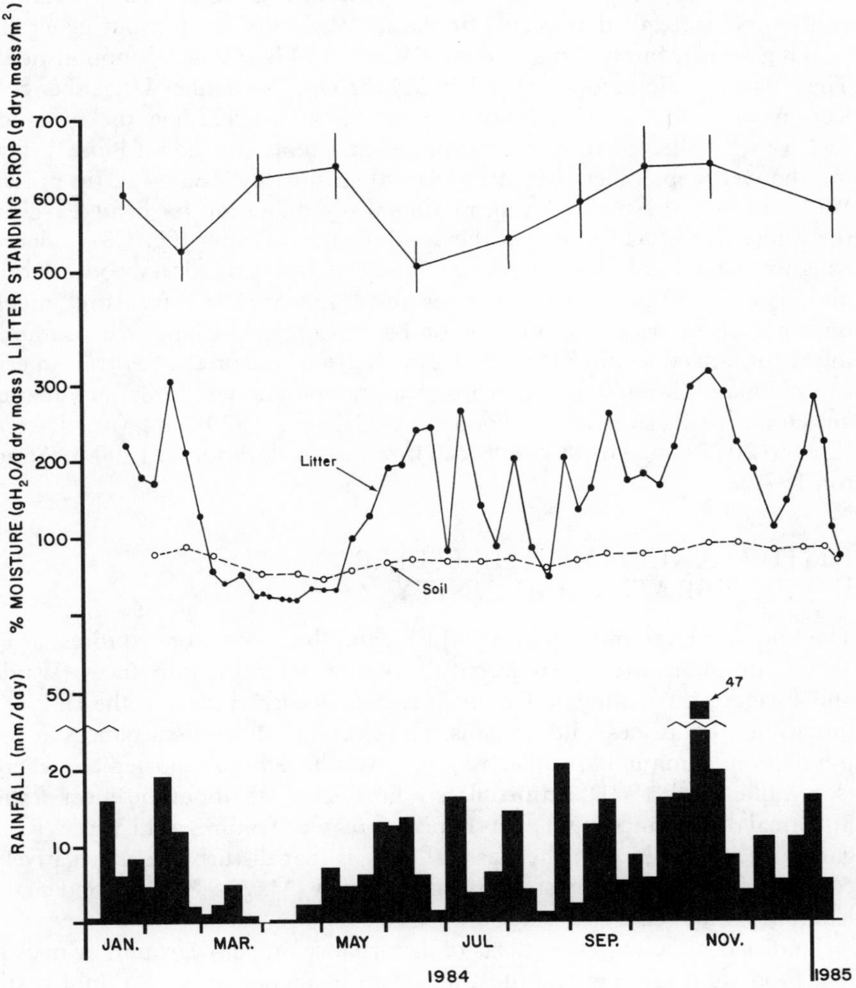

Figure 5.1. Rainfall distribution (bottom panel), litter and soil (0–5 cm) moisture (middle panel), and standing crop of litter (top panel) during the annual cycle that litter arthropods were sampled in tabonuco rain forest at El Verde.

Windsor (1982). Moisture levels in the El Verde soil cores, on the other hand, displayed a damped oscillation relative to the Panamanian site, perhaps partially reflecting the shallower core depths (0 to 3 cm) taken at the latter site. The annual rainfall in the tabonuco forest during the study period was 36% greater than the three-year average for Barro Colorado Island during the Levings and Windsor study.

Seasonal trends in the dynamics of litter standing crops from D-Vac collections (see app. 5) were not readily apparent (fig. 5.1). This reflects the continuous leaf fall that occurs at the El Verde site, with input generally peaking from March through June (Wiegert 1970a; Zucca unpublished). The mean standing crop for the January through November-December collections was 588 ± 18 (standard error) g m^{-2}, exceeding the mean of 392 g m^{-2} collected from four experimental plots (including Plot 3) during the wet (September) and dry (March) seasons by Zou et al. (in press). Wiegert (1970a) estimated a mean annual standing crop for dead leaves in the litter layer of 511 g m^{-2} at the South Control Center (fig. 1.3), a figure roughly comparable to the present study, if one excludes woody debris and fruits from this estimate. Levings and Windsor (1982) measured much higher levels of litter accumulation on Barro Colorado Island, where standing crops ranged from 731 to 1973 g m^{-2}. Elsewhere on the Central American mainland, litter standing crops in a premontane wet forest in Panama amounted to 2116 g m^{-2} (Woods and Gallegos 1970) and varied from 1160 to 2015 g m^{-2} in a Costa Rican forest at an elevation of 1300 m (Stanton 1979).

DISTURBANCE AND THE LITTER INVERTEBRATE COMMUNITY

The Luquillo Experimental Forest (LEF), and thus the tabonuco forest at El Verde, are intermittently subjected to intense natural disturbances (Waide and Reagan, this volume). The main causes of disturbance in the LEF are hurricanes, landslides, and treefalls. The effects of these disturbances at the population, community, and ecosystem levels have been reported elsewhere (see Walker et al. 1991). Unfortunately, little is known about the effect of the historical disturbance regime on shaping food web features at El Verde (Reagan et al., this volume). The classical view is that disturbance is negatively correlated with ecosystem stability and diversity (May 1974) and food chain length (Pimm 1982).

Studies dealing with the effects of disturbance on belowground or microbial food webs are few and most are from temperate systems (Hunt et al. 1989; Moore and de Ruiter 1991; Lenz 1992). These systems do not conform to conventional wisdom about food webs. Feeding loops, also known as cross predation, are common (Moore et al. 1988; Lenz 1992);

omnivory is prevalent at all trophic levels (Hunt et al. 1989; Moore and de Ruiter 1991; Lenz 1992); and the presence of compartments, as well as longer than expected food chain lengths, were evident even though disturbance was a common factor (Moore et al. 1988; Moore and de Ruiter 1991; Lenz 1992).

Theoretical studies indicate that the stability of ecosystems is tied to recycling of materials through the decomposer food web (Moore et al. 1993). Models based on primary productivity alone (Lotka-Volterra–based) were less resilient and less feasible than models that included recycling via the decomposer circuit. The prevalence of disturbance in ecosystems is an accepted paradigm in ecology (Pickett and White 1985). Thus, ecosystem persistence in ecological time will be tied to the recycling of materials through the decomposer circuit (Moore et al. 1993).

The effects of disturbance on the microarthropod community and its interaction with decomposition processes are only now being explored. At the Coweeta Hydrological Laboratory (North Carolina) litter decomposition was highly correlated with increased microarthropod densities (Blair and Crossley 1988). Both, in turn, were negatively associated with clear-cutting. Eight years after clear-cutting, densities of microarthropods in litter bags were still 20 to 54% lower in the experimental site when compared to the control site. In the LEF, wood decomposition was also highly correlated with increased arthropod densities as well as diversity (Torres 1994). The number of trophic guilds, as well as densities within guilds, also increased as decomposition progressed.

A main difference between natural and anthropogenic disturbance in forests is the removal of biomass, namely in the form of wood. The El Verde ecosystem seems to be well adapted to the natural disturbance regime (Doyle 1981; Walker et al. 1981). This regime, which includes hurricanes, landslides, and treefalls (Doyle 1981; Brokaw 1985; Guariguata 1990) produces steady, as well as pulse, fluxes of wood and litter down to the forest floor. Evidence points to the fact that increased complexity in the decomposer food web is positively correlated with increased decomposition rates in the LEF (Torres 1994). Management of forestry resources must take into consideration the effects of wood removal on microarthropod communities in particular and decomposer food webs in general.

LITTER AND SOIL FORAGING GUILDS

The functional foraging groups that occur in litter and soil often include a wide variety of invertebrate taxa. Information on the invertebrates occuring in litter at El Verde is organized below by these functional foraging groups and includes discussions of the abundance, biomass, and diet of each of the principal taxa constituting the group.

Seprotrophic and Microbitrophic Arthropods

Arthropods falling into either or, more likely, both of these trophic categories include organisms that feed directly on dead plant tissue (leaves, reproductive structures, fruits, woody tissue, and belowground parts) or animal remains (saprovores), organic material derived from the parent plant substrate (such as fecal pellets), and microbes (microbivores). These two trophic categories are difficult to distinguish in many cases since, for example, an arthropod consuming leaf litter is likely to assimilate microbes colonizing the leaf surface as well as the infiltrating fungal hyphae.

Luxton (1982b) attributed the degradation of dead organic tissue to four principal processes: (*a*) oxidation of inorganic carbon, (*b*) leaching, (*c*) microbial activity, and (*d*) catabolic breakdown and physical comminution due to faunal activity. Microflora (fungi and bacteria) are responsible for the vast bulk of carbon mineralization in temperate forest soils (Macfadyen 1963; Crossley 1977; Reichle 1977; Luxton 1982c), in large part due to high mass-specific metabolic rates, rapid population turnover, and substrate specificity. Experimental studies, mostly employing litterbag techniques, have generally indicated that the presence of microarthropods accelerates plant litter disappearance, with an average rate enhancement of 23% in the nineteen studies reviewed by Seastedt (1984). The contribution of soil and litter invertebrates to the process of plant litter decomposition often has been considered primarily from the perspective of substrate comminution (Edwards et al. 1970; Kitchell et al. 1979). Fragmentation of litter by primary decomposer fauna has been assumed to enhance leaching and microbial catabolism of organic substrate by increasing its surface/volume ratio (van der Drift and Witkamp 1960; Englemann 1961; Kevan 1962; Ausmus and Witkamp 1973; Jensen 1974).

However, Webb (1977) has argued that the pelletization of unassimilated material passed from some macroarthropod guts may actually decrease the surface/volume ratio. Nicholson et al. (1966) and Webb (1977) measured decomposition rates for millipede fecal pellets with low surface/volume ratios that were similar to the parent litter. Although the minute fecal pellets of microarthropods (such as mites and collembolans) individually offer greatly enhanced surface/volume ratios for microbial catabolism, they have a lesser tendency to fragment and are even prone to reaggregate (Webb 1977; Kitchell et al. 1979). The relative contribution of arthropod activity towards decomposition rates may be greatest for woody tissue through the production of fecal pellets that are heavily colonized by microflora, and particularly as a consequence of tunneling activities by wood-boring arthropods (Fager 1968; Edwards et al. 1970; Ausmus 1977; Kitchell et al. 1979).

The feeding activities of arthropods may also significantly influence the structure of microbial communities (see a more complete discussion in the section on diplopods and isopods). Persson et al. (1980) estimated that despite their relatively low standing stocks, microbivores removed 30 to 60% of the annual microbial production in the litter and humus layers of Scandinavian pine forest. Newell (1984) has also presented evidence that the presence of a preferentially feeding collembolan grazer can alter the competitive relationship between a pair of basidiomycetes and, owing to differential degradative abilities of the two fungal species, thereby indirectly affect the rate of litter decomposition.

Acari

Mites usually comprise the numerically dominant arthropod taxon in forest litter (see Petersen 1982a), although studies of energy flow in litter communities have found that mites and collembola combined generally ingest less than 1% of the annual litter input to the forest floor (Mitchell 1977; Thomas 1979; Luxton 1982b; Norton 1985). However, other estimates using analysis of alkali element flow indicate that saprovore consumption in forest litter communities is an order of magnitude higher than indirectly estimated by energy flow studies (Gist and Crossley 1975a; Crossley 1977).

Two acarid orders are significant as saprovores and microbivores: the Cryptostigmata (Oribatida) and the Prostigmata. Cryptostigmatids, which usually compose over 50% of the mites in forest soils (Petersen 1982a, Wallwork 1983), are generally considered fungivorous particulate feeders that produce fecal pellets contributing to soil structure (Wallwork 1983). Prostigmatids, on the other hand, exploit intracellular resources through modified appendages (stylettiform chelicerae, Norton 1985). Mesostigmatid mites are generally considered to be predaceous, with Collembola serving as a significant prey resource (Christiansen 1964; Wallwork 1967).

Mites comprised 69% of all arthropods recovered from high-gradient extractions of litter collected from the South Control Center by Wiegert (1970a). The mean densities of 15,960 mites m^{-2} in the upper 5 cm of soil from Wiegert's three collection dates (January 1964, May 1965, and May 1966) lie near the low end of the range reported from forested habitats and considerably below levels found in temperate forests, where 50×10^3 to 10^6 mites m^{-2} commonly occur (Petersen 1982a). Although the litter and soil of some tropical and subtropical forests harbor populations exceeding 50×10^3 mites m^{-2} (Fittkau and Klinge 1973; Goffinet 1975, 1976; Plowman 1979), other sites, notably various Neotropical locations (see table 5.1), yield similarly sparse densities as were extracted from tabonuco forest litter and soil. Fungal standing stocks in the litter and upper soil layers of the El Verde forest appear sufficient to support much denser mite populations (see Lodge, this volume), but the thin litter layer may limit densities through a relative scarcity of habitat space and refuges from high densities for predaceous arthropods.

Table 5.1. Densities of arthropod taxa from the litter and soil of Neotropical forests

		Puerto Rico		Mexico		Trinidad		Panama		Brazil	
	(A) Litter	(B) Litter	(B) Soil	(C) Litter	(C) Soil	(D) Litter	(D) Soil	(E) Litter	(F) Litter	(G) Litter	(G) Soil
Acari	—	15,960	1,390	—	—	10,759	8,033	1,715	—	61,200	11,500
Collembola	1,292	1,200	—	—	—	1,574	934	2,288	—	10,300	1,680
Diplopoda	300	435	—	107	180	208	119	200	50	280	10
Isopoda	549	—	—	3	3	32	10	146	10	70	5
Araneae	356	70	—	54	19	—	—	—	20	70	20
Opiliones	146	—	—	—	—	84	82	146	—	—	—
Pseudoscorpionida	262	340	—	—	65	—	—	—	30	210	9
Dictyoptera and Orthoptera	115	—	—	—	36	69	2	20	5	0	7
Isoptera	374	130	—	—	897	591	3,865	12	—	90	40
Homoptera-Hemiptera	521	770	—	12	5	272	314	410	5	3,940	0
Coleoptera	236	—	—	249	193	198	259	171	40	140	7
Diptera	618	—	—	30	34	517	259	52	—	—	—
Formicidae	1,209	1,120	2,253	880	2,344	1,196	1,814	1,641	320	720	140
Arthropods less Acari & Collembola	4,928	6,040	—	1,846	4,563	3,259	6,790	2,469	600	6,700	1,500
Total Arthropods	—	22,317	4,210	—	—	15,592	15,758	6,472	—	78,200	14,680

Sources and notes: (A) densities for present study expressed in terms of mean annual density, the calculations for which excluded the January 1985 collections, (B) entries for litter Acari, Collembola, and total arthropod densities as well as for soil Acari, Formicidae, and total arthropods derived from Wiegert 1970a; remaining entries from Wiegert (1965); entry for litter Diplopoda under Wiegert also includes other Myriapoda (i.e., Chilopoda); soil cores sampled the upper 5 cm of mineral soil, (C) Mexican data derived from a humid tropical forest at Bonampak (Lavelle and Kohlmann 1984); entries based on estimates obtained from a wet sieving technique (their table 3) adjusted for apportionment of the taxa between the litter and soil horizons (their table 5); entries between columns indicate sum of densities for both litter and soil components; soil cores sampled the upper 20 cm of soil, (D) entries for Trinidad from rain forest (Strickland 1944); arthropods extracted in Tullgren funnels; soil cores sampled the upper 7.6 cm of soil, (E) Williams (1941) also employed Tullgren funnel extraction for samples collected during the months of July and August only, (F) entries for Levings and Windsor (1982) from Barro Colorado Island based on rough and liberally biased visual estimates of average taxal densities from figures in Levings and Windsor showing density fluctuations during their thirty-month study, (G) entries for the Brazilian rain forest based on summarization of Beck's work at the Manaus site by Fittkau and Klinge (1973); soil cores sampled the upper 10 cm of soil. Densities given as individuals m^{-2}.

Collections during 1984–85 yielded acarid densities of only 2000 to 3500 mites m^{-2}, most of which were fairly large cryptostigmatids. The mesh size of the muslin collection bags associated with the D-Vac (see app. 5) is sufficiently coarse to allow the passage of small mites while suction was being applied in the field. For this reason, mite densities calculated from the present study are considered to be underestimates.

Taxonomic work on El Verde mites has thus far been limited to a small subset of the collections. Thirteen species of cryptostigmatids have thus far been recognized (Dr. R. Norton pers. comm.; see app. 6 in Garrison and Willig, this volume). Wallwork (1983) considers thirty to forty species of cryptostigmatid mites as typical for most forest litter and soil systems, although Plowman (1981) recovered eighty-three and 112 species, respectively, from the litter layer of subtropical rain forest and wet sclerophyll forest in Australia.

Collembola

Collembola are primarily regarded as consumers of decaying plant material and fungal tissue (Christiansen 1964; Butcher et al. 1971; Swift et al. 1979; Luxton 1982a, 1982c). Exploitation of these resources may vary seasonally with their availability (Anderson and Healey 1972; McMillan 1975). Consumption of decomposing organic matter by Collembola appears to enhance the growth of associated bacterial populations during gut passage, whereas fungal standing stocks are reduced (Hanlon and Anderson 1979).

In both the surveys of Wiegert (1970a) and the 1984–85 study, low collembolan densities were recovered from tabonuco forest litter (table 5.1). Low densities in the upper 5 cm of mineral soil at the El Verde control site were also apparent in Wiegert's study. Of the approximately 200 published estimates of collembolan densities collated by Petersen (1982a), only a handful of desert and steppe habitats supported levels similar to or lower than those of the El Verde site. A recent study of collembolans from three tropical rain forest sites in northern Queensland (Australia) reported densities in the litter layer less than half of those at El Verde, but densities range from 1.1 to 12.0×10^3 in the upper 4 cm of soil (Holt 1985). In general, densities in Neotropical litter and soil systems (table 5.1) also fall near the low end of the range of estimates from a variety of habitats.

Entomobryidae constitute the dominant family of Collembola in tabonuco forest litter, representing 69% of the mean annual collembolan density and containing about half of the fifteen or so species extracted from the 1984–85 samples. Dominant entomobryid species included *Dicranocentropha* sp., *Dicranocentruga* sp., and *Lepidocyrtus* ap., while *Ptenothrix* sp. and *Sphyrotheca* sp. comprised nearly all of the Sminthuridae and *Proisotoma* sp. the majority of the Isotomidae. As would be expected by their prominence, much of the dynamic pattern in the collembolan community

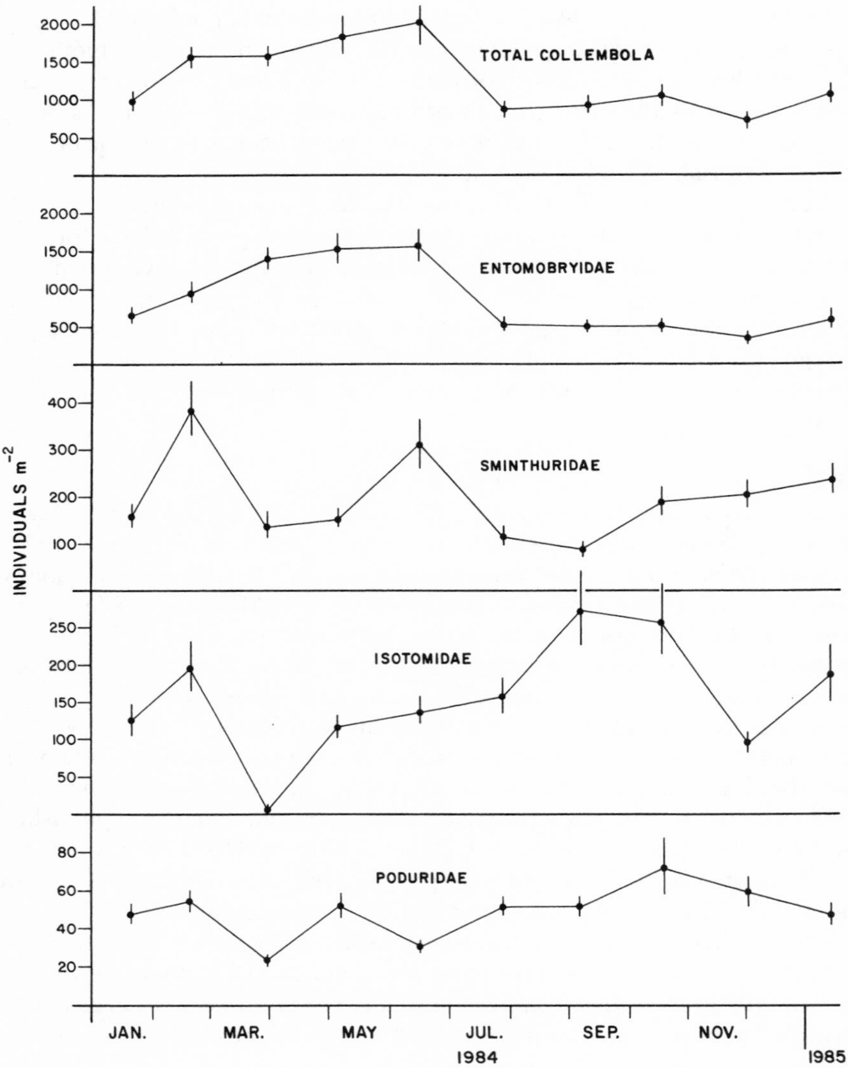

Figure 5.2. Density dynamics of collembola (top panel) and constituent families in tabonuco rain forest litter. Arithmetic means were converted to a m^2 basis; upper and lower limits of standard error bars were established through back-transformation of confidence limits calculated after logarithmic transformations were performed on the original data (see Elliot 1977).

was generated by fluctuations in Entomobryidae numbers. Highest densities occurred from February through June (fig. 5.2), which coincided with the onset of the dry season through the beginning of the 1984 wet season.

The unusually low densities of Collembola in the soil and the litter layer of the El Verde rain forest could be attributed to the low levels of resources afforded by the relatively sparse litter layer. However, Lodge (this volume) has demonstrated the presence of large fungal standing stocks. As suggested above for the circumstance of low mite densities, the abundant array of predators—spiders, pseudoscorpions, opilionids, *Strumigenys* spp. ants as well as other predaceous ants, dipsocorid hemipterans, pselaphid beetles, the gekko *Sphaerodactylus klauberi,* etc.—concentrated in a habitat with limited refuges argues that predation plays a key role in the suppression of collembolan populations.

Diplopoda and Isopoda

Millipedes and isopods are well represented in the litter layer of tabonuco forest, with mean annual densities of 300 and 549 individuals m^{-2}, respectively, estimated for the 1984 samples. The value for millipedes surpasses all of the twenty-eight estimates from a variety of habitats presented in Petersen (1982a), although that compilation is restricted to communities for which both density and standing stock estimates are available. Peak millipede densities occurred in September 1984 (624 individuals m^{-2}) and January 1985 (561 individuals m^{-2}). Wiegert (1965) also reported high millipede densities, with 435 individuals m^{-2} present in litter samples collected during late January and early February. Two deciduous forests not included in Petersen's compilation support millipede densities that were more than double the El Verde annual mean (Lebrun cited in Geoffroy 1981, Blower 1979). Additionally, some of the millipede estimates for tropical and subtropical forests in table 5.2 surpass the El Verde millipede densities, largely due to the contribution of soil-dwelling individuals. Millipede numbers in the mineral soil layer of the tabonuco forest appear to be negligible.

Millipedes appear to represent the predominant taxa in tabonuco forest litter in terms of standing stocks. A preliminary estimate for the mean annual standing stock of millipedes (0.6 g m^{-2}), however, falls into the mid-range of values plotted in Petersen (1982a). On the other hand, the potential for high turnover, which is associated with relatively small species residing under favorable conditions for year-round growth, exists in El Verde. Resource turnover and production by this assemblage may be high relative to the temperate communities that dominate the data base of Petersen's analysis.

The species richness (eleven species) of millipedes recovered from tabonuco forest litter in the 1984–85 samples (table 5.3) corresponds to that found in European forests (e.g., eleven species from a sycamore-ash woodland

Table 5.2. Densities of arthropod taxa of African and Australasian tropical forests

	Australia				Zaire		Sarawak	
	Rain forest		Wet sclerophyll		Dense forest	Gallery forest	Mixed diptocarp	Lower montane
	Litter	Soil	Litter	Soil	Litter/Soil	Litter/Soil	Litter/Soil	Litter/Soil
Acari	5,170	50,940	17,400	49,560	357,308	74,019	—	—
Collembola	2,480	27,750	1,430	12,690	38,888	19,100	—	—
Diplopoda	106	1,380	100	1,736	4	0	12	20
Isopoda	145	104	110	554	9	37	19	24
Araneae	117	129	183	291	33	40	32	22
Opiliones	15	69	30	45	—	—	6	5
Pseudoscorpionida	63	578	9	55	1	3	3	2
Dictyoptera-Orthoptera	5	0	31	100	18	8	4	18
Isoptera	0	0	0	0	102	9	1,058	154
Homoptera-Hemiptera	57	250	71	709	15,022	47	10	8
Coleoptera	622	2,036	568	1,917	469	96	82	89
Diptera	405	2,407	609	3,499	—	—	14	78
Hymenoptera: Formicidae	74	923	352	2,326	770	440	619	449
Arthropods other than Acari and Collembola	1,934	16,590	3,185	15,162	18,095	2,443	2,043	1,048
Total Arthropods	9,580	95,280	22,015	77,862	414,291	95,562	—	—

Sources and notes: (A) soil cores in Australian forests taken to a depth of 5 cm (Plowman 1979), (B) soil cores in African study taken to depths of 10 to 15 cm (Goffinet 1975), (C) soil cores in Sarawak study taken to depth of 20 cm (Collins 1980). For all sites listed in tables 5.1 and 5.2, formicid densities do not include the larval and pupal stages. Densities given as individuals m^{-2}.

Table 5.3. Millipede species in tabonuco forest litter

Species	Jan	Feb	Apr	May	Jun	Jul	Sep	Oct	Nov–Dec	Jan	MAD	CV
Docodesmus maldonadoi	72	43	5	25	167	199	354	155	269	412	143	82.7
Prostemmiulus heatwoli	93	73	72	104	53	117	154	135	115	76	102	31.9
Spirobodellus richmondi	11	14	17	16	5	13	55	41	11	28	20	80.8
Glomeridesmus marmoreus	6	32	10	22	7	12	35	11	22	22	17	61.8
Agenodesmus reticulatus	1	3	0	0	6	2	19	19	9	8	6.7	
Lophoturus niveus	8	7	3	8	13	2	3	8	2	6	5.9	
Liomus obscurus	1	1	1	1	3	3	4	9	0	0	2.1	
Liomus ramosus	0	3	0	1	0	1	1	0	0	6	0.6	
Styraxodesmus juliogarciai	3	0	0	1	1	0	0	0	0	0	0.6	
Siphonophora portoricensis	2	0	0	1	0	0	1	0	0	0	0.4	
Ricodesmus stejneri	0	0	0	0	0	0	0	1	0	0	0.1	
All species	196	178	108	178	263	348	624	381	428	561	300	53.8

Notes: Mean annual densities (MAD) and annual coefficients of variation (CV) are calculated for the January 1984 through November–December 1984 collections, where meaningful. Densities are given as individuals m^{-2}.

Blower 1970; twelve species from a mixed hardwood-pine forest, Geoffroy 1981; and thirteen species from an oak woodland, Blower and Gabbutt 1964), but appears impoverished for a tropical site, judging from the sixty species recovered from a clear cut portion of humid tropical forest in the Ivory Coast (Africa) by Maurin and Levieux (1984). Two species, *Docodesmus maldonadoi* (Polydesmidae) and *Prostemmiulus heatwoli* (Prostemmiulidae), composed 82% of the millipede assemblage recovered from the litter layer at El Verde. Millipede densities fluctuated nearly six-fold, with the lowest densities recorded during the dry portion of the year. Fluctuations in *D. maldonadoi* densities contributed greatly to that of the community as a whole. The population structure of *D. maldonadoi* consisted almost exclusively of adult and late instar nymphs during the dry season, and peak densities corresponded to pulsed reproductive activity. *Prostemmiulus heatwoli* population dynamics exhibited a more muted oscillation, as indicated by its relatively low annual coefficient of variation (table 5.3). In contrast to *D. maldonadoi,* this prostemmiulid maintained a much more consistent age distribution structure throughout the study year.

Only four species of isopods have been collected from litter at the El Verde site, two of which occurred very infrequently. Although *Porcellionides* (?) sp. accounts for 85% of the mean annual density of isopods, preliminary analysis suggests that *Philoscia richmondi* probably contributes the bulk of an annualized isopod standing stock of approximately 0.1 g m^{-2}. Isopod densities and standing stocks in tabonuco forest litter fall in the upper range of the fifteen studies presented by Petersen (1982a). Much lower densities of isopods have been reported from Neotropical forests (table 5.1), other than from Brazilian rain forest (Fittkau and Klinge 1973). Plowman (1979), however, measured densities comparable to those in tabonuco forest from Australian rain forest and wet sclerophyll forest (table 5.2). Isopod densities at El Verde displayed seasonal fluctuations similar to those of millipedes, with highest numbers occurring during the wettest portions of the study year.

Millipedes and isopods are generally viewed as primary decomposers that obtain nutrition from the litter-microbe complex. Several investigators have suggested that some millipede and isopod species preferentially forage for fungal mycelia (Heeley 1941; Gist and Crossley 1975b; Soma and Saitô 1983), and Beck (1971) has further contended that fungal mycelia and spores serve as the exclusive food resource for millipedes in central Amazonian rain forest. Even species feeding directly on forest litter usually prefer leaves that have aged on the forest floor, presumably at the stage where the concentration of secondary compounds have been reduced through leaching and microbial degradation (Edwards et al. 1970; Satchell 1974; Edwards and Heath 1975; Cameron and LaPoint 1978; Sutton 1980; Rushton and Hassall 1983; Hassall and Rushton 1984) and/or microbial colonization has enriched the substrate (Anderson 1973; Neuhauser and Hartenstein 1978; Swift et al. 1979). Litter infiltrated with fungal hyphae may further provide millipedes and isopods with a concentrated source of calcium, since high levels of calcium bound to oxalate ions frequently occur in fungal hyphae (Cromack et al. 1977; Swift et al. 1979; Lodge, this volume). Exoskeletons of many species of millipedes are highly calcareous (Reichle et al. 1969; McBrayer 1973), with concentrations of calcium in some species approaching 50% of their dry biomass. The numerically dominant millipede in tabonuco forest litter, *D. maldonadoi,* very likely falls into the upper range of this distribution. At some sites the low availability of calcium may drastically limit the numbers of large decomposer invertebrates, such as diplopods, isopods, and gastropods, which possess substantial calcium requirements for their exoskeletons (Albert 1979).

Substantial increases in bacterial numbers and standing crops occur during passage of litter through diplopod and isopod guts (Reyes and Tiedje 1976; Anderson and Bignell 1980; Hanlon and Anderson 1980; Hanlon 1981b), whereas, as has been demonstrated in collembolan gut passage (Hanlon and Anderson, 1979), fungal standing crops are reduced (Anderson and Bignell 1980; Hanlon and Anderson 1980). In addition, finely comminuted and compacted fecal pellets appear to favor bacterial colonization, whereas fungal colonization may initially be retarded (Hanlon 1981a). The initial enhancement of bacterial activity on diplopod and isopod feces is soon exhausted (van der Drift and Witkamp 1960; Nicholson et al. 1966; Hanlon 1981b), presumably with the depletion of readily utilizable carbohydrate fractions (Nicholson et al. 1966), leaving more recalcitrant fractions (such as cellulose and lignin) for fungal attack. Hassall and Rushton (1985) measured peak standing crops of fungal hyphae on isopod feces three weeks after their deposition and reasoned that products resulting from fungal catabolism of complex carbohydrates should enrich these particles, making them attractive resources for further bacterial exploitation and reingestion by arthropods. Coprophagy appears to be widespread among many primary decomposers, and several authors (Mason and Odum 1969; Crossley 1977; Hassall and Rushton 1985) have evoked the image of an external rumen when discussing such recycling of feces. In some circumstances this process may be crucial for arthropods: McBrayer (1973) reported that restricted access to fecal pellets resulted in severely reduced growth rates and, in some cases, mortality in various diplopod species.

Diptera

The greatest seasonal fluctuation of numbers for major arthropod taxa in tabonuco forest litter was displayed by Diptera (fig. 5.3) and Lepidoptera larvae (fig. 5.4). The decline in dipteran densities during the dry period from mid-February to early May was particularly steep for larval densities, and the rise of larval densities during the first month of the wet season was over twenty-fold (fig. 5.3). The consequent pulse of adult Diptera emerging from the litter may have been largely responsible for the increase in reproductive activity that was apparent in the understory spider community during the summer, particularly for the orb-weaver *Leucauge regnyi* (Pfeiffer, chap. 7, this volume). Chironomidae were particularly abundant components of the larval dipteran community, and members of the Tipulidae, Mycetophilidae, Phoridae, and Drosophilidae also contributed significant numbers. Garrison

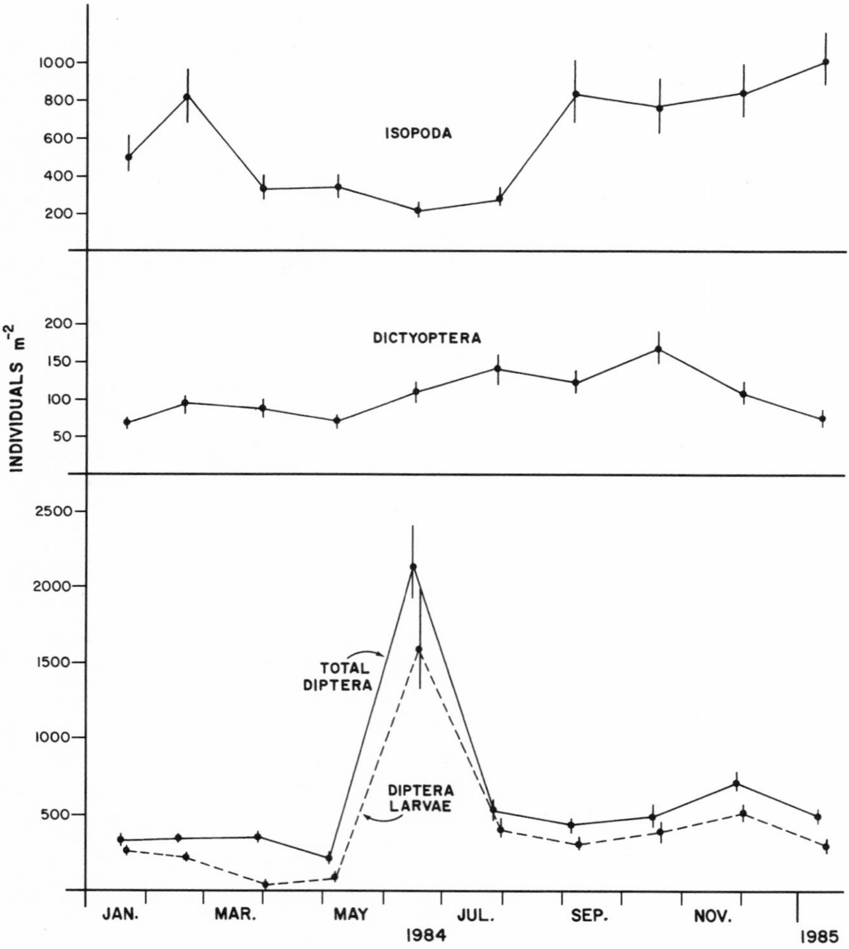

Figure 5.3. Density dynamics of Diptera larvae and adults (bottom panel), Dictyoptera (middle panel), and Isopoda (top panel) in tabonuco forest litter. Upper and lower limits of standard error bars derived as indicated in the legend of figure 5.2.

(in Reagan et al. 1982) also reported representatives of Psychodidae, Ceratopogonidae, Sciaridae, Cecidomyiidae, Stratiomyiidae, and Dolichopodidae among the more frequent dipterans occupying litter bags placed on the floor of tabonuco forest.

Tullgren extraction of fly larvae from mull and moder soils from a temperate deciduous forest greatly underestimated densities relative to wet fun-

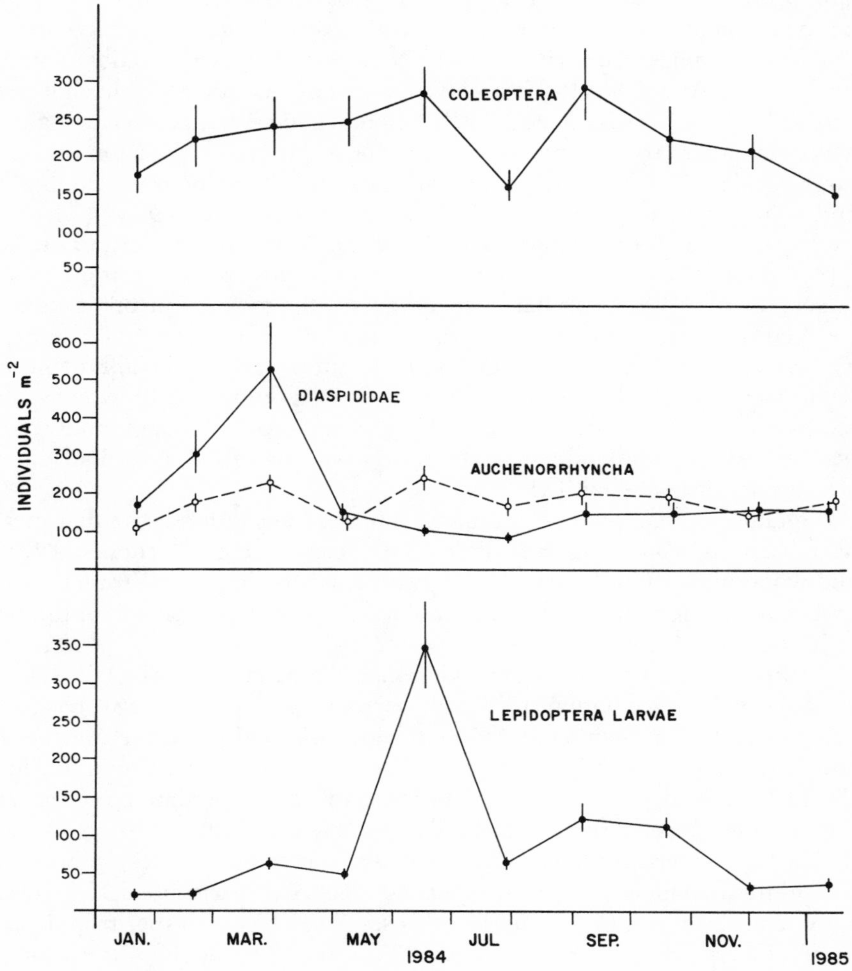

Figure 5.4. Density dynamics of Lepidoptera larvae (bottom panel), diaspididid scale and auchenorrhynchous Homoptera (middle panel), and Coleoptera (top panel) in tabonuco rain forest litter. Upper and lower limits of standard error bars derived as indicated in the legend of figure 5.2.

nel and flotation techniques (Healey and Russell-Smith 1970). Thus, the density estimates for dipteran larvae obtained during this study may be undervalued, although Tullgren extraction of larvae from the relatively meager litter and virtually nonexistent humus layer typical of tabonuco forest probably does not lead to an undesirably low efficiency of extraction.

As a group, larval Diptera exhibit a broad range of nutritional proclivities. Their placement in this section is based on the inclusion of dipteran larvae among the major litter-fragmenting arthropods by Edwards et al. (1970) and on the abundance of Chironomidae and Mycetophilidae in tabonuco forest litter, two families that Swift et al. (1979) regarded as being predominantly fungivorous. Anderson (1975) observed fungal hyphae and, to a much lesser extent, finely pulverized detritus in chironomid and mycetophilid larval guts. Mycetophilidae larvae primarily inhabit fungal sporophores, whereas the allied Sciaridae reside primarily in the soil and only exploit fungal sporophores and wood as ancillary resources (Binns 1981). These two families (collectively known as fungus gnats) often form the dominant proportion of the dipteran community in temperate forest litter (Binns 1981). Chironomidae larvae, often viewed as primarily aquatic, were the most frequently encountered larval group in the present study and in litter bag colonization studies by Garrison (in Reagan et al. 1982). This family can also be abundant at temperate sites as indicated by Healey and Russell-Smith (1970), who found that Chironomidae larvae represented the second most abundant family in a beech-chestnut woodland where larval flies averaged 4900 individuals m^{-2} (Healey and Russell-Smith 1970).

Tipulidae larvae, present in tabonuco forest litter in moderate densities, have been associated with leaf litter decomposition (e.g., Perel et al. 1971) and wood boring (Swift et al. 1979). Prior conditioning of substrate by microbes and the presence of their metabolites are considered critical for tipulid utilization of decomposing leaves (Lawson et al. 1984).

Phoridae display the broad trophic range characteristic of the Diptera as a whole. Sustenance within this family varies from apparent Aufwuchs foraging on leaf litter (Swift et al. 1979), to close association with termite nests and *Eciton* ants (Brues 1932), to endoparasitism of a Trinidadian millipede (Brues 1932) and, specific to the El Verde food web, to predation on eggs of the frog *Eleutherodactylus coqui* (Villa and Townsend 1983).

The feeding relationships of the larval dipterans associated with tabonuco forest litter remain largely unexplored and reliance on literature data derived from temperate sites has dominated this section. The only other trophic information regarding larval Diptera specific to the El Verde site was reported by Lavigne 1970c) in a preliminary study of arthropods associated with fallen fruit. Lavigne reared representatives of the eleven dipteran families (all mentioned at the beginning of this section) from such fruits. In an Australian subtropical rain forest Atkinson (1985) discerned twenty-eight species of Diptera as exploiters of fallen fruit, with broad resource overlap among the species. Pipkin (1965) has also documented the utilization of fallen fruits and blossoms by a wide array of drosophilid species in Panamanian forests.

Dictyoptera, Orthoptera, and Phasmatodea

Cockroaches (Dictyoptera) are particularly diverse and abundant in tropical forests as compared with temperate sites. Wolda (1983a), for instance, recovered 164 species at black-light traps placed at just 6 locations in Panama, nearly three times the number of species found in all of North America above the Mexican border (Arnett 1985). Thus far, twenty-five species of Blattidae have been identified at the El Verde site (see invertebrate species list in app. 6).

Cockroaches achieve unusually high densities (108 individuals m^{-2}) in tabonuco forest litter, in excess of that reported from any other tropical forest litter (tables 5.1 and 5.2) and vastly greater than from temperate sites (see summary in Gist and Crossley 1975a). Lowest densities were evident during the drier portion of the year, increasing to a peak density of 169 individuals m^{-2} in the October collections (fig. 5.3). Two species of *Cariblatta* (*C. hebardi* and *C. suave*) are the dominant constituents of the roach community.

The trophic impact of cockroaches in the litter layer of tabonuco forest may be far less than is implied by their high densities. A significant portion of resource consumption by these cockroaches may be derived from the understory layer. Schal and Bell (1986) documented the crepuscular vertical migration of cockroaches from the litter to understory foliage in a Costa Rican lowland forest. Avoidance of exposure to diurnally foraging vertebrates and insect parasitoids was postulated as a benefit of the occupation of litter during daylight hours. Negligible diurnal locomotor activity was recorded in outdoor insectaries for most of the species in the Costa Rican community in marked contrast to nocturnal observations. Low levels of feeding activity in the litter layer would be expected as a consequence of this apparent diurnal quiescence.

Cockroaches are commonly observed at night, but only rarely during the day, on low understory foliage at the El Verde site. The large contribution of cockroaches (20% of prey volume) to the diet of the nocturnal *Eleutherodactylus coqui* (Stewart and Woolbright, this volume) compared with a lower proportion in stomachs of the diurnally foraging *Anolis* spp. (Reagan, this volume) supports the contention of increased aboveground exposure at night. Nocturnal examination of the litter layer at night, however, reveals that a significant proportion of the dictyopteran community is present throughout the night. The relative contribution of cockroaches to litter and understory food webs has not been examined at El Verde, nor the factors that influence the selection of aboveground or epigeic foraging sites, such as resource availability and quality, microclimatic conditions, and life history.

Refuge selection by *Cariblatta* spp. represents one apparent distinction between the El Verde and Costa Rican roach communities. Whereas the *Car-*

Table 5.4. Miscellaneous arthropod taxa in tabonuco forest litter

Arachnida		Insecta	
Schizomida	3.1	Entotrophi	24
Amblypygi	1.8	Microcoryphia	1.5
		Phasmatodea	7.9
		Orthoptera: Gryllidae	6.8
Myriapoda		Psocoptera	15
Chilopoda	5.6	Thysanoptera	34
Pauropoda	5.6	Neuroptera	1.6
		Hymenoptera (wasps)	49
		Lepidoptera	96

Notes: Mean annual densities given as individuals m^{-2}. Mean annual densities for major taxa can be found in tables 5.1, 5.3, 5.5, and 5.7.

iblatta spp. in Costa Rica sought refuge in furled dead leaves above ground rather than in the litter (Schal and Bell 1986), the El Verde species (particularly *C. hebardi* and *C. suave*) are abundant in the litter layer during the day. The scarcity of diurnal refuges in the low understory at El Verde, other than the attached dead fronds of the sierra palm (*Prestoea montana*) and suspended dead *Cecropia* leaves, may necessitate daily vertical migration to and from the litter layer for those cockroaches utilizing aboveground resources at night.

Gryllidae (crickets) were relatively minor constituents of the litter arthropod community of tabonuco forest (table 5.4). The abundance of species, such as *Amphiacusta caraibea,* which are common in crevices beneath large rocks by day, may have been greatly underestimated with the present sampling regime. The prevalence of *A. caraibea* in *E. coqui* guts (Stewart and Woolbright, this volume) and frequent observations of this gryllid on understory foliage at night (Stewart pers. comm.) indicates that at least one species occupying El Verde litter by day migrates vertically at dusk. Another possibility is that *A. caraibea* has a vertical habitat range that emcompasses both litter and understory independent of time of day.

The eggs and early instars of Phasmatodea (walking sticks) were frequently encountered in litter samples. The eggs, which resemble small seeds, are preyed upon by parasitoid wasps and ants (Bedford 1978; Carlberg 1986).

Other Saprotrophs and Microbitrophs

Wiegert (1965) recovered high densities of Diaspididae (armored scale insects, 770 individuals m^{-2}) from fifty Tullgren extractions of litter from the El Verde rain forest during late January and early February. Densities of less than 200 individuals m^{-2} of Diaspididae occurred from May through December of the present study, with sharp increases observed in late February and early April collections (fig. 5.4). Armored scales were particularly observed in association with leaves of *Cecropia schreberiana* in the litter; Zucca (unpublished) measured peak leaf fall for *C. schreberiana* in litter basket collections from March through May.

The termite *Parvitermes discolor* maintains small colonies in and consumes the decaying wood of standing trees and stumps as well as fallen woody debris (McMahan 1970b, this volume). Manual sorting or litter following Tullgren extraction revealed that this species also occupies hollow twigs. A mean annual density of 374 termites m^{-2} was calculated for the litter layer at the El Verde site, which falls within the middle of the broad range of densities reported from other tropical forest litter and soil layers (tables 5.1 and 5.2). Colonies were encountered with such infrequency that no clear seasonal pattern could be discerned. Like *Nasutitermes costalis,* the other common termite in tabonuco forest, *P. discolor* constructs carton tunnels (McMahan 1970b, this volume) between nesting sites and resources that often extend well above ground. The ability to forage during dry seasons may be one selective advantage of this behavior (Swift et al. 1979); protection from predation may be another. Despite the abundance of *P. discolor* and the substantial prey biomass this species potentially represents, *P. discolor* forms only a negligible portion of the prey consumed by *Anolis* and *Eleutherodactylus* species (Reagan, this volume; Stewart and Woolbright, this volume). However, Thomas and Gaa Kessler (this volume) found that *Parvitermes* contributed 52% of the stomach contents of the blind snake *Typhlops richardi platycephalus,* and that the more common species of blind snake (*Typhlops rostellatus*) at El Verde also probably consumed substantial numbers of termites.

The dynamics of lepidopteran larvae (fig. 5.4) displayed a pattern similar to that of dipteran larvae (fig. 5.3), with an abrupt increase in densities in the June collections. Lepidopteran larvae were often observed in association with floral tissue in the litter samples. Lavigne (1970c) recovered lepidopteran larvae from sierra palm seeds and reared an unidentified microlepidopteran from *Inga vera* fruits. Owen (1983) observed frugivory among various genera of adult Lepidoptera in a Sierra Leone rain forest, in addition to the consumption of flowers and seeds by some larval species.

Densities of Thysanoptera in tabonuco forest litter varied from 14 to 88 individuals m^{-2} during the 1984–85 study, with peak densities occurring in the May and June collections. In the litter of a Panamanian tropical moist forest, Thysanoptera abundance was positively correlated with the flower input (Levings and Windsor 1982, 1985) along with which the thysanopterans may have been transported to the forest floor. Because there is no pronounced pattern of flower fall among tree species in tabonuco forest (Estrada Pinto 1970), a similar relationship is difficult to establish at El Verde without more detailed knowledge of specific Thysanoptera flower trophic linkages. In addition, resource consumption by litter Thysanoptera probably encompasses a broader range than just floral tissue. Strickland (1944) considered that litter and soil Thysanoptera, present in a Trinidadian rain forest at densities of 64 individuals m^{-2}, included mostly fungivorous forms and possibly some species that are predators of immature mites.

An apterous species of Aradidae (flat bugs) was very patchily distributed in the litter collections due to their association with similarly distributed chunks of woody debris. A mean annual density of 25 individuals m^{-2} was maintained by this species. Aradids consume fungi on dead wood and under the bark of dead trees (Picchi 1977; Slater and Baranowski 1978; Arnett 1985).

Seed Predators and Root-feeders

Hemiptera and Homoptera

Two families of the true bug order (Hemiptera), whose food resources include seeds and roots, were represented in tabonuco forest litter: Lygaeidae, by the genus *Ozophora,* and Cydnidae. Fallen seeds of *Ficus* sp. are eaten by Neotropical members of the widely distributed lygaeid genus *Ozophora,* including the Puerto Rican form *O. atropicta.* Sweet (1960) maintained that the members of the lygaeid subfamily encompassing *Ozophora* display no predaceous tendencies. In tabonuco forest litter, *Ozophora* was represented by a species whose density ranged from 5 to 39 individuals m^{-2} (mean annual density, 22 individuals m^{-2}). Cydnidae, a hemipteran family largely comprised of burrowing forms that feed on the roots of plants, maintained a mean annual density of 6.5 individuals m^{-2} in the litter.

Homoptera of the suborder Auchenorrhyncha, represented by the families Cicadellidae (leafhoppers) and Cixiidae (cixiid planthoppers), appeared in collections throughout the study in substantial numbers (fig. 5.4). Cixiids are known to feed on roots, and may be tended by ants (O'Brien and Wilson 1985). Mueller-Dumbois and Howarth (1981) noted an unusual situation in which a sapsucking cixiid was associated with tree roots suspended in lava tubes in a Hawaiian forest. At least two cixiid species, *Bothriocera undata* and *Pintalia alta,* have been recovered from the litter at El Verde. The cicadellid *Xestocephalus maculatus* was also collected with suction sampling, and although these individuals may have been associated with seedlings in the study plots, this species and four other cicadellid species were recovered from within litterbags placed on the forest floor by Garrison (Reagan et al. 1982).

Coleoptera

The bark beetle *Cocotrypes carpophagus* (Scolytidae) consumes the endocarp of sierra palm seeds after the exocarp of the fallen fruit has rotted away or has been removed by *Rattus rattus* (Janzen 1972). Over 99% of *P. mon-*

tana seeds in a palm forest at 680 m elevation in the Luquillo Mountains were discovered by Janzen to have been attacked by this beetle. In tabonuco forest at El Verde, Bannister (1970) observed heavy infestations of sierra palm seeds by adults and larvae by this bark beetle (note Janzen's correction of Bannister's erroneous reference to a weevil, rather than bark beetle, infestation). Janzen reported fruiting synchrony among trees in the higher altitude palm forest such that there was almost no seed fall from early May to early January. Due to seed germination and the destruction of seeds by the bark beetles, resources for these seed predators grew scarce during Janzen's (1972) study by August. During the 1984–85 sampling, peak bark beetle densities of 75 to 125 individuals m^{-2} sere observed in the April through June collections and declined sharply thereafter. Bannister (1970) observed that some palms produced flowers and fruits throughout the year, which may explain the persistence of low bark beetle populations throughout the study year.

Scarabaeidae larvae were commonly uncovered in soil excavations both at the El Verde site and a nearby secondary growth forest. Arnett (1973) remarked on a wide range of resources utilized by scarabaeid larvae, including roots.

Euryphagic Arthropods

Formicidae

Although many ant species are predominantly predatory, this diverse hymenopteran family will be considered here for the sake of continuity. Because the ant community is dominated by ponerines in tabonuco forest, consumption of arthropod prey by ants in the litter layer is significant and may rival the level attained by spiders and opilionids.

The Puerto Rican ant fauna displays an impoverishment similar to that observed for other major terrestrial invertebrate taxa on the island. Smith's (1936) compilation of the Puerto Rican ant fauna included sixty-six species, of which eighteen were regarded as being introduced. Work in progress on the ant fauna of Puerto Rico puts the number of species over 100 (J. Torres and R. Snelling pers. comm.). Although this diversity exceeds the forty-two species occurring in the British Isles (Brian 1977) and the forty-seven species found in the Netherlands (Drift 1963), far greater diversity has been demonstrated at tropical sites. For example, Wilson (1959) recognized a minimum of 172 species from a few square kilometers of New Guinea rain forest, and Levings (1983) reported that well over 200 species of ants have been collected from the 1500 ha of Barro Colorado Island, Panama. From a roughly 5 km^2 region in the Peruvian Amazon, Wilson (1987) identified 135 arboreal species alone, including forty-three species dislodged from a single leguminous tree by insecticidal fogging.

Approximately 40% of the thirty-six species that have been reported from tabonuco forest in the vicinity of El Verde (Lavigne 1970b, 1977) were recovered in the 1984–85 suction samples of the litter layer. *Solenopsis* spp. (mostly *S. azteca pallida*) and *Pheidole moerens* accounted for over half of the recovered adults (table 5.5). Little variation in community numbers was apparent through the study year, with the annual coefficient of variation of 13.6% representing the lowest for any major arthropod taxon.

As has been observed at other Neotropical sites, ants are one of the numerically dominant taxa in the litter layer of the tabonuco forest at El Verde (table 5.1). Excluding mites and collembolans, which were not efficiently sampled, ants made up about half of the arthropods in the litter layers of forests on Barro Colorado Island (Levings and Windsor 1982) and at Bonampak, Mexico (Lavelle and Kohlmann 1984; table 5.1), as well as in forest soil and litter at Laguna Verde, Mexico (Lavelle 1984). The proportional representation by ants in the litter arthropod community increased to 70% during the wet seasons and declined to 20 to 40% during the dry seasons in the thirty months of collections of the Levings and Windsor (1982) study although, as at El Verde, the overall worker densities remained fairly constant. In other Neotropical forests ants ranked only behind mites and collembolans in abundance (table 5.1). Ant densities in the soil exceeded those in the overlying litter layer at all tropical sites except Manaus, Brazil. Ants represented over 50% of the arthropods recovered from the upper 5 cm of mineral soil in El Verde, with *S. azteca pallida* contributing 90% of the workers (Wiegert 1970a).

The Ponerinae, a subfamily of predominantly predaceous ants, represented by four species in the tabonuco forest litter, composed the bulk of the ant standing stocks. Lévieux (1983) regarded the Ponerinae as the prevailing ant subfamily in the tropical forests of Africa, where they contributed as much as 60 to 85% of ant species at specific sites. The largest and most dominant ant species in tabonuco forest litter, in terms of standing stocks, was *Odontomachus brunneus*. This genus is characterized as a generalist predator (Levings 1983; Deyrup et al. 1985). This ponerine accounts for approximately two-thirds of the 0.2 g m^{-2} mean annual standing stock of formicids in the litter of this forest. Levings (1983) also categorized Panamanian species of *Pachycondyla* as generalist predators, although Wheeler (1936) noted three Central and South American species of *Pachycondyla* as being specialized predators of termites. *Hypoponera* species occupying Ivory Coast savannas were classified as Collembola specialists (Lévieux 1983), and Levings (1983) categorized species of this genus as specialist predators of unstated preference. *Hypoponera opacior* may be regarded as a collembolan predator, and *Pachycondyla stigma* as a potential termite predator. No specific trophic information for the remaining ponerine (*Anochetus mayri*) recovered from tabonuco forest litter has been reported in the literature, and it is likely that this species represents a generalist predator of small litter arthropods, perhaps also with a preference for Collembola.

Table 5.5. Ant species in tabonuco forest litter

Species	Jan	Feb	Apr	May	Jun	Jul	Sep	Oct	Nov–Dec	Jan	MAD	CV
Solenopsis spp.	529	399	500	276	261	596	363	564	361	650	428	29.0
Pheidole moerens	135	225	381	218	272	122	218	159	258	166	221	36.0
Brachymyrmex heeri	121	105	289	198	249	37	92	73	118	104	142	59.2
Paratrechina near *vividula*	120	77	121	199	95	53	223	87	120	121	122	45.8
Ochetomyrmex auropunctata	210	23	101	139	52	225	22	0	23	0	88	96.6
Strumigenys gundlachi	28	42	47	85	56	47	95	18	35	23	50	50.2
Strumigenys rogeri	49	36	7	17	29	34	46	60	95	14	42	62.2
Odontomachus brunneus	20	9	24	135	20	18	24	31	58	39	38	103.4
Iridomyrmex melleus	6	5	46	52	24	33	10	35	7	5	24	75.0
Cyphomyrmex minutus	0	1	4	3	23	13	15	91	24	9	19	146.8
Anochetus mayri	14	23	11	24	20	12	21	15	8	8	16	34.8
Hypoponera opacior	1	6	7	15	16	4	20	5	3	2	7.3	
Paratrechina microps	0	4	1	2	3	1	1	22	0	0	3.7	
Paratrechina steinheili	2	4	0	3	1	1	7	1	7	3	2.9	
Pheidole sp.	2	5	4	0	0	0	0	0	0	0	1.2	
Pachycondyla stigma	0	3	0	3	2	0	2	0	0	0	1.1	
Monomorium floricola	2	0	0	0	0	0	0	0	0	0	0.2	
All species	1237	969	1543	1369	1125	1199	1159	1161	1116	1142	1209	13.6

Notes: Mean annual densities (MAD) and annual coefficients of variation (CV) are calculated for the January 1984 through November–December 1984 collections, where meaningful. Larval and pupal stages are not included. Densities are given as individuals m^{-2}. Totals do not sum to "All species" sum because of rounding.

Wilson (1959) considered *Strumigenys* spp. (Myrmicinae) to be specialist predators of Collembola. Observations by E. O. Wilson of Cuban workers of *Strumigenys gundlachi* and *Strumigenys rogeri* in laboratory colonies led Brown (1954, 1959) to conclude that members of this genus were ambush predators characterized by a preference for entomobryid collembolans and a nearly universal rejection of podurid collembolans. Other preferred prey included Campodeidae and sminthurid collembolans, whereas termites, isopods, and adult staphylinid beetles were usually ignored or rejected. Smith (1936) and Torres (1984a) classified the ants *Paratrechina steinheili, Pheidole moerens,* and *Ochetomyrmex* (*Wasmannia*) *auropunctata* as being primarily insectivorous, although Torres was uncertain as to the predatory proclivity of *O. auropunctata.* Lavigne (1977) considered *P. moerens* to be omnivorous and observed workers feeding on fruits in the El Verde forest. *Ochetomyrmex auropunctata* workers exploit both dead and living insects as well as a variety of decomposing fruits at El Verde. In Maricao forest (western Puerto Rico) these ants were observed tending a *Pseudococcus* scale insect inside a hollow fruit (Lavigne 1977). *Pheidole* spp. and *O. auropunctata* primarily employ mass recruitment to exploit resources (Levins et al. 1973). Mass recruitment by the *Paratrechina* spp. in tabonuco forest litter may be reserved for large prey items, if their behavior is consistent with that observed by Levins et al. (1973) for *Paratrechina longicornis* elsewhere on the island.

Cyphomyrmex minutus (= *C. rimosus*), considered by Levins et al. (1973) to be the most abundant fungus-growing ant on the island, was reputed by these investigators to depend largely on insect feces for use as a substrate in their fungus gardens. Other Neotropical *Cyphomyrmex* spp. also use insect carcasses and pieces of fruit as substrates for their fungus gardens (Hölldobler and Wilson 1990). The remaining species that appear in table 5.5 are likely omnivores that rely on mass recruitment. Levins et al. (1973) observed that mass-recruiting species in the Maricao forest were distinguished by the range over which they could respond to resources. *Paratrechina longicornis* recruited to resources located 5 to 10 m from its nest, *Solenopsis geminata* could only respond to much more closely placed resources, and the response of *Brachymyrmex heeri* was normally limited to resources within centimeters of its nest. Torres (1984a) noted that, in addition to granivory, *S. geminata* tended root-feeding insects. Such tending of sapsucking homopterans may be an important trophic link in tabonuco forest, considering the abundance of both auchenorrynchous Homoptera and *Solenopsis* in the litter and upper soil layers. Lavigne (1970c) recovered several species of *Solenopsis* (principally *S. azteca pallida*), *P. moerens, Pheidole subarmata borinquensis, Paratrechina* sp., *C. minutus, O. auropunctata, Myrmelachista ramulorum,* and *Mycocepurus smithi* from fruits on the floor of the El Verde rain forest. The latter species was also reported by Levins et al. (1973) to use bat dung as a substrate for their fungal gardens.

One of the more salient contrasts between the ant community occupying tabonuco forest litter in Puerto Rico and those of the Central and South American mainland is the absence of army ants in the island's forests. Smith (1936) observed that the Dorylinae are absent throughout the West Indies, except for the southernmost Lesser Antillean islands near the coast of Venezuela. In continental litter communities, army ant foraging can have pronounced effects on arthropod community structure. Franks and Bossert (1983) concluded that the Central American doryline *Eciton burchelli* acted as a keystone predator (Paine 1969) on the ant community residing in forest litter on Barro Colorado Island, removing competitively dominant ant species and continually opening patches of disturbed litter for recolonization. A simulation model suggested that at any given time 50% of the litter on Barro Colorado Island is undergoing "secondary succession" by the ant community. Initially, highly opportunistic species, such as those of the genera *Paratrechina* and *Pheidole,* flourish in these gaps, but as succession proceeds more competitively able species begin to dominate. The action of *E. burchelli* probably not only enhances species diversity, but also maintains a fairly constant mosaic. Moreover, the effects of the foraging behavior of this army ant likely extends to the arthropod community as a whole.

The absence of army ants in tabonuco forest litter may have led to a distinctly different composition in the ant assemblage and the underlying mechanisms influencing this structure as compared with mainland Neotropical sites. If litter in the El Verde rain forest were subjected to dorylinid raids, then larger ponerine species, especially *O. brunneus* which so dominates the biomass of this ant community, might well be reduced to lower equilibrium densities. Levings (1983) observed a much-diminished presence of an *Odontomachus* species at bait traps subsequent to the passage of an *Eciton* swarming raids. In the Amazon region of Ecuador *Eciton rapax* obtains a significant portion of its prey by raiding ponerine nests, among them *Odontomachus* (Brown 1976).

In contrast to Central American forests where competition for nesting sites may be lessened as a consequence of doryline swarming raids, competition for nesting sites in tabonuco forest may be more severe, with the input of fruit and woody debris to the forest floor a major determinant of local community composition. Lavigne (1970b) found that the vacant seed coats of *Dacryodes excelsa, Prestoea montana,* and *Sloanea berteriana,* as well as the pith chambers of decomposed *Cecropia peltata* limbs, frequently served as nesting sites, particularly for colonies of *S. azteca pallida, S. rogeri, Paratrechina* sp., and *O. auropunctata.* It is a moot question as to whether the absence of army ants has played an ancillary role in the historic development of a comparatively low species richness of ants in the tabonuco litter community through the absence of opportunities for competitively inferior immigrant species to gain a foothold. Levins et al. (1973) and Culver (1974) considered the component species of Puerto Rican ant communities to be broad-niched relative to, respectively, the tropical mainland and an Appalachian hardwood forest. Culver attributed the lower species packing in the Puerto Rican at Maricao to the greater aggressiveness of the tropical species. Clearly, however, an explanation of the relatively impoverished arthropod diversity in tabonuco forest must begin with island biogeographic considerations.

Coleoptera

Little seasonal variation was evident in the densities of beetles (fig. 5.4). Adults contributed 72% of the beetles recovered from samples collected from January through December 1984. Of these, 44% were from the family Pselaphidae and another 22% were from the family Scolytidae. Other frequently encountered adults included representatives from the Nitidulidae, Curculionoidea, Ptilidae, Staphylinidae, and Scydmaenidae. Carabidae and Scarabaeidae were rarely encountered in the litter samples. Of the immature stages, Scotylidae larvae and a case-bearing Chrysomelidae occurred most frequently.

Lavigne (1970c) reared a variety of larvae in the families Staphylinidae and Nitidulidae from fallen fruits in tabonuco forest, with the scarabaeid *Canthochilum borinquensis* and a curculionid appearing less frequently. Whether the staphylinids occurring on these fruits are predaceous or mycophagous is not clear; several species from the mycophagous subtribe Gyrophaenina are known from the West Indies (Ashe 1984). The majority of the Nitidulidae exploit decaying fruit and fungi, and the Ptilidae and Scydmaenidae also appear to primarily exploit fungi (Borror and DeLong 1971; Arnett 1973), although Penny and Arias (1982) considered that the scydmaenids of a central Amazonian rain forest were mite predators.

Predaceous Arthropods

The quantitative assessment of the relative strength of trophic linkages between arthropod prey and predators in litter systems is severely restricted by the small size and cryptic habits of many of the component species. Furthermore, for the arachnid and hemipteran predators so abundant in tabonuco forest litter, prey composition cannot be discerned from examination of gut contents because prey tissue is digested externally (preoral digestion) and then ingested as fluid by these arachnids or the bodily fluids of impaled prey are sucked out by the predaceous Hemiptera. Serological techniques may be applied in such cases to delineate feeding preferences among broad taxonomic categories such as insect families and arachnids orders (e.g., Adams 1984), as well as finer discriminations (Stuart and Greenstone 1990) for the estimation of trophic links and trophic level (Walter et al. 1991). At present only broad dietary composition among the arthropod predators in tabonuco forest litter can be assumed, with the possible exception of the ponerine ant species previously mentioned. Many of the minute species of arachnid and hemipteran predators so well represented in this system are likely constrained to a more restricted universe of microarthropods and the early developmental stages of macroarthropods.

Predaceous Arachnids

Araneae

Spiders are a dominant predatory group in the litter of many forests, both in terms of density and standing stocks. Current evidence suggests that spiders may exert a significant impact on arthropod prey populations in forest litter. Moulder and Reichle (1972) calculated that the biomass of litter invertebrates lost annually through mortality to spiders in the litter of a southern Appalachian hardwood forest amounted to 2.3 times the annual mean standing stock of prey, although their collection and extraction techniques may have severely underestimated collembolan densities and standing stocks (cf. with those of similar sites from McBrayer and Reichle 1971 and Gist and Crossley 1975a). Experimental reductions of spiders in Canadian beech-maple forest litter led to increases in collembolan densities of about 25% (Clarke and Grant 1968), although the lack of replication in the experimental design was a critical shortcoming.

The arachnid order Araneae encompasses a highly speciated group of euryphagic predators that are significant insectivores in most terrestrial systems. The ability of spiders to control the growth of insect populations, however, is in doubt. The vast majority of spiders are opportunistic predators that exhibit lower potential rates of population increase than most insect prey species. Compared with vertebrate predators and arthropod parasitoids, spiders display a limited ability to detect and aggregate in patches of dense prey concentrations that are often ephemeral. Hence, the capacity of spiders to regulate prey species in a density-dependent manner appears limited (Kuno and Hoyko 1970; Riechert 1974; Riechert and Lockley 1984). Rather, the significance of spider communities in ecosystem processes appears to lie chiefly in their potential, in combination with other generalist predators, to suppress insect populations below outbreak levels (Riechert and Lockley 1984).

Arthropod prey densities can be subject to rapid and radical change. Intrinsic and extrinsic factors, however, appear to buffer the densities of spider communities from wide fluctuations in prey availability. At one extreme, behavioral traits, such as territoriality (Riechert 1978, 1981), intraspecific predation (cannabilism), and the structural features of the habitat may restrain spider densities below levels dictated solely by prey productivity. On the other hand, metabolic rates of spiders are low relative to poikilotherms of similar mass (Anderson 1970; Greenstone and Bennett 1980; Anderson and Prestwich 1982), and these rates can be further reduced by as much as 50% during starvation (Miyashita 1969; Anderson 1974; Tanaka and Ito 1982). This physiological trait should favor the persistence of relatively stable densities in spider communities during periods of low prey availability (Riechert and Lockley 1984), contributing to the maintainance of a sufficient predatory counterbalance to check the growth of prey populations from low densities to outbreak levels, beyond which effective control by a community of polyphagous predators may be precluded.

The mean annual density of spiders (356 nymphs and adults m^{-2}) in the litter of the tabonuco forest at El Verde exceeds those reported for other forests, with the exception of combined litter and soil estimates from two Finnish spruce stands and an Australian subtropical forest (table 5.6). Spider densities in the soil of the El Verde rain forest appear to be very low, although the contribution to the overall standing stocks by soil-dwelling *Psalistops corozali* (Barychelidae) and *Ischnocolus culebrae* (Theraphosidae) may be of some significance. The two numerically dominant species, *Modisimus montanus* (Pholcidae) and *Theotima* sp. (Ochyroceratidae), comprise 61% of the mean annual density of the litter spider community (table 5.7), but their impact on the standing stock of this community is much less due to their small size (fully gravid females range up to 500 and 50 μg, respectively). *Pseudosparianthus jayuyae* (Sparassidae) and *Corinna jayuyae* (Clubionidae), ranging up to 12 and 15 mg, respectively, appear to make the greatest contributions to community standing stocks.

Twenty-three species of spiders have been recognized in the samples collected from litter at El Verde (table 5.7). Three additional species associated with the ground layer include a Clubionidae (*Trachelas borinquensis*), known only from an *Eleutherodactylus coqui* gut, a Scytodidae (*Loxosceles* sp.) that occupies the recesses of pitted boulders, and a very patchily distributed Linyphiidae (*Centromerus ovigerus*). Two species represented in the collections, *Clubiona portoricensis* (Clubionidae) and *Stasina portoricensis* (Sparassidae), forage primarily on aboveground vegetation but may seek diurnal refuge in the litter layer. In addition, females of both species have been collected manually from the litter layer while associated with egg clutches.

The species richness of the spider assemblage in tabonuco forest litter falls near the lower end of the range reported for communities in temperate forest litter (table 5.8). This paucity of spider species in the litter is more

Table 5.6. Spider densities in forest litter and soil

Locality	Habitat	Layer(s)	Density	Source
Norway	birch: dominated by *Dryopteris*	litter	238*	Hauge 1977
	Vaccinium-Empetrum	litter	176	
Finland	spruce: *Hylocomium-Myrtillus* type	above mineral horizon	428*	Huhta and Koskenniemi 1975
	Myrtillus type	above mineral horizon	431*	
	Oxalis-Myrtillus type	above mineral horizon	309*	
Denmark	beech		58	Bornebusch 1930
	oak		34	
	spruce		66	
England	oak: site 1	F and H horizons	36*	Gabbutt 1956
	site 2	F and H horizons	74	
Netherlands	beech	F and H horizons	163	Drift 1951
		F, H, and A horizons	231	
France	chestnut coppice	L, F, and H horizons	75*	Christophe and Blandin 1977
Tennessee	*Liriodendron*-dominated deciduous	litter	126*	Moulder and Reichle 1972
Japan	beech	litter & mineral horizons	186	Kitazawa 1967
Australia	rain forest	litter	117	Plowman 1979
		soil	129	
	wet sclerophyll	litter	183	
		soil	291	
Puerto Rico	tabonuco forest	litter	356	this study
Mexico	moist tropical forest	litter and soil	73	Lavelle and Kohlmann 1984
Panama	moist tropical forest	litter	<25	Levings and Windsor 1982

Notes: Densities are given as individuals m^{-2}. Densities followed by asterisks are mean annual estimates.

Table 5.7. Spider species in tabonuco forest litter

Species	Jan	Feb	Apr	May	Jun	Jul	Sep	Oct	Nov–Dec	Jan	MAD	CV
Modisimus montanus	113	93	128	85	74	173	253	190	171	201	142	41.4
Theotima sp.	74	96	87	101	55	43	77	74	59	166	74	25.9
Masteria petrunkevitchi	36	33	9	14	28	33	52	69	52	49	36	52.5
Pseudosparianthus jayuyae	35	17	14	13	14	25	18	38	14	19	21	45.9
Corythalia gloriae	16	17	19	25	19	27	16	21	17	25	20	20.2
Heteroonops spinimanus	11	15	26	8	18	15	8	16	10	19	14	40.6
Ochyrocera sp.	8	25	23	8	7	1	6	8	3	11	9.9	
Phrurolithus sp.	7	13	7	12	8	11	16	14	7	3	9.4	
Corinna jayuyae	14	8	8	8	2	8	12	20	7	10	8.7	
Lygromma sp.	8	2	8	2	4	4	3	8	5	5	4.9	
Oligoctenus ottleyi	6	3	1	5	2	3	1	6	0	1	2.9	
Caponina spp.	2	1	5	2	1	2	3	2	5	3	2.4	
Clubiona portoricensis	1	4	4	1	0	0	3	3	0	2	1.7	
Ischnocolus culebrae	1	2	0	2	1	3	4	1	0	0	1.5	
Stasina portoricensis	2	2	0	0	2	5	2	2	1	3	1.1	
Psalistops corozali	1	1	2	2	1	3	1	1	0	0	1.1	
Dysderina sp.	0	2	1	0	0	2	0	2	1	1	0.8	
Nops sp.	0	2	1	0	1	1	2	0	0	1	0.7	
Triaeris stenaspis	0	1	0	1	3	0	0	0	0	1	0.6	
Emanthis sp.	1	2	0	0	1	1	0	0	0	0	0.5	
Oonops sp.	1	1	1	0	0	1	0	0	0	1	0.4	
Hahnia ernesti	0	0	0	0	0	2	0	0	0	0	0.2	
All species	336	338	343	286	239	362	478	473	352	521	356	21.8

Notes: Densities are given as individuals m^{-2}. Mean annual densities (MAD) and annual coefficient of variation (CV) are calculated for the January 1984 through November–December 1984 collections. Totals do not sum to “All species” sum because of rounding.

Table 5.8. Species richness of spiders in litter habitats

Locality	Forest type	Number of Species	Collection technique	Area sampled (m^2)	Source
Finland	pine: *Calluna*	24	pitfall	8.75	Huhta 1965
	Vaccinium	31	pitfall	7.5	
	spruce: *Myrtillus*	24	pitfall	7.0	
	Oxalis-Myrtillus	29	pitfall	11.25	
Finland	pine: *Vaccinium*	92	pitfall		Huhta 1971
	spruce: *Hylocomium-Myrtillus*	23	pitfall		
	Oxalis-Myrtillus	65	pitfall		
England	oak: site 1	49	manual	37.5	Gabbutt 1956
	site 2	30	manual	25.0	
England	oak-hawthorn	39	pitfall		Williams 1962
England	chestnut coppice	40	pitfall		Russell-Smith and Swann 1972
Belgium	beech	56	pitfall		Jocqué 1973
	mixed coppice	61	pitfall		
West Germany	beech	35	pitfall		Albert 1976
Yugoslavia	beech	38	pitfall		Polenec 1964
	oak-hornbeam	46	pitfall		Polenec 1974
France	chestnut coppice	47	Tullgren	46.0	Christophe and Blandin 1977
Michigan	oak-hickory	61		1.8	Bohnsack 1954
Michigan	beech-maple climax	21	pitfall		Bultman et al. 1982
	oak subclimax	11	pitfall		
Illinois	streamside forest	26	pitfall		Uetz 1979
	upland hardwood forest	21	pitfall		
Delaware	oak-tuliptree-maple	33	pitfall		Uetz 1975, 1979
Ohio	beech-maple	58	manual		Bultman and Uetz 1982
Tennessee	pine	13	Tullgren	0.3	Crossley and Bohnsack 1960
Tennessee	*Liriodendron* hardwood	40	manual	79.2	Moulder and Reichle 1972
North Carolina	pine-hardwood forest	40	various		Coyle 1981
Puerto Rico	tabonuco forest	23	Tullgren	11.8	this study
Botswana	mopane deciduous woodland	87	pitfall		Russell-Smith 1981

remarkable considering that the species richness at some of the sites listed in table 5.8 was probably substantially underestimated because of an exclusive use of pitfall trapping as a sampling technique. This collection technique is ineffective for the acquisition of sedentary web-building forms, which generally account for nearly 50% of the spider species in litter communities.

Studies in temperate forests have indicated that the species richness of spider communities increases with the depth of the litter layer and to a lesser extent with its structural complexity (Huhta 1971; Uetz 1975, 1979; Bultman and Uetz 1982). Thus, the relatively thin litter layer in tabonuco forest may be a contributing factor to the low species richness of the litter spider community. Considerable structural complexity, however, does arise on the floor of this forest from the diversity of dead leaf morphology as well as from the presence of vines, buttresses, and rocks of various sizes, which should ameliorate the influence of shallow litter depth. A relatively sparse spider diversity is also apparent in the arboreal strata (Pfeiffer, chap. 7, this volume), suggesting that an extrinsic factor such as the island effect (i.e., distance from the mainland and island size) may be an overriding barrier to the development of a rich litter spider assemblage in Puerto Rican forests. Comparative data for mainland Neotropical forests are not available.

The diversity of the spider community of tabonuco forest litter is, however, relatively rich if considered at the familial level, a taxonomic category that generally corresponds to distinctiveness of foraging mode in spiders. Fifteen families of spiders are represented from the forest floor of the El Verde rain forest, comparable to or in excess of the number present in other forest communities (table 5.8). The species-to-family ratio (1.7) may represent a crude index of the diminished redundancy within feeding guilds of this community. This, in turn, suggests that the overlap along the food resource dimension may be reduced, especially when compared to other spider assemblages.

Comparison of density estimates for litter spiders with those extrapolated for aboveground populations provides a striking illustration of the intensity of faunal activity in the litter layer. Although a vertical cross-section of tabonuco forest from the litter layer to the top of the canopy is roughly 500 times thicker than the litter layer itself, the density of spiders in the litter is fourteen times higher than that estimated for the entire vertical profile of the arboreal layers (Pfeiffer, chap. 7, this volume). If considered on the basis of spider density per cm of height, the concentration of spiders is approximately 7000 times higher in the litter layer than in aboveground habitat. It should be noted, however, that the population estimate for arboreal spiders does not consider strictly nocturnal foragers or web-building spiders whose webs were not deployed during the diurnal censuses. Thus, the disparity between the size of spider populations in the litter and arboreal layers is less than stated above. Still, the difference in relative concentrations is noteworthy. In terms of standing stocks the difference appears to be somewhat less, but the aboveground values for the relatively massive tarantulas (Theraphosidae) and giant crab spiders (Sparassidae), are inadequately known.

Other Predaceous Arachnids

Five other predominately or exclusively predaceous orders of arachnids occupy tabonuco forest litter. Three of these orders were represented by single species in the D-Vac collections: Schizomida by *Schizomus yunquensis,* Amblypygi by *Charinides* sp., and Scorpiones by *Tityus obtusus.* All three species maintained mean annual densities in litter of less than 5 individuals m^{-2}. Field observations by the author and by M. Stewart and L. Woolbright (this volume) indicate that *T. obtusus* is primarily an arboreal forager, and that diurnal refuge may be taken by some individuals in furled dead leaves and palm fronds on the forest floor. Both of the schizomid and amblypygid species are minute, and what predation pressure they exert on the litter arthropod community is likely directed towards the microarthropods and early developmental stages of macroarthropods. The size of *Charinides* sp., with maximum body length of less than 5 mm, is markedly smaller than the other amblypygid (*Phrynus longipes*) occupying tabonuco forest. *Phrynus longipes,* whose mature body length is at least 40 mm, normally resides during daylight hours in crevices under large boulders in rarely flooded streambeds and forages at night in the vicinity of these retreats and on trunks of trees in the lower portion of this forest. This large arachnid species appears to have little interaction or impact on the litter arthropod community.

The remaining two orders of arachnids (Pseudoscorpionida and Opiliones) represented in tabonuco forest litter achieve densities much higher than any reported from other Neotropical humid forests (table 5.1). On a worldwide basis for tropical and subtropical sites only Plowman's (1979) estimates of combined soil and litter pseudoscorpions in Australian rain forest exceed the El Verde numbers. Petersen (1982a) notes that few population estimates have been made for these groups, even in temperate forests. Generally, higher densities would be expected in the thicker organic layers of temperate forests, especially for pseudoscorpions. Densities well in excess of those present in tabonuco forest litter have been reported in litter and soil layers of English oak and beech forests (Gabbutt 1967; Wood and Gabbutt 1978).

The highest densities of pseudoscorpions and opilionids were observed during the wet season (fig. 5.5), although no obvious seasonal bursts of reproduction were apparent. Field observations of predation by pseudoscorpions and opilionids are rare not unexpectedly because of their small size and cryptic habits. Pseudoscorpions in laboratory cultures accept a wide variety of soft-bodied prey, including collembolans and holometabolous insect lar-

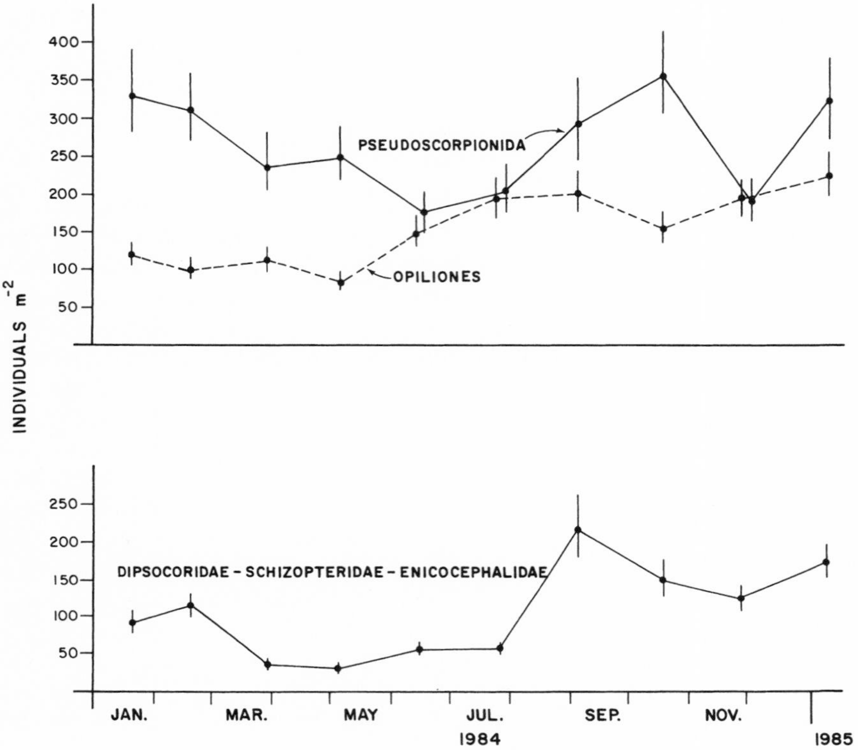

Figure 5.5. Density dynamics of combined hemipteran families Dipsocoridae, Schizopteridae, and Enicocephalidae (bottom panel) and the arachnid orders Opiliones and Pseudoscorpionida (top panel). Upper and lower limits of standard error bars derived as indicated in the legend of figure 5.2.

vae (Levi 1953). Considering their abundance, these taxa probably contribute substantially to pseudoscorpion diets in tabonuco forest litter. Serological analyses of six opilionid species in English woodlands by Adams (1984) found that dipteran larvae (28% of diet), isopods (12%), and collembolans (10%) represented the major prey items.

Two species of pseudoscorpions were recovered from tabonuco forest litter: 63% of individuals belonged to a species of the family Menthidae, possessing robust chelae and a maximum body size of 350 μg, whereas the remainder were Ideoroncidae with long, thin, and less scleroticized chelae and a maximum body size of about 150 μg. Differences in size and chelate structure may determine the proportions of prey types captured by these two species. *Stygnoma spinulata,* armed with heavily spined pedipalps, dominated an opilionid assemblage consisting at least six species. Because of the much larger size of the opilionids, these arachnids probably consume at least several times the biomass of arthropods than do the pseudoscorpions and likely rank behind spiders and possibly ants as the dominant predatory group in tabonuco forest litter. Estimates for mean annual standing stocks of pseudoscorpions amount to roughly 0.015 g m^{-2}; for opilionids the corresponding value is roughly 0.090 g m^{-2}.

Other Predaceous Arthropods

Although little is known of the natural history of the hemipteran families Dipsocoridae, Schizopteridae, and Enicocephalidae, they are presumed predaceous by various authorities (Emsley 1969; Borror and DeLong 1971; Slater and Baranowski 1978; Kritsky 1979; Arnett 1985). Together these minute hemipterans maintained mean annual densities of 83 individuals m^{-2} in tabonuco forest litter, exhibiting peak densities during the wetter months and minimum numbers during the dry period from March to May (fig. 5.5).

A variety of diminutive wasps were collected in fairly constant numbers during the study period, ranging from 33 to 71 adults m^{-2}. Pselaphidae, the most numerous of the coleopteran families in the litter layer, are classified as nocturnal predators of small arthropods such as collembolans, insect larvae, and mites and are often associated with ant or termite nests (Park 1942; Park 1964; Arnett 1973; Penny and Arias 1982). Predaceous beetles were uncommon except for the Staphylinidae, whose trophic habits are unclear in this community. Rove beetles (Carabidae), in particular, are poorly represented. Neuropteran larvae of the family Ascalaphidae, although present in densities of only 1 to 2 individuals m^{-2}, are capable of impaling the largest of the litter arthropods and, along with the largest of the arachnid species and centipedes, may be responsible for much of the invertebrate predation on large prey such as cockroaches.

The mean annual density of centipedes of 8 individuals m^{-2} probably represents a severe underestimate, since during daylight hours many centipedes may remain firmly secluded under rocks and in other crevices. Centipedes can achieve considerable densities in forest litter. Mean annual densities of lithobiid centipedes alone amounted to 51 to 94 individuals m^{-2} in English forest litter (Wignarajah and Phillipson 1977). In German beech forest litter lithobiid densities averaged 94 individuals m^{-2} (Albert 1983). Centipedes are very abundant in the litter and soil layers of two Australian subtropical forests examined by Plowman (1979), where densities exceed 850 individuals m^{-2}.

Onychophora, sometimes classified within the phylum Arthropoda, is represented in tabonuco forest by a species of *Peripatus* that is occasionally found foraging on moss- and algae-covered rocks on the forest floor. Prey immobilization is accomplished by directing an adhesive spray at the victim; large prey such as crickets and roaches may be captured by this strategy.

Annelida

Earthworms appear to represent the dominant faunal group in tabonuco forest, contributing 36% (4.2 g m^{-2}) of faunal standing stocks (Odum 1970c). On this basis, the apparent proportional contribution of earthworms to the litter and soil faunal community at El Verde is comparable to several Nigerian and Mexican tropical forests (Lee 1983; Lavelle 1984). However, this conclusion is based on a very limited data set from tabonuco forest, consisting of only six soil monoliths (Moore and Burns 1970).

Trophic behavior of earthworms falls along a spectrum between geophagous species consuming soil humus and detritivorous species exploiting litter and roots (Lavelle 1983b), with the exception of isolated cases of species adapted for carnivory (e.g., Lavalle 1983a). In general, the incorporation of earthworm biomass into higher trophic levels appears insignificant compared with mortality related to stressful physical conditions (Lavelle 1983b).

TROPHIC LINKS BETWEEN LITTER AND ARBOREAL FOOD WEBS

Vertical movement corresponding to light/dark cycles constitute a prominent faunal interchange between litter and arboreal habitats in tabonuco forest, as indicated most dramatically for *E. coqui* (Stewart and Woolbright, this volume), the dominant nocturnal predator in the forest. As suggested previously, crepuscular movement from the litter layer to understory foliage also appears to form part of the behavioral repertoire of cockroaches and crickets, as well as for some of the large arachnid predators (e.g., the amblypygid *Phrynus,* the sparassid *Stasina,* the scorpion *Tityus*). In all of these instances, food resources are being harvested above ground and refuge from diurnal predators (anoles and birds) is apparently sought in the litter layer. All of these species may exploit prey in the litter or rocky areas of the forest floor, but to what extent their diet is composed of litter arthropods remains unclear.

The synchronized emergence of adult insects from the pupal stage, particularly dipterans, releases large quantities of potential prey for aboveground predators, especially for those dwelling at the lower levels of the understory. A potentially large pulse of dipteran adults appears to have been released near the beginning of the 1984 wet season (fig. 5.3). This may have contributed greatly to the midsummer rise in the understory densities of the orbweaver *Leucauge regnyi* (Pfeiffer, chap. 7, this volume).

VERTEBRATE PREDATION ON LITTER ARTHROPODS

Despite the abundance of arthropods in the litter layer, exploitation of this resource concentration *in situ* by vertebrate predators appears limited except for nonanoline reptiles. The dominant herpetofauna at El Verde appear to restrict the majority of their foraging to the arboreal layers. *Anolis* lizards, in particular, limit their activity to aboveground vegetation and on rare occasions to large boulders. The dense frog populations in tabonuco forest actively seek prey above ground at night, although individuals seeking refuge in the litter layer by day may consume prey that venture into their retreats (Stewart and Woolbright, this volume). *Eleutherodactylus portoricensis* forages at night near the ground, and its diet reflects some consumption of litter arthropods (Stewart and Woolbright, this volume). Little foraging in the litter layer by insectivorous birds is apparent (Waide, this volume).

Nonanoline reptiles may account for the majority of vertebrate consumption of litter arthropods. The gekko *Sphaerodactylus klauberi* was recovered in D-Vac collections at a density of 1 individual m^{-2} and appears to represent the most abundant of the nonanoline reptiles. The predominant arthropod taxa recovered from *S. klauberi* stomachs include (in descending order of frequency): Acari, Araneae, Collembola, Isopoda, and Coleoptera (Thomas and Gaa Kessler, this volume). The anguid *Diploglossus pleei,* on the other hand, consumes mostly Diplopoda, Coleoptera, Dermaptera, and Gastropoda. Thomas and Gaa Kessler (this volume) suggest that this anguid may be a millipede specialist. Stemmiulid millipedes (probably *Prostemmiulus heatwoli*) numerically composed 86% of the diplopod prey and 24% of the total stomach contents.

The remaining two groups represented in tabonuco forest litter also may concentrate prey selection on a few arthropod taxa, although this conclusion is drawn on relatively small sample sizes (Thomas and Gaa Kessler, this volume). Half of stomach contents of the amphisbaenid *Amphisbaena caeca* consist of Coleoptera and Formicidae. The latter group, although abundant in the litter layer, formed less than 2% of the stomach contents of *S. klauberi* and *D. pleei.* Blind snakes (*Typhlops*) also exploit ants, but termites may provide the greatest part of their diet (Thomas and Gaa Kessler, this volume).

CONSUMPTION RATES

Tabonuco forest in the Luquillo Mountains of Puerto Rico is apparently subject to a level of aboveground herbivory similar to that reported for most of the world's forested ecosystems. Measurements have generally indicated that 5 to 10% of green leaf tissue is shunted into the folivore-based food web (e.g., Bray 1964; Reichle and Crossley 1967; Misra 1968; Woodwell and Whittaker 1968; Gosz et al. 1972; Funke 1973; Reichle et al. 1973; Leigh

1975; Leigh and Smythe 1978; Nielsen 1978). Odum and Ruiz-Reyes (1970) estimated that chewing folivores had consumed 6% of the surface area of leaves collected from litter baskets at control sites at El Verde. This value reflects a minimum estimate for total aboveground herbivory, since neither consumption by sapsucking insects nor the loss of foliage totally consumed *in situ* or dropped to the litter layer through herbivore-induced greenfall (see Coley 1982; Lowman 1985b; Risley and Crossley 1988) was assessed.

Faunal inventories in El Verde (Odum et al. 1970a; Odum 1970a) demonstrated that approximately half of the total faunal biomass was concentrated in a relatively thin layer of soil and litter. A similar distribution of faunal biomass was reported by Fittkau and Klinge (1973) for central Amazonian rain forest. Although such inventories are laborious exercises prone to a variety of measurement errors, they illustrate the relatively intense faunal activity on and below these tropical forest floors, particularly when considered in conjunction with the potentially rapid turnover rates of tropical litter and soil invertebrates.

The quantitative analysis of arthropods associated with the litter and soil of tropical forests has only begun to receive widespread attention in recent years. Reflecting the wide diversity of tropical forests and sampling techniques, estimates for arthropod densities in tropical and subtropical soils and litter vary greatly. At some sites (e.g., Fittkau and Klinge 1973 in Brazil; Goffinet 1975, 1976 in Zaire; Plowman 1979 in Australia) arthropod densities approach the range commonly seen in temperate forests (Petersen 1982a). Other studies (e.g., Williams 1941 in Panama; Strickland 1944 in Trinidad; Wiegert 1970a in Puerto Rico; Lasebikan 1974 in Nigeria) have reported arthropod densities that were nearly an order of magnitude less. Although Swift et al. (1979) suggested that arthropod standing stocks in the soil and litter of tropical forests may be similar or even exceed those of temperate sites, Petersen (1982b) tentatively indicated that biomass accumulations for tropical soil and litter fauna to be the lowest of seven major biomes. Realistically, however, biomass estimates for whole arthropod communities in litter and soils of tropical forests (Petersen 1982b; Lavelle 1984; Lavelle and Kohlmann 1984) are still too scarce to allow for meaningful comparisons with temperate forests, especially when the diversity of tropical forests, widely appreciated only recently, is considered.

Knowledge of food web structure in litter and soil arthropod communities in tropical forests exists at an even more rudimentary stage. Even in temperate habitats with longer histories of research activity, trophic relationships at specific sites have at best been detailed for small subsets of these communities (Moore et al. 1988; Moore and de Ruiter 1991). The difficulty of establishing the relative strength of trophic links cannot be overstated (Walter et al. 1991). For primary decomposers, the major obstacle lies in the unequivocal demonstration of assimilation for specific plant and microbial constituents ingested into their guts. At the other end of the trophic spectrum, visual examination of gut contents, indispensable for determining prey selection in vertebrates, is fruitless for the dominant arthropod predatory group (arachnids), where prey digestion occurs prior to ingestion. A complete delineation of the feeding relationships for the many species of euryphagic habits and whose trophic links potentially extend throughout the litter subsystem poses an overwhelming undertaking. The most practical approach would entail rigorous examination of the trophic habits of the key species occupying major taxa. In this respect the arthropod community associated with tabonuco forest litter offers an attractive opportunity for this type of approach, because so few species dominate the densities and standing stocks for many of the taxa.

SUMMARY

The interactions between the arthropod community associated with litter and soil in tabonuco forest and other components of the food web extend beyond the physical boundaries of the litter-soil subsystem. The ultimate resource base (plant material) for the litter arthropods originates in the arboreal zone of the forest and population sizes of both decomposer arthropods and their *in situ* arthropod predators are greatly influenced by the nature and quantity of litterfall. In turn, biomass is exported out of the soil-litter system in the form of predation by arboreally based predators that exploit the litter habitat or from the movement of arthropods residing or developing in litter to the arboreal zones where they are preyed upon. Two major examples of the latter case are the nightly migration by cockroaches and crickets from the litter and the emergence of flies and moths as winged adults from the litter during the wet season.

One of the important aspects of the litter-soil subsystem from a food web perspective is the habitat that litter provides for the larval life stages of arthropods which later occupy other forest compartments. This function intertwines the dynamics of litter, understory, and canopy prey populations and creates the possibility of competition for resources among predator species differing in habitat and foraging structure.

The arthropod community occupying tabonuco forest litter manifests a community structure more typical of temperate forests within a habitat structure characteristic of many humid tropical forests. The insular setting has shaped a biotic community in which the species richness of the dominant taxa are more comparable to those of temperate forests in North America and Europe than to mainland tropical forests. The eleven species of millipedes thus far collected at the El Verde site, for instance, fall within the range of species found in several European forests, but well below that recovered from an African site (Ivory Coast). The twenty-three species of spiders occupying the litter and soil habitats at El Verde represent a somewhat impoverished diversity even when compared to many temperate sites (table 5.8). Likewise, the twenty-five species of cockroaches and thirty-six species of ants collected from tabonuco forest litter at El Verde are impoverished in comparison to collections made elsewhere in the Neotropics.

On the other hand, the physical setting that the arthropod community inhabits in tabonuco forest litter is meager compared with litter habitats in temperate forests. The rapid turnover of fallen organic matter on the forest floor results in the accumulation of only a thin layer of litter and virtually none of the humus layer that is typical of temperate sites. The thin layer does not provide the ample spatial refuges that are available to microarthropod populations occupying the thick organic layers of many temperate and boreal forests. The scarcity of refuges from predation may limit the populations of mites and collembolans to relatively low densities in tabonuco forest litter, despite standing stacks of fungal hyphae that are comparable to temperate sites (see Lodge, this volume). The mean annual density (MAD) of collembolans at El Verde (1292 collembolans m^{-2}) lies near the low portion of the range reported from various tropical sites and far below the densities measured at temperate sites. Mite densities (roughly 17,000 individuals m^{-2}) in tabonuco litter and soil are roughly an order of magnitude lower than many forests in North America and Europe.

The abundance of arthropod predators in tabonuco forest litter may be a major factor limiting the number of mites and collembola. Many of the numerically dominant members of this assemblage are minute, with maximum dry masses of around 500 μg or less. These small predators include the spiders *Modisimus montanus, Theotima* sp., *Corythalia gloriae,* and *Heteroonops spinimanus;* two species of pseudoscorpions; predaceous Hemiptera in the families Dipsocoridae, Schizopteridae, and Enicocephalidae; various species of pselaphid beetles; and the ants *Strumigenys gundlachi, Strumigenys rogeri,* and *Anochetus mayri.* Collectively, these predators along represent MAD values of approximately 850 individuals m^{-2}.

Some taxonomic groups are present in tabonuco forest litter in densities that are high relative to those reported for most forest litter habitats, be they temperate or tropical. Macrodecomposers such as millipedes (MAD = 300 individuals m^{-2}), isopods (MAD = 549 individuals m^{-2}), and cockroaches and crickets (collectively 113 individuals m^{-2}) manifest densities well in excess of those reported at other Neotropical sites. Similarly, the major predaceous arachnid taxa—spiders (MAD = 356 individuals m^{-2}), pseudoscorpions (MAD = 262 individuals m^{-2}), and the largely raptorial assemblage of opilionids (MAD = 146 individuals m^{-2})—appear far more abundant at El Verde than other Neotropical sites and even fall into the upper range of densities reported from temperate forest litter.

Ants are second to mites as the most numerous major taxa occupying the litter and soil of tabonuco rain forest. In litter, ant densities amounted to a

MAD of about 1200 workers m^{-2}, while in the soil Wiegert (1970a) reported densities of about 2250 workers m^{-2}. *Solenopsis azteca pallida* accounted for 90% of the individuals recovered from soil samples. In the litter the large predaceous ponerine *Odontomachus brunneus* comprises about 67% of total ant biomass (roughly 0.2 g m^{-2}) in tabonuco forest litter. Notably absent from Puerto Rico are army ants, which exert a strong influence on arthropod community structure in Central and South American forests.

With a less pronounced dry season than in Central American forests, seasonal fluctuation in arthropod densities was less apparent in tabonuco forest at El Verde. Ant densities, for example, exhibited a coefficient of variation of only 13.6% for the eight sampling periods over the course of an annual cycle. The greatest seasonal variation in numbers was exhibited by armored scales, flies, and moth larvae. Armored scale densities rose dramatically in the April samples, possibly due to their input in the litter with the falling leaves of certain trees. Large increases in fly (to about 2100 individuals m^{-2}) and moth (to about 350 individuals m^{-2}) densities in mid-June 1984 were likely a response to the resumption of moderate rainfall following a period of greatly reduced rainfall that began in late February. The ensuing emergence of adults from the dense larval populations present in June may have been the major influence in the near-doubling of web-building spider populations in low understory from late June to late August of 1984 (Pfeiffer, chap. 7, this volume).

Some tantalizing but meager data suggest that earthworms may represent the dominant taxon in tabonuco forest in terms of biomass. Odum (1970c) concluded that earthworms contribute 36% (4.2 g m^{-2}) of the total faunal biomass in the forest, but the evidence for this rests on but six soil monoliths that were exhumed by Moore and Burns (1970). This is clearly an area in need of further research.

Little specific information on the food web linkages exists for the invertebrates of the litter and soil habitats in tabonuco forest. Despite the relatively low species richness of the litter arthropod community, the determination of specific trophic linkages within and without this assemblage remains a formidable task. Determination of prey taken by the major predatory group in the litter subsystem is greatly complicated by the external digestion of prey by arachnids. Resource selection by the saprotrophic and/or microbitrophic arthropods residing in tabonuco forest litter is likely generalized and opportunistic, further complicating an inventory of trophic linkages. However, some species that tend to be primarily saprotrophic may preferentially select decomposing leaves of particular tree species or organic complexes with particular combinations of dead leaves and microbial species. Primary decomposer species, such as millipedes and isopods, may enhance the decomposition of certain plant species through their comminutory activities and thus enhance nutrient cycling on the forest floor, particularly if dominant litter arthropods and tree species are involved.

Appendix 5 Study Site and Sampling Methods

Litter arthropods were sampled from a 1 ha plot in the Luquillo Experimental Forest designated as Plot 3 (see chap. 1). The center of this experimental plot is situated within 200 m of the Quebrada Sonadora, and elevation within the plot ranges from 350 to 380 m above sea level (fig. 1.1). Basal areas, densities, and leaf fall for the eight dominant tree species within this plot are listed in descending order of species importance value in table A5.1.

The floor of tabonuco forest is typically covered with only a thin, often discontinuous litter layer that usually lies directly on mineral soil without an intervening humus layer. Where the litter layer accumulates in depressions and around protruding boulders and vegetation, a relatively thin humus layer may form. Stones and boulders, root mats, vines, and spreading tree buttresses frequently protrude from or underlie the soil surface. As an indication of rock coverage, 10.5% of the 400 collection sites in Plot 3 that were initially selected on a random basis during this study fell on bare rock surface. The complex forest floor structure of tabonuco forest creates formidable barriers to a randomized sampling of soil and litter arthropods if core tubes are used, particularly for core diameters appropriate for the accurate quantification of macroarthropod populations. Manual collection of litter for Tullgren extraction was not considered to efficiently sample many of the minute arthropods that numerically dominate the litter community.

Consequently, a sampling methodology involving the use of a D-Vac suction sampler (Dietrick 1961) and extraction funnels was adopted. The D-Vac apparatus consists of a small gasoline motor mounted on a backpack frame. Suction generated through the air intake is transferred to the ground via flexible tubes and a plastic cone. At approximately five-week intervals from 20 January 1984 to 20 January 1985, coordinates for forty plots were established from a random number table (Rohlf and Sokal 1969). An alternate site was sampled when the original plot coor-

Table A5.1. Basal area, density, and leaf fall of tree species in Plot 3, El Verde

Species	Basal Area (m^2 ha^{-1})	Density (individuals ha^{-1})	Leaf Fall (g m^{-2} y^{-1})
Dacryodes excelsa	5.8	108	102.7
Prestoea montana	2.5	145	—
Sloanea berteriana	2.8	83	—
Guarea trichilioides	5.1	38	60.2
Manilkara bidentata	1.7	64	36.8
Buchenavia capitata	3.0	19	—
Homalium racemosum	2.6	19	—
Croton poecilanthus	1.1	51	—
Total all species	34.4	815	509.3

Source: From Zou et al. (in press).

Note: Basal areas and densities include individuals of at least 10 cm dbh (diameter at breast height)

dinates coincided with a bare rock surface. The "November-December" collections mentioned in previous sections refer to a single series of samples that were collected over a period of several days because of inclement weather and consequently encompassed both months.

During the collection of litter and associated arthropods from these plots, the nozzle of the D-Vac (sampling area = 295 cm^2) was pressed against the forest floor and suction was applied for 10 seconds. Material sucked from the enclosed plot were trapped in a nylon organdy collection bag, for which the average mesh diameter equaled 200 μm. Arthropods were later extracted into 75% ethanol from litter supported in Tullgren funnels by screens with a coarse (5 mm) mesh size. Following funnel extraction, arthropods retained above the screen in the litter fraction were sorted out. Recovery of macroarthropods through this manual sorting indicated that extraction efficiencies by the Tullgren funnel technique generally ranged from 70 to 90%.

Litter samples and 2 cm diameter soil cores, taken to a depth of 5 cm, were removed from the perimeter of the study site and dried at 60° C for 10 days to determine moisture content. Rainfall was recorded on a nearly daily basis at the El Verde field station, approximately 400 m from the study site.

6

Arboreal Invertebrates

Rosser W. Garrison and Michael R. Willig

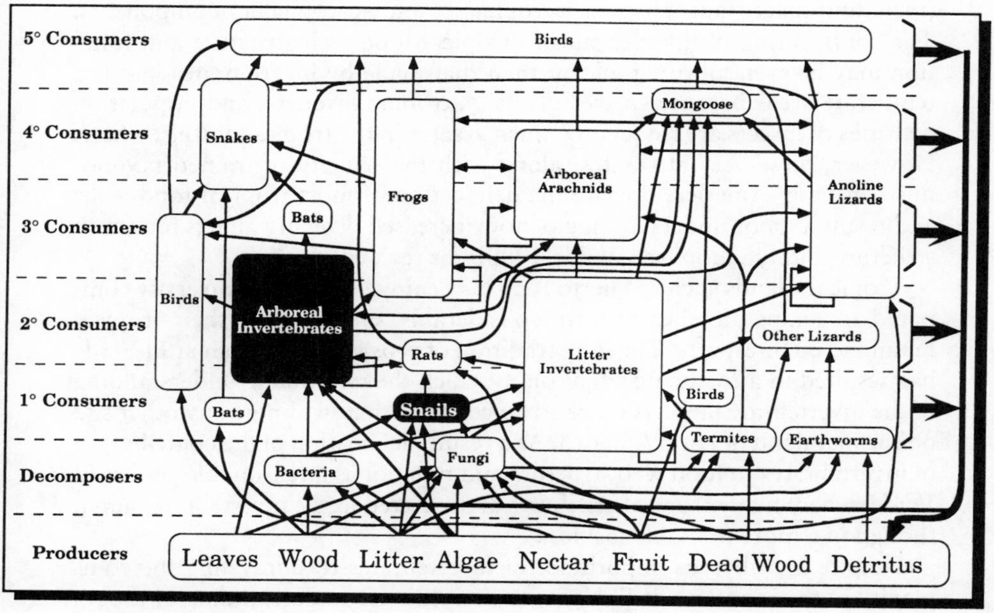

INVERTEBRATES make critical contributions to the structure and function of most ecosystems. Their dominance among consumers is derived from their high diversities, densities, and reproductive rates, as well as from their occupation of most consumer trophic categories within communities. Indeed, invertebrate size spans several orders of magnitude, with feeding specializations that include herbivore (folivore, granivore, frugivore, nectarivore), carnivore (predators, parasites, and parasitoids), and detritivore (including macrosaprophagic, necrophagic, and scatophagic) components. The contribution of invertebrates to tropical food web structure and function may be even more significant than that made by invertebrates in more temperate areas because species diversity, trophic diversity, and population densities of insects are frequently much greater in the tropics than elsewhere. However, these very attributes, along with the paucity of trained taxonomists in Latin America, obviate the delineation of most tropical food webs and result in poor understanding of how increased diversity affects food web structure and function (see Wolda 1983b for review).

Tropical islands such as Puerto Rico may enjoy high species diversity compared to temperate sites but are considerably simpler than their tropical mainland counterparts. The long tradition of ecosystem research at El Verde has resulted in a fortunate situation in which the taxonomic understanding of the invertebrate fauna is more advanced than that in almost any other site on the mainland of Latin America. As a result, delineation of the contribution of invertebrates to food web structure and function is a reasonable goal at El Verde, where many tropical attributes are reflected in the invertebrate fauna, though in a more tractable fashion.

Despite the obvious importance of invertebrates to unraveling the complexity of the food web at El Verde, comprehensive understanding of the relationships of most species with their environment is elusive. The ecology of even the more abundant species is poorly known compared to that of other consumer groups in the forest. Although Drewry (1970b) recorded over 1,200 species of insects from El Verde, few of them are seen by casual observers. Many species are rare or infrequent, and most are known from few specimens. The true number of species of invertebrates at El Verde must be considerably higher than the number reported in the literature (Owen 1983).

The most readily seen invertebrates at El Verde are the large camaenid snails (*Caracolus caracolla*), walking sticks (*Lamponius portoricensis*), cockroaches, arboreal crickets, various fulgoroid plant hoppers, and the pierid butterfly, *Dismorphia spio* (fig. 6.1).

The complexity of tropical food webs may be underestimated because species remain undetected or unnamed, or two or more taxa masquerade under

Figure 6.1. Prominent invertebrates in the forest at El Verde. (*a*) camaenid snail (*Polydontes acutangula*), (*b*) walking stick (*Lamponius portoricensis*), (*c*) pierid butterfly (*Dismorphia spio*), (*d*) cockroach (*Epilampra wheeleri*), and (*e*) butterfly (*Siproeta stelenes*).

a single name. Biased sampling procedures also compromise ecological and taxonomic studies of invertebrates, resulting in lower values for species richness, simplified views of taxonomic composition, and lower estimates of population densities. This bias is often habitat- or microhabitat-specific. Near ground sites are more adequately sampled (e.g., Beebe 1916; Williams 1941; Schubart and Beck 1968; Beck 1971; Willis 1976; Pfeiffer, chap. 5, this volume) than are canopy sites (e.g., Wolda 1979; Erwin and Scott 1980; Erwin 1983a, 1983b; Shelly 1988). Preliminary data suggest that the canopy harbors greater numbers of invertebrate species (e.g., Erwin 1983a, 1983b) and individuals (Rees 1983; Sutton 1983; Shelly 1988) than near ground-level sites, but great difficulties characterize sampling such locations. Erwin (1983b) described the difficulties of sampling arboreal insects.

In discussing arboreal invertebrates, we include in this chapter those that are aerial or occur in the canopy, understory, or shrubbery above the forest litter. Some invertebrates (e.g., springtails) have life stages which occur in the litter as well as in the understory. Others, such as millipedes, centipedes, and certain ants, occur in both litter and understory. Because most of these animals are primarily inhabitants of litter, they are dealt with in chapter 5. Termites (Isoptera) are considered in chapter 4 and arboreal arachnids in chapter 7.

HISTORY

Most biological data on invertebrates at El Verde are qualitative. Martorell (1975) listed plant-insect relationships gleaned from literature referring to insects in Puerto Rico, and Velez (1979) provided a bibliography of the entire literature dealing with the invertebrate fauna of the island.

The first field studies on the invertebrate fauna of El Verde are probably those of McMahan and Sollins (1970), who assessed species diversity of soil microarthropods within and outside of the Radiation Center. They concluded that low-level irradiation did not result in lower diversity of organisms compared with control (non-irradiated) areas. Drewry (1970b) provides a list of over 1,200 insect species known from El Verde. The list was the result of several years of work by George Drewry and Robert Lavigne, who collected or reared thousands of specimens, many of which were sent to specialists for identification. Voucher specimens were deposited at the El Verde Field Station insect collection, which continues to be a resource for other entomologists. Appendix 6 is an expansion of their work, based on more recent collections and identifications of several groups by specialists.

Some common groups have received attention because of their conspicuousness or abundance. Heatwole and Heatwole (1978) described the biology of the large camaenid snail, *Caracolus caracolla*. Lavigne (1970b, 1977) provided information on ant ecology and diversity at El Verde. The foraging activities of some Puerto Rican ants (not from El Verde) are discussed by Torres and Canals (1983) and Torres (1984a, 1984b); Willig et al. (1986) detailed the population dynamics of the common walking stick, *Lamponius portoricensis*. Lister (1981), in an analysis of niche relationships among three species of *Anolis* lizards in the tabonuco forest near El Verde, used sticky traps and collections by sweep net to measure arthropod diversity, abundance, and biomass. He found no significant differences among arthropod diversity, numbers, and biomass between wet and dry seasons. Araneida (spiders), Orthoptera (crickets, etc.), Coleoptera (beetles), and Diptera (flies) were dominant components, both in numbers and biomass during wet and dry seasons. However, Lister's analysis was focused at the ordinal level and most likely would underestimate differences in these categories, compared to analyses to familial, generic, or specific levels. Liebherr (1988) edited a volume on the zoogeography of insects of the Caribbean, in which ten authors reviewed the biogeographic history of various insect groups. Willig and Camilo (1991) documented declines of six invertebrate species at El Verde and Torres (1992) described outbreaks of various lepidopteran species (especially *Spodoptera eridana*) following Hurricane Hugo.

TAXONOMIC STATUS OF INVERTEBRATES AT EL VERDE

The systematics of many insect groups are well known for Puerto Rico compared to other islands of the Greater Antilles (Cuba, Hispaniola, Jamaica). Principal works on systematics of Puerto Rican invertebrates that are of importance to the El Verde fauna are given in table 6.1. This list, although not exhaustive, gives an idea of the uneven taxonomic coverage for invertebrate groups. For example, taxonomic treatment of auchenorrhynchous Homoptera is considered reasonably complete (Ramos 1988), but virtually nothing has been published on the microlepidoptera.

An updated and expanded list of known invertebrate taxa from El Verde appears in appendix 6, and incorporates the data in Drewry (1970b). The update incorporates nomenclatural changes, where possible, and comprises over 1,560 species, an increase of approximately 30%, compared to the over 1,200 species listed by Drewry (1970b).

The higher classification of some insect groups is in dispute. For example, the order Orthoptera (grasshoppers) is considered by some (e.g., Borror et al. 1989) to include walking sticks, cockroaches, mantids, and others, whereas we generally follow the classification of the Insects of Australia (CSIRO 1991). Unfortunately, the systematics of many holometabolous insects (e.g., Coleoptera, Lepidoptera, Diptera, Hymenoptera) has lagged far behind that of the hemimetabolous forms (e.g., Odonata, Hemiptera, Homoptera, Orthoptera). Dissimilar morphology, and differences in habits, habitat use, and species richness of these groups render a more complete knowledge of the

Table 6.1. Taxonomic references to invertebrate fauna of the El Verde Field Station and surroundings

Taxon	Reference	Comments
General	Martorell 1945a,b	Thorough account of insect pests of forests
General	Wolcott 1948	Anecdotal account of insects of Puerto Rico, mostly out of date
General	Martorell 1975	Listing of food plants of insects recorded in literature
General	Velez 1979	Bibliography of taxonomic works for Puerto Rico
Aquatic biota	Hurlbert and Villa-Figueroa 1982	Compilation of taxonomic references by many specialists
Mollusca	van der Schalie 1948	Keys, descriptions
	Aguayo 1961	Puerto Rican snails
	Enrique de Jesus 1987	Taxonomy of land snails of Caribbean National Forest
Araneae	Petrunkevitch 1929, 1930a,b	Descriptive treatment of the spiders of Puerto Rico, with keys
	Archer 1965; Banks 1896, 1914; Bryant 1940, 1942, 1943, 1945, 1947a,b, 1948; Chickering 1945, 1964, 1968a,b,c,d, 1969a,b, 1970, 1972a,b,c; Coddington 1986; Exline & Levi 1962; Lehtinen 1967; Levi 1955a,b, 1957, 1959, 1962, 1963a,b,c, 1971, 1977, 1978, 1980, 1981, 1986a,b; Levi & Randolph 1975; Opell 1979, 1981, 1984; Platnick 1974; Roewer 1951; Shear 1978	Descriptions of spiders
Acarina	Cromroy 1958	Plant mites
Collembola	Wray 1953	New species described
	Mari Mutt 1976, 1987	Keys to genera, descriptions
Ephemeroptera	Travers 1938	Keys
	Peters 1971	Revision of West Indian Leptophlebiidae
Odonata	Klots 1932	Keys to adults, larvae
	García-Diaz 1938	Ecology of adults, larvae
	Garrison 1986	Updates Puerto Rican Odonata
Blattodea	Rehn & Hebard 1927	Keys, descriptions
Orthoptera	Otte 1981	Monograph of North American Orthoptera
Thysanoptera	Medina-Gaud 1961, 1963	Keys, descriptions
Hemiptera	Barber 1939	Keys
	Drake & Maldonado 1954	Waterstriders
	Martorell 1955	Describes new Tingid
	Maldonado-Capriles 1969	Keys to Puerto Rican Miridae
Homoptera	Davis 1928	Cicadas of Puerto Rico
	Osborn 1935	Keys, descriptions
	Caldwell & Martorell 1950a	Keys, descriptions of Cicadellidae
	Caldwell & Martorell 1950b	Keys, descriptions of Fulgoroidea
	Caldwell & Martorell 1951b	Descriptions of new leafhoppers
	Young 1953	Descriptions of *Empoasca* leafhoppers

	Ramos 1957	Keys, descriptions of Membracidae, Cercopidae, Kinnaridae
	Smith 1960, 1970	Describes new aphids
	Smith et al. 1963, 1971	Keys, descriptions of Aphididae
	Caldwell 1942	Describes psyllids
	Caldwell & Martorell 1951a	Review Psyllidae of Puerto Rico
	Nakahara & Miller 1981	Lists Coccoidea of Puerto Rico
Coleoptera	Hlavac 1969	Treatment of genus *Scarites*
	Peck 1970, 1972	Leiodidae
	Matthews 1965, 1966	Keys, descriptions, biogeography of Scarabaeinae
	Chalumeau 1978, 1982, 1985	Descriptions, notes of Antillean Scarabaeidae
	Chalumeau & Gruner 1976	Descriptions of melolothines and rutelines of Antilles
	Chapin 1940	Revision of West Indian Aphodiinae
	Ratcliffe 1976	Revision of West Indian *Strategus*
	Blake 1941, 1943, 1948, 1950, 1951, 1952, 1953, 1964, 1970	Descriptive treatment of various Chrysomelidae
	Bright 1985	Descriptions of new Scolytidae
	Equihua-Martínez & Atkinson 1987	Catalog of North and Central American Platypodidae
Trichoptera	Flint 1964, 1992	Keys, descriptions
Lepidoptera	Forbes 1930	Microlepidoptera
	Schaus 1940a,b	Macromoths
	Comstock 1944	Butterflies
	Riley 1975	Butterflies
	Ramos 1982	Checklist of Puerto Rican butterflies
Diptera	Curran 1928, 1931	Flies, keys, and descriptions
	Alexander 1932	Descriptions of Tipulidae
	Fox 1946	Culicidae
	Snyder 1957	Keys, descriptions of *Neodexioopsis*
	Wheeler & Takoda 1963	Revision of *Mycodrosophila*
	Drewry 1969b,c	Keys to Dolichopodidae and Muscidae of El Verde
	Borgmeier 1969	Phoridae
	Romero & Ruppel 1973	New species of Lonchaeidae
	Telford 1973	Syrphidae, keys
	Thompson 1981	Syrphidae of the West Indies
Hymenoptera	Wheeler 1908	Formicidae of Puerto Rico
	Smith 1936	Formicidae of Puerto Rico
	Smith & Lavigne 1973	New species of ants
	Lavigne 1970b, 1977	Key, Formicidae of El Verde
	Bohart & Strange 1965	Revision of *Zethus*

Note: This list is not intended to be exhaustive.

immature stages of the holometabolous forms almost impossible. A more thorough knowledge will be acquired only by careful rearing of immature stages to adult (see Janzen 1988).

DIVERSITY

The arboreal invertebrate fauna of El Verde in particular, and of Puerto Rico in general, can be considered depauperate compared to mainland tropical forests (Martorell 1945a; Allan et al. 1973; Waide 1987). Examples from four insect groups illustrate this statement.

Odonata

Dragonflies and damselflies are large, predaceous insects which are most speciose in the tropics. Their taxonomy in the Antilles is well known, and few species likely remain to be discovered in Puerto Rico. Garrison (1986) cited forty-nine species for the island, but only ten species (app. 6) occur at El Verde. In contrast, Paulson (1982) listed 228 species from Costa Rica, and the probability of finding new records and species there seems high. Infrequent collecting at one site in Rondônia State, Brazil, has revealed over 130 species (pers. observation). As with the Lepidoptera, Puerto Rico and the other Antilles lack characteristically Neotropical families. No calopterygids of the genus *Hetaerina* (ruby spots) are known from the Antilles, although thirty-seven species are known from the United States south through South America (Garrison 1989). Other families known from the mainland tropics include Polythoridae, Platystictidae, and Perilestidae, all unknown from the Antilles. A dominant genus of the American mainland tropics is *Argia* (Coenagrionidae). Over 110 species are known, with at least forty to fifty new taxa to be described. However, only one endemic Lesser Antillean species, *A. concinna,* is known from the Caribbean.

Homoptera

Another relatively well known insect group at El Verde is the flying auchenorrhynchous Homoptera, including cicadas (Cicadidae), tree hoppers (Membracidae), leaf hoppers (Cicadellidae), and plant hoppers (Fulgoroidea). We have recorded only sixty-four species for El Verde, compared with at least 120 species from a site in central Sulawesi (Rees 1983).

Puerto Rico has an even lower diversity of sternorrhynchous forms (Aphididae, Coccidae, Diaspididae). The aphids (Aphididae) comprise only one species at El Verde, although Smith et al. (1963) cite several records from El Yunque. Dixon et al. (1987) addressed the problem of low species diversity of this family in the tropics. They attributed nonuse of most rare host plants to constraints in aphid biology (i.e., short life cycles, inability to live long without food, high degree of host specificity, and low efficiency in locating proper host plants). Aphids are not favored in tropical communities, they argue, because high plant diversity and low numbers of plants per species are generally the rule.

The same reasoning may apply to coccids and diaspidids. Adult females in these families are completely sessile. The only coccid thus far found at El Verde is *Ceroplastes rubens,* a widespread species which is highly polyphagous (Gimpel et al. 1974). Similarly, only one diaspidid has been found at El Verde: a heavy infestation of the black thread scale, *Ischnaspis longirostris* on *Guarea guidonia* during March 1942 (Martorell 1945a).

Coleoptera: Cerambycidae

Longhorned beetles are attractive insects and favorites with collectors. The family is speciose; all species are phytophagous. Hovore (1989a,b) provided a list of all species of the family from the Monteverde Cloud Forest, Costa Rica, collected from 1974 to 1989, and from the Turrialba region of Costa Rica. The Monteverde region contains at least 225 species, and Hovore speculates that perhaps 25 to 50% more will be found at the site. A total of 348 species has been recorded from Turrialba, and Giesbert and Hovore (pers. comm.) record about 400 species from eight years of collecting 10 to 15 km north of El Llano, Pana Province, Panama. In stark contrast, there are thus far only nineteen species of cerambycids known from El Verde.

Lepidoptera

About 1,560 species of invertebrates are recorded from El Verde (app. 6), yet Janzen (1988) records 3,142 species of Lepidoptera alone from an approximately 100 km^2 area within Santa Rosa National Park, Costa Rica. Even a continental temperate site such as Ithaca, New York, is credited with 1,577 species of Lepidoptera (Janzen 1988), yet we found only 234 species of Lepidoptera at El Verde. Admittedly, our investigation was not as thorough as those for Costa Rica or New York, but our fauna is only about 7% as rich as that of Costa Rica, and about 15% as rich as that of New York. Perhaps a more accurate comparison can be made when comparing the butterflies, including skippers (Hesperiidae). These showy diurnal insects are well known, and estimates of species numbers are probably more accurate. We have found only twenty-six species at El Verde, compared with 345 in Costa Rica and 105 in New York (Janzen 1988). Again, species richness at El Verde is only 8% that of Costa Rica and 25% that of New York. Only 106 species of butterflies occur in Puerto Rico (Ramos 1982), and only about 300 species occur throughout the Antilles (Riley 1975), compared with over 600 species in Trinidad alone (Barcant 1970) and 1,500 to 1,600 species from a 750 ha tract in Rondônia, Brazil (Emmel and Austin 1990). The unusually high number of butterfly species recorded from Trinidad is probably due to its proximity to Venezuela.

Puerto Rico also contains no unusual elements in its lepidopteran fauna, and some characteristic, predominantly New World taxa are lacking. Owl butterflies (Brassolinae), morphos (Morphinae), and ithomiid butterflies (Ithomiinae) are absent, and the island has only one species of satyr (Satyrinae). Similarly, only one species of saturniid moth (Saturniidae) is found in Hispaniola (Ferguson 1971), compared with thirty-five species in Santa Rosa National Park, Costa Rica, and eleven species in Ithaca, New York (Janzen 1988).

DENSITY

Data on the density of arboreal invertebrates at El Verde are known for only two groups, snails and the large walking stick, *Lamponius portoricencis*.

Snails

Recent work (Alvarez 1991; Cary 1992; Alvarez and Willig 1993; Alvarez and Willig unpublished; Willig et al. unpublished) focuses on the population and community ecology of common snails (*C. caracolla, Nenia tridens, Austroselenites alticola, Megalomastoma croceum,* and *Subulina octana*) and the community ecology of all terrestrial snails at El Verde. Alvarez (1991) and Alvarez and Willig (1993) identified seven species of snails at El Verde that were not previously recorded for the tabonuco forest (*Lamellaxis micra, Opeas pumilum, Nesovitrea subhyalina, Guppya gundlachi, Habroconus ernsti, Striatura meridionalis,* and *Chondropoma riisei*). These taxa may have been absent from earlier inventories at El Verde in part because of their small size (diameter or length less that 5 mm) and in part because of their soil or litter microhabitat associations during the day. In the tabonuco forest at Bisley, the densities of the common snails *Nenia tridens, Gaeotis nigrolineata,* and *C. caracolla* were 6.2, 0.7, and 3.8 individuals 100 m^{-2}, respectively; moreover, each species is significantly hyperdispersed (Willig and Camilo 1991).

Based on quadrats arranged along transects that bisected thirteen light gaps at El Verde, Alvarez (1991) and Alvarez and Willig (1993) could evaluate the density response of the five common snail species to changes in cover. Three species (*A. alticola, M. croceum,* and *S. octana*) did not significantly differ in density between light gaps and the surrounding forest matrix. In contrast, two species did respond to light gaps created by treefalls. The abundance of *N. tridens* was significantly higher in gaps, whereas that of *C. caracolla* was significantly higher in the surrounding forest. Differences in microhabitat distribution may be attributable to factors related to diet and body water loss rates.

Substrate selection by each of the five common snail species was compared separately in the wet and dry seasons by Alvarez (1991) and Alvarez and Willig (1993). Substrate was classified into four categories: litter or topsoil, rock, live plant material, and dead plant material. Differences among species in substrate selection were identical in both seasons in that two statistically distinguishable groups of snails were produced. The first group comprised *A. alticola, M. croceum,* and *S. octana:* these snails may be considered forest floor specialists because they were collected in litter or topsoil over 85% of the time in each season. The second group comprised *C. caracolla* and *N. tridens:* these snails were captured more frequently in plant material above the forest floor. In particular, *N. tridens* was highly associated with dead plant material, and was collected from this substrate more than 70% of the time in either season. *Caracolus caracolla* exhibits a seasonal change in substrate associations: 53% of individuals were collected from the litter or topsoil in the dry season, whereas 45% were collected from live plant material in the wet season.

Two levels of community analysis were undertaken by Alvarez (1991) and Alvarez and Willig (unpublished). The first focused on the five common species and the second examined the entire assemblage of snail species. In the former case, they were able to distinguish between quadrats occurring in gaps and those occurring in the undisturbed forest based upon the joint densities of the common taxa. This suggests that these habitats harbor different assemblages of snails and may represent different spatial compartments within the detrital food web. Nonetheless, distinctions between gaps and undisturbed forest were not obtained when the entire snail fauna was considered in community-level analyses, in part because many rare species overwhelmed any pattern based upon the common taxa. It may also have been related to variation among gaps in microhabitat attributes which change during secondary succession. Such variation may obviate the production of distinct assemblages for such a diverse fauna.

Walking Sticks

Willig et al. (1986) examined the population structure of one deme of the only common walking stick, *Lamponius portoricensis,* in a small light gap (100 m^2) at El Verde. They found an average of 0.4 to 1.0 walking sticks m^{-2} during the wet season. Individuals moved an average of 0.5 m day^{-1} and were generally restricted to their host plants. In a nearby part of the tabonuco forest (Bisley), Willig and Camilo (1991) estimated the densities of *L. portoricensis* and *Agamemnon iphemedia,* based upon minimum numbers known alive in each of forty circular quadrats (78.54 m^2), to be 0.034 and <0.001 individuals m^{-2}, respectively. The lower density at Bisley than at El Verde for *L. portoricensis* is attributable to two methodological differences between the studies. First, the survey regime at Bisley predominantly sampled undisturbed forest with some light gaps, whereas at El Verde the entire grid was located in a single light gap. Second, the minimum number known alive technique used at Bisley is likely an underestimate of ecological density because it is based on a single survey; in contrast, the multiple mark and recapture rates used at El Verde adjust population estimates based on recapture rates during repeated surveys and is a more accurate measure of ecological density. Finally, each species of walking stick was significantly hyperdispersed at Bisley and El Verde (Willig et al. 1986; Willig and Camilo 1991).

POST-HURRICANE EFFECTS

On 18 September 1989, Hurricane Hugo with sustained winds of 166 km h^{-1} (Scatena and Larsen 1991) passed over the eastern end of Puerto Rico resulting in extensive damage to the rain forest at El Verde. This hurricane had a dramatic effect on the invertebrate fauna because mature forest was heavily damaged and there was, soon after, a luxuriant growth of secondary or early successional vegetation. Documentation of response to this disturbance is given by Willig and Camilo (1991) for five species of forest snails and two species of walking sticks, by Torres (1992) for larvae of various lepidopteran species, by Schowalter (1994) for canopy phytophagous insects, and by Perfecto and Camilo (in press), for ants.

Willig and Camilo (1991) noted significant decreases in population densities of two walking sticks (*Lamponius portoricensis* and *Agamemmon iphimedia*) and three of four species of snails in the season before Hurricane Hugo (July–August 1989) as compared to similar samplings ten to eleven months after the hurricane. All of these species suffered reductions of up to 75%. The large reductions were a consequence of direct effects of the hurricane (e.g. dislodging of snails and walking sticks by strong winds), as well as of indirect effects (substantial alteration of habitat manifested by reduction of food sources and increased insolation due to the destruction of forest canopy). The dramatic proliferation of low, early successional plant species can present extremely favorable conditions for rapidly reproducing phytophagous insects. Torres (1992) reported substantial increases in population densities of the noctuid moth, *Spodoptera eridania* by April 1990. Larvae of this moth are known to feed on at least fifty-six species of plants from thirty-one families (Torres 1992). Four plant species, *Phytolacca rivinoides, Impatiens wallerana, Ipomoea tiliacea,* and *Cestrum macrophyllum,* were especially abundant, and many sites with these plants suffered moderate to complete defoliation by *S. eridania.* After Hurricane Hugo, canopy lepidopterans, predaceous beetles, and decomposers were more abundant in standing trees than in gap areas, whereas sapsucking insects were more abundant in canopy gap areas (Schowalter 1994). One introduced species of ant, *Wasmania auropunctata,* became the dominant ant species representing 94% of the individual ants collected at transects at El Verde in March 1990, less than one year after the hurricane (Perfecto and Camilo in press). Pre-hurricane sampling in the summer of 1989 yielded eighteen species of ants from 120 sites. Two common species, the endemic *Linepithema mellea* (formerly *Iridomyrmex melleus*), an associate of the sierra palm, *Prestoea montana,* and the epigaeic *Pheidole moerens* were the most commonly found species before Hurricane Hugo. Stomach analysis of *E. coqui* collected in June 1990 revealed that *Wasmania* comprised a major component of the ant diet (pers. observation).

A comprehensive two-year study of the autecology of *C. caracolla* (Cary 1992; Willig and Cary unpublished) was conducted at El Verde on three grids (374.68 m^2), beginning two years after Hurricane Hugo. Grids were selected based on hurricane damage. In general, the grid less affected (based on tree damage and canopy openness immediately after the hurricane) consistently had higher snail densities (141 to 182 individuals per grid) than did either of the other two more disturbed sites (91 to 139 and 88 to 126 individuals per grid). Nonetheless, survivorship (between seasons) on the three small grids was indistinguishable during the course of the study (survivorship averaged 0.56). Snails grew more slowly on the less disturbed grid (mean growth rate, 2.38 mm y^{-1}) than on the disturbed grids (mean growth rate, 4.99 mm y^{-1}), in part as a consequence of increased resource levels derived from fallen trunks and limbs, as well as because of increased density of early successional shrubs in disturbed grids. Simulation analyses indicated that snails exhibited site fidelity and have home ranges (Minimum Convex Polygon Method; Cary 1992) that are significantly smaller than those expected by chance alone on all three grids. However, after controlling for the effects of season, snail size, and number of captures, analysis of covariance detected a significant difference between the two disturbed grids as a group (mean, 4.30 m^2) and the less disturbed grid (mean, 9.50 m^2), but no difference between the two disturbed grids (4.06 versus 4.53 m^2) The same statistical results were obtained when attention was restricted to foraging home range (day retreats were not included in calculations of home range). Hence, snails traverse a smaller range in search of forage and retreat sites in disturbed grids; if this translates to reduced energy costs, it likely contributes to the higher growth rates enjoyed in disturbed sites.

AGE STRUCTURE

Virtually no data on age structure of invertebrates exist for El Verde. Some species, such as ants, are undoubtedly continuously brooded, while others are synchronously brooded. We compared results gathered from the two-week sticky trap survey during 9–22 June 1981. Nineteen 5 oz. plastic cups were covered with Tanglefoot® sticky trap adhesive and suspended at 1 m intervals on a string parallel to the El Verde Tower. Samples were collected at 0900 and 1800 hrs. over a two-week period (ten days, eight nights), excluding weekends. Our results indicated that one taxon, phorid flies, is probably synchronously brooded. The data indicate a sudden mass emergence or mass flight over a short period of time. Only one specimen was collected on 9 June, one on 10 June, two on 11 June, six on 12 June, and 285 on 15 June (the next sampling period). Peak density of 913 was reached on 16 June, but numbers fell to eighty-four on 17 June. It is not known if several broods occur throughout the year, or whether mass emergence is restricted to the wet season.

Other invertebrates are long-lived. The large snail, *Caracolus caracolla,* apparently lives an average of three to six years (Heatwole and Heatwole 1978). One adult specimen was recaptured seven years and four months after initial marking. Because the individual was at least three years old when marked (minimum time to reach maturity), its total age was over ten years. As mentioned earlier, the life span of the Central American chrysomelid beetle, *Chelobasis perplexa,* is probably about two years (Strong 1983), and some species of adult *Heliconius* butterflies are known to live six months (Ehrlich and Gilbert 1973).

SEASONALITY

Several studies have stressed the differences in invertebrate abundance throughout the year in the tropics (e.g., Janzen and Schoener 1968; Allan et al. 1973; Janzen 1973a,b; Wolda 1978a,b, 1979, 1980a,b, 1983b; McElravy et al. 1981; Penny and Arias 1982; McElravy et al. 1982; Wolda and Flowers 1985), but fewer studies have been conducted on island faunas (Allan et al. 1973; Frith 1975; Janzen 1973a,b; Tanaka and Tanaka 1982; Snyder et al. 1987; Stewart and Woolbright, this volume). In general, these studies indicate seasonality in size of many invertebrate populations and corresponding increases in species diversity, abundance, and biomass during the rainy season. Studies reported by Snyder et al. (1987) and Stewart and Woolbright (this volume) have shown this to be true at El Verde. Janzen and Schoener (1968) report high habitat specificity in many insect species during the dry season, but with the onset of rains they leave moist riparian areas to repopulate previously dry areas (Janzen 1983b). However, not all insects follow a seasonal trend. The Panamanian cicadellid, *Polana scinna,* showed no detectable differences in numbers trapped throughout the year, whereas other species, even other congeners, did (Wolda 1980a).

In tropical areas where the dry season is not marked, fluctuations in invertebrate populations numbers may be less evident. This may be the case for certain mayflies (Wolda and Flowers 1985). In contrast, many dragonflies and damselflies are highly seasonal. Adult platystictids, *Palaemnema desiderata* and *P. paulitoyaca,* are present as adults only during the rainy season in Mexico (Garrison and Gonzalez pers. observation). The onset of rain can trigger emergences of various stream and lake species (pers. observation). Tanaka and Tanaka (1982), in their study of arthropod abundance in Grenada, found that most species increased in number about two weeks after rainfall. In the tabonuco forest at El Verde, increased insect abundance was detected in February in the middle of the dry season (Snyder et al. 1987). Stewart and Woolbright (this volume) provide further comparisons of availability of invertebrate prey between wet and dry seasons.

Species diversity may correspondingly be expected to increase during the rainy season. This has been shown for Hemiptera (Janzen 1973b) and Coleoptera in Costa Rica (Janzen 1973b; Buskirk and Buskirk 1976), but such was not the case for Coleoptera in Grenada (Tanaka and Tanaka 1982). Janzen (1973b) and Tanaka and Tanaka (1982) argue that tropical island faunas, being depauperate compared with mainland tropical ecosystems, comprise more generalist species, that is, species more polyphagous than their mainland counterparts. Such island generalists respond to the onset of the rainy season with an increase in numbers of individuals, whereas species previously not present at mainland sites appear during the rainy season.

FEEDING GUILDS

Herbivores

We categorize phytophagous invertebrates into two broad groups: polyphages and monophages. Some polyphagous species may be limited to only a few species of hosts and represent a special category, oligophages. Many, if not most, hemimetabolous insects can be considered polyphagous, as many do not seem to be host-specific and sample a wide variety of plant species. A detailed analysis of one common herbivore, *Lamponius portoricensis* (Willig et al. 1986; Sandlin-Smith 1989; Sandlin and Willig 1993; Willig et al. 1993), and host records gleaned from Martorell (1975) for various Orthoptera and Homoptera support this designation. Some Homoptera are extremely host specific, although we have no evidence for the common species recorded at El Verde.

Polyphagous Forms

Snails. Prior to 1990, the only substantive ecological work on snails at El Verde was that of Heatwole and Heatwole (1978), and they focused primarily on the large common camaenid, *Caracolus caracolla.* They report that

C. caracolla is quite polyphagous. It has been observed eating dead brown leaves (one leaf was identified as an introduced *Hibiscus*), unidentified green leaves, large seeds (one was *Ormosia krugi*), wet discarded paper, arum roots, and inflorescences of *Inga vera*. In the laboratory, these snails fed on carrots, paper, and *Hibiscus* leaves (Heatwole and Heatwole 1978). A macroscopic analysis of fecal material of snails at El Verde revealed that 54% was leaf material, 18% thin fibers, 14% wood, and 10% bark. A microscopic examination of fecal material showed that *C. caracolla* primarily ingested diatoms (42%), wood cells (34%), plant hairs (11%), and calcium oxalate crystals (5%). Ratios of these items differed in fecal samples collected at El Yunque, a site east of El Verde and at a higher elevation in the Luquillo Mountains. However, Lodge (this volume) reports that epiphyll composition on leaves adjacent to snail feeding trails was 77% fungi and 23% algae; fungi do not appear in fecal samples.

BLATTODEA. About twenty to twenty-five species of cockroaches occur at El Verde (Drewry 1970b, app. 6), some of which are common and reach high densities in the forest litter (Pfeiffer, chap. 5, this volume). One of these, *Epilampra wheeleri,* is large (25 mm), but nothing is known of its foraging ecology. A preliminary study of food habits of *Eleutherodactylus coqui* (Woolbright and Garrison unpublished) indicates that about 18% by volume of the diet comprises these insects.

ORTHOPTERA. Grasshoppers are considered to be generalized feeders (Mulkern 1967; but see Rowell et al. 1983 for exceptions). Orthoptera at El Verde are large (5–45 mm). A few species are common, and many are consumed regularly by frogs and *Anolis* lizards (Lister 1981; pers. observation). All species are probably important herbivores at El Verde. Two common katydids and nine gryllids (table 6.2) are the dominant forms at El Verde. *Orocharis* contains two small species (*O. vaginalis, O. terebrans;* about 15 mm) which occur from the understory to the canopy. Martorell (1975) records several host plants for *Cyrtoxipha* and *Orocharis,* but these plants are primarily monocultural crops. No definite host relationships have been recorded for any of these species at El Verde. Over 50% of the total volume of food consumed by *E. coqui* comprises these insects (Woolbright and Garrison unpublished), so they provide an important link in the food chain.

PHASMATODEA. Four species of walking sticks occur at El Verde, but only one, *Lamponius portoricensis,* is common. At El Verde, *Lamponius* commonly consumes leaves from four plant taxa (Willig unpublished; Sandlin and Willig 1993): *Piper treleaseanum* and *P. hispidum* (herbaceous shrubs in the Piperaceae), *Urera baccifera* (a woody shrub in the Urticaceae which grows from prostrate stems), and *Dendropanax arboreus* (a mid-successional canopy tree in the Araliaceae). In areas with appreciable human modification in the tabonuco forest, *Hibiscus rosa-sinensis* is a common ornamental and forage species for *Lamponius* as well.

Lamponius is the largest common insect at El Verde, and is the only one for which food consumption data have been quantified in any detail (Willig unpublished; Sandlin-Smith 1989; Sandlin and Willig 1993). In controlled experiments where five foods were offered for consumption in equal amounts by wet weight (Willig unpublished), *U. baccifera* was the most preferred food regardless of total availability; however, all other foods were consumed as well, even though the supply of *U. baccifera* was not exhausted during feeding trials. *Urera baccifera* had the lowest caloric content but the highest or second highest content of phosphorus, sulfur, zinc, manganese, potassium, calcium, and magnesium. Moreover, despite significant changes in consumption patterns which accompanied alterations in total abundance of foods offered, the ratio of nutrients (calories, ash, elements listed above, nitrogen, and sodium) remained constant in the diet of the experimental population of *Lamponius*.

Willig et al. (1993) subsequently evaluated microhabitat selection by multiple regression analysis. The percentage of all captures (618) during which *Lamponius* was found on or consuming its natural forage plants was 62% *P. treleaseanum*, 12% *D. arboreus*, 9% *P. hispidum*, 4% *U. baccifera*, and 13% other taxa (*Ruellia coccinea, Panicum adspersum, Hippocratea volubilus, Inga vera, Palicourea barvineira,* and *Prestoea montana*). Walking sticks were associated with areas characterized by high apparency (foliar development in the understory) of *P. treleaseanum* and *Symplocos martinicensis,* and low apparency of *Dryopteris deltoides*. The total development of the understory, regardless of taxonomic composition, at 2.5 feet and 3.5 feet above the ground also contributed to high density of walking sticks. In addition, *Lamponius* occurred twice as often on *P. treleaseanum* as expected based on its total contribution to the understory flora. The authors hypothesized that the disproportionate occurrence on *P. treleaseanum* was related to the production by *Piper* of aromatic attractants that act as proximate cues in patch selection.

To understand why *Lamponius* disproportionately occurs on its least preferred forage plant, a number of experiments were conducted that evaluated the manner in which forage attributes (e.g., nutrient content) or herbivore characteristics (age, sex, or previous foraging experience of walking sticks) interact to affect food preference. Multivariate repeated measures analysis of variance revealed that at different ages, males and females exhibit different patterns of food consumption when offered *P. treleaseanum*, *P. hispidum*, *U. baccifera*, and *D. arboreus*. Likewise, preexposure to only one food influences subsequent diet composition differently, depending on walking stick sex and which of the four plants were preexposure foods during a particular

Table 6.2. Common phytophagous insects at El Verde

Taxon	Foraging Time	Food Plants (if known)
POLYPHAGOUS FORMS:		
Blattodea: Blattellidae		
Aglaopteryx facies	Night	—
Cariblatta hebardi	Night	—
Cariblatta suave	Night	—
Plectoptera infulata	Night	—
Epilampra wheeleri	Night	—
Neoblattela vomer	Night	—
Orthoptera: Tettigoniidae		
Anaulocomera laticauda	Night	*Inga vera*[a]
Turpilia rugosa[a]	Night	Probably *Inga vera*[b]
Orthoptera: Gryllidae		
Anaxipha sp.	Night	—
Cyrtoxipha gundlachi	Night	*Citrus* spp., *Musa sapientum, Saccharum officinarum, Solanum melongena, Zea mays*[a]
Laurepa krugii	Night	*Coffea arabica, Rhizophora mangle*[a]
Orocharis terebrans	Night	*Citrus* spp., *Coffea arabica*[a]
Orocharis vaginalis	Night	*Citrus* spp., *Saccharum officinarum, Coffea arabica, Dracaena fragrans, Gossypium hirsutum*[a]
Orocharis spp. (4 undescribed)	Night	—
Phasmatodea: Phasmatidae		
Lamponius portoricensis	Night	*Dendropanax arboreus*[c], *Hibiscus rosa-sinensis*[c], *Piper hispidum*[c], *Piper treleaseanum*[c], *Urera baccifera*[c], probably *Lobelia portoricensis*[a]
Psocoptera (13 spp.)	?	Fungal spores, lichen[d]
Homoptera: Cicadidae		
Borencona aguadilla	Day, Dusk?	*Coffea arabica*[a]

Homoptera: Cicadellidae		
Sibovea coffeacola	Day	*Castilla elastica, Coffea arabica, Inga fagifolia, Manilkara bidentata, Pothomorphe peltata, Rubus rosifolius, Solanus* spp.[a]
Homoptera: Cixiidae		
Bothriocera undata	Day, Night?	22 plants listed by Martorell (1975)
Homoptera: Delphacidae		
Ugyops occidentalis	Day, Night	*Coffea arabica, Inga fagifolia*[a]
Homoptera: Derbidae		
Dawnaria sordidulum	Day, Night	*Musa sapientum*[a]
Dysmia maculata	Day, Night	—
Homoptera: Achilidae		
Catonia cinera	Day, Night	*Dendropanax arboreus, Guarea guidonia, Hibiscus rosa-sinensis, Inga vera, Montezuma speciosissima*[a]
Catonia dorsovittata	Day, Night	—
Catonia arida	Day, Night	—
Homoptera: Tropiduchidae		
Ladellodes stali	Day, Night	*Panicum muticum*[a]
Homoptera: Flatidae		
Melormenis magna	Day, Night	—
MONOPHAGOUS FORMS:		
Lepidoptera (larvae): Papilionidae		
Papilio pelaus	Night?	*Zanthoxylum martinicense*[a]
Lepidoptera: Pieridae		
Dismorphia spio	Night?	*Ruellia coccinea*
Lepidoptera: Nymphalidae		
Siproita stelenes	Night?	*Ruellia coccinea*[a]

Sources and notes: [a]Martorell (1975). Food plants gleaned from Martorell (1975): does not indicate that those plants occur at El Verde.

[b]Martorell (1975: 144) records a similar tettigoniid, *Anaulacomera laticauda,* as feeding on *Inga vera;* thus this tree is probably used by other orthopterans at El Verde.

[c]From Willig (1989)

[d]Feeding habits for this order summarized by Broadhead (1983).

experiment. In addition, preferences were shown for different qualities of leaves within single forage species (old, intermediate-aged, or young leaves). In particular, older leaves of *P. treleaseanum* were preferred, whereas intraspecific differences in consumption based on leaf age or position did not occur for *D. arboreus* or *U. baccifera*. In summary, walking sticks distinguish among plant species, recognize differences in plant quality associated with age or position for some taxa, and modify diet content to reflect past experience.

Studies of insect consumption by birds, anoles, and frogs at El Verde indicate that walking sticks constitute only a minor part of their diet. However, their large size, abundance, and ability to defoliate forage plants indicate that *Lamponius* may be important in returning nutrients to the soil during early successional stages.

PSOCOPTERA. The thirteen species of Psocoptera from El Verde are common elements in the understory. We have no data regarding their feeding habits, but Broadhead (1983) characterizes the group as microepiphyte feeders, of which some species are primarily bark inhabitants and others are foliage inhabitants. Broadhead classifies species as fungal spore and lichen feeders. However, the two groups are not obligatory in their feeding habits: both can switch to the alternate food source when their preferred host is scarce or absent. Our fauna (thirteen species) is depauperate compared with that of the Canal Zone, Panama, from which Broadhead (1983) cites 219 species.

HOMOPTERA. About eighty species occur at El Verde, of which at least sixteen (table 6.2) are common there. These species, according to records (Martorell 1975), sample a wide array of food plants. As with the Orthoptera, many of the food plants are primarily of agricultural importance. However, many species of Homoptera are known to be extremely host specific, and several less frequently encountered species at El Verde may be monophagous or oligophagous. Measurements of herbivory rates by this group are scarce, because damage to plant tissues is difficult to assess. Laboratory experiments have been conducted to measure the feeding rate of few sucking insects. Most subjects involved relatively sessile aphids (e.g., Auclair 1958, 1959; Mittler 1958, 1970; Van Hook et al. 1980); we know of no studies conducted for more vagile tropical Homoptera. Their feeding habits may cause reduced viability in certain plants, and several homopterans are economically important pests. Another deleterious and far-reaching consequence of homopteran herbivory is the ability of some species to transmit plant viruses and other diseases. Homoptera are an important food source for arboreal *Anolis* lizards (*A. evermanni, A. stratulus,* Garrison and Reagan unpublished).

Monophagous Forms

We include here members of the large orders Coleoptera and Lepidoptera, which are more speciose than the more primitive Orthoptera and Homoptera. The Lepidoptera generally are considered to have a narrow foodplant range (Gilbert 1984), and evidence suggests that Coleoptera follow this trend also. For example, Linsley (1961), in discussing host selection of the North American Cerambycidae, indicated that the more primitive tribes tend to be more polyphagous than are more advanced tribes. Linsley (1959) noted that nearly seventy-five species of the cerambycid genus *Plagithmysis* (confined to the Hawaiian Islands), are highly host specific. Abundance of ecological niches and diversity of hosts have probably contributed to the abundance of species in this genus. Exceptions occur, and many species of both orders are important pests, feeding on a wide array of plant species. Examples include the sugar cane weevil, *Diaprepes abbreviatus,* and the Melolonthine scarab, *Phyllophaga portoricensis.* Many species of these orders are known from El Verde, but virtually nothing is known of their food habits or importance in the food web.

COLEOPTERA. The most common beetles encountered in our sampling program have been bark beetles (Scolytidae). These insects are injurious to stressed or unhealthy trees in the temperate zone, but little is known of their feeding habits at El Verde, though some species are polyphagous and probably play an important role in the food web. At least three pantropical species collected at El Verde (app. 6) are introduced: *Xyleborus ferrugineus, X. affinus,* and *Coccotrypes carpophagus.* Both species of *Xyleborus* are among the most important tropical tree pests in the world. Wood (1982) cites records of over 150 hosts for *X. ferrugineus* and over 250 for *X. affinus. Xyleborus ferrugineus* is known to be a principal vector of *Ceratoeystis fimbriata,* which causes wilt disease in cacao trees (Saunders 1965). *Coccotrypes carpophagus,* as its specific name indicates, breeds in large seeds, especially those of palms, on the ground.

Some bark beetles have been reared from decaying seed pods of *Inga vera.* Larvae of the common Middle American bark beetle, *Scolytodes atratus panamensis* (Cecropia Petiole Borer) feed only on petioles of recently fallen *Cecropia* leaves (Wood 1983). Wolcott (1948) lists thirty species from Puerto Rico, but there must certainly be more than that. Wolcott (1948) lists *Xyleborus affinus* as attacking healthy *Inga vera. Xyleborus* (*Ambrosiodmus*) *lecontei* has been collected from dying terminals of *Cedrela mexicana,* and a species of *Pterocyclon* has been found attacking *Dacryodos excelsa.*

Only three species of chrysomelid beetles have been listed for El Verde, but a more realistic number is probably eighty-five to ninety species (E. Sleeper pers. comm.). Strong (1983) describes the biology of rolled-leaf hispines of the tribes Cephaloliini and Arescini. Members of the genus *Cephaloleia,* of which there are 182 species, are specific to families of plants of the order Zingiberales.

Lepidoptera. Little is known of the host plant range for the 234 species of Lepidoptera at El Verde. The larval forms of almost all Lepidoptera are phytophagous. The adults are nectar feeders, but some species (e.g., *Gonodonta* spp., Noctuidae, Todd 1959) have short tongues for piercing fruit and imbibing fruit juices. Two butterflies at El Verde, the green and white *Siproeta stelenes* and the pierid *Dismorphia spio,* are known to feed on *Ruellia coccinea* (Wolcott 1948 pers. observation). The large swallowtail butterfly, *Papilio pelaus,* feeds on *Zanthoxylum martinicense* Wolcott (1948).

Janzen (1988) documented the host range of most of the Lepidoptera at Santa Rosa National Park in Costa Rica. He provided good evidence of the narrow host range for larvae of the order. At least half of the caterpillar species studied are monophagous; and he speculated that at least 80% of the remainder are oligophagous. He believed that about twenty of 3,142, or less than 1%, of the fauna are polyphagous.

Total defoliation of various plant species in mature ecosystems by caterpillars or by other insects is apparently rare in the Neotropics. Janzen (1988) records forty such episodes over nine years at Santa Rosa National Park, and Wolda and Foster (1978) document an outbreak of the dioptid moth, *Zunacetha annulata.*

Carnivores

Two broad classes are defined here, predators and parasitoids/parasites. Predators, which attack and consume other invertebrates, are usually indiscriminate in prey acquisition and therefore sample a wide array of organisms. Parasitoids are usually specific to one kind of organism; their larvae feed on and destroy the host. In contrast, parasites may be host specific but do not usually kill the host. The holometabolous orders Diptera and Hymenoptera (table 6.3) represent this feeding guild.

Predators

The only common group of predators appears to be the beetle family, Lampyridae (table 6.3). Seven species occur at El Verde. The sickle-shaped mandibles of the larvae are used to stab and suck dry their prey. Females of some lampyrids are predatory on other similar species (Lloyd 1965), but it is not known if this phenomenon occurs at El Verde. Lampyrids are occasionally found in the stomachs of frogs (*Eleutherodactylus*) and lizards (*Anolis*), and

Table 6.3. Common carnivorous insects of El Verde

Taxon	Foraging Time	Feeding Guild	Host (Parasitoids)
Coleoptera			
Lampyridae (about 6 spp.)	Night	Predator	
Diptera (adults)			
Culicidae (9 spp.)	Day, Night	Parasite	
Ceratopogonidae (about 34 spp.)	Day, Night	Parasite	
Dolichopodidae	Day	Predator	
Diptera (larvae)			
Phoridae (about 65 spp.)	Day, Night	Predator[a]	
Muscidae			
Philornis spp.	Day, Night	Parasite	Aves[b]
Hymenoptera			
Mymaridae (about 13 spp.)	Day, Night	Parasitoid	Eggs of Lepidoptera, Coleoptera, and other insects depending on species
Eulophidae (about 14 spp.)	Day, Night	Parasitoid	Homoptera, Lepidoptera, and other insect larvae
Scelionidae (about 17 spp.)	Day, Night	Parasitoid	Eggs of insects, spiders, depending on species
Formicidae			
Linepithema mellea	Day	Predator[c]	
Myrmelachista ramulorum	Day	Predator[b]	
Vespidae			
Mischocyttarus phthisicus	Day	Predator	
Polistes crinitus	Day	Predator	

Notes: These data are based on general knowledge of the biology of various insect groups. Primary hosts are included for host specific forms.

[a] Larvae of one species have been observed eating eggs of *E. coqui* (Woolbright pers. observation).

[b] Snyder et al. (1987).

[c] Are also scavengers.

one was found in the stomach of a juvenile Puerto Rican boa, *Epicrates inornatus* (Reagan 1984).

Curiously, members of the large coleopterous family Carabidae appear to be absent from El Verde. A few species are found at higher elevations near El Yunque, but none has been collected at El Verde.

Parasites and Parasitoids

The Hymenoptera and Diptera compose these groups. Generally, they are small to very small insects and, according to preliminary sticky trap sampling, are represented abundantly at El Verde. Their precise role as potential regulators of other insect and vertebrate groups is largely unknown, although the general biology of the groups indicates that most species are extremely host specific. Janzen (1988) reports over 300 species of tachinid, ichneumonid, and brachonid parasitoids from Santa Rosa National Park, Costa Rica. Many of these are monophagous, and others limit their host selection to clusters of closely related species.

Many nematocerous Diptera (Culicidae, Ceratopogonidae) adults suck the blood of vertebrate and invertebrate hosts, but no quantitative data have been gathered for these insects. Snyder et al. (1987) provide data on avian parasitism by larvae of the warble fly, *Philornis* spp. They observed a 26% to almost 47% death rate of nestling pearly eyed thrashers due to infestations of these flies.

The ectoparasite fauna of bats is reported by Willig and Gannon (this volume). Levels of parasitism based on the age and sex of both bat host and invertebrate parasite species are documented elsewhere (Gannon 1991; Gannon and Willig 1994b, in press). The level of infestation by *P. iheringi* on *S. rufum* depends upon the age and sex of the host, but not upon season of capture (wet versus dry season). In particular, subadult bats harbored significantly higher numbers of this wing mite than did adult males or adult females. The same pattern obtains for *P. iheringi* on *A. jamaicensis*. In contrast, levels of infestation by the other ectoparasites (*M. aranea, Aspidoptera* sp., *Trichobius* sp., and *Spelaeorhynchus* sp.) of *A. jamaicensis* are not influenced by the age or sex of the host. For the other bat taxa, the number of captured hosts was too small to conduct powerful tests for differences in ectoparasite infestation levels.

Two species of ectoparasite (*P. iheringi* and *Trichobius* sp.) occurred on all three common bat taxa; differences in infestation by each of these ectoparasites, as well as by all ectoparasites, could be compared among host taxa and between seasons. In all three cases, season-independent, host-specific differences in ectoparasite infestation were detected in statistical analyses. In particular, infestation levels by *Trichobius* sp. were the same on *S. rufum* and *A. jamaicensis,* but levels of infestation on *M. redmani* differed from that on each of the other bats. In contrast, levels of infestation by *P. iheringi* and all ectoparasites were the same on *S. rufum* and *M. redmani,* but each of these bat taxa differed from *A. jamaicensis*.

The distribution of ectoparasites on hosts differs among host age-sex groups and may be related to behavioral attributes of each bat species. The number of ectoparasites per host was randomly distributed in *A. jamaicensis* and *M. redmani,* whereas the distribution of ectoparasites on *S. rufum* was significantly hyperdispersed (even). Both *A. jamaicensis* and *M. redmani* roost in colonies where ectoparasite transmission among bats may be facilitated. In the case of bats which roost in a solitary fashion, such as *S. rufum,* barriers to interhost transmission may give rise to the clumped distribution of ectoparasites.

Comparisons of ectoparasite community composition can be evaluated based upon the proportional representation of ectoparasite species. Because the sample size of hosts was large, the effect of host age and sex on community composition was determined for *A. jamaicensis*. Differences in ectoparasite community composition were detected among adult males, adult females, and subadults. Less powerful *a posteriori* tests were unable to identify pairwise differences. Nonetheless, the contrast between adult females and subadults approached significance ($p = .052$) and most likely contributed to overall differences.

FORAGING ACTIVITY

Most foraging probably occurs at night, because most invertebrate activity is observed during that time. Data pertinent to the day-night comparison and vertical stratification of flying insects were accumulated from the 9 to 22 June 1981 study mentioned above. After we identified all invertebrates, we tabulated mean numbers and subjected the data to a one-way analysis of variance (ANOVA) and sum of squares simultaneous testing procedure (SS-SSTP) (Sokal and Rohlf 1969). Stewart and Woolbright (this volume) provide further data on day-night activity and abundance of forest dwelling arthropods at El Verde. Data on seasonal abundance of flying insects were gleaned from unpublished data accumulated by Kepler and summarized by Snyder et al. (1987) and by Lister's (1981) work with *Anolis* lizards.

Table 6.4 lists invertebrates collected over ten days and eight nights, and table 6.5 gives the percentage contribution of each order. No significant differences in mean numbers of invertebrates were detected between day and night, but Blattodea, Orthoptera, and Lepidoptera showed a nocturnal preference (table 6.5). The dipteran suborders Brachycera and Cyclorrhapha (except Phoridae) were strongly diurnal (table 6.4). Willig (unpublished) has found that *Lamponius* feeds only at night. During the day, Orthoptera and Blattodea remain hidden, while *Lamponius* remains quiescent. The sticky trap survey and personal observation indicate that Homoptera are active day and night, but it is not known if they feed during both times. Heatwole and Heatwole (1978) have observed *Caracolus caracolla* feeding only at night; however, these snails become active when humidity is high or during frequent showers throughout the year (Cary 1992; Willig pers. observation).

Evidence of diel cycles among ants comes from gut analysis of nocturnal frogs and diurnal *Anolis* lizards. The major ant eaten by *E. coqui* appears to be *Paratrachina* spp., but these species seldom appear in diets of *Anolis* lizards. The most common ant components of the diet of these animals are *Pheidole moerens, Linepithema mellea* (formerly *Iridomyrmex melleus*), and *Myrmelachista ramulorum*.

Table 6.4. Flying or wind-drifting arthropods trapped 9–22 June 1981

Taxon	Day [a]	Night [b]
Class ARACHNIDA		
Order ARANEAE		
unidentifiable to family	1	2
Pholcidae		
unidentifiable to species	1	
Modisimus sp.	2	
Linyphiidae	1	
Clubionidae	1	5
Araneidae		
unidentifiable to species	1	1
Leucauge regnyi	1	2
Thomisidae		
Epicaudus mutchleri	1	
Salticidae	3	1
Order ACARINA		
Suborder Cryptostigmata	1	
Class ELLIPURA		
Order COLLEMBOLA		
Entomobryidae		
Lepidocyrtinus sp.?	1	
Class INSECTA		
Order EPHEMEROPTERA		
Leptophlebiidae?	1	
Order BLATTODEA		
Blattidae	1	
Blattellidae		
Cariblatta hebardi		2
undetermined species		4
Order ORTHOPTERA		
Gryllidae		
Cyrtoxipha gundlachi		3
undetermined Trigonidinae		1
Orocharis vaginalis or *terebrans*	4	
Order ISOPTERA		
Termitidae		
Nasutitermes sp.	1	
Kalotermitidae		
Glyptotermes ?pubescens (winged)	3	2
Order PSOCOPTERA		
Polypsocidae		2
Epipsocidae		2
Psocidae		3
Lepidopsocidae	12	7
Order THYSANOPTERA		
Phlaeothripidae	8	10
Thripidae	5	2
Order HEMIPTERA		
Dipsocoridae	1	
Miridae		
undetermined species	1	
Polymerus pallidus	1	
Lygaeidae		6
Cydnidae		
?*Amnestus* sp.	1	

Table 6.4. (*continued*)

Taxon	Day [a]	Night [b]
Order HOMOPTERA		
Membracidae		
Nessorchinus esbeltus	1	
Cicadellidae		
undetermined species	1	
undetermined species (larva)	1	
Sibovea coffeacola		1
Xestocephalus maculatus		2
Ponana insularis		3
Superfamily FULGOROIDEA		
Cixiidae (larva)	1	
Delphacidae		
Ugyops occidentalis		1
Derbidae		
undetermined species	1	
Dysimia maculata		1
Dawnaria sordidulum		9
Patara albida		1
Achilidae		
Amblycratus striatus? (larvae)	8	7
Catonia cinerea		1
Catonia dorsovittata (larvae)	3	5
Quadrana punctata? (larva)		1
undetermined species (larvae)	2	
Tropiduchidae		
Ladellodes stali	1	8
Issidae		
Thionia borinquensis	3	
Colpoptera maculifrons	6	
Colpoptera brunneus	9	5
Neocolpoptera monticolens	1	
Kinnaridae		
Quilessa fasciata	2	
Psyllidae	1	3
Superfamily COCCOIDEA	1	1
Order COLEOPTERA		
Ptiliidae		
Actinopteryx sp.	1	
Scaphidiidae		1
Staphylinidae		
undetermined species	4	3
Palaminus sp.		1
Pselaphidae	3	
Histeridae	4	1
Elateridae		1
Throscidae	4	2
Anobiidae		3
Trogositidae (Tribe Tenebroidini)	1	
Cucujidae		2
Coccinellidae	3	3
Tenebrionidae		1
Colydiidae		1
Melandryidae		1
Mordellidae	1	

Table 6.4. (*continued*)

Taxon	Day [a]	Night [b]
Euglenidae	3	3
Chrysomelidae	1	2
Scolytidae	7	7
Anthribidae	1	
Curculionidae		2
Order LEPIDOPTERA		
Gracillariidae?		1
Cosmopterygidae?		1
Gelechiidae?		1
Order DIPTERA		
Tipulidae	5	9
Mycetophilidae	18	4
Sciaridae	35	20
Cecidomyiidae	8	11
Psychodidae	9	11
Scatopsidae	7	1
Ceratopogonidae	102	80
Chironomidae	125	44
Asilidae		1
Empididae	14	4
Dolichopodidae	28	12
Phoridae	1,520	1,879
Pipunculidae	1	
Lonchaeidae	1	
Tephritidae	1	
Odiniidae	4	
Agromyzidae	1	
Lauxaniidae	2	
Chamaemyiidae?	1	
Heleomyzidae	3	2
Drosophilidae	3	3
Ephydridae	1	
Chloropidae	3	
Muscidae	7	
Calliphoridae	4	1
Sarcophagidae	1	
Tachinidae	39	5
Order HYMENOPTERA		
Braconidae	2	5
Ichneumonidae		1
Mymaridae	11	1
Trichogrammatidae	1	4
Eulophidae	6	2
Encyrtidae	54	18
Eupelmidae	8	
Agaonidae	2	
Torymidae	1	1
Pteromalidae		1
Cynipidae	1	
Ceraphronidae	3	
Diapriidae	4	
Scelionidae	29	14
Platygasteridae	5	

Table 6.4. (*continued*)

Taxon	Day [a]	Night [b]
Bethylidae	4	
Dryinidae	1	1
Formicidae		
Monomorium floricola	1	
Linepithema mellea (formerly *Iridomyrmex melleus*)	3	1
Myrmelachista ramulorum	1	4
Brachymyrmex heeri	1	
undetermined workers	7	3
winged males	6	5
Sphecidae (Cabroninae)	4	

[a] Ten days.
[b] Eight nights.

Table 6.5. Invertebrates collected by day and by night

Order	Day [a]	Night [b]	Total	% Overall Total
Acarina	0	1	1	0.02
Araneidae	9	14	23	0.51
Collembola	0	1	1	0.02
Ephemeroptera	1	0	1	0.02
Blattodea	0	7	7	0.16
Orthoptera	0	8	8	0.18
Isoptera	4	2	6	0.13
Psocoptera	12	14	26	0.58
Thysanoptera	13	12	25	0.55
Hemiptera	4	6	10	0.22
Homoptera	42	49	91	2.02
Coleoptera	33	28	61	1.35
Lepidoptera	0	3	3	0.07
Diptera	1,943	2,087	4,030	89.42
Hymenoptera	152	62	214	4.75
Totals	2,213	2,294	4,507	100%

[a] Ten days.
[b] Eight nights.

VERTICAL STRATIFICATION

Our brief trapping survey and data gleaned from the literature (e.g., Lister 1981) confirm that certain kinds of invertebrates are not equally distributed vertically at El Verde. A total of 4506 invertebrates representing fifteen orders and 105 families (table 6.4) was collected over the ten-day, eight-night sampling period. Diptera constituted the most abundant insect group (89%), followed by Hymenoptera (5%), Homoptera (2%), and Coleoptera (1%) (table 6.5). Phorid flies representing several species made up 75% of the en-

Height (m)	1	2	3	11	5	19	14	15	4	9	13	7	17	6	12	16	8	10	18
$\bar{x}$	839.5	340.5	177.5	70	69.5	69	61	60	59	55.5	55	53.5	52	52	51.5	50	50	49	40.5

Figure 6.2. Sum of Squares-Simultaneous Test Procedures (SS-STP) for differences of mean numbers of insects along 1 m height intervals. Horizontal lines indicate ranges over which differences are nonsignificant.

tire invertebrate fauna and were obviously the dominant group during the sampling period. Diptera comprised the most families (twenty-seven), followed by Coleoptera (twenty) and Hymenoptera (nineteen).

There were significant differences among the mean numbers of invertebrates collected at the nineteen heights ($F_{[0.01](18,19)} = 6.68$, $p < .001$). An SS-STP test (fig. 6.2) showed the first 2 m to contain a significantly greater number of invertebrates than did the upper 17 m. The Phoridae likewise showed significant differences in mean numbers collected along the 19 m ($F_{[0.01](18,19)} = 6.68$, $p < .001$) and were the major factor contributing to the differences observed among total invertebrate groups. An SS-STP test of phorids showed the same results as for all trapped invertebrates. When all invertebrates minus the Phoridae were compared no mean differences were detected.

Members of the superfamily Fulgoroidea (Delphacidae through Issidae, app. 6), or plant hoppers, are conspicuous herbivores in the rain forest. Though they are often seen and collected in sweep-net and D-Vac samples near the ground, more of these insects were found near the canopy than below. When the 19 m strata were divided into three equal samples of 6 m (the first meter sample was deleted because it had so few specimens, and to equalize sample sizes, i.e., numbers of cups), a significant difference was observed between the top 6 m and the lower 12 m (fig. 6.3).

The small, inconspicuous Diptera appear to be the most abundant insects on a regular basis. Studies reported by Drewry (1969a) and Snyder et al. (1987) showed Diptera to make up 91% and 63% of all insects collected in their mosquito light trap and sticky-trap samples at El Verde and El Yunque, respectively. Similarly, Penny and Arias (1981), after a year of light and trap sampling in the Amazonian rain forest, found 84 to 91% of the invertebrates to be Diptera, primarily *Luzomyia* spp. (Psychodidae). Phorid flies were the most abundant Diptera trapped at the tower. Phorids are a large group with varied habits. Adults and larvae probably feed on decaying organic matter, which explains their greater numbers near the ground. Phorids collected at 1 m above ground during separate twenty-four-hour periods ranged from zero to 913. Next to phorids, the nematoceran families (Tipulidae through Cecidomyiidae) were the most common insects trapped. Their numbers were relatively constant throughout the 19 m. Large invertebrates, such as drag-

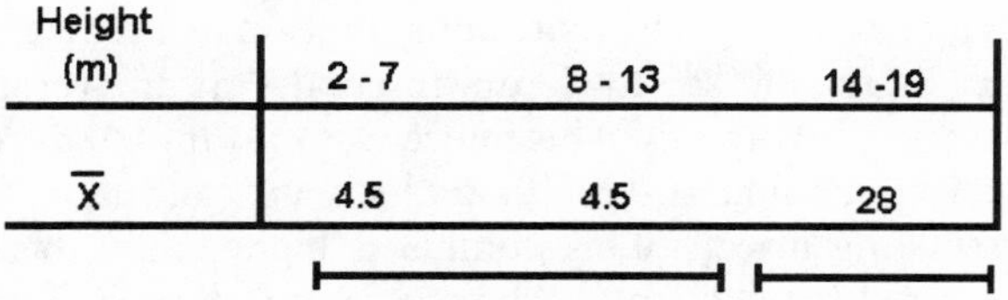

Figure 6.3. Sum of Squares-Simultaneous Test Procedures (SS-STP) for differences of mean numbers of fulgoroids per 6 m height interval (above lowest 1 m). Horizontal lines indicate ranges over which differences are nonsignificant.

onflies and butterflies, were absent from our samples and may have avoided the traps or escaped.

Sutton (1983), in conducting a vertical census in a rain forest in Sulawesi, found higher numbers of Homoptera, Hemiptera, Lepidoptera, Diptera, Hymenoptera, and Coleoptera in the upper canopy than in the understory. Erwin (1982, 1983a) believes that tropical canopies, when adequately sampled, will yield dramatic increases in the number of species.

Erwin (1983b) found low degrees of vagility among canopy Coleoptera in the Amazonian rain forest. This resulted in each forest type harboring its own assemblage of species. Even if this is true for most Coleoptera, we suspect that more vagile insects, such as macrolepidoptera and Homoptera, are more widely distributed. Dispersion of such insects also may be influenced by wind and rugged terrain patterns such that neighboring tree crowns are not in close proximity to each other. Sutton (1983), for example, found intercrown diversity to be fairly uniform for Homoptera.

Among ants, *Pheidole moerens* appears to be primarily a litter species, as we have taken it commonly in litter traps. It is consumed most commonly by the ground-dwelling anole, *A. gundlachi. Myrmelachista ramulorum,* on the other hand, appears to be primarily an arboreal species. It constitutes the greatest number of Formicidae consumed by the arboreal anole, *A. stratulus,* but is rarely found in stomachs of *A. gundlachi* (see Reagan, this volume). Further, we have collected specimens of *M. ramulorum* from birds which use them to smear formic acid over their feathers to help ward off ectoparasites. The most common ant species, *Linepithema mellea* (formerly *Iridomyrmex melleus*), is found from litter to canopy. We have taken them in litter traps and in sticky traps in the canopy, and they are consumed by anoles that occur from ground to canopy.

ENERGY FLOW AND NUTRIENT CYCLING

Estimation of energy and nutrient flow in the forest canopy is hindered by the difficulty of sampling that assemblage. We are aware of no studies that provide nutrient cycling data for arboreal neotropical ecosystems. Schowalter et al. (1981), in assessing herbivore consumption in a temperate zone forest, used 2.5 ± 3.2 mg dry sap mg^{-1} dry insect d^{-1} as an average consumption rate for sucking herbivores. This figure was extrapolated from previous papers. They estimated that 100 to 200 kg ha^{-1} yr^{-1} of foliage biomass was consumed by sucking insects. This number is higher than the 60 to 70 kg ha^{-1} yr^{-1} measured for chewing herbivores and indicates that Homoptera can be important primary consumers.

SUMMARY

Aboveground invertebrates occupy many trophic roles in the El Verde food web. Their contribution to nutrient cycling roles within the forest is emphasized by their great diversity compared to vertebrates. Although over 1,500 invertebrate species have been recorded at El Verde, diversity is poor compared to comparable mainland ecosystems. For example, ten species of Odonata are found at El Verde compared to over 130 species recorded from a rain forest site in Brazil; sixty-four species of auchenorrhynchous Homoptera are found at El Verde compared to 120 from a site in Sulawesi; and 234 species of Lepidoptera are found at El Verde compared to 3,142 and 1,577 species from sites in Costa Rica and New York, respectively. Although the rain forest at El Verde is superficially similar to mainland tropical rain forests, Puerto Rico is lacking in characteristic families and subfamilies present in these mainland tropical ecosystems. Well-known Neotropical families such as Calopterygidae, Polythoridae, Platystictidae, and Perilestidae (all Odonata) and Brassolinae, Morphinae, Ithomiinae, and Saturniidae (all Lepidoptera) are lacking at El Verde. Intensive, long-term studies have been conducted for only two groups of invertebrates. Snails and the walking stick, *Lamponius portoricencis,* allow evaluation of the effects of disturbance on population dynamics. Three species of snails, *Austroselenites alticola, Megalomastoma croceum,* and *Subulina octana,* were found to be primarily ground dwellers that were equally suited to forest cover and light gaps. Two others, *Nenia tridens* and *Caracolus caracolla,* occurred primarily on plant material above the forest floor during the wet season but respond differently to light gaps. *Caracolus caracolla* was more common in the forest, but *N. tridens* was more likely to be associated with light gaps. The arboreal walking stick, *L. portoricencis,* was found distributed among their food plants where they moved little (0.5 m d^{-1}). Hurricane Hugo had a profound effect on the densities of these organisms. Most suffered population reductions of up to 75%.

Seasonal variation in invertebrate populations occurs at El Verde. Sticky-trap studies conducted over a two-week period at El Verde documented an abrupt increase and decrease of phorid flies indicating a synchronous emergence. Greater insect abundance was detected during the onset of the rainy season. Increases in numbers were detected from passive (sticky-trap) collections as well as from an assessment of *E. coqui* and *Anolis* stomach contents.

Trophic relationships for most of the invertebrate taxa at El Verde can only be generalized from comparison with known feeding habits of similar taxonomic categories. El Verde has a diversity of polyphagous herbivores (snails, cockroaches, crickets, katydids, walking sticks, bark lice, and various sucking insects), monophagous herbivores (many beetles, moths, and butterflies), generalized predators (Lampyrid beetles), and parasites and parasitoids (parasitic wasps and flies, blood sucking flies). Data gleaned primarily from sticky-trap samples, gut analyses of nocturnal frogs and diurnal *Anolis* lizards, and personal observations at El Verde indicate variation in diurnal activity patterns for some groups of invertebrates. Cockroaches, tree crickets, walking sticks, and Lepidoptera larvae appear to be more common at night, when they probably feed; other invertebrates, such as various flies, are more active during the day. Some species such as the snail, *C. caracolla,* and the walking stick, *L. portoricensis,* are usually inactive during the day and feed primarily at night, although they may feed during frequent showers during the day. Vertical stratification of invertebrates occurs for some groups of insects. Plant hoppers were more abundant in the upper story of the forest, whereas others, such as adult phorid flies, were most common in the understory. Of three abundant ants, one, *Myrmelachista ramulorum,* is arboreal, another, *Pheidole moerens,* occurs in the litter on the forest floor, and *Linepithema mellea* (formerly *Iridomyrmex melleus*) occurs from canopy to forest floor.

ACKNOWLEDGMENTS

We appreciate constructive criticisms of F. Hovore, E. Masteller, D. Reagan, E. Sleeper, J. Torres, and R. Waide. For their help and input on the taxonomy of certain groups, especially for appendix 6: J. de Jesus (land snails), R. Edwards, W. Pfeiffer, H. Levi, V. Roth (spiders and most soil arthropods); R. Norton (Cryptostigmata mites); G. Eichwort (Mesostigmata mites); G. Camilo (Opiliones and Schizomida); M. Velez (Diplopoda, Chilopoda, other terrestrial noninsect groups); E. Masteller (aquatic insects); O. Flint, Jr. (Trichoptera); A. Evans (Scarabaeidae); E. Giesbert, F. Hovore (Cerambycidae); E. Sleeper (Curculionidae); J. Sorensen (Aphididae and literature); S. Wood (Scolytidae); C. Hogue (Blepharoceridae); E. Fisher (Asilidae); R. Snelling (ants). Special appreciation is due R. Lavigne and G. Drewry who, through the years, worked assiduously to collect and rear most of the invertebrates at the El Verde Field Station. M. Gannon, J. Alvarez, E. Sandlin, and J. Torres provided access to unpublished manuscripts. The University of Puerto Rico,

U.S. Department of Energy and National Science Foundation Long-Term Ecological Research Program provided M. Willig with research facilities and logistical support and funding during research at El Verde. Finally, we thank the editors for their input, help, and for inviting us to contribute to this volume.

Appendix 6 Foraging Status of Invertebrates at El Verde

Taxon	Foraging location (see notes on p. 245)			
Class ADENOPHOREA (Nematoda)				
(many species) *	S	L	U	C?
Class SECERNENTEA (Nematoda)				
(many species) *	S	L	U	C
Class GASTROPODA				
Order Archaeogastropoda				
Helicinidae				
Alcadia alta			U	
Alcadia n. sp. 1			U	
Alcadia striata			U	
Alcadia n. sp. 2			U	
Order Mesogastropoda				
Cyclophoridae				
Megalomastoma croceum *	S	L		
Megalomastoma verruculosum		L		
Pomatiasidae				
Chondropoma riisei			U	
Chondropoma yunquei			U	
Order Systellommatophora				
Veronicellidae				
Vaginulus occidentalis		L	U	
Order Stylommatophora				
Camaenidae				
Caracolus caracolla *		L	U	
Caracolus marginella		L	U	
Polydontes lima		L?	U	
Polydontes luquillensis		L	U	
Polydontes acutangula *			U	
Cepolidae				
Cepolis squamosa		L	U	
Helicarionidae				
Habroconus ernsti		L	U	
Zonitidae				
Nesovitrea subhyalina		L	U	
Sagdidae				
Hyalosagda selenina		L	U	
Platysuccinea portoricensis		L	U	
Thysanophora plagioptycha		L		
Yunquea denselirata		L		
Bulimulidae				
Gaeotis nigrolineata *			U	
Clausiliidae				
Nenia tridens *		L	U	
Subulinidae				
Lamellaxis gracilis		L		
Lamellaxis micra		L		
Leptinaria unilamellata		L		
Obeliscus terebraster		L		
Obeliscus swiftianus		L		
Obeliscus hasta		L		
Opeas alabastrinum		L		

Taxon	Foraging location			
Opeas pumilum		L		
Subulina octana		L		
Haplotrematidae				
Austroselenites alticola		L		
Oleacinidae				
Oleacina glabra			U	
Oleacina playa			U	
Oleacina interrupta			U	
Zonitidae				
Glyphyalinia indentata		L		
Euconulidae				
Guppya gundlachi		L		
Gastrodontidae				
Striatura meridionalis			U	
Zonitoides arboreus		L		
Limacidae				
Deroceras laeve		L		
Pupillidae				
Pupisoma minus			U	
Pupisoma dioscoricola			U	
Vertigo hexodon		L	U	
Class OLIGOCHAETA				
Order Haplotaxida				
Megascolecidae				
*Pheretina hawayana**	S			
Class ONYCHOPHORA				
Peripatidae				
Peripatus juanensis		L		
Class ARACHNIDA				
Order SCORPIONIDA				
Buthidae				
Tityus obtusus		L	U	
Order PSEUDOSCORPIONIDA				
Menthidae				
Menthus sp.*		L		
Ideoroncidae (1 sp.)*		L		
Order SCHIZOMIDA				
Schizomidae				
Schizomus portoricensis		in termite mounds		
Schizomus yunquensis		L		
Order AMBLYPYGIDA				
Charinidae				
Charinides sp.		L		
Phrynidae				
*Phyrnus longipes**	understory (tree trunks), rock substrate			
Order ARANEAE				
Dipluridae				
*Masteria petrunkevitchi**			U	
Barychelidae				
Trichopelma corozali		L	U	
Theraphosidae				
Avicularia laeta				C
Ischnocolus culebrae		L	U	
Scariidae				
Loxosceles carribbaea			U	

Taxon	Foraging location		
Ochyroceratidae			
Ochyrocera sp.		U	
Theotima sp. (possibly *radiata*) *	L		
Pholcidae			
Micromerys dalei			C
Modisimus montanus *		U	C (rarely)
Modisimus signatus *		U	
Caponiidae			
Caponina sp.		U	
Nops sp.	L		
Oonopidae			
Dysderina sp.	L		
Oonops spinimanus	L		
Triaeris stenopsis	L		
Mimetidae			
Mimetus portoricensis		U	
Uloboridae			
Miagrammopes animotus *			C
Theridiidae			
Achaearanea porteri		U	
Argyrodes caudatus		U	
Argyrodes exiguus		U	
Argyrodes nephilae		U	
Theridiosomatidae			
Ogulnius gloriae	L		
Theridiosoma nechdomae	L		
Wendilgarda clara	L		
Wendilgarda theridionina	L		
Linyphiidae			
Centromerus ovigerus	L	C	
Tetragnathidae			
Leucauge moerens		U	
Leucauge regnyi *		U	C
Tetragnatha tenuissima		U	
Araneidae			
Agryiognatha gloriae		U	
Alcimosphenus borinquenae		U	
Capichameta hamata		U	
Cyclosa caroli		U	
Cyclosa walckenaeri		U	
Edricus crassicauda		U	
Eriophora edax		U	
Eustala sp.	L		
Gasteracantha cancriformis		U	C
Micrathena militaris		U	
Nephila clavipes		U	
Verrucosa arenata		U	
Hahniidae			
Neohahnia ernesti	L		
Anyphaenidae			
Hibana tenuis		U	
Wulfila macropalpus	L	U	
Wulfila tropica		U	
Wulfila sp.	L		
Clubionidae			
Clubiona portoricensis	L		

Taxon	Foraging location		
?Liocranidae			
Phrurolithus sp.	L		
Corinnidae			
Corinna jayuyae	L		
Trachelas borinquensis	L		
Gnaphosidae			
Lygromma sp.	L		
Ctenidae			
*Oligoctenus ottleyi**	L		
Selenopidae			
Selenops sp. (probably *lindborgi*)	L		
Heteropodidae			
Olios antiguensis		U	C
Pseudosparianthus jayuyae	L		
*Stasina portoricensis**	L	U	
Thomisidae			
Epicaudus mutchleri		U	
Misumenops bulbulcus		U	
Salticidae			
Corythalia gloriae		U	
Emanthis portoricensis		U	
Emanthis tetuani		U	
Lyssomanes portoricensis		U	
Order OPILIONES			
Cosmetidae			
Neocynortoides obscura		U	
*Cynorta v-album**		U	
Phalangoididae			
Stygnomma spinula		U	
Pseudomitraceras minutus		U	C
At least 4 other sp.		U	C
Order ACARINA			
Suborder Metastigmata			
Argasidae			
Ornithodoros sp.	on bat host: *Erophylla sezekorni*		
Suborder Mesostigmata			
Ameroseiidae? (at least 1 sp.)	L	U	
Ologamasidae (at least 1 sp.)	L	U	
Phytoseiidae (at least 1 sp.)		U	
Podocinidae (at least 1 sp.)	L		
Spelaeorhynchidae			
Spelaeorhynchus monophylli	on bat host: *Monophyllus redmani*		
Spelaeorhynchus sp.	on bat hosts: *Artibeus jamaicensis, Monophyllus redmani*		
Spinturicidae			
Periglischrus iheringi	on bat hosts: *Artibeus jamaicensis, Stenoderma rufum*		
Periglischrus vargasi	on bat host: *Artibeus jamaicensis*		
Periglischrus sp.	on bat hosts: *Erophylla sezekorni, Monophyllus redmani*		
Spinturnix sp.	on bat host: *Eptesicus fuscus*		
Uropodidae	L		

Taxon	Foraging location		
Suborder Prostigmata			
Eupodidae (at least 1 sp.)		L	
Suborder Astigmata			
Labidocarpidae			
Paralabidocarpus artibei	on bat hosts: *Artibeus jamaicensis, Stenoderma rufum*		
Paralabidocarpus foxi	on bat host: *Stenoderma rufum*		
Paralabidocarpus stenodermi	on bat host: *Stenoderma rufum*		
Suborder Cryptostigmata			
Cymbaeremaidae			
Scapheremaeus sp.		L	
Dampfiellidae			
Beckiella sp.		L	U
Eremulidae			
Eremulus sp.		L	U
Galumnidae			
Acrogalumna/Allogalumna group		L	U
Haplozetidae			
Haplozetes sp.		L	U
Rostrozetes sp.		L	U
Malaconothridae			
Trimalaconothrus sp.		L	U
Oppiidae			
Oppia (*sensu latu*) sp.		L	U
Phthiracaridae			
Hoplophorella sp.		L	U
Plasmobatidae			
Orbiculobates sp.		L	U
Scheloribatidae			
Scheloribates sp.		L	U
Trhypochthoniidae			
Afronothrus sp.		L	U
Allonothrus sp.		L	U
Class CRUSTACEA			
Order ISOPODA			
Oniscidae			
Philoscia richmondi *		L	U
Porcellionides sp. ? *		L	U
Sphaeroniscus portoricensis		L	
Synuropus granulatus		L	
Order PODOCOPA (1 sp. ?) *	S	L	
Subclass COPEPODA (1 sp. ?) *	S	L	
Order DECAPODA			
Pomamonidae			
Epilobocera situatifrons		L	
Class CHILOPODA			
Order Scutigeromorpha			
Scutigeridae			
Antillora portoricensis		L	
Order Lithobiomorpha			
Henicopidae (1 sp.)		L	
Order Scholopendromorpha			
Cryptopidae			
Scolopocryptops ferrugineus		L	

Taxon	Foraging location			
Scolopendridae				
*Scolopendra alternans**		L	U	C
Class DIPLOPODA				
Order Polyxenida				
Lophoproctidae				
Lophoturus niveus		L		
Order Glomeridesmida				
Glomeridesmidae				
*Glomeridesmus marmoreus**		L		
Order Polydesmida				
Cryptodesmidae				
*Docodesmus maldonadoi**		L		
Liomus obscurus		L		
Liomus ramosus		L		
Stylodesmidae				
Styraxodesmus juliogarciai		L		
Chelodesmidae				
Ricodesmus stejneri		L		
Vanhoeffenidae		L		
Agenodesmus reticulatus				
Paradoxosomatidae				
Orthomorpha coarctata		L		
Order Spirobolida				
Spirobolellidae				
*Spirobelellus richmondi**		L		
Order Spirostreptida				
Epinannolenidae				
Epinannolina trinidadensis		L		
Order Stemmiulida				
Stemmiulidae				
*Prostemmiulus heatwoli**		L		
Order Siphonophorida				
Siphonophoridae				
Siphonophora portoricensis		L		
Class PAUROPODA (1 sp.)		L		
Class ELLIPURA				
Order COLLEMBOLA				
Sminthuridae				
Ptenothrix sp.	S	L		
Sphyrotheca sp.	S	L		
1 other sp.	S	L		
Entomobryidae				
Drepanocyrtus sp.		L		
Dicranocentropha sp.*	S	L		
Dicranocentruga sp.	S	L		
Entomobrya sp.		L		
Liepdocyrtinus sp.		L		
Salina sp.		L		C
Isotomidae				
Proisotoma sp.	S	L		
1 other sp.	S	L		
Poduridae				
Pseudachorutes sp.	S	L		
1 other sp.	S	L		

Taxon	Foraging location		
Class INSECTA			
Order ARCHAEOGNATHA			
Machilidae (1 sp.)		U	
Order EPHEMEROPTERA			
Leptophlebiidae (1 sp.)		U	C
Order ODONATA			
Coenagrionidae			
Enallagma coecum		U	C
Telebasis vulnerata		U	C
Aeshnidae			
Aeshna psilus		U	C
Coryphaeschna viriditas		U	C
Gynacantha nervosa		U	C
Triacanthagyna septima		U	C
Triacanthagyna ?trifida		U	C
Libellulidae			
Erythrodiplax umbrata		U	C
Macrothemis celeno		U	C
Micrathyria didyma		U	C
Orthemis ferruginea		U	C
Scapanea frontalis		U	C
Order BLATTODEA			
Blattidae			
Pelmatosilpha coriacea	L	U	
Periplaneta australasiae	L	U	C?
Blattellidae			
Aglaopteryx facies *	L	U	C
Aglaopteryx sp.			C
Cariblatta craticulata	L	U	
Cariblatta hebardi *	L	U	C
Cariblatta plagia	L	U	
Cariblattoides suave *	L	U	
Cariblattoides sp.	L	U	
Epilampra wheeleri *	L	U	
Eurycotis sp.	L	U	
Neoblattella borinquensis	L	U	
Neoblattella vomer	L	U	
Neoblattella sp. a	L	U	
Neoblattella sp. b	L	U	
Plectoptera dorsalis	L	U	
Plectoptera infulata *	L	U	
Pseudosymploce personata	L	U	
Pseudosymploce sp.	L	U	
Blaberidae			
Panchlora sagax	L	U	
Order ISOPTERA			
Termitidae			
Nasutitermes costalis *		U	C
Nasutitermes nigriceps		U	C
Parvitermes discolor	L		
Kalotermitidae			
Glyptotermes ?pubescens		U	C
Order MANTODEA			
Mantidae			
Gonatista grisea		U	C

Taxon	Foraging location			
Order DERMAPTERA				
Carcinophoridae (1 sp.)		L	U	C
Labiidae (1 sp.)		L	U	C
Order ORTHOPTERA				
Acrididae				
Schistocerca colombina			U	
Tettigoniidae				
Phaneropterinae				
Anaulocomera laticauda			U	C
Microcentrum triangulatum			U	C
Turpilia rugosa			U	C
Copiphorinae				
Erioloides sp.			U	
Neoconocephalus triops			U	
Agraecinae (1 sp.)			U	
Conocephalinae				
Conocephalus cinereus			U	
Gryllacrididae				
Gryllacridinae				
Abelona sp.		L	U	
Gryllidae				
Phalangopsinae				
Amphiacusta caraibea *		L		
Gryllinae				
Anurogryllus muticus *		L		
Gryllus assimilis		L		
Trigonidiinae				
Cyrtoxipha gundlachi *			U	C
Anaxipha sp.*			U	C
Eneopterinae				
Orocharis vaginalis			U	C
Orocharis terebrans *			U	C
Orocharis sp. a			U	C
Orocharis sp. b			U	C
Orocharis sp. c			U	C
Orocharis sp. d			U	C
Laurepa krugii *			U	C
Tafalisca lurida			U	C
Nemobiinae (1 sp.)	S	L		
Gryllotalpidae				
Scapteriscus vicinus	S	L		
Order PHASMATODEA				
Heteronemiidae				
Pseudobacteria yersiniana			U	
Phasmatidae				
Agamemnon iphimedeia			U	
Diapherodes achalus			U	
Lamponius portoricensis *			U	
Order EMBIOPTERA			U	
Teratembiidae (1 sp.)				
Order PSOCOPTERA			U	C
Polypsocidae (1 sp.) *		L	U	C
Epipsocidae (3 spp.) *		L	U	C
Psocidae (2 spp.) *		L	U	C

Taxon	Foraging location		
Pseudocaeciliidae			
Pseudocaecilius pretiosus	L	U	
1 other sp.	L	U	
Psyllopsocidae (1 sp.)	L	U	C
Lepidopsocidae (1 sp.)	L	U	C
Pachytroctidae (1 sp.)	L	U	C
Liposcelidae			
Liposcelis divinatorius *	L		
Myopsocidae (1 sp.)	L	U	C
Order HEMIPTERA			
Veliidae (2 spp.)	water surface		
Belostomatidae			
Belostoma subspinosum	in water		
Schizopteridae (3 spp.)	L		
Dipsocoridae (1 sp.)	L	U	C
Enicocephalidae (1 sp.)	L	U	
Phymatidae (1 sp.)		U	
Miridae			
Itacoris trimaculatus		U	
Itacoris nigroculus		U	
Antias miniscula		U	
Pycnoderes heidemanni		U	
Pycnoderes quadrimaculatus		U	
Fulvius anthocorides		U	
Dagbertus sp.		U	
Collaria oleosa		U	
Rhinacloa pusilla		U	
Rhinacloa pallida		U	
Diphleps unica		U	
Phytocoris ricardoi		U	
Polymerus pallidus		U	C
Cyrtopeltis modesta		U	
Parthenicus nigrosquamis		U	
1 other sp.		U	
Reduviidae			
?*Ploiaria* sp.*	L	U	
Oncerotrachelus sp.		U	
Empicoris sp.		U	
Nabidae			
Neogorpis neotropicalis		U	
Lygaeidae			
Ozophora atropicta		U	C
Ozophora subimpicta		U	C
Ozophora sp.	L		C
Pachybrachius sp.		U	
Coreidae			
Phthia rubropicta		U	
1 other sp.		U	
Aradidae (2 spp.)	L		
Saldidae (1 sp.)	shores of streams		
Cydnidae			
?*Amnestus* sp.	L	U	C
Scutelleridae			
Pachycoris fabricia		U	

Taxon	Foraging location		
Pentatomidae			
Piezosternum subulatum		U	
Loxa pilipes		U	
Acrosternum marginatum		U	
Edessa cornuta		U	
Edessa parvinula		U	
Fecelia minor		U	
Order HOMOPTERA			
Superfamily CICADOIDEA			
Cicadidae			
*Borencona aguadilla**		U	C
Membracidae			
Nessorchinus esbeltus			C
Cicadellidae			
*Sibovea coffeacola**		U	C
*Xestocephalus maculatus**	L	U	C
Xestocephalus sp. a		U	
Xestocephalus sp. b		U	
Cicadulina tortilla		U	
Hortensia similis		U	
Krisna insularis		U	
Deltocephalus flavicosta		U	
Ponana insularis			C
Protalebrella braziliensis		U	
Macrosteles fascifrons		U	
Tylozygus fasciatus		U	
Protalebra sp.		U	
Empoasca sp.		U	
Balclutha sp.		U	
Osbornellus sp.		U	
Graminella sp.		U	
Idiocerus parvulus		U	
Hybla maculata		U	
Superfamily FULGOROIDEA			
Cixiidae			
*Bothriocera undata**		U	C
Oliaris slossonae		U	C
*Pintalia alta**		U	C
*Pintalia supralta**		U	C
*Pintalia nemaculata**		U	C
Pintalia sp. nr. *nemaculata*		U	C
Pintalia martorelli		U	C
*Pintalia osborni**		U	C
Pintalia sp.		U	C
Cubana tortriciformis		U	C
Delphacidae			
Ugyops osborni		U	C
*Ugyops occidentalis**		U	C
Neomalaxa flava		U	C
Nilaparvata sp.		U	C
Abbrosoga sp.		U	C
Euidella sp.		U	C
Punana sp.		U	C
Derbidae			
*Dysimia maculata**		U	C

Taxon	Foraging location		
*Dawnaria sordidulum**		U	C
Dawnaria sp.		U	C
Patara albida		U	C
Cedusa wolcotti		U	C
Cedusa sp.		U	C
Otiocerus schonherri		U	C
Achilidae			
?*Amblycratus striatus*		U	C
*Catonia cinerea**		U	C
*Catonia dorsovittata**		U	C
*Catonia arida**		U	C
Martorella puertoricensis		U	
Quadrana punctata		U	C
Tropiduchidae			
*Ladellodes stali**		U	C
Ladellodes nepallata		U	C
Ladellodes or *Neurotmeta* sp.		U	C
Flatidae			
Petrusa epilepsis		U	C
Petrusa pivota		U	C
Petrusa torus		U	C
Petrusa rocquensis		U	C
Flatormenis pseudomarginata		U	C
Ilesia nefuscata		U	C
Puertormenis virgina		U	C
Melormenis antillarum		U	C
Melormenis basalis		U	C
*Melormenis magna**		U	C
Pseudoflatoides albus		U	C
Issidae			
Thionia borinquensis		U	C
Colpoptera maculifrons		U	C
Colpoptera brunneus		U	C
Neocolpoptera monticolens		U	C
Neocolpoptera puertoricensis		U	C
Acanalonidae			
Acanalonia agilis		U	C
Acanalonia vivida		U	C
Kinnaridae			
Quilessa fasciata		U	C
Superfamily PSYLLOIDEA			
Psyllidae (5 spp.)		U	C
Superfamily APHIDOIDEA			
Aphididae (1 sp.)		U	
Superfamily COCCOIDEA			
Coccidae			
Ceroplastes rubens		U	C
Ortheziidae (1 sp.)	L		
Diaspididae (1 sp.)	L		
Order THYSANOPTERA			
Phlaeothripidae			
At least 2 spp.*	L	U	C
Thripidae			
At least 8 spp.	L	U	C

Taxon	Foraging location			
Order NEUROPTERA				
Coniopterygidae (1 sp.)			U	
Mantispidae				
Mantispa sp.			U	
Climaciella cubana			U	
Hemerobiidae				
Nusalalia cubana			U	
Chrysopidae				
Chrysopa collaris			U	
Chrysopa nr. *cubana*			U	
Chrysopa sp. a			U	
Chrysopa sp. b			U	
Nodita sp.			U	
Ascalaphidae				
Ululodes opposita			U	
Order COLEOPTERA				
Suborder Adephaga				
Carabidae (5 spp.)		L		
Dytiscidae				
Copelatus posticatus	in water			
Suborder Polyphaga				
Hydrophilidae				
Enochrus debilis	in water			
Ptiliidae				
Actinopteryx sp.		L	U	C
Scydmaenidae (1 sp.)		L		
Silphidae (1 sp.)	S	L	U	
Scaphidiidae (1 sp.)				C
Staphylinidae				
Palaminus sp.				C
6 other spp.*	S	L		
Pselaphidae (4 spp.)*		L		
Histeridae				
Ormalodes ruficlavis			U	C
Opalides sp.			U	C
Passalidae (1 sp.)				
Paxillus crenatus		L	U	
Scarabaeidae				
Strategus oblongus		L	U	C
Phyllophaga portoricensis		L	U	C
Phyllophaga sp. a		L	U	C
Phyllophaga sp. b		L	U	C
Phyllophaga sp. c		L	U	C
Phyllophaga sp. d		L	U	C
Chalepides barbata		L	U	C
Canthonella parva		L	U	C
Canthochilum borinquensis		L	U	C
Canthochilum histeroides		L	U	C
Ataenius floridanus		L		
Dascillidae (3 spp.)			U	
Ptilodactylidae (at least 5 spp.)			U	
Chelonariidae (1 sp.)			U	
Limnichidae				
Limnichoderus insularis	in water			

Taxon	Foraging location		
Elmidae			
Neoelmis sp.	in water		
Phanocerus sp.	in water		
Elateridae			
Dicrepidius ramicornis		U	C
Pyrophorus luminosus	L	U	C
Platycrepidus sp.		U	C
2 other spp.		U	C
Throscidae (1 sp.)		U	C
Telegeusidae (1 sp.)		U	C
Lampyridae			
Callopisma borencona		U	C
Photinus triangularis		U	C
Photinus vittatus		U	C
Photinus dubiosus		U	C
Photinus sp.*	L	U	C
Cantharidae			
Tylocerus barberi		U	C
Lycidae (4 spp.)		U	
Dermestidae (1 sp.)		U	
Anobiidae (2 spp.)		U	C
Trogositidae (1 sp.)		U	C
Cleridae (2 spp.)		U	
Nitidulidae			
Europs maculata	L	U	
1 other sp.	L	U	
Rhizophagidae (1 sp.)		U	
Cucujidae (at least 3 spp.)	L	U	C
Cryptophagidae (1 sp.)		U	
Phalacridae (1 sp.)		U	
Coccinellidae			
Curinus sp.		U	
2 other spp.	L	U	C
Endomychidae (2 spp.)		U	
Tenebrionidae (at least 3 sp.)	L	U	
Colydiidae (1 sp.)	L	U	C
Oedemeridae (1 sp.)		U	
Melandryidae (1 sp.)		U	C
Mordellidae (1 sp.)		U	C
Euglenidae (at least 2 spp.)		U	C
Cerambycidae			
Parandrinae			
Parandra cribrata	L	U	C
Prioninae			
Callipogon proletarius	L	U	C
Derancistrus thomae	L	U	C
Stenodontes exsertus	L	U	C
Lepturinae			
Bellamira scalaris		U	C
Cerambycinae			
Brittonella chardoni		U	C
Chlorida festiva	L	U	C
Elaphidion tomentosus		U	C
Methia necydalea		U	C
Neoclytus araneiformis	L	U	C

Taxon	Foraging location		
Lamiinae			
Batocera rubis	L	U	C
Leptostylus antillarum		U	C
Leptostylus longicornis		U	C
Leptostylus oakleyi		U	C
Lagochirus araneiformis		U	C
Nanilla sp.		U	C
Oreodera prob. *glauca*		U	C
Proecha spinipennis		U	C
Typanidius nocturnus	L	U	C
Distenidae			
Distenia darlingtoni		U	C
Chrysomelidae			
Diabrotica sp.		U	
2 other spp.	L	U	C
Platypodidae (1 sp.)		U	C
Scolytidae			
Ambrosiodmus hagedorni		U	C
Coccotrypes carpophagus	L	U	C?
Corthylus papulans		U	C
Xyleborus affinis *	L	U	C
Xyleborus ferrugineus *	L	U	C
Xlyeborus sp.		U	
at least 2 other spp.	L	U	C
Brentidae			
Stereodermus sp. 1	probably U on tree trunks		
Belophorus maculatus	U, on tree trunks		
Belophorus sp.		U	
Brentus volvulus	U, on tree trunks and branches		
Anthribidae			
Homocloeus? conspersus?		U	
Ormiscus sp. 1		U	
Ormiscus sp. 2		U	
Phaenotheriopsis conciliatus		U	
Phaenotheriopsis sp.		U	
Euxenus? sp. 1	L	U	
Genus? (was *Neanthribus*)		U	C
Attelabidae			
Attelabinae			
Euscelus biguttatus		U	
Euscelus dentipes		U	
Euscelus sexmaculatus		U	
Rhynchitinae			
Auletobius sp. 1		U	C
Pselaphorhynchites sp. 1		U	C
Apionidae			
Cylas formicarius elegantulus		U	
Apion martinezi		U	
Apion oakleyi		U	
Apion salarium		U	
Apion subaeneum		U	
Apion sp. 1		U	
Apion sp. 2		U	
Curculionidae			
Apodrosus argentatus		U	C
Apodrosus wolcotti		U	C

Taxon	Foraging location		
Polydrosinae			
Polydacrys depressifrons		U	
Artipus sp. 1		U	
Menoetius coffeae montanus		U	
Menoetius curvipes		U	
Menoetius trilineatus		U	
Menoetius yaucona?		U	
Pachnaeus psittacus		U	
Compsus luquillo		U	C
Compsus mariacao		U	C
Diaprepes abbreviatus		U	
Diaprepes maugei		U	
Exophthalmus quindecimpunctatus		U	
Exophthalmus roseipes		U	C
Exophthalmus sphacelatus?		U	
Exophthalmus sp. 1		U	C
Molytinae			
Heilipus elegans		U	C
Conotrachelus seniculus		U	C
Conotrachelus sp. 1		U	
Conotrachelus sp. 2		U	
Conotrachelus sp. 3		U	
Anchonus sp. 1	L	U	
Anchonus sp. 2	L	U	
Smicronyx sp. 1		U	
Smicronyx sp. 2		U	
Pantoteloides sp. 1	L	U	
Derelomus? albidus?		U	
Notolomus sp. 1		U	
Phyllotrox? pallidus		U	
Nanus uniformis	L	U	
Micromyrmex pulicarius		U	
Sicoderus sp. 1		U	C
Anthronomus albocapitis		U	C
Anthonomus alboannulatus		U	C
Anthonomus annulipes		U	
Anthonomus convexifrons		U	C
Anthonomus costulatus		U	
Anthonomus dentipes		U	
Anthonomus flavus		U	
Anthonomus incanus		U	
Anthonomus nigrovarigatus?		U	
Anthonomus sp. 1		U	
Anthonomus sp. 2		U	
Pseudanthonomus sp. 1		U	
Tychiinae			
Lygnyodes sp. 1		U	
Sibinia aliguantula		U	
Sibinia pulcherrima		U	
Sibinia setosa		U	C
Sibinia sp. 1		U	
Pyropinae			
Pyropus sp. 1		U	
Cryptorhynchinae			
Pseudomus militaris		U	C
Pseudomus sp. 1		U	

Taxon	Foraging location		
Tyloderma danforthi	L	U	
Euscepes porcellus		U	C
Euscepes postfasciatus		U	C
Neoulosomus sp. 1		U	
Neoulosomus sp. 2		U	
Cryptorhynchus? sp. 1		U	
Cryptorhynchus? sp. 2		U	
Pseudomopsis cucubano?		U	C
Pseudomopsis sp. 1		U	C
Pseudomopsis sp. 2		U	C
Pseudomopsis sp. 3		U	C
Pseudomopsis sp. 4		U	C
Pseudomopsis 4–5 more spp.	L	U	
Macromerus sp. 1		U	C
Sternocoelus armipes		U	C
Eubulus sp. 1		U	C
Zygopinae			
Lechriops psidii		U	C
Lechriops sp. 1		U	C
Ceutorhynchinae			
Hypurus bertrandi		U	
Auleutes insepersus		U	
Panophthalmus puertoricanus		U	
Baridinae			
Peridinetus concentricus		U	C
Peridinetus signatus		U	
Baris torquata		U	
Ampeloglypter cissi		U	
Geraeus? montanus		U	
Anacentrinus sp.		U	
Rhynchophorinae			
Sphenophorus sp.		U	
Metamasius hemipterus	? Taken on bananas in market in El Verde		
Cosmopolites sordidus	? Taken with *Metamisius* above		
Sitophilus granarius		U	
Stiophilus linearis	U, in grain at pet store		
Sitophilus oryzae	In grain and flour products		
Cossoninae			
Cossonus impressus	U, under bark		
Decuanellus pecki	L		
Decuanellus sp. 2	L		
Decuanellus sp. 3	L		
Caulophilus oryzae	L		C
Stenotrupis acicula	L		
Stenancylus sp. 1		U	C
Micromimus sp. 1	L	U	
Dryophthorinae			
Dryophthorus sp. 1	L		
Order TRICHOPTERA			
Philopotamidae			
Chimarra albomaculata		U	C
Chimarra maldonadoi		U	

Taxon	Foraging location		
Psychomyiidae			
Antillopsyche tubicola		U	
Xiphocentron borinquensis		U	
Polycentropodidae			
Cernotina mastelleri		U	
Polycentropus zaneta		U	
Hydropsychidae			
Smicridea protera		U	
Rhyacophilidae			
Atopsyche trifida		U	
Glossosomatidae			
Cariboptila orophila		U	
Cariboptila trispinata		U	
Hydroptilidae			
Hydroptila martorelli		U	
Ochrotrichia juana		U	
Ochrotrichia marcia		U	
Ochrotrichia squamigera		U	
Ochrotrichia ceer		U	
Oxyethira janella		U	
Oxyethira puertoricensis		U	
Alisotrichia circinata		U	
Alisotrichia hirudopsis		U	
Alisotrichia setigera		U	
New genus, sp.		U	
Helicopsychidae			
Helicopsyche minima		U	
Helicopsyche ramosi		U	
Calamoceratidae			
Phyllocius pulchrus		U	
Leptoceridae			
Nectopsyche sp.		U	
Order LEPIDOPTERA			
Tineidae			
Tiquadra aeneonivella		U	
Acropholus sp. a*		U	
Acropholus sp. b*		U	
Acropholus sp. c*		U	
Acropholus sp. d*		U	
Acropholus sp. e*		U	
Acropholus sp. f*		U	
Acropholus sp. g*		U	
Acropholus sp. h*		U	
Acropholus sp. i*		U	
?Gracillariidae (at least 1 sp.)		U	C
Oecophoridae			
Ethmia zanthorrhoa		U	
Blastobasidae (1 sp.)		U	
?Cosmopterygidae (1 sp.)		U	C
Gelechiidae			
Dichomeris sp.	L	U	C
Alucitidae			
Orneodes sp.		U	
Cossidae			
*Psychonoctua personalis**		U	

Taxon	Foraging location	
Tortricidae		
Eulia sp. a	U	
Eulia sp. b	U	
Bactra sp.	U	
1 other sp.	U	
Hesperiidae		
Panoquina nero	U	
Perichares phocion	U	
Perichares philetes	U	
Choranthus vittellius	U	
Urbanus dorantes	U	
Proteides mercurius	U	
Epargyreus zestis	U	
Pyrgus syrichtus	U	
Wallengrenia otho druryi	U	
Papilionidae		
Papilio pelaus	U	C
Pieridae		
Dismorphia spio *	U	C
Phoebis sennae	U	C
Phoebis phileae	U	C
Phoebis trite	U	C
Phoebis argante	U	C
Eurema portoricensis	U	C
Nymphalidae		
Heliconiinae		
Heliconius charitonius	U	C
Dryas julia	U	C
Satyrinae		
Calisto nubila	U	C
Charaxinae		
Prepona antimache	U	C
Apaturinae		
Adelpha gelania	U	C
Nymphalinae		
Marpesia petreus	U	C
Hypanartia paullus	U	C
Anartia jatrophe	U	C
Siproeta stelenes	U	C
Lycaenidae		
Chlorostrymon maesites	U	C
Electrostrymon angelica	U	C
Megalopygidae		
Megalopyge krugii *	U	
Pyralidae		
Pyraustinae		
Sparagmia gigantalis *	U	
Pantographa limata	U	
Terastia meticulosalis	U	
Azochis rufidiscalis	U	
Margaronia flegia	U	
Margaronia costata	U	C
Margaronia elegans *	U	
Margaronia nitidalis	U	
Margaronia marginepuncta	U	
Margaronia sibillalis	U	

Taxon	Foraging location
Margaronia sp.	U
Sylepta onophasalis	U
Sylepta elevata	U
Sylepta ceresalis	U
Sylepta silicalis	U
Sylepta sp. a	U
Sylepta sp. b	U
Pycnarnon receptalis	U
Mesocondyla concordalis	U
Mesocondyla sp.	U
Crocidolomyia palindalis	U
Neoleucinodes elegantalis	U
Pyrausta cerata	U
Pyrausta cardinalis	U
Pyrausta sp.	U
Phostria humeralis	U
Phostria simialis	U
Phostria prolongalis	U
Desmia tages *	U
Desmia ufeus	U
Maruca testulalis	U
Pilocrocis ramentalis	U
Pilocrocis infuscalis	U
Pilocrocis lauralis	U
Epipagis mopsalis	U
Syngamia florella	U
Syngamia cassidalis	U
Syngamia sp.	U
Herpetogranma phaeropteralis	U
Herpetogranma perusialis	U
Lygropia lelex	U
Bradina hemingalis	U
Hileithia ductalis	U
Diasemia ramburialis	U
Samea carrelalis	U
Lamprosema zoilusalis	U
Lamprosema indicata	U
Lamprosema stenialis	U
Lineodes metagrammalis	U
Argyractis serapionalis	U
Argyractis sp. a	U
Argyractis sp. b	U
Cataclysta sumptiosalis	U
Cataclysta miralis	U
Scoparia sp.	U
Gonopionea sp.	U
Condolorrhiza sp. a	U
Condolorrhiza sp. b	U
Undulambia sp.	U
Pyralinae	
Pyralis manihotalis	U
Epipaschiinae	
Jocara ferrifusalis	U
Jocara sp.	U
Tetralopha scabridella	U

Taxon	Foraging location
Tetralopha sp.	U
Pococera atramentalis	U
Crambinae	
Argyria lacteela	U
Diatraea saccharalis	U
Crambus sp.	U
Chrysauginae	
Pachymorphus subductellus	U
Caphys bilinea	U
Parachma sp.	U
Schoenobiinae	
Rupela sp. a	U
Rupela sp. b	U
Rupela sp. c	U
Thyrididae	
Rhodoneura leuconotula	U
Rhodoneura thiastoralis	U
Rhodoneura myrsusalis	U
Pterophoridae	
Oidaematophorus basalis	U
Sphenarches caffer	U
Adaina sp. a	U
Adaina sp. b	U
Platyptilia sp.	U
Geometridae	
Microgonia vesulia	U
Sphaecelodes vulneraria	U
Semaeopus perletaria	U
Drepanodes hamata	U
Cambogia mexicaria	U
Racheospila sanctae-crucis	U
Racheospila gerularia	U
Racheospila herbaria	U
Racheospila sp.	U
Pleuroprucha rudimentaria	U
Hammaptera chloronotata	U
Tricentrogyna vinacea	U
Tricentrogyna floridora	U
Cloropteryx paularia	U
Sterrha sp.	U
Scopula sp. a	U
Scopula sp. b	U
Scopula sp. c	U
Psaliodes sp.	U
Bronchelia sp.	U
Phrygionis sp.	U
Semiothisa sp.	U
Sphingidae	
Manduca sexta	U
Erinnyis alope	U
Erinnyis ello	U
Pholus fasciatus	U
Xylophanes tersa	U
Pachylia ficus	U
Aellopos fadus	U

Taxon	Foraging location
Aellopos sp.	U
1 other sp.	U
Notodontidae	
Rifargia distinguenda	U
Proelymniotis aequipars	U
Disphragis baracoana	U
Disphragis sp.	U
Arctiidae	
Pericopinae	
Ctenuchida virginalis	U
Hyalurga vinosa	U
Arctiinae	
Eupseudosoma involutum	U
Ecpantheria icasia	U
Ecpantheria sp.	U
Utethesia ornatrix	U
Phegoptera bimaculata	U
Tricypha proxima	U
Lomuna negripuncta	U
Talaria sp.	U
Ctenuchinae	
Cosmosoma auge	U
Cosmosoma achemon	U
Lymire flavicollis	U
Correbida terminalis	U
Nyridela chalciope	U
Eunomia colombina	U
Euceron sp.	U
Noctuidae	
Blosyris mycerina	U
Ophisma tropicalis	U
Gonodonta sicheus	U
Gonodonta incurva	U
Mocis diffluens	U
Mocis megas	U
Prodenia pulchella	U
Prodenia rubrifusa	U
Prodenia eridania	U
Heliothis virescens	U
Eulepidotis addens	U
Heterochroma berylliodes	U
Heterochroma sp.	U
Ephrodes cacata	U
Sylectra erycata	U
Condica cupentia	U
Messala obvertens	U
Speocropia scriptura	U
Mastigophorus demissalis	U
Phlyctaina irregularis	U
Mamestra soligena	U
Gonodes liquida	U
Metalectra analis	U
Lascoria phormisalis	U
Anepischetos porrectalis	U
Anepischetos mactatalis	U
Phalaenophana eudorealis	U

Taxon	Foraging location	
Carteris oculatalis	U	
Callipistra floridensis	U	
Callipistra jamaicensis	U	
Plusia admonens	U	
Araeoptera vilhelmina	U	
Afrida tortriciformis	U	
Nymbis garnoti	U	
Leucania rosea	U	
Leucania sp.	U	
Plusiodonta sp. a	U	
Plusiodonta sp. b	U	
Calpe sp.	U	
Diptherigia sp.	U	
Bleptina sp. a	U	
Bleptina sp. b	U	
Antiblemma sp.	U	
Diomyx sp.	U	
Tortricoides orneodalis	U	
Tortricoides sp. a	U	
Tortricoides sp. b	U	
Thursania sp.	U	
Physula sp.	U	
Pseudaletia sp.	U	
Lascoria sp.	U	
Zale sp.	U	
Nola bistriga	U	
Order DIPTERA		
Suborder Nematocera		
Tipulidae*		
Tipulinae		
Dolichopeza puertoricensis	U	C?
Brachypremna unicolor	U	C?
Limoniinae		
Helius albitarsus	U	C?
Limonia diva	U	C?
Limonia gowdeyi	U	C?
Limonia cinereinota	U	C?
Limonia tibialis	U	C?
Limonia myersiana	U	C?
Limonia subrecisa	U	C?
Limonia rostrata antillarum	U	C?
Limonia tetraleuca	U	C?
Limonia domestica	U	C?
Limonia sp. k	U	C?
Limonia sp. hh	U	C?
Limonia willistoniana	U	C?
Limonia schwarzi	U	C?
Limonia sp. aa	U	C?
Limonia hoffmani	U	C?
Limonia divisa	U	C?
Limonia trinitatis	U	C?
Limonia sp. t	U	C?
Limonia sp. bb	U	C?
Atarba sp.	U	C?
Elephantomyia westwoodi	U	C?
Polymera geniculata pallipes	U	C?

Taxon	Foraging location	
Hexatoma sp. a	U	C?
Hexatoma sp. b	U	C?
Psiloconopa portoricensis	U	C?
Psiloconopa caliptera	U	C?
Teucholabis sp. gg	U	C?
Gonomyia pleuralis	U	C?
Gonomyia puer	U	C?
Gonomyia subterminalis	U	C?
Trentepohlia nivetarsis	U	C?
Trentepohlia sp. kk	U	C?
Shannonomyia leonardi	U	C?
Shannonomyia sp. p	U	C?
Shannonomyia sp. m	U	C?
Blephariceridae		
Paltostoma argyrocincta	U	
Mycetophilidae *		
Leia sp.	U	C?
Manota sp.	U	C?
Platyura sp. d	U	C?
Platyura sp. n	U	C?
Platyura sp. o	U	C?
Platyura sp. x	U	C?
Platyura sp. y	U	C?
Megopthalmida sp.	U	C?
Boletina incompleta	U	C?
Boletina sp.	U	C?
Neompheria sp.	U	C?
Zygomyia sp. h	U	C?
Zygomyia sp. aa	U	C?
Zygomyia sp. bb	U	C?
Zygomyia sp. cc	U	C?
Exechia sp. a	U	C?
Exechia sp. c	U	C?
Exechia sp. q	U	C?
Exechia sp. u	U	C?
Exechia sp. dd	U	C?
Rhymosia sp.	U	C?
Mycetophilia sp. f	U	C?
Mycetophilia sp. m	U	C?
Mycetophilia sp. p	U	C?
Mycetophilia sp. r	U	C?
Mycetophilia sp. s	U	C?
Mycetophilia sp. w	U	C?
Sciaridae (33 spp.) *	U	C
Cecidomyiidae (30 spp.)	U	C
Psychodidae (37 spp.) *	U	C
Scatopsidae		
Rhegmoclema sp.	U	C
Aldrovandiella sp.	U	C
Dixidae		
Dixa sp.	U	C?
Chaoboridae		
Chaoborus brasiliensis	U	C
Chaoborus sp. e	U	C
Chaoborus sp. c	U	C

Taxon	Foraging location	
Culicidae		
Toxorhynchites portoricensis	U	C
Aedes mediovittatus	U	C
Aedes taeniorhynchus	U	C
Aedes serratus	U	C
Culex nigripalpus	U	C
Culex pipiens quinquefasciatus	U	C
Culex sp.	U	C
Wyeomia sp.	U	C
Uranotaenia sp.	U	C
Mansonia flaveolus	U	C
Simuliidae (2 spp.)	U	C
Ceratopogonidae		
Monohelea johannseni	U	C
Culicoides hoffmani	U	C
Polpomyia sp. n	U	C
Atrichopogon sp. r	U	C
Atrichopogon sp. s	U	C
Atrichopogon sp. t	U	C
Stilobezzia bimaculata	U	C
Stilobezzia sp. h	U	C
Stilobezzia sp. q	U	C
Dasyhelea sp. b	U	C
Dasyhelea sp. e	U	C
Dazyhelea sp. ee	U	C
Dasyhelea sp. ff	U	C
Dasyhelea sp. jj	U	C
Forcipomyia glauca	U	C
Forcipomyia fuliginosa	U	C
Forcipomyia genualis	U	C
Forcipomyia corsoni	U	C
Forcipomyia pluvialis	U	C
Forcipomyia sp. a	U	C
Forcipomyia sp. c	U	C
Forcipomyia sp. d	U	C
Forcipomyia sp. f	U	C
Forcipomyia sp. g	U	C
Forcipomyia sp. j	U	C
Forcipomyia sp. k	U	C
Forcipomyia sp. l	U	C
Forcipomyia sp. v	U	C
Forcipomyia sp. x	U	C
Forcipomyia sp. y	U	C
Forcipomyia sp. aa	U	C
Forcipomyia sp. bb	U	C
Forcipomyia sp. cc	U	C
Forcipomyia sp. hh	U	C
Chironomidae		
Orthocladiinae		
Corynoneura sp. (near water)	U	
Cricotopus sp. (near water)	U	
Diplosmittia sp. (near water)	U	
Limnophyes sp. (near water)	U	
Parametriocnemus sp. (near water)	U	
Thienemanniella sp. (near water)	U	
Pseudosmittia sp. (near water)	U	

Taxon	Foraging location		
Unknown Orthoclad genus 1 (near water)		U	
Unknown Orthoclad genus 2 (near water)		U	
Chironomini			
Nilothauma? n. sp. (near water)		U	
Paralauterborniella? sp. (near water)		U	
Polypedilum sp. 1 (near water)		U	
Polypedilum sp. 2 (near water)		U	
Polypedilum sp. 3 (near water)		U	
Polypedilum sp. 4 (near water)		U	
Stenochironomus cf. *innocuus* (near water)		U	
Stenochironomus sp. 1 (near water)		U	
Xestochironomus furcatus (near water)		U	
Xestochironomus cf. *nebulosus* (near water)		U	
Tanytarsini			
Tanytarsus sp. 1 (near water)		U	
Tanytarsus sp. 2 (near water)		U	
Rheotanytarsus sp. (near water)		U	
Tanypodinae			
Ablabesmya sp. (near water)		U	
Dialmabatista sp. (near water)		U	
Labrundinia sp. (near water)		U	
Larsia sp. (near water)		U	
Pentaneura sp. (near water)		U	
About 50 unidentified spp.*	L	U	C
Suborder Brachycera			
Tabanidae			
Stenotabanus brunettii		U	
Rhagionidae (1 sp.)		U	
Xylophagidae (1 sp.)		U	
Stratiomyidae			
Hermetia illucens		U	
Hermetia sexmaculata		U	
Nothomyia nigra		U	
3 other spp.		U	
Asilidae			
Andrenosoma chalybeum		U	C
Empididae (6 spp.)		U	C
Dolichopodidae*			
Condylostylus graenicheri		U	C
Condylostylus flavicornis		U	C
Condylostylus sp. r		U	C
Condylostylus sp. d		U	C
Pelastoneurus sp.		U	C
Neurigona sp.		U	C
Thrypticus sp.		U	C
Chrysotus flavohirtus		U	C
Chrysotus sp. a		U	C
Chrysotus sp. c		U	C
Chrysotus sp. h		U	C
Chrysotus sp. j		U	C
Chrysotus sp. g		U	C
Chrysotus sp. l		U	C
2 other spp.		U	C
Phoridae			
Chaetopleurophora formosa		U	C?
Diploneura picea		U	C?

Taxon	Foraging location		
Macrocerides brevicornis		U	C?
Megaselia basichaeta		U	C?
Megaselia defecta		U	C?
Megaselia fausta		U	C?
Megaselia perspicua		U	C?
Megaselia subfava		U	C?
Megaselia subinflava		U	C?
Megaselia violata		U	C?
Metopina reflexa		U	C?
Pseudacteon simplex		U	C?
Puliciphora borinqensis		U	C?
Puliciphora parvula		U	C?
51 other spp.		U	C?
Syrphidae			
Meromacrus cinctus		U	
Ornidia obesa		U	
Eristalis cubensis		U	
Baccha capitata		U	
Baccha latiuscula		U	
Baccha deceptor		U	
Baccha parvicornis		U	
Baccha gracilis		U	
Baccha cylindrica		U	
Eristilis albifrons		U	
?*Mesograpta arcifera*		U	
Mesograpta sp. *verticalis* or *floralis*		U	
Mesograpta violacea		U	
Mesograpta sp.		U	
?*Parapenium banksi*		U	
Volucella tricincta		U	
Pipunculidae (1 sp.)		U	
Micropezidae			
Taeniaptera lasciva		U	
Taeniaptera sp.		U	
Systellapha scurra		U	
Systellapha sp.		U	
Neriidae (3 spp.)		U	
Lonchaeidae			
Lonchaea sp.		U	C
Silba sp.		U	C
Otitidae			
Euxesta thomae		U	
Euxesta sp.		U	
Tephritidae			
Anastrepha sp.		U	C
1 other sp.		U	
Clusiidae (3 spp.)	L	U	
Odiniidae			
Odinia biguttata	L	U	C
Agromyzidae			
Melanagromyza sp.		U	C
Milichiidae (2 spp.)		U	
Sepsidae			
Paleosepsis scabra		U	
Lauxaniidae			
Pseudogriphoneura albovittata		U	C

Taxon	Foraging location		
Pseudogriphoneura octopunctata		U	C
Pseudogriphoneura sp.		U	C
Neogriphoneura sordida		U	C
Poecilominettia picticornis		U	C
Sapromyza sp.		U	C
Minetta octopuncta		U	C
?Chamaemyiidae (1 sp.)		U	C
Heleomyzidae (1 sp.)		U	C
Sphaeroceridae (4 spp.)	L	U	
Curtonotidae (1 sp.)		U	C
Drosophilidae			
Drosophila sp. d		U	C
Drosophila sp. e		U	C
Drosophila sp. f		U	C
Aulacigaster sp.		U	C
4 other spp.		U	C
Ephydridae (1 sp.)		U	C
Chloropidae			
Oscinella lutzi		U	C
Pentanotaulax sp.		U	C
Muscidae			
Neomuscina sp.		U	C
Neodexiopsis ditiportus		U	C
Neodexiopsis rex		U	C
Neodexiopsis cavalata		U	C
Neodexiopsis discolorisexus		U	C
Neodexiopsis crassicrurus		U	C
Neodexiopsis maldonadoi		U	C
Bithoracochaeta sp.		U	C
4 other spp.		U	C
Calliphoridae			
Phaenica rica	L	U	C
Sarcophagidae			
Paraphrissopoda capitata		U	
Sarcophaga sp. a		U	
Sarcophaga sp. d		U	
11 other spp.		U	
Tachinidae			
Euphasiopteryx dominicana		U	
Eucletoria armigera		U	
Tachinophyta sp.		U	
At least 17 other spp.		U	C
Hippoboscidae (1 sp.)	On bats		
Streblidae			
Aspidoptera sp.	On bat host: *Artibeus jamaicensis*		
Megistropoda aranea	On bat host: *Artibeus jamaicensis*		
Icterophilia sp.	On bat host: *Monophyllus redmani*		
Trichobius sp. nr. *sparsus*	On bat host: *Monophyllus redmani*		
Trichobius sp.	On bat hosts: *Artibeus jamaicensis, Monophyllus redmani*		

Taxon	Foraging location		
Order HYMENOPTERA			
Suborder Apocrita			
Braconidae			
Apanteles carpatus		U	C
Heterospilus sp.		U	C
Xenarcha sp.		U	C
Ecphylus sp.		U	C
Orthostigma sp.		U	C
Spathius sp.		U	C
Macrocentrus sp.		U	C
Clinocentrus sp.		U	C
Ichneumonidae (4 spp.)		U	C
Mymaridae (13 spp.)*		U	C
Trichogrammatidae (3 spp.)		U	C
Eulophidae (14 spp.)*		U	C
Encyrtidae (7 spp.)		U	C
Eupelmidae (1 sp.)		U	C
Agaonidae			
Blastophaga sp.		U	C
Torymidae (1 sp.)		U	C
Pteromalidae (at least 1 sp.)		U	C
Cynipidae			
Hypolethria sp.		U	C
Kleidotoma sp.		U	C
2 other spp.		U	C
Ceraphronidae			
Ceraphron sp. a		U	C
Ceraphron sp. b		U	C
Ceraphron sp. c		U	C
Ceraphron sp. d		U	C
Ceraphron sp. h		U	C
Aphanogmus sp.		U	C
Aphanogmus sp.		U	C
Aphanogmus sp.		U	C
Aphanogmus sp.		U	C
Diapriidae (12 spp.)		U	C
Scelionidae (17 spp.)*	L	U	C
Platygasteridae (9 spp.)		U	C
Bethylidae (4 spp.)		U	C
Dryinidae (1 sp.)	L	U	C
Formicidae			
Ponerinae			
Amblyopone falcata	L		
Anochetus mayri	L		
Hypoponera puntatissima	L		
Hypoponera opacior	L		
Odontomachus brunneus	L		
Odontomachus bauri	L		
Pachycondyla stigma	L		
Platythyrea punctata	L		
Myrmicinae			
Cyphomyrmex minutus	L		
Macromischa leptothorax	L		
Monomorium ebeninum	L		
Monomorium floricola	L	U	C
Mycocepurus smithi	L		

Taxon	Foraging location		
Myrmelachista ramulorum *		U	C
Wasmannia auropunctata	L		
Pheidole moerens	L		
Pheidole subarmata	L		
Pheidole sp.	L		
Solenopsis azteca	L		
Solenopsis corticalis	L		
Solenopsis geminata	L		
Solenopsis sp. a	L		
Solenopsis sp. b	L		
Strumigenys eggersi	L		
Strumigenys gundlachi	L		
Strumigenys rogeri	L		
Tetramorium bicarinatum	L		
Dolichoderinae			
Linepithema mellea (formerly *Iridomyrmex melleus*)	L	U	C
Tapinoma littorale	L		
Tapinoma melanocephalum	L		
Formicinae			
Brachymyrmex heeri	L	U	C
Paratrechina cisipa		U	
Paratrechina longicornis	L	U	
Paratrechina myops	L	U	
Paratrechina steinheili	L	U	
Paratrechina sp. a	L	U	
Paratrechina sp. b	L		
Paratrechina sp. o	L		
Paratrechina sp. r	L		
Camponotus sp. 1	L?		
Camponotus sp. 2	L?		
Scoliidae			
Campsomeris atrata	L	U	
Pompilidae			
Pepsis ruficornis		U	
Vespidae			
Mischocyttarus phthisicus *		U	
Polistes crinitus *		U	
Sphecidae			
Sphex ichneumoneus		U	
2 other spp.		U	
Halictidae (2 spp.)		U	
Apidae			
Apis mellifera *		U	C
Centris haemorrhoidalis		U	
Xylocopa mordax *		U	

Sources: Assignment of taxa to the foraging locations is based on personal observations by J. Alvarez, R. W. Garrison, W. J. Pfeiffer, and M. R. Willig, and on knowledge of general habitat associations pertaining to certain arthropod groups. The list augments that of Drewry (1970b).

Note: This list includes arthropods in soil (S), litter (L), understory (U), and canopy (C) strata.

* Major component of the food web based on abundance and/or size.

7

Arboreal Arachnids

William J. Pfeiffer

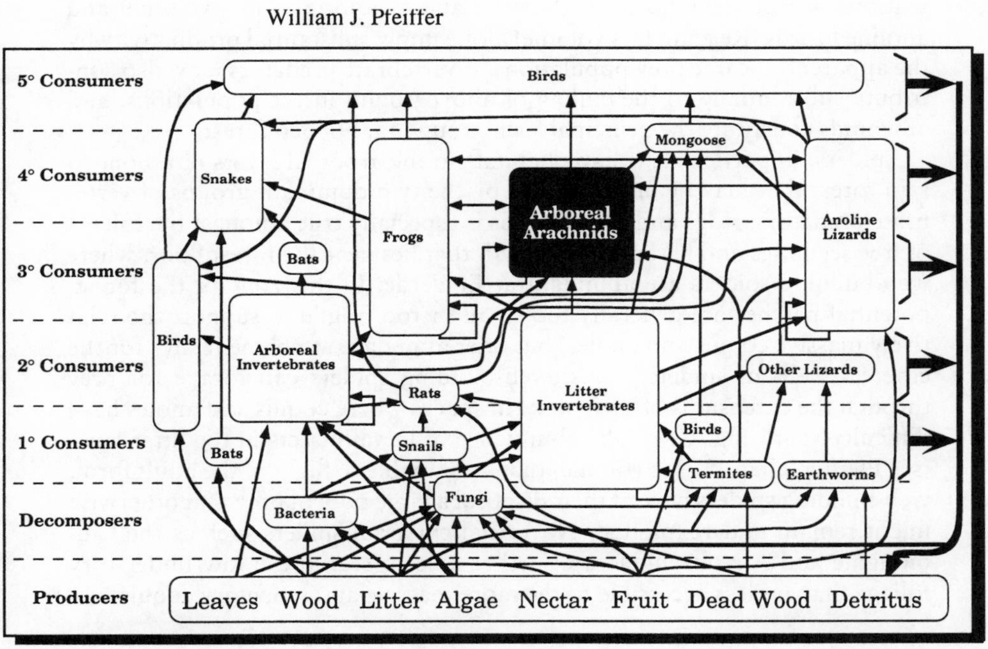

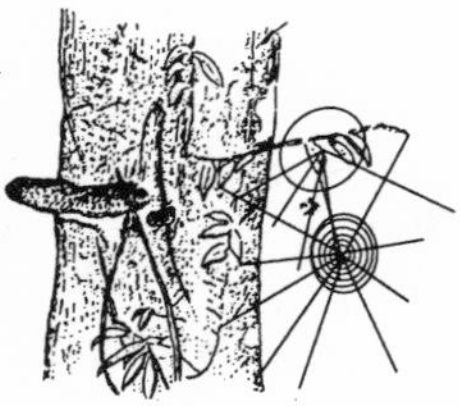

A STRIKING aspect of the tabonuco rain forest at El Verde is the apparent scarcity of insects in the understory and canopy, especially during daylight hours. This phenomenon is not unique to tabonuco rain forest; Elton's (1973, 1975) surveys indicated a paucity of invertebrates in rain forest understories in Brazil and Panama, which he attributed to heavy predation pressure. Predators are the most conspicuous trophic segment of the arboreal faunal community in the tabonuco rain forest, where the dense populations of leptodactylid frogs (Stewart and Woolbright, this volume) and anoline lizards (Reagan, this volume) alone imply substantial productivity by the apparently scarce prey populations. Invertebrate predators may also contribute substantially to the culling of aboveground insect populations, and arachnids constitute the principal such group in tabonuco forest.

Spiders, in particular, occupy habitats in the arboreal layers of tabonuco rain forest that may be underutilized by the two dominant groups of vertebrate predators: frogs and lizards. This is especially true amongst the foliage of tree seedlings and herbaceous growth that lies near the forest floor, where web-building spiders are abundant at El Verde. In this zone of the forest, potential perches are generally too small or too fragile to support the relatively massive coquis and anoles, but offer abundant attachment sites for the ethereal webs of spiders. Thus, web-building spiders can forage for prey through the extensions of their webs in spaces where coquis and anoles have difficulty venturing. Given the abundance of larval insects in the litter layer (see Pfeiffer, chap. 5, this volume) that are capable of flight in the adult form, web-building spiders have at their disposal a large pool of prey that otherwise might remain underexploited. Nocturnal cursorial spiders such as the Clubionidae and Anyphaenidae are further able to search the low understory foliage that are less accessible to the more massive and sedentary coquis.

COMMUNITY STRUCTURE

Representatives of five arachnid orders have been observed in the arboreal strata of the tabonuco rain forest at the El Verde site. Aboveground activity of the scorpion *Tityus* obtusus (Scorpiones), the tailless whip scorpion *Phrynus longipes* (Amblypigida), and at least one species of harvestman (Opiliones) is largely restricted to tree trunks, although *P. longipes* is occasionally sighted on boulders. One Pseudoscorpionida species has been recovered only from the undersurface of bark on a single dead tree at a height of 10 m, but it may be widespread and abundant in such inaccessible habitats. Spiders, as the most numerous and most readily quantifiable arachnid predators in the arboreal strata, have been studied in greatest detail at El Verde and thus will be the focus of this chapter.

Forty-one species of spiders have been collected from the arboreal strata of the tabonuco forest at El Verde (see app. 6 in Garrison and Willig, this volume). Of these, twenty-two species appeared in the understory transects along with one species that was unidentifiable (and not collected) during the censuses (table 7.1). Four species of web-building spiders dominated the diurnally active community in these transects, accounting for 95.7% of the mean annual density of 36,194 individuals ha^{-1}. Principally diurnal species that capture prey without the assistance of snares were of little numerical significance and were largely represented by several small jumping spider (Salticidae) species and a cryptic crab spider (Thomisidae) closely resembling tree bark. The high densities of anoles in the arboreal layers of this forest may preclude the presence of moderate to large species of diurnal hunting spiders that possess slow rates of maturation and hence a lesser likelihood of avoiding predation prior to reaching reproductive age. Cursorial and ambush predators are of much greater significance at night, when giant crab spiders (Sparassidae), ghost spiders (Anyphaenidae), and sac spiders (Clubionidae) forage on understory vegetation, and bark spiders (Selenopidae: *Selenops* spp.) forage on tree trunks.

Table 7.2 compares the community composition of diurnal web-building spiders observed in El Verde transects with two other tropical sites where such assemblages have also been surveyed in the lower 2 m of the understory over an annual cycle. In each of these other sites only one transect, out of three on New Guinea and four on Barro Colorado Island, bear close enough resemblance to the El Verde site in terms of topography, understory shrub and the ground growth, and absence of ecotone habitat to be suitable for comparison. The Barro Colorado Island transect (#2) was situated in a moist lowland tropical forest subject to a distinct and extended dry season. Located in a moist secondary growth forest on the site of a former coffee plantation at an altitude of 1150 m, the New Guinea transect (#1) supported reduced understory growth and sustained a less seasonal precipitation pattern than the Panamanian site (Robinson et al. 1974; Lubin 1978).

The species impoverishment of invertebrates in the El Verde forest reflected elsewhere (e.g., litter arthropods in Pfeiffer, chap. 5, this volume, and arboreal insects in Garrison and Willig, this volume) is also apparent in the diurnal web-building spider community of the lower understory. Compared with the forest transects on Barro Colorado Island, the El Verde transects

Table 7.1. Spiders from El Verde understory transects

Species	March	May	June–July	August	October	Nov–Dec	MAD
Leucauge regnyi	16,433	10,833	12,467	28,933	18,767	20,800	18,039
	2,327	1,020	1,933	6,238	2,693	3,214	2,661
Miagrammopes animotus	3,767	4,633	9,200	10,967	9,267	9,767	7,934
	658	545	1,120	792	1,894	937	1,214
Modisimus signatus	6,233	5,700	4,900	5,000	4,500	5,500	5,306
	1,697	1,423	1,075	911	1,105	1,175	256
Theridiosoma nechdomae	2,967	2,600	2,433	4,900	3,767	3,567	3,372
	1,090	242	654	912	765	875	373
Wendilgarda clara	833	533	567	1,300	867	933	839
Alcimosphenus borinquenae	133	33	400	300	233	167	211
Witica crassicauda	333	267	133		100	33	144
Micromerys dalei			100	100	200	67	78
Emanthis spp.	67		33	233		33	61
Corythalia gloriae	33	67		133		67	50
Stasina portoricensis				200			33
Argyrodes spp.				67	33		17
Clubiona portoricensis			33		33		11
Micrathena militaris	33				33		11
Mimetus portoricensis	33			33			11
Aysha tenuis				33			6
Chrysometa hamata				33			6
Epicaudus mutchleri				33			6
Eriophora edax						33	6
Gasteracantha cancriformis					33		6
Nephila clavipes				33			6
Unidentified			133	33	67	33	44
All species	30,867	24,667	30,400	52,333	37,867	41,033	36,194
	4,574	2,039	4,051	7,774	6,477	6,486	4,006

Notes: Densities are given as individuals ha^{-1}. The standard error is given below the mean density estimate. MAD = mean annual density.

Table 7.2. Web-building spider communities in understories of tropical forests

	Puerto Rico	Panama	New Guinea
Web-building species	14	35	22
Non-orb-building species	5	13	5
% In most abundant species	50	34	—
% In 4 most abundant species	96	61	—
% In following families:			
Pholcidae	15	7	0
Uloboridae	22	18	0
Tetragnathidae	0	0	17
Theridiosomatidae	12	0	0
Araneidae	51	31	70
Theridiidae	0	41	10

Sources: El Verde, Puerto Rico (this study); Barro Colorado Island, Panama (Lubin 1978); Wau, New Guinea (Robinson et al. 1974).

contained only 40% as many species of diurnal web-building spiders, in addition to a greater degree of dominance by a few species (table 7.2). Considering the total aboveground assemblage of forty-one spider species that have been collected at El Verde, the species richness in this forest also pales in comparison with the English oak stand studied by Turnbull (1960b), where virtually all of the ninety-eight species collected in that forest occurred during some time of the year in the arboreal layers. Vegetation clipped at heights of 7.5 to 16 m in Polish oak, pine, and spruce stands also yielded a greater diversity of, respectively, sixty-eight, seventy, and sixty-four species (Dziabaszewski 1976).

The relative numerical contribution by diurnally foraging individuals from major web-building families, which tend to employ distinctive styles of web architecture, also varied among the three tropical sites. Individuals constructing non-orb webs (69.0%) appear to dominate the diurnal community at the Panamanian site, whereas orb-webs represent the most common snare structure in the Puerto Rican (60.6%) and New Guinea (87.9%) rain forests. The omission of a minute orb-weaving *Theridiosoma* in the Panamanian counts, however, understates to some extent the relative contribution of the orb-weaving guild. Lubin (1978) attributed the low proportion of three-dimensional webs in the New Guinea secondary forest to reduced understory growth that offered less dense structural support for the construction of such web-forms.

Leucauge regnyi (Araneidae) was the dominant species in the understory, stairwell, and canopy transects in tabonuco forest (app. 7), where this orb-weaver accounted for approximately half of all spiders recorded during the censuses (table 7.3). At the extremes of the vertical profile the species compositions were largely dissimilar. Only one of the five next most abundant species in the understory, representing an additional 48.8% of individuals

Table 7.3. Species composition of diurnal spider communities in tabonuco forest at El Verde

	Understory	Stairwell	Walkway
Leucauge regnyi	49.8	60.4	49.3
Miagrammopes animotus	21.9	5.7	1.5
Modisimus signatus	14.7	28.3	0.0
Theridiosoma nechdomae	9.3	0.0	0.0
Wendilgarda clara	2.3	0.0	0.0
Alcimosphenus borinquenae	0.6	0.0	0.0
Gasteracantha cancriformis	0.02	3.8	28.4
Cyclosa walckenaeri	0.0	0.0	3.7
Argyrodes spp.	0.05	0.0	3.7
Total individuals in census	6515	53	134

Note: Species composition given as percentage of individuals in the census.

from this strata, appeared in the canopy, and that species accounted for only 1.5% of the total observed. Conversely, *Gasteracantha cancriformis* (Araneidae: variety *tetracantha,* see Petrunkevitch 1930a), an associated kleptoparasite *Argyrodes nephilae* (Theridiidae), and *Cyclosa walckenaeri* (Araneidae) were numerically important components of the canopy community that only rarely occupied the understory transects.

ARACHNIDS IN THE INVERTEBRATE COMMUNITY

Few estimates of arachnid populations in the aboveground strata of forests have appeared in the literature, primarily because of the difficulty in sampling such heterogeneous and largely inaccessible habitats. Nevertheless, the existing estimates suggest the importance of arachnids, especially spiders, in the invertebrate predator community occupying the arboreal layers of forests. Crossley et al. (1976) determined that spiders comprised 52% and phalangids (= Opiliones) another 22% of the predatory arthropod biomass associated with the summer canopy of a mixed hardwood forest in North Carolina. Summer standing stocks of spiders accounted for 24 to 29% of the total arthropod community associated with the canopy foliage of a Tennessee growth forest dominated by *Liriodendron tulipfera* (Reichle and Crossley 1967), compared with 11% in the North Carolina forest. Within a black spruce stand in Alaskan taiga, spiders contributed 22% of the aboveground arthropod biomass during the summer (Werner 1976).

Surveys in the tropics indicate a similar contribution by arachnids to the arthropod community occupying low understory vegetation, although relative population estimates for the entire macroinvertebrate community have been limited to just two sites. In the Mocambo Forest on the Amazon delta, spiders accounted for 30% of all invertebrates swept from vegetation at heights from 0.15 to 1.8 m above the forest floor (Elton 1973). Visual censuses conducted on Barro Colorado Island during the wet season indicated an even greater dominance of the invertebrate community by predaceous arachnids, with 42% of all invertebrates identified as spiders and another 8% as opilionids (Elton 1973).

SPIDER POPULATIONS OF THE ARBOREAL STRATA

Population Dynamics

Densities of the diurnal spider community fluctuated in the same general pattern for the six transects of tabonuco understory (fig. 7.1). Mean spider densities ranged from a minimum of 24,667 ± 2,039 (standard error) ha $^{-1}$ just prior to the termination of the 1984 dry period (compare fig. 7.1 with rainfall pattern in fig. 5.1 of Pfeiffer, chap. 5, this volume) to a peak of 52,333 ± 7,774 ha $^{-1}$ in the August censuses. Analysis of variance revealed significant differences among spider densities over time (F = 3.107, d.f. = 5,30; $p < .025$), although a multiple comparison test (least significant difference method) clearly distinguished only the August mean from the others.

Robinson et al. (1974) found that densities of immature spiders at Wau, New Guinea, increased markedly during the months of greatest rainfall and suggested that the variation in the numbers of both immature and adult spiders was linked to the influence of precipitation patterns on prey productivity and activity. In the El Verde transects the increase in overall densities from the June-July to August censuses was largely attributable to large increases in the number of early instar *L. regnyi.* The apparent burst of reproductive activity could be related to the mass emergence of adult Diptera and Lepidoptera from the litter layer during the early summer, which may also have increased the survivorship of the early nymphal stages of *L. regnyi.* Litter collections made during the census period indicated densities of 21 × 10^{6} Diptera ha $^{-1}$ in mid-June (Pfeiffer, chap. 5, this volume).

Lubin (1978) reported a similar seasonal pattern for the numerical dynamics of a web-building spider community occupying rain forest undergrowth on Barro Colorado Island. Minimum densities of web-building spiders were supported in Transect 2 during the latter portion of a four-month dry season that commenced in January. Lubin noted that virtually none of the web-building species on Barro Colorado Island achieved peak abundance during the dry season and suggested that population sizes are limited at this time due to reduced prey populations and/or desiccation. Although the rain forest at El Verde was subject to a less severe and extended dry season during the study period than the Panamanian and New Guinea sites, particularly so in the former case, a similar divergence of spider densities between wet and dry seasons was evident. Maximum densities during the wet season surpassed minimum values during the dry season by only a slightly lower factor than the 2.5- to three-fold difference exhibited at the other two sites.

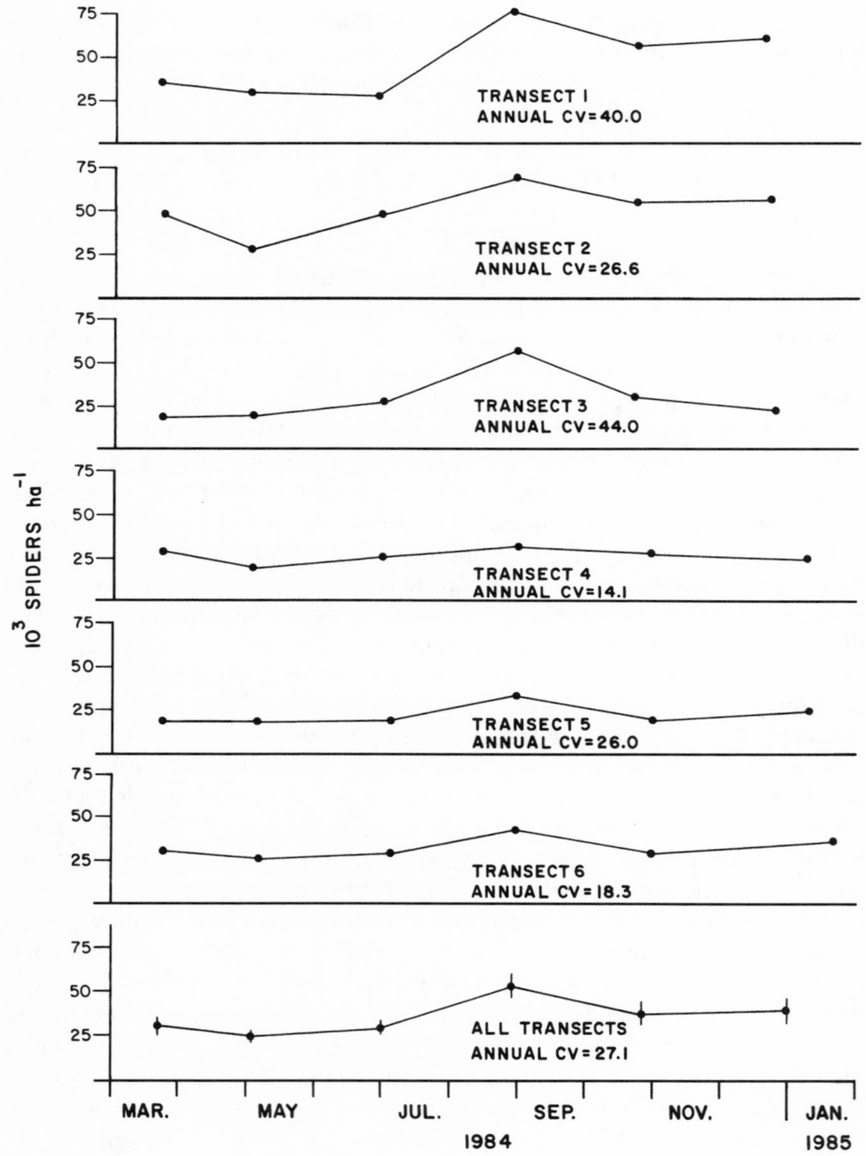

Figure 7.1. Spider densities observed in diurnal censuses of transects during an annual cycle in the tabonuco forest at El Verde. Annual CV refers to the annual coefficient of variation for the six census periods. Error bars associated with the means for all transects represent ±1 standard error.

The densities of spiders in the lower 2 m of understory in tabonuco forest (annual mean = 36,194 individuals ha $^{-1}$; range = 24,667 to 52,333) exceed those reported from other tropical forests. In Lubin's (1978) Transect 2, densities ranged from approximately 6700 to 21,000 spiders ha $^{-1}$, roughly two-thirds of the levels recorded in a nearby forest site bordering a grassy clearing. Elton's (1973) understory transects at Barro Colorado Island yielded density estimates of nearly 6000 spiders ha $^{-1}$ during the rainy season, whereas the mean density of weekly censuses at Wau was 8255 spiders ha $^{-1}$ (Robinson et al. 1974). Mean densities of web-building spiders during the rainy season in the understories (to undisclosed heights) of a Peruvian subtropical floodplain rain forest and a Gabonian equatorial rain forest were, respectively, 32,867 spiders ha $^{-1}$ and 23,467 spiders ha $^{-1}$ (Rypstra 1986).

In contrast to these tropical sites, Rypstra (1986) reported mean densities for web-building spiders in the understory of a Pennsylvania chestnut oak forest of 15,333 individuals ha $^{-1}$. Turnbull (1960b) consistently recorded peak densities exceeding 150,000 spiders ha $^{-1}$ over three years in the lower 1.8 m of understory in an English oak forest. However, his collection techniques included the use of such quantitatively unreliable methods as sweep nets and beating trays (see Turnbull 1973).

Standing Stocks

A preliminary estimate of mean annual standing stock for the four major species in the lower 2 m of the tabonuco forest undergrowth that make up 95.7% of the mean annual density of understory spiders amounts to approximately 10 g ha $^{-1}$ (all values are in dry mass). *Leucauge regnyi* accounted for nearly half of this biomass. Minimum standing stocks for these species were exhibited during the May census, when approximately 6 g ha $^{-1}$ were present. Standing stocks during the March, August, and December-January censuses all fell in the range of 11 to 12 g ha $^{-1}$. Applying the same proportions that were used in generating a density estimate for the entire vertical profile of tabonuco rain forest (see the following section on vertical distributions), the corresponding standing stock estimate works out to a minimum of 70 g ha $^{-1}$.

Energy Flows

Given the conservative estimate for the mean annual biomass for the portion of the arboreal arachnid community for which data exists (i.e., the diurnal web-building spiders), the estimates of energy flow for the arachnid community are mere ballpark figures. This is compounded by the unknown contribution of cursorial species (such as the giant crab spider *Stasina por-*

toricensis, the arboreal tarantula *Avicularia laeta,* and the scorpion *Tityus obtusus*) as well as nocturnal web-building spiders.

Using the conservative estimate of 70 g ha $^{-1}$ as a point of departure, an initial step towards estimating the annual production of the arboreal arachnid community can be made. First, assume an overall production/standing stock (P/B) ratio of 3.6 (not adjusted for cohort production interval, see Benke 1979, 1984). The estimate of production is approximately 250 g ha $^{-1}$ yr $^{-1}$. This ratio of 3.6 is derived from a detailed energetics study of arachnids in a Georgia salt marsh (Pfeiffer 1988). In addition to accurate standing stock estimates for the arboreal arachnids in tabonuco forest, an accurate assessment of the generation time for the dominant species (especially *Leucauge regnyi*) is critical. It is likely, given the climatic conditions, that *L. regnyi* produces at least two generations per year. Therefore, a P/B ratio of 6.0 for this species would not be unreasonable to assume. This would boost the estimate of arboreal arachnid production to 420 g ha $^{-1}$ yr $^{-1}$ or, assuming a mid-range estimate of 25 kJ g $^{-1}$ (Hagstrum 1970; Edgar 1971; Van Hook 1971; Moulder and Reichle 1972; Humphreys 1977), equivalent to 10.5 $\times$ 10^3 kJ ha $^{-1}$.

Values for net production efficiencies (production/assimilation) for spiders range from 20 to 37% (Humphreys 1978; Pfeiffer 1988). Assuming an average efficiency of 30% for the arboreal arachnid community in the tabonuco forest, the assimilation estimate is 3.5 $\times$ 10^4 kJ ha $^{-1}$ yr $^{-1}$. Estimates for assimilation efficiencies (assimilation/ingestion) of spiders consistently range around 90% (Van Hook 1971; Moulder and Reichle 1972; Steigen 1975; Humpreys 1978); thus a preliminary ingestion estimate for the El Verde arboreal arachnid community is ca. 4.0 $\times$ 10^4 kJ ha $^{-1}$ yr $^{-1}$. Humphreys (1977) estimated a prey killed/ingestion ratio of 1.37 for a wolf spider (Lycosidae) which, if applied to El Verde data, estimates the energetic content of prey killed by the arboreal community to be approximately 5.5 $\times$ 10^4 kJ ha $^{-1}$ yr $^{-1}$. The ratio of prey killed to the energy ingested by web-building spiders is likely to be considerably higher than that for the hunting (non-web-spinning) species studied by Humphreys, and thus this may lend a conservative estimate for prey killed. In any event, these conservative estimates suggest that prey ingestion by the arboreal arachnid community may be at least an order of magnitude less than that for *Eleutherodactylus coqui* (Stewart and Woolbright, this volume).

Vertical Distribution

In the lower 2 m of the understory the vertical distribution of the four dominant species was skewed towards the forest floor (fig. 7.2). For all species, 50.3% of the observed spiders occurred within 30 cm of the litter layer (fig. 7.3). The general pattern of this vertical distribution remained consistent

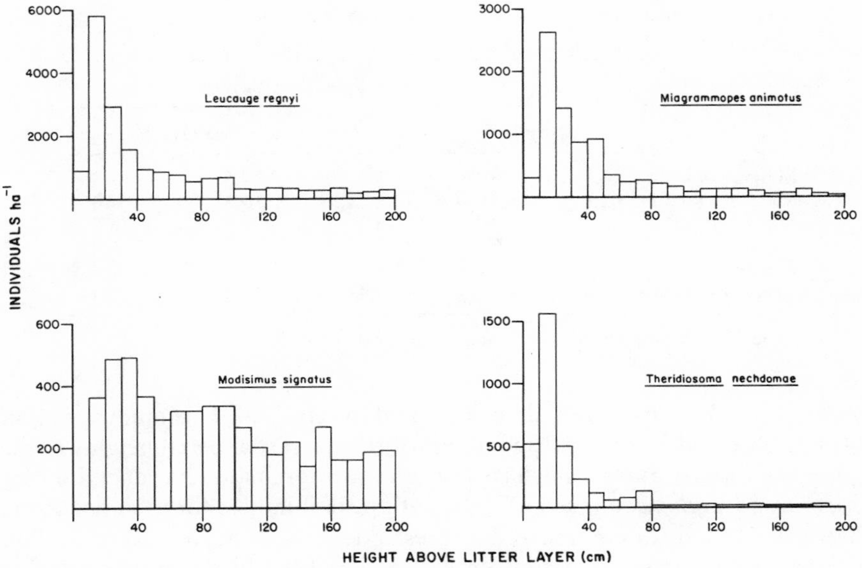

Figure 7.2. Vertical distribution of the dominant spider species observed during the six census periods

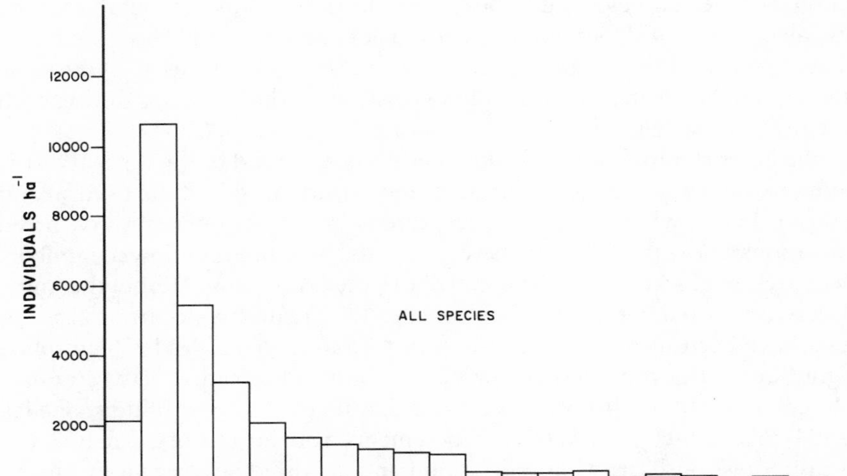

Figure 7.3. Vertical distribution for all spiders observed during the six census periods

Table 7.4. Vertical distribution of spiders in understory transects, 1984–1985 study

	Height above litter layer (cm)			
	<30	30–50	50–100	100–200
March	45.9	15.2	21.6	17.3
May	59.1	16.6	16.9	7.4
June–July	47.4	16.2	19.4	16.9
August	55.0	12.9	15.8	16.3
October	47.9	15.9	20.2	16.0
December–January	46.9	13.4	20.5	19.4

Note: Distribution given as percentage of individuals in the census.

throughout the census period, as indicated in table 7.4. Formanowicz et al. (1981) observed a similar distribution for the nocturnal perch heights of the giant crab spider *Stasina portoricensis* and the dominant leptodactylid frog *Eleutherodactylus coqui* at El Verde, with peak numbers for each occurring between 10 and 20 cm above the forest floor. Sticky traps placed at 1 m height intervals over the entire vertical profile of the El Verde forest during June 1982 also indicated that the greatest concentration of insects occurs near the ground, with the number of insects trapped at 1 m being 2.5 times greater than that at 2 m and nearly equal to the cumulative total recorded from heights of 4 to 19 m (Reagan et al. 1982). Most of the difference in numbers collected below and above the 2 m station was attributable to the prevalence of phorid flies in the lower portion. Stewart and Woolbright (this volume), on the other hand, report that a similar vertical transect at the same site yielded the greatest quantity of insects at the highest stations during both wet and dry seasons.

The latter results from El Verde generally correspond to the vertical distribution pattern reported for a semi-deciduous rain forest in Zaire (Sutton and Hudson 1980), where the greatest capture of insects in sticky traps occurred at stations above and within the canopy and substantially lower numbers were present at stations 1 m and 8 m above the forest floor. Within the lower understory of temperate and tropical sites the greatest concentration of insects have been measured near the forest floor. Uetz et al. (1978) trapped higher quantities of insects below 60 cm than from heights between 60 to 140 cm and 140 to 200 cm in a mature oak-maple forest in Illinois. Shelley (1984) placed sticky traps at 0.3 m increments from heights of 0.3 m to 2.1 m in rain forest on Barro Colorado Island and captured nearly twice as many Diptera and Coleoptera at 0.3 m than any of the other stations.

The abundance of web-building spiders near the forest floor at El Verde corresponded to the presence of moderate to dense patches of tree seedlings and herbaceous plants, especially in Transects 1, 2, and 3. Such vegetation supplied abundant attachment sites for the minute webs of *Miagrammopes animotus, Theridiosoma nechdomae,* and early instars of *L. regnyi.* Vegetational architecture exerts a strong influence on spider community structure, potentially limiting population densities and species richness in communities of web-building spiders as well as shaping vertical distribution patterns (Barnes 1953; Colebourn 1974; Gertsch and Riechert 1976; Hatley and MacMahon 1980; Greenstone 1984; Shelley 1984; Wise 1984; Riechert and Gillespie 1986). Moreover, vegetational architecture may also be at least partly responsible for the vertical distribution of other predators in tabonuco forest. Formanowicz et al. (1981), for instance, observed that *S. portoricensis* and *E. coqui* primarily utilize the upper surfaces of understory leaves as perch sites, suggesting the possibility that the concentration of these nocturnal predators at low heights in the understory are correlated with the vertical distribution of potential perch sites.

Vertical stratification according to size classes within orbweaving species has been reported by Enders (1974) and Pacala and Roughgarden (1984), including a Lesser Antillean species of *Leucauge* by the latter authors. This phenomenon was distinctly apparent for *L. regnyi* populations in the understory transects at El Verde, with early instars concentrated in vegetation near the forest floor. Webs of late instar and mature individuals are positioned higher in the understory, their snare placement sites likely being restricted most frequently by lack of sufficient open space to deploy their webs or, where the forest floor is sparsely carpeted by vegetation, by attachment sites that are too widely spaced.

The vertical transects along the stairwell from 2 to 19.5 m above the forest floor yielded an average density of 23,257 spiders ha^{-1} per 2 m vertical increment. This is less than the mean annual density (36,194 $\pm$ 4006 spiders ha^{-1}) of the 0 to 2 m understory transects or, to consider comparable months, of the average densities (39,450 spiders ha^{-1}) for the October and November-December understory transects. The average density per 1 m increment (11,628 spiders ha^{-1}) for the vertical transects does exceed the levels measured from 1 to 2 m in the understory transects (5,811 spiders ha^{-1}), where the densities appeared to be leveling off (fig. 7.3). The canopy walkway censuses (from 19.5 to 21.5 m) yielded an average of 14,381 spiders ha^{-1} in these 2 m high transects, considerably lower than the overall densities in the understory transects, but still in excess of the mean annual density of the upper meter of those transects.

Combining the results of these censuses generates an estimate of 2.54×10^5 diurnally active spiders ha^{-1} for a 21.5 m vertical profile of tabonuco rain forest. The lack of spatial replication and the few individuals counted in the vertical transects ($N = 53$) and the walkway transects ($N = 134$) must be underscored, and caution is required in the extrapolation of these results. The above estimate for arboreal spider densities is particularly sensitive to the results of the stairwell censuses, which unfortunately are also the censuses with the least volume of habitat examined.

Previous estimates of spider densities in the arboreal layers of temperate forests exceed those estimated above for the El Verde rain forest. Turnbull (1960b) estimated that spider densities in the arboreal layers of an English oak forest surpassed 500×10^3 individuals ha^{-1}, although the potential errors associated with his collection techniques should be considered. Extrapolating Reichle and Crossley's (1967) data for a *Liriodendron*-dominated hardwood forest in North Carolina to a whole canopy estimate indicates 410×10^3 spiders ha^{-1}, whereas Werner (1976) arrived at mean aboveground densities for summer populations of 950×10^3 spiders ha^{-1} in Alaskan spruce stands.

Disturbance

Hurricanes and long-term droughts are the two major large-scale natural disturbances to which the tabonuco forest in El Verde is subjected. The effect of an extended drought on the magnitudes of understory spider populations would be expected to be negative, the result of both prey shortfalls and mortality through desiccation. The short dry period, experienced during the 1984 sampling period of February–April, coincided with the lowest densities of understory spiders recorded for this study. It has been suggested that a pulsed production of adult Diptera from the litter, with the onset of the wet season, resulted in the eventual rise of understory spider populations in the summer months. A more extended drought should cause understory spider densities, in particular those of the web-building species, to decline correspondingly.

The effect of moderate to severe hurricane damage to the forest on understory spider populations is more complex because of the physical transformation of the forest that is involved. With the passage of Hurricane Hugo over Puerto Rico in September 1989, extensive damage was done to the vegetation structure of the tabonuco forest at El Verde. Detached foliage and foliage-bearing branches and twigs from the upper and mid-canopy portions of the forest littered the forest floor along with felled trees. The resulting reallocation of vegetation from the upper reaches of the forest towards the forest floor should have been advantageous for the understory spider community as a whole from a short-term perspective with respect to both prey availability and useful habitat structure. For many web-building spiders the existence of portions or whole tree crowns near the forest floor dramatically increased the availability of firm attachment sites from which to support their webs. Rypstra (1986) found that abundance of spiders was most highly correlated with the density of vegetation. For nocturnal cursorial (primarily Clubionidae, Anyphaenidae, and Sparassidae) and web-building species the eventual abundance of furled dead leaves suspended from fallen branches and twigs would supply diurnal refuges from reptilian and avian predators. And for web-building spiders in particular, the initially much-thickened litter layer should sharply increase the resources for many larval fly populations that would subsequently provide increased prey populations. To the debit side of the equation, mortality during the passage of the hurricane resulting from the fierce winds and impact of the falling debris would initially depress understory spider populations, especially for the exposed web-building species.

Transects examined in the El Verde forest from 22 to 25 December 1989 (see app. 7 for procedural details) revealed diurnally exposed densities of 90,833 ± 9,112 spiders ha^{-1}. This density was 151% greater than the mean annual density (36,194 spiders ha^{-1}) measured during the 1984 censuses, 74% greater than the peak density (52,333 spiders ha^{-1}) observed during that census year, and 114% greater than the most comparable census date (November-December) when 41,033 spiders ha^{-1} were estimated.

The enhancement of spider densities in the post-Hugo transects were virtually solely attributable to increases in those of the orb-weaver *Leucauge regnyi*. Present at densities of 78,500 ± 8,671 individuals ha^{-1}), this species accounted for 86.4% of all individuals observed in the December 1989 transects. During the 1984 censuses, *L. regnyi* accounted for only 49.8% of all spiders or maximally at a given census only 55.3% in August. All remaining species totaled 12,333 ± 989, a reduction of 32.1% from the corresponding mean annual density and 39.0% from the November-December 1984 census.

Particularly dramatic was the apparent decline in the densities of the pholcid *Modisimus signatus* (833 ± 307 individuals ha^{-1}), 81.5% lower than the mean annual density observed in the 1984 transects. The primary attachment site for webs of this species are the undersides and margins of live leaves of understory shrubs and saplings, vegetation that was extensively damaged or destroyed by the high winds and falling debris associated with Hurricane Hugo. Of the other two species that maintained mean annual densities exceeding 1000 individuals ha^{-1} during the 1984 series, the densities of *Miagrammopes animotus* (9,333 ± 955 individuals ha^{-1}) in December 1989 fell within the range of the earlier censuses, while those of *Theridiosoma nechdomae* (1333 ± 422) were 60.5% lower than the mean annual density of the 1984 transects.

The concentration of web-building spiders towards the forest floor was even more pronounced in the post-Hugo transects, with 72.8% of all individuals in the 2 m high profile occurring within 30 cm of the forest floor. During the entirety of 1984 censuses this figure was 50.3%, with the highest concentration (55.3%) appearing in the August 1984 census. That census marked the peak densities observed during the 1984 censuses, a peak associated with presence of high densities of early instar *L. regnyi,* similar to the situation observed during the December 1989 census.

Because the species of *Eustala* present in the El Verde forest occupies its web predominantly from dusk to dawn, this orb-weaver was not directly observed in the transect censuses. Nocturnal traversals of the forest at night in December 1989 revealed a dramatic increase in the presence relative to observations during the previous six years. The increase in potential diurnal retreat sites associated with the increased concentration of furled dead leaves suspended from fallen branches and twigs was a favorable outcome from Hurricane Hugo for this species. Primarily observed previously in edge habitats around the forest, the opening of the canopy and understory in parts of the forest also favored the population rise of *Eustala* sp.

Trophic Relations

Web Structure and Prey Capture

The architecture of a web not only influences the vulnerability of certain prey types to capture, but also sets constraints on where the web may be deployed, since suitable scaffolding and open space must exist for the deployment of the snare. Selection of the web site itself may affect the composition of prey that may potentially be entangled in a particular web (Turnbull 1960a; Craig 1987).

The dominant diurnal species of spiders in tabonuco understory exhibit distinct modes of snare construction. *Leucauge regnyi* constructs orb-webs up to 30 cm in diameter, usually at angles between 45° and 90° to the forest floor. *Theridiosoma nechdomae* spins a small orb-web up to 7 cm in diameter that is drawn into a conical shape by force applied along a guide line attached to the center of the snare. Tension on the guide line is released when the web is disturbed or prey is sensed near the web, thus perhaps enhancing the entanglement of the prey as the web springs back to equilibrium. *Modisimus signatus* (Pholcidae) resides inverted within an irregular, bowl-shaped web that is usually positioned below the foliage of woody vegetation. The positioning of webs beneath leaves confers the advantage of reduced web damage from heavy rainfall that is deflected by the foliage. *Miagrammopes animotus* depends on the very reduced type of snare characteristic of the genus (Lubin et al. 1978; Opell 1984), one consisting of single strands containing sticky, cribellar silk that are held under tension diagonally away from the long axis of the body by the anterior pair of legs.

Wendilgarda clara (Theridiosomatidae) and *Alcimosphenus borinquenae* (Araneidae) construct snares very near the surface of small streams. The widespread Neotropical theridiosomatid strings horizontal nonsticky lines from which hang vertical lines containing sticky, cribellar silk and which cling to the surface film of small streams (Coddington and Valerio 1980;

Coddington 1986). *Alcimosphenus borinquenae* constructs orb-webs up to 16 cm in diameter very near the ground in damp places as well as over small brooks. The web of late instars and mature females, although only about half of the diameter as that for a *L. regnyi* of similar mass, still requires substantial space for placement and hence when not positioned above small streams this species is found in areas with low to moderate density of ground vegetation.

Systematic study of prey composition captured by the web building spiders in tabonuco rain forest have not yet been undertaken. Casual observations of *L. regnyi* webs indicate that Diptera and Homoptera constitute the major prey categories. The intimate association of *M. signatus* webs with live leaves may increase the frequency of herbivorous insects with short flight capacity (e.g., Homoptera) and ambulatory arthropods in the diet of this species. Moreover, nocturnal predators establishing perches on the upper surfaces of leaves may, on occasion, divert prey into the web of this pholcid. Likewise the diet of *W. clara* and *A. borinquenae* should reflect the microhabitats in which their snares are deployed and accordingly should be dominated by adult aquatic insects.

Miagrammopes animotus may derive a significant portion of its diet from certain species of nematocerous Diptera that have frequently been observed hanging, often in large aggregations and particularly at night, from nonsticky silk threads of other spiders in several tropical forests (Robinson and Robinson 1976; Eberhard 1980; Lahmann and Zúñiga 1981) as well as in the El Verde tabonuco forest (pers. observation). The value that this behavior may confer to the hanging flies arises from the isolation it affords from ambulatory predators and the early warning given to the approach of potential predators by resting on a structure very sensitive to disturbance (Eberhard 1980; Lahmann and Zúñiga 1981). Robinson and Robinson (1976) have suggested that portions of *Miagrammopes* capture lines covered with sticky, cribellar silk may ensnare flies that mistake the lines for nonadhesive draglines and frameworks left by other species.

Predation on juvenile *E. coqui* treefrogs by *Stasina portoricensis* and *Oligoctenus ottleyi* (Cteniidae) in the tabonuco forest has been reported by Formanowicz et al. (1981), who further suggested that the arboreal tarantula *Avicularia laeta* also preys on juvenile frogs. Although *O. ottleyi,* which like *S. portoricensis* employs stoutly spined anterior tibia to immobilize prey, appears to confine most of its activity to the litter layer and surface rocks, individuals have been observed in association with bromeliads in the understory (Stewart and Woolbright, this volume).

Predators of Spiders

Analyses of stomach contents indicate that spiders contribute only 5.1% to the prey volume consumed by the dominant amphibian (*Eleutherodactylus*

Figure 7.4. The giant crab spider, *Stasina portoricensis,* a common predator of frogs in the tabonuco forest, Luquillo Experimental Forest. (Photograph by D. Reagan)

coqui) in the tabonuco rain forest (Stewart and Woolbright, table 8.3). *Stasina portoricensis* (fig. 7.4), which occurred in densities of 200 to 1,700 individuals ha $^{-1}$ in nocturnal censuses (Formanowicz et al. 1981), accounts for nearly 75% of this volume of spider prey, whereas only two web-building arboreal species (the pholcid *M. signatus* and the araneid *G. cancriformis*) were recovered from *E. coqui* stomachs (Stewart and Woolbright, this volume). The presence of *G. cancriformis* in these stomachs supports the possibility that some *E. coqui* treefrogs migrate nocturnally towards the canopy (Stewart and Woolbright, this volume), since this araneid appeared only once during all of the transect censuses in the understory, but was commonly observed in the canopy walkway transects. *Leucauge regnyi,* recovered from guts of *Eleutherodactylus portoricensis,* is the only other web-building species identified from stomach contents of any of the other *Eleutherodactylus* species examined. With a foraging strategy best described as "sit-and-wait" (Schoener 1971b), it is not surprising that these frogs consume a proportionately small quantity of web-building spiders, whose own sit-and-wait strategy and relatively isolated suspension in their snares should lead to only rare encounters between these two abundant predator groups. Understory species most vulnerable to predation by *E. coqui* are members of the cursorial spider families Anyphaenidae and Clubionidae as they forage for prey on vegetation at night, and Sparassidae (Formanowicz et al. 1981), particularly as they move between their diurnal retreats and their nocturnal perch sites.

Spiders form a more substantial proportion of the prey in lizard stomachs, especially in *Anolis evermani* and *Anolis gundlachi* (Reagan, this volume). In contrast to their virtual absence in leptodactylid frog guts, web-building species numerically comprised over 50% of the identified spider prey. *Leucauge regnyi* was the most frequently recorded species among the spider prey in anole guts. Several studies have indicated that anole predation may severely constrain communities of web-building spiders. Pacala and Roughgarden (1984) found that the virtual exclusion of two *Anolis* species from the understory of a riverine deciduous xeric forest on St. Eustatius (Lesser Antilles) led to a twenty- to thirty-fold increase in the densities of three species of orb-weaving spiders (including *G. cancriformis* and a species of *Leucauge*) over the six-month experimental period. These authors attributed the increased orb-weaver densities to higher survival rates for immature stages as a consequence of both lessened predation by lizards and increased Diptera populations, which more than doubled in the absence of anoles.

Lizards have also been demonstrated to strongly influence the abundance and structure of web-building spider communities in the Bahamas (Schoener and Toft 1983; Toft and Schoener 1983). At the extreme, small islands lacking lizard populations supported ten-fold higher densities of web-spiders and 50% more species when other parameters (island area and distance from presumed mainland species pools) were factored out. In contrast to what may be inferred from Pacala and Roughgarden (1984), the only species whose densities were apparently not adversely affected by the presence of lizards was *Gasteracantha cancriformis,* whose well-armored dorsum and remotely placed orb-web apparently reduce the vulnerability of this species to lizard predation. Whether the dense lizard populations in the Puerto Rican tabonuco rain forest have historically provided significant resistance to the establishment of immigrant web-building spider species is unknown, but this may be one ancillary factor that has led to the relatively low species richness of this community at El Verde. The isolation of the island from presumed mainland species pools and its small area are likely dominant influences in the reduced species diversity in this forest.

OTHER UNDERSTORY ARACHNIDS

The tailless whip scorpion, *Phrynus longipes* (fig. 7.5) has been collected from caves and from under rocks in moist forests in the Greater Antilles and from some of the United States and British Virgin Islands (Quintero 1981). In the tabonuco forest individuals of *P. longipes* take refuge under rocks by day, particularly aggregating under boulders strewn along dry stream beds.

Figure 7.5. The tailless whip scorpion, *Phrynus longipes,* a common arboreal arachnid predator in the tabonuco forest, Luquillo Experimental Forest

Nocturnal foraging primarily takes place on the ground near their diurnal refuges, where *Amphiacusta* cave crickets are often abundant and presumably constitute a major portion of prey items. *Phrynus longipes* also forages on the lower portions of tree trunks, where they have been observed preying on *E. coqui* (Stewart and Woolbright, this volume). Foraging activity, particularly on tree trunks, appears more prevalent during rainy periods. As is characteristic of other amblypygids, *P. longipes* tactually samples a large volume of the environment in proportion to its body size by means of elongate antenniform legs. Amblypygids snare and restrain prey within clasped spiny pedipalps while serrated chelicerae shred and masticate the prey prior to preoral digestion (Cloudsley-Thompson 1968; Quintero 1981).

Very little is known of the natural history for most of the other eleven recognized species of *Phrynus* other than habitat occurrences. Only for *Phrynus marginemaculatus* has detailed information on reproductive and developmental biology been extensively documented (Weygoldt 1969, 1970, 1975). Prey acquisition in natural habitats has rarely been observed for phrynids; individuals have been maintained in captivity for extended periods on diets of crickets, roaches, and termites (Muma 1967; Quintero 1981). No demographic studies have been undertaken for phrynids; mark-recapture approaches appear best suited in light of their cryptic habits.

Other than their occurrence in the stomach contents of frogs and anoles (see Stewart and Woolbright, this volume; Reagan, this volume), little is known of the ecology of the scorpion *Tityus obtusus* and four opilionid species occasionally observed at night on tree trunks. The recovery of *T. obtusus* from litter samples indicates that some portion of the population resides at least by day in the litter layer. Stewart and Woolbright (this volume) suggest that the aboveground occurrence of this scorpion may be limited by the availability of diurnal retreats. Analyses of stomach contents has also revealed the rare occurrence of the pseudoscorpion *Menthus* sp. in the diet of two anole species and three *Eleutherodactylus* species (Stewart and Woolbright, this volume; Reagan, this volume). Whether these instances result from litter foraging by these vertebrates or the presence of *Menthus* sp. in the arboreal layers is unclear.

SUMMARY

Arachnids are the major arboreal invertebrate predators in tabonuco forest and spiders the numerically dominant predators within this class. Conservative estimates of densities (2.54×10^5 spiders ha^{-1}) and standing stocks (70 g ha^{-1}) for a 21.5 m vertical profile in tabonuco forest suggest that the contribution of spiders as predators in the arboreal strata is minor relative to anole and coqui populations (see Reagan, this volume; Stewart and Woolbright, this volume). However, the turnover rate of spider populations is potentially much higher than their considerably more massive vertebrate counterparts, making their impact on prey populations more substantial than biomass comparisons alone would suggest.

In order to support the standing stocks of vertebrate and invertebrate predators observed in the arboreal zones of tabonuco forest, a high level of productivity must be maintained by the apparently sparse prey populations residing in the arboreal zone and, additionally, supplemented by prey from the litter layer. Faced with a high risk of mortality because of a dense predator array, selective advantage is gained by potential prey individuals that maximize maturation rates. The resulting high rates of population turnover has been proposed by Elton (1973) as a resolution to the seeming paradox of low apparent prey populations supporting large and conspicuous predator populations in Neotropical forests of Brazil, Costa Rica, and Panama. In addition, more rapid maturation rates lead to smaller body size of prey species at maturity—a phenomenon observed by Elton (1973) in the Neotropical forests he studied and apparent as well in tabonuco forest.

Elton further suggested that the dominance of ectothermic organisms (such as spiders, anoles, and frogs) in the predator community of Neotropical forests allows for the presence of a more productive and stable predator array. Without the additional cost of metabolically maintaining an elevated body temperature, an ectotherm can make more efficient energetic use of ingested prey and better survive periods of prey scarcity, as might occur during seasonal droughts. The low metabolic rates of spiders and their ability to withstand long periods of starvation have been well documented (Miyashita 1969; Anderson 1970, 1974; Greenstone and Bennett 1980; Anderson and Prestwich 1982; Tanaka and Ito 1982).

Censuses following the same transects in the low understory (0–2 m) over an annual cycle in tabonuco forest indicated seasonal dynamics that may be related to rainfall patterns. Spider densities ranged from 25 to 52 $\times$ 10^3 individuals ha^{-1}, with a mean annual density of 36 $\times$ 10^3 individuals ha^{-1}. These densities exceed those observed in similar transects of forests in the Neotropics and New Guinea.

Minimum spider densities were present in the early May censuses, which coincided with the resumption of rain at El Verde following an eight-week period of reduced precipitation (see Pfeiffer, fig. 5.1, this volume). By mid-June Diptera populations in litter, predominately in the larval stages, alone accounted for densities as high as 21 $\times$ 10^6 individuals ha^{-1} (see fig. 5.3 in Pfeiffer, this volume). The subsequent emergence of adult flies early in the wet season preceded the near-doubling of the web-building spider populations from early July to late August during 1984, mainly due to increases in densities of the orb-weaver *Leucauge regnyi.* The seasonal dynamics of densities for the community as a whole was similar to that observed at Barro Colorado (Lubin 1978), where most species had population peaks during the wet season. The dynamical pattern of the spider community in the low understory of tabonuco forest was largely determined by that of *L. regnyi,* which accounted for 50% of all spiders observed in the transects.

The highest concentrations of arboreal spider populations in the the transects were observed within 30 cm of the litter layer. Two major contributors to this vertical distribution are the abundance of vegetation near the forest floor that can serve as scaffolding for spider webs and the substantial quantity of potential prey that is harbored in and emerges from the litter layer.

In censuses conducted in the lower 2 m of tabonuco forest understory during December 1989, densities of *L. regnyi* were nearly triple that of the August 1984 peak. The dense populations of this orb-weaver in 1989 were likely a consequence of the extensive canopy destruction from high winds associated with passage of Hurricane Hugo over Puerto Rico in September. The fallen vegetation created a thick litter layer that favored the development of dense prey populations and enhanced the structural complexity near the forest floor on which webs could be supported.

Leucauge regnyi and three other web-building species accounted for 95% of all spiders observed in low understory transects during daylight hours. These four species construct webs with distinctly different architecture: a planar orb-web (*L. regnyi,* an orb-web deformed into a conical shape (*Theridiosoma nechdomae*), a three-dimensional bowl-shaped web (*Modisimus signatus*), and a web structure reduced to single strands (*Miagrammopes animotus*). Whether these different foraging structures result in dissimilar diets is unknown. The external digestion of prey by spiders precludes the straightforward analyses of gut contents as a means of prey identification and field observation of predation in progress is costly in terms of time. Thus, quantitative tracing of food web links to generic and specific levels (i.e., link strength and per capita predator effect) remains unexplored.

Anolis lizards appear to be the principal vertebrate predators of spiders in tabonuco forest at El Verde. *Anolis* spp. capture a broad variety of spiders including orb-building species and especially *L. regnyi* (Reagan, this volume). Predation by these lizards may be a significant population regulating agent for *L. regnyi* as has been suggested for other web-building spiders on small islands elsewhere in the West Indies. In contrast, arboreal arachnids are of minor importance in the diet of the frog *Eleutherodactylus coqui,* with the giant crab spider *Stasina portoricensis* being the predominant arachnid consumed.

Appendix 7 Methods

Census Techniques

Populations of tropical web-building spiders have been enumerated previously by visual census techniques from fixed forest transects in New Guinea (Robinson et al. 1974), Panama (Elton 1973, Lubin 1978), and Costa Rica (Liebermann-Jaffe 1981) and from random plots in Gabon and Peru (Rypstra 1986). The fixed transect approach creates negligible disturbance to both populations and habitat structure. This enables repeated censuses within the same habitat space through time and thus avoids the sampling error introduced with the destructive sampling of different subsets of a highly heterogeneous habitat at each census period. The spatial distribution of individuals can also be determined during visual censuses, especially with sedentary organisms such as web-building spiders.

Diurnal censuses of spiders were taken from six transects in the lower 2 m of understory vegetation in the El Verde rain forest at approximately eight-week intervals spanning 20 March 1984 to 10 January 1985. Each transect measured 50 m by 1 m. At each census, the vertical position of spiders occurring within 200 cm of the litter layer was recorded, along with indirect (based on web dimensions) or direct estimates of body size.

On 29 December 1986 and 17 September 1987 censuses were conducted in transects associated with a canopy walkway located at the edge of Plot 3. A vertical transect 1.3 × 1.0 × 19.5 m was surveyed along the stairwell from ground level to the uppermost stairwell platform 19.5 m above. Two-meter-high horizontal transects, extending 1 m out from both sides of the canopy walkway and running 22.6 m of its length were also examined on these dates.

Subsequent to the passage of Hurricane Hugo over Puerto Rico on 18 September 1989, six transects were examined from 22 to 25 December 1989. Because of the extensive damage to the tabonuco forest (Walker et al. 1991), the boundaries of the original 1984 transect could not be established accurately. New transects, near the location of the originals, were established. As a consequence of the considerable input of woody debris, the width of the transects were reduced to 0.5 m, in order to facilitate as unobstructed a view into the transects as possible. Time constraints also forced a reduction of the transect length to 20 m. Because of hurricane damage to the canopy walkway, censuses could not be taken above 2 m.

The accuracy of visual censuses is sensitive to environmental conditions. Heavily overcast skies further reduce the already low ambient light conditions near the forest floor and intense precipitation may damage fragile webs. Such conditions may bias against the perception of small and cryptic individuals or of species with reduced web architecture. Consequently, overcast days and periods following heavy precipitation were avoided for the censuses and transects were run during midday to late afternoon whenever possible. Although the diurnal censuses excluded species that forage or occupy snares solely at night and are cryptically positioned by day, nocturnal observations suggested that such species are proportionately insignificant in the understory except for the giant crab spider, *Stasina portoricensis*. In addition, some individuals of species normally constructing diurnal webs may not have them deployed during a census interval. These individuals will be unaccounted for in a given census, although Robinson et al. (1974) did not consider the failure to include such web-building spiders to be a critical source of error in their New Guinea transects if censuses were not conducted shortly after heavy rains.

The densities presented here represent snapshots of the number of spiders that are actively foraging (in the sense that web-building spiders forage with their webs) during daylight hours rather than estimates of absolute population densities. These density estimates may serve as a basis for extrapolating diurnal insect consumption by spiders, although the timing of the censuses to avoid periods during and shortly after heavy precipitation lend an upward bias to such estimates.

8

Amphibians

Margaret M. Stewart and Lawrence L. Woolbright

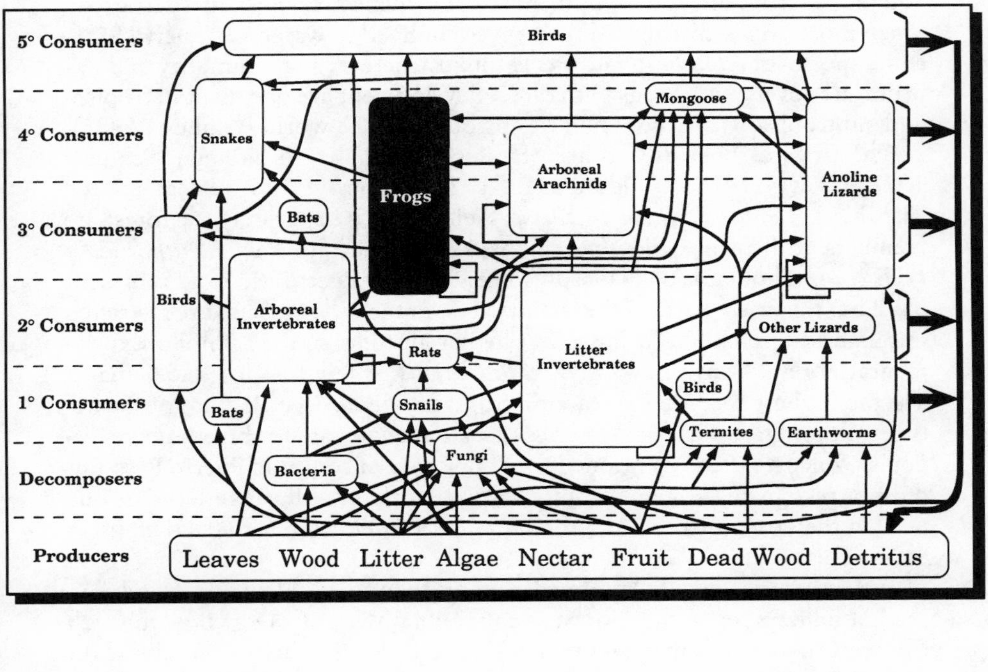

THE NATIVE Puerto Rican amphibian fauna (anurans only), like that of other Antillean islands, consists primarily of members of the family Leptodactylidae (Rivero 1978). The predominance of the genus *Eleutherodactylus* is characteristic of the entire West Indies (Schwartz 1978; Schwartz and Henderson 1991) and results from complex phenomena. These include climate, insularity, and proximity to Central and South America, the centers of diversity of the genus (Pregill and Olson 1981). By having nonaquatic eggs and direct development (no free-living tadpoles), these frogs are much more evenly distributed than those confined to water. On Puerto Rico, some species of *Eleutherodactylus* is found wherever high humidity and appropriate cover exist. Frogs with direct development are prevalent in tropical and subtropical island communities throughout the world (Gibbons 1985).

The frogs of Puerto Rico are obvious in virtually all habitats (Schmidt 1928; Drewry 1970a,c; Odum et al. 1970c; Rivero 1978). Except for the islandwide *Leptodactylus albilabris,* all eleven native species in the forest are members of the genus *Eleutherodactylus* (table 8.1, fig. 8.1). *Eleutherodactylus coqui* Thomas is the widespread generalist of Puerto Rico and is present in all but the driest parts of the island. Most of the other species are restricted in range and in habitat, primarily at higher altitudes in the Luquillo Experimental Forest. Densities of *Eleutherodactylus* in Puerto Rico and Jamaica are the highest recorded so far for frogs anywhere (see Population Structure), although comparisons with species that aggregate for breeding are difficult. Wolcott (1924) mentioned the abundance of coquis in Puerto Rico but did not recognize their importance as predators; he still suggested that one needed the large introduced *Bufo agua* (= *B. marinus*) for effective insect control.

We summarize available information on the role of *Eleutherodactylus* in the Luquillo Experimental Forest as a major pathway of energy flow through the system. We present a comprehensive study of the diet of the abundant coqui, *E. coqui,* along with data on the diets of other common species, and we compile a list of known frog predators from field observations. We evaluate the importance of frogs to the forest ecosystem in terms of biomass and numbers of prey harvested by coquis and the number of coquis harvested by predators. We also estimate energy intake and expenditure by adult coquis.

Table 8.1. Species of *Eleutherodactylus* at El Verde

Species	SVL max. (mm)	Microhabitat	Height off ground (m)	General abundance	Density (adults/100 m^2)	Maximum density (individuals/ha)
*E. coqui**	55	ground-canopy	0–18.5	abundant throughout	6–29	24,800
*E. portoricensis**	39	ground-low canopy	0–3.0	common locally	0–8	800
*E. wightmanae**	20	litter-low shrubs	0–0.4	common	0–4	400
E. antillensis	30	open grass-shrub not in forest	0.5–1.8	common locally on roadsides to 750 m	0–4	400
E. hedricki	33	canopy-subcanopy treeholes	0–18.5	widely scattered and patchy	0–2	200
*E. richmondi**	45	fossorial-low boulders, litter	0–0.6	sparsely clumped	0–1	100
E. brittoni	20	forest edge; low open vegetation	0–0.5	common along road-sides only	rare	—
E. locustus	19	litter, grassy patches	0–0.3	higher elevations	rare	—
E. eneidae	30	mossy boulders	0–0.5	not seen in study area	rare	—
E. gryllus	16	low canopy; forest edges	2–15	above 450 m	rare	—
E. karlschmidti	80	riparian; wet boulders	0–0.2	extirpated?	—	—

Notes: Species are listed in order of abundance. Density estimates are based on nine years of nocturnal counts at 2000 to 2300 h. Per hectare estimates are derived from the maximum count per 100 m^2 of all frogs for *E. coqui* and of adults only for other species. Additional size data are from Drewry (1970a,c) and Rivero (1978).

* Species treated in detail.

METHODS

Most of the data reported herein are based on seventeen years of field studies at El Verde. Our information on *E. coqui* is based on a much larger sample size than that for other species. Because of the difficulty in aging amphibians, we have size-indexed populations and most discussions deal only with size. The size dimension used is snout-vent length (SVL), or the length from tip of the snout to the posterior end of the backbone. That distance approximates the length of the trunk (head + body) of a frog. Most stomach contents were obtained at dawn by the stomach pumping method of Legler and Sullivan (1979). We present our methods of density estimates and sampling for prey availability in appendix 8.B. We show the locations of specific field sites and plots in figure 8.2.

The climate in the Luquillo Mountains is mildly seasonal. Seasons, defined by temperature and rainfall, follow the same pattern as more temperate areas

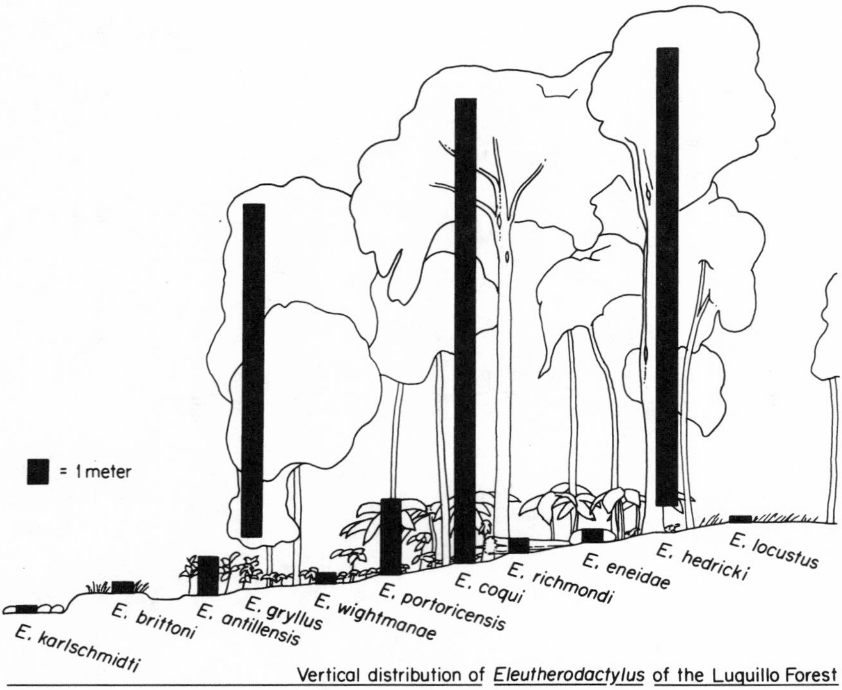

Figure 8.1. Generalized profile of Luquillo Forest habitats showing vertical distribution (black bars) of *Eleutherodactylus* species that live in and around the forest. *Eleutherodactylus gryllus* and *E. locustus* are found at higher elevations. Except for *E. karlschmidti, E. eneidae,* and *E. gryllus,* distributions are derived from field measurements by the authors.

of southeastern North America. The months of January through March average one-third less rainfall per month than the rest of the year and lower temperatures prevail (Briscoe 1966; Odum et al. 1970b; fig. 1.2 in Waide and Reagan, this volume). We refer to those months as the "dry season" or "winter." Temperatures we recorded in the forest range from 18° to 26° C in winter, and 20° to 28° C during summer months.

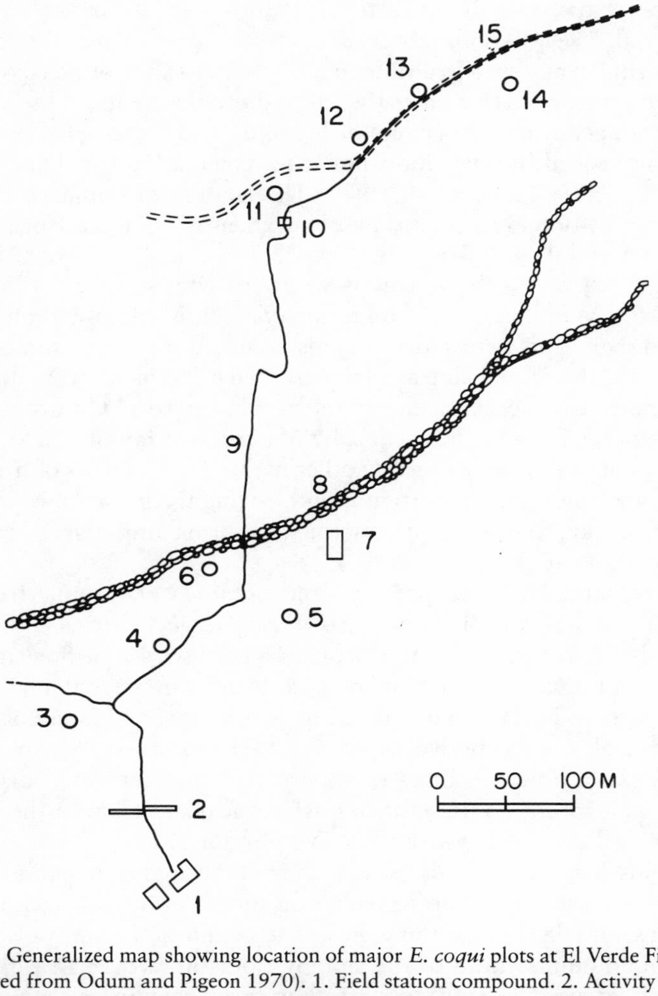

Figure 8.2. Generalized map showing location of major *E. coqui* plots at El Verde Field Station (map adapted from Odum and Pigeon 1970). 1. Field station compound. 2. Activity Transect. 3. River Experimental Plot. 4. River Control Plot. 5. Bridge Experimental Plot. 6. Bridge Control Plot. 7. Woolbright Mark-recapture Plot. 8. Quebrada Sonadora. 9. Main Trail. 10. Walk-up Tower. 11. Tower I Experimental Plot. 12. Tower I Control Plot. 13. Tower II Experimental Plot. 14. Tower II Control Plot. 15. Old Ox Trail.

NATURAL HISTORY

Activity Patterns and Behavior

All Puerto Rican *Eleutherodactylus* are nocturnal, although activity levels during the night differ with species (Drewry 1970c; Drewry and Rand 1983; Stewart 1985; Woolbright 1985a; Stewart 1995). During the day frogs seek shelter in litter, in leaf axils or tree holes, in crevices in the ground, under roots, rocks, bark, or in human artifacts (Cintron 1970; Drewry 1970a; van Berkum et al. 1982; Woolbright 1985a). We have no evidence of active diurnal foraging in any species although it is possible that frogs feed opportunistically on arthropods that enter the frogs' diurnal retreats.

The entire behavioral repertoire of *E. coqui,* and presumably other forest frogs, is influenced by their moisture requirements (Heatwole et al. 1969; Pough et al. 1983; Taigen et al. 1984). The amount of rainfall is critical to tree climbing, foraging success, and movements of larger frogs (Stewart 1985; Townsend 1985; Woolbright and Stewart 1987). Also, frogs move more frequently when the foliage is wet than when it is dry (Woolbright 1985a). At El Verde, periods of up to ten consecutive days without rain are not uncommon, and longer dry periods sometimes occur (Stewart 1995). Foraging (and therefore being available to predators) is moisture dependent; that dependence is especially important to smaller frogs. On dry nights few juveniles emerge from the litter. In spite of abundant rain in the forest, temporary dry conditions may override other major determinants of frog behavior. Moisture is not only important in influencing the behavior of individual frogs, but appears to be the major factor structuring amphibian assemblages (e.g., Heatwole 1982).

Breeding patterns, hence presence and numbers of juvenile frogs, vary with species and season. Typically, *Eleutherodactylus* species lay terrestrial eggs that hatch directly into tiny froglets (Townsend and Stewart 1985). Since there is no tadpole stage, there is no major ontogenetic niche shift as individuals grow. Rather, in preadults, there is an increase in use of horizontal and vertical space as body size increases (Townsend 1985). Space use by adults varies from home ranges of apparently a few meters in diameter in litter-dwelling species to use of the entire vertical spectrum of the forest by tree-climbing *E. coqui* (Stewart 1985; Woolbright 1985a).

As coquis increase in body length, they climb higher in the understory vegetation; all size classes forage higher off the ground when conditions are wet (Townsend 1985). Depending on weather conditions, many of the larger preadults and adults climb to the canopy, sometimes over 18.5 m, to forage at dusk (Stewart 1985). Most breeding males remain on the understory vegetation near calling and nest sites although a few call and nest in the canopy if appropriate sites are available (Pough et al. 1983; Stewart 1985; Townsend 1989).

Habitat

Habitat and microhabitat requirements of the species at El Verde vary (table 8.1, fig. 8.1). The entire horizontal and vertical spectrum of the forest and forest edges is used by members of the anuran assemblage. *Eleutherodactylus coqui* is the anuran generalist of Puerto Rico (Rivero and Segui Crespo 1992); other species are much more restricted in their habitat use. The coqui frequents human artifacts and tolerates disturbance more than other species. It uses all levels of the forest from ground to canopy. Another principal occupant of the understory, *E. portoricensis,* occurs in most moist, cooler sites in the forest. Its distribution is patchy, and it is usually much less abundant than its sibling *E. coqui.*

Litter and ground cover species include the semifossorial *E. richmondi* that occurs in boulder-strewn sites or around large fallen logs. It calls only in wet weather from under or atop boulders and forest debris. The few egg clutches observed were in rotting logs (Drewry 1970c; Rivero 1978). Also living around mossy boulders is *E. eneidae,* now rare at El Verde. The most common litter species is the small *E. wightmanae* that forages and calls from the litter or low shrubs. Canopy species include *E. hedricki* and *E. gryllus.* The treehold dwelling *E. hedricki* (Rivero 1963) occurs in more mature tree stands where holes in the upper portion of the trunks of larger broad-leaved trees provide appropriate nest sites. It is rarely seen on or near the ground (Kepler 1977; pers. observation) but we have found it in bamboo frog houses in experimental plots (Stewart unpublished). *Eleutherodactylus gryllus* lives in bromeliads and under moss on trees at higher elevations and is heard rarely at El Verde.

The largest native anuran, *E. karlschmidti,* lives around boulders in stream beds (see Association with Streams). Apparently it has disappeared after having been common in the recent past (Drewry 1970a,c; Rivero 1978). Also living along streams, rivulets, and wet seeps is *Leptodactylus albilabris,* which calls day and night in wet weather. Adults range widely on the forest floor on wet nights; they deposit foam nests in cavities at the water's edge where their aquatic tadpoles develop.

Several species are adapted to open sites and occur in forest openings and edges. Common at lower altitudes is *Eleutherodactylus antillensis* that forages from grasses, forbs, and shrubs. It has invaded the forest along highways up to 750 m. A small population is active in warmer months in the field station compound. *Eleutherodactylus brittoni* also occurs along forest edges and openings, using primarily herbaceous vegetation. At higher elevations *E. locustus* lives in and near the litter in more open grassy areas. *Bufo marinus,* the only introduced anuran in the forest (Rivero 1978), is sparsely distributed through the forest (0 to 2 in eight 100 m^2 experimental plots).

Forest floor litter is a feature of prime importance to the life of most species of *Eleutherodactylus* in the forest. Litter provides several essentials such as moist refugia and rehydration sites (van Berkum et al. 1982), predator protection, and food (Pfeiffer, chap. 5, this volume). The litter provides nest sites for most species, as well as reliably moist retreats. Moist shelter and predator protection, rather than food, are important provisions of the litter for *E. coqui,* especially for juveniles. Although the litter is not an important foraging site for coquis, it is so for *E. wightmanae* and other ground-dwelling species (see Diets of other *Eleutherodactylus* species; Lavigne and Drewry 1970). Litter characteristics have a major effect on population densities of resident species. Litter depth varies spatially and temporally (LaCaro and Rudd 1985; Lawrence, this volume); it is deeper in April and May, the time of major leaf fall, or after hurricanes. In plots with dependably heavy litter, juveniles are more numerous. Only two species, the arboreal *E. hedricki* (tree holes) and *E. gryllus* (bromeliads and arboreal moss), do not use the litter for refugia.

Population Structure of *Eleutherodactylus*

Sex Ratio and Size Structure of E. coqui *Populations*

The sex ratio of *E. coqui* appears to be 1:1 (Stewart 1985). Since many females often forage in the canopy, a cursory examination of the population in the understory will yield more males than females. Long-term marked populations, however, suggest similar numbers of males and females (Stewart unpublished).

At El Verde, body sizes (SVL) range from 6 to 7 mm at hatching to 55 mm as adult females. We categorized the population into three size classes: juveniles 6 to 17 mm, subadults 18 to 23 mm, adults 24 to 55 mm. Preadults include juveniles and subadults. Results of counts in the Activity Transect (fig. 8.2; app. 7.B) show the size structure of *E. coqui* populations (fig. 8.3). Numbers of adults remain fairly constant at any one site (Stewart and Pough 1983; Stewart unpublished). Juvenile populations fluctuate widely from census to census because of seasonality of breeding, localized hatching of egg clutches, and weather. The high frequency of preadults compared to adults in populations in summer (up to 15:1) suggests that much predation occurs at small body sizes. Among adults, females are about 30% longer (mean = 46 mm, SVL) than males (mean = 36 mm, SVL; fig. 8.4; Woolbright 1989).

Eleutherodactylus coqui grows to reproductive size in approximately one year. Growth rates for froglets are difficult to obtain, but rates appear constant among juveniles. Mean growth rate suggests that hatchlings require approximately 316 days to reach 28 mm SVL, the minimum size of reproductive males (Woolbright unpublished). Growth rates of marked adults that were later recaptured at one site (Woolbright mark-recapture plot, fig. 8.2) are shown in figure 8.5. The growth rate of males slows dramatically when reproduction begins (Woolbright 1989) due to the energetic costs of calling and parental care (Townsend et al. 1984; Woolbright 1985b; Townsend 1986a; Woolbright and Stewart 1987).

Total prey mass per frog increases with body size. Body sizes within *E. coqui* populations differ with geographic location. Coquis increase dramatically in body size with increase in altitude (Beuchat et al. 1984; Narins and Smith 1986; Stewart and Woolbright unpublished). Since population densities decrease with altitude, increased body size in highland populations may

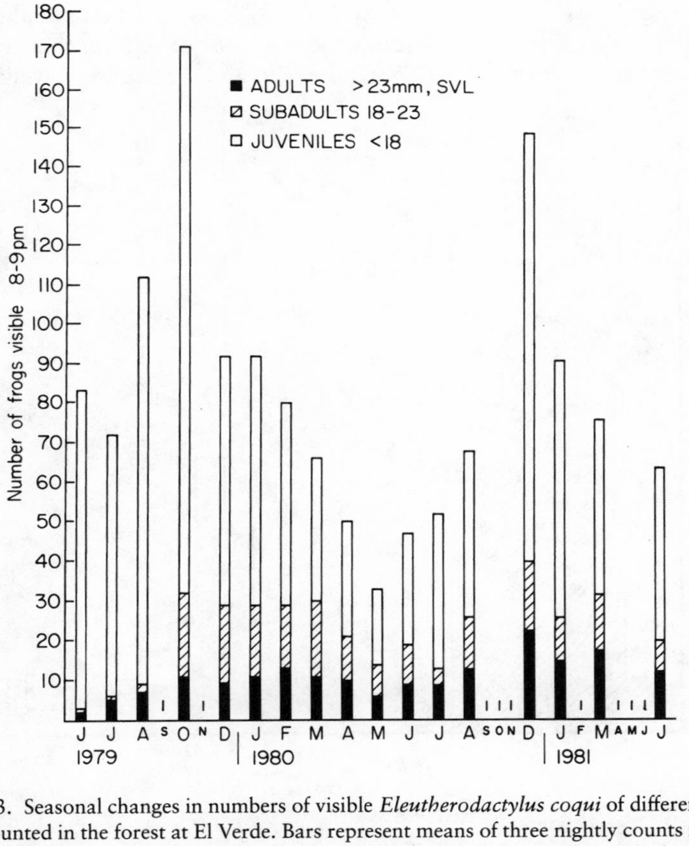

Figure 8.3. Seasonal changes in numbers of visible *Eleutherodactylus coqui* of different size classes counted in the forest at El Verde. Bars represent means of three nightly counts made in a 2 × 50 m transect (Activity Transect) between 2000 and 2100 h every fortnight from June 1979 to July 1981. Bar totals are cumulative. Dashes in place of bars indicate no data for those months.

not result in greater total biomass of coquis in those populations. From lowlands to mountain tops, males increase from 27 to 44 mm SVL, females from 32 to 58 mm SVL. The total population density decreases from an estimate 45,000 ha $^{-1}$ at 30 m altitude to 2,500 ha $^{-1}$ at 750 m (Stewart unpublished). *Eleutherodactylus antillensis* and *E. portoricensis* show similar trends in body size but to a lesser degree. The reasons for the clinal increase have not been determined, although we are investigating several aspects of the question.

Comparisons of Body Size

In most communities, the smallest species are the most numerous (Wallace 1858; Elton 1927). This is so for many anuran assemblages (e.g., Barbault 1967, 1976; Milstead et al. 1974; Stewart 1974) as well as for snakes (Fitch 1982), but not for frogs in the Luquillo Mountains. Except for the introduced *Bufo marinus*, the extirpated *E. karlschmidti*, and the ground-dwelling *Lep-*

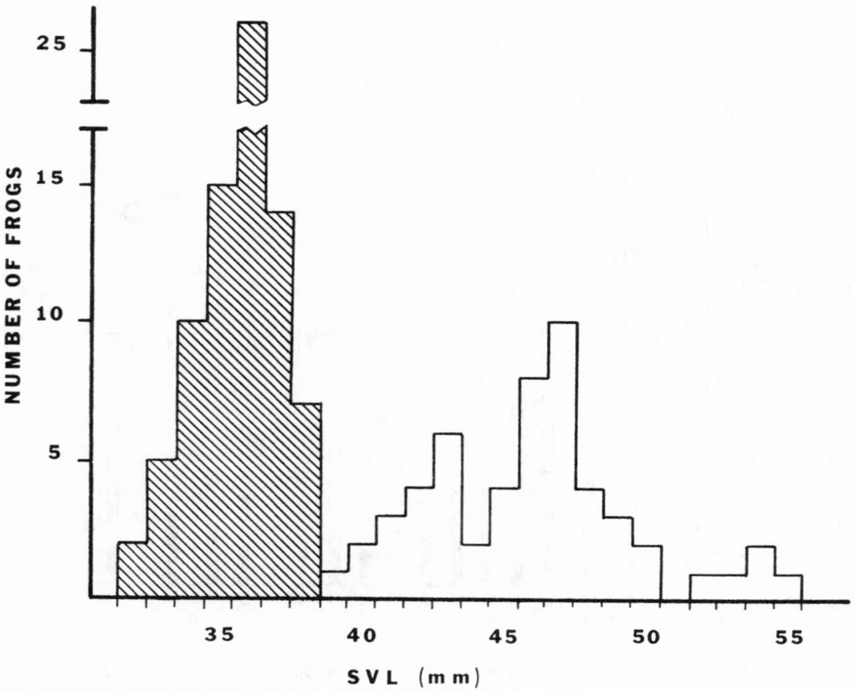

Figure 8.4. Body sizes (SVL) of reproductively mature male (shaded bars) and female (open bars) *Eleutherodactylus coqui* examined during mark-recapture study at one site. Data were taken only from the first capture of each individual. Lack of overlap is an artifact of the sample, not a characteristic of the entire population. (Reprinted from Woolbright 1989).

todactylus albilabris (53 mm, SVL), *E. coqui* is the largest frog in the forest, yet its populations exceed all others in all sites examined. Factors determining body size are complex (Cotgreave 1993). Physiological limits of frogs to periods of hydric stress may be a contributing factor (Pough et al. 1983). Even though rain occurs throughout the year, there are periods of five to

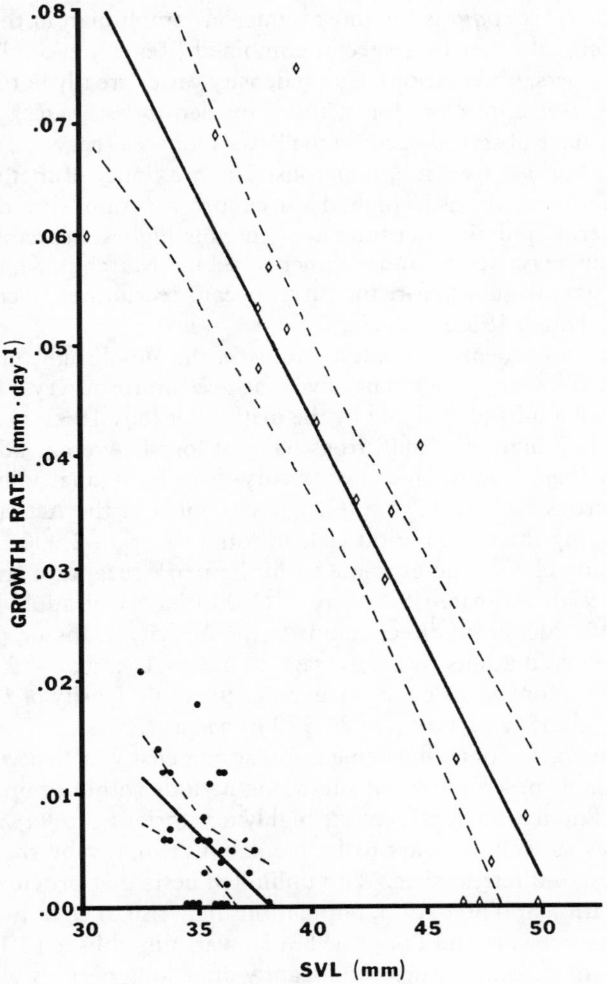

Figure 8.5. Growth rates of adult male (circles) and female (diamonds) *Eleutherodactylus coqui* collected during a mark-recapture study between 1980 and 1984. Least squares regression lines are shown with 95% confidence intervals. Data include some small females identified as such only by subsequent recaptures. (Reprinted from Woolbright 1989).

twenty-eight days without rain when the understory vegetation wilts and the litter dries. Even with a broad diet available, foraging is often limited. Refugia for small frogs appear to be limited due to thin, rapidly decaying litter and dense clay soil. The habitat is therefore unpredictable for small frogs, especially juveniles.

Population Densities of E. coqui

Eleutherodactylus coqui is the most numerous amphibian in the forest, far outnumbering all other frog species combined (Reagan et al. 1982; Woolbright 1991; pers. observation). Coqui density varies greatly both temporally and spatially (see app. 8.B.1 for methods for density estimates). The highest densities we have observed occur in the River Plot area (fig. 8.2), where sierra palms and *Cecropia* trees are numerous. The proximity of that site to Quebrada Sonadora as well as its high, dense canopy no doubt slow drying of the soil and litter. Population densities are generally highest September to January following heavy spring and summer breeding (March to September) and lowest February to June before the current year's recruitment (Schmidt 1928; Stewart and Pough 1983; fig. 8.3).

The population density of adult coquis in the Woolbright plot (fig. 8.2) was estimated by means of twenty-five mark-recapture surveys. Estimates of the number of adults (calculated by the method of Jolly 1965) ranged from a low of 988 to a high of 11,440 frogs ha $^{-1}$ of forest. Average adult population density (based on the mean of twenty-three individual Jolly estimates) was 3,265 frogs ha $^{-1}$ (S.E. = 556). Direct counts in the Activity Transect (fig. 8.2) during the same period yielded totals of up to 2,900 adults ha $^{-1}$, approximating closely the estimate from the mark-recapture study. Turner and Gist (1970) estimated 800 acre $^{-1}$ (2,000 ha $^{-1}$) of adult *E. coqui* at 450 m at El Verde. From direct counts in the Activity Transect, preadult coquis outnumbered adults by an average of 5.3 to 1 (range = 2:1 to 15:1, $N = 72$). Therefore we calculate an average preadult density of 17,305 frogs ha $^{-1}$ (5.3×3265) and a total of 20,570 frogs ha $^{-1}$.

Several major factors influencing population density in *E. coqui* are numbers of available nest and retreat sites, long periods without rain, and major hurricanes. Population density of the highly territorial *E. coqui,* and possibly other species as well, appears to be regulated primarily by the number of available nest and retreat sites. With unlimited nests that provide protection from desiccation and predation, populations may rise to eight times those in control plots (Stewart and Pough 1983; Stewart unpublished). The population density of *E. coqui* drops significantly after long periods without rain. In the Activity Transect from 1984 to 1989 adult density dropped by as much as 50% of densities during the prior five years. In larger plots monitored by Woolbright (1991), juveniles dropped by 60% over that same time period. That drop was inversely correlated with an increased number of periods of days with ≤3 mm of rain (Stewart 1995). Evidently juveniles cannot survive extensive drought. Populations of other species that prefer moist condition, such as *E. portoricensis* and *E. richmondi,* also decreased during that period.

Hurricane Effects

The greatest impact on population densities of frogs occurred following Hurricane Hugo, September 1989 (Woolbright 1991; Reagan, this volume). The major change in *E. coqui* populations was a large drop in numbers of juveniles, but a six-fold increase in density of adults both in the Activity Transect (Stewart 1995) and in larger plots monitored by Woolbright (1991). Numbers of adults attained higher levels than in the five years prior to Hugo. The decrease in juveniles probably resulted from severe droughts that followed the hurricane. Increase in retreat sites from fallen litter, as well as a decrease in invertebrate predators, warmer temperatures and increased insect abundance, probably favored coquis. The warmer, dryer conditions were unfavorable for species that prefer cool moist conditions, such as *E. portoricensis* and *E. richmondi,* and populations of those species crashed. By January 1992, juvenile numbers have recovered to historic levels. Adult numbers have declined from their immediate post-hurricane peak numbers but are still three times their pre-hurricane levels (Woolbright unpublished).

Amphibian Density on Islands Compared to Continents

Comparisons of any group between continents and islands are not easy, and generalizations are often naive. Environmental conditions for communities or guilds that are used for comparisons often differ greatly. In general, species diversity is much greater on mainland than on island sites, but numerous authors have noted just the opposite for vertebrate population densities on islands compared with densities on continents (Darwin 1859; Wallace 1880; Crowell 1962; Grant 1968; Diamond 1970; Karr 1971; MacArthur 1972; MacArthur et al. 1973; Case 1975; Andrews 1979). These patterns appear to hold for amphibians, although specific comparisons are limited. We know of only two investigations (Anderson 1960; Townsend 1979) that compared directly mainland and island population densities of amphibians, although some information exists for individual groups on a few islands (Stewart 1979; Stewart and Martin 1980; Townsend 1986b). Anderson estimated salamander densities to be 4,000 to 20,000 ha^{-1} on islands off the California coast while mainland populations of the same or similar species were 500 to 4,500 ha^{-1} (Anderson 1960; conversions by Vial 1968). Townsend (1979) found significantly more red-backed salamanders on Beaver Island in Lake Michigan than at either of two mainland sites.

Densities of anurans in continental tropical areas vary greatly depending on conditions. Explanations for differences vary, but center around moisture levels and seasonal droughts, nutrient levels of soils that influence the amount and rate of decomposition of litter, elevation, and insect abundance. Inger (1980b) found 12 to 131 frogs ha $^{-1}$ at different Asian sites. By comparison, Neotropical sites produced 1160 to 1470 frogs ha $^{-1}$ in Costa Rica and 2,980 ha $^{-1}$ in Panama (Inger 1980a). Several studies show much greater density as well as diversity of frogs in Neotropical sites compared to Asia and Africa (Heyer and Berven 1973; Barbault 1976; Scott 1976, 1982; Toft 1982). Mid-level continental Neotropical forests support greater amphibian abundance than those at either high or low elevations (Scott 1976), a situation that pertains in Puerto Rico as well. Other studies show a decrease in density with elevation (Fauth et al. 1989).

Jamaica, which is approximately the same size and latitude as Puerto Rico but has greater elevational differences, is biogeographically more similar to Puerto Rico than the localities mentioned above. Unfortunately, Jamaica's low- and mid-elevation rain forests have been destroyed, so habitats comparable to the Luquillo Forest no longer exist. Stewart (1979) and Stewart and Martin (1980) found combined densities of 300 to 21,400 ha $^{-1}$ of all common species of *Eleutherodactylus* in disturbed agricultural lowland habitats of Jamaica. Townsend (1986b) estimated combined densities of 8,250 ha $^{-1}$ of the three most numerous of seven species of *Eleutherodactylus* occurring at 1250 m in Jamaican highlands.

Biomass

The relationship between snout-vent length and frog dry mass is shown in Fig. 8.6. Mean SVL of adult females is 45.9 mm (S.E. = 0.50, N = 54), that of adult males is 35.6 mm (S.E. = 0.16, N = 79; Woolbright 1989), and that of preadults is 13.5 mm (N = 75; Townsend 1985). By multiplying population densities obtained in mark-recapture plots by mass estimates from the regression equation, and assuming a 1:1 adult sex ratio, we estimate a total coqui biomass of 3.28 kg ha $^{-1}$ of forest (table 8.2). In experimental plots enriched with 100 bamboo nest sites where population density reached a maximum of 4,800 adult coquis ha $^{-1}$, the calculated maximum biomass is 4.1 kg ha $^{-1}$.

We estimated biomass (dry mass) per hectare of forest for adults of four additional species (table 8.2). Because we have direct measurements for *E. coqui* only, we have assumed that the SVL:dry mass relationship for these species is the same as for *E. coqui.* Lacking detailed population density studies of those species, we used the maximum counts of other species found in coqui surveys. The figure estimated for the total biomass of these smaller forest dwelling species is 0.42 kg ha $^{-1}$. A conservative total figure for all *Eleutherodactylus* species in the forest is an impressive 3.70 kg ha $^{-1}$. Our incomplete information on *Leptodactylus albilabris* (<1/900 m ^{2}) and the highly localized *Bufo marinus* (1/900 m ^{2}) prevents our including those large species in our calculations. The ratio of dry mass to wet mass for adult coquis is approximately 0.19.

FROGS AS PREDATORS

Diets of several species of Puerto Rican anurans have been determined by previous investigators (Dexter 1932; Rivero et al. 1963; Cintron 1970; La-

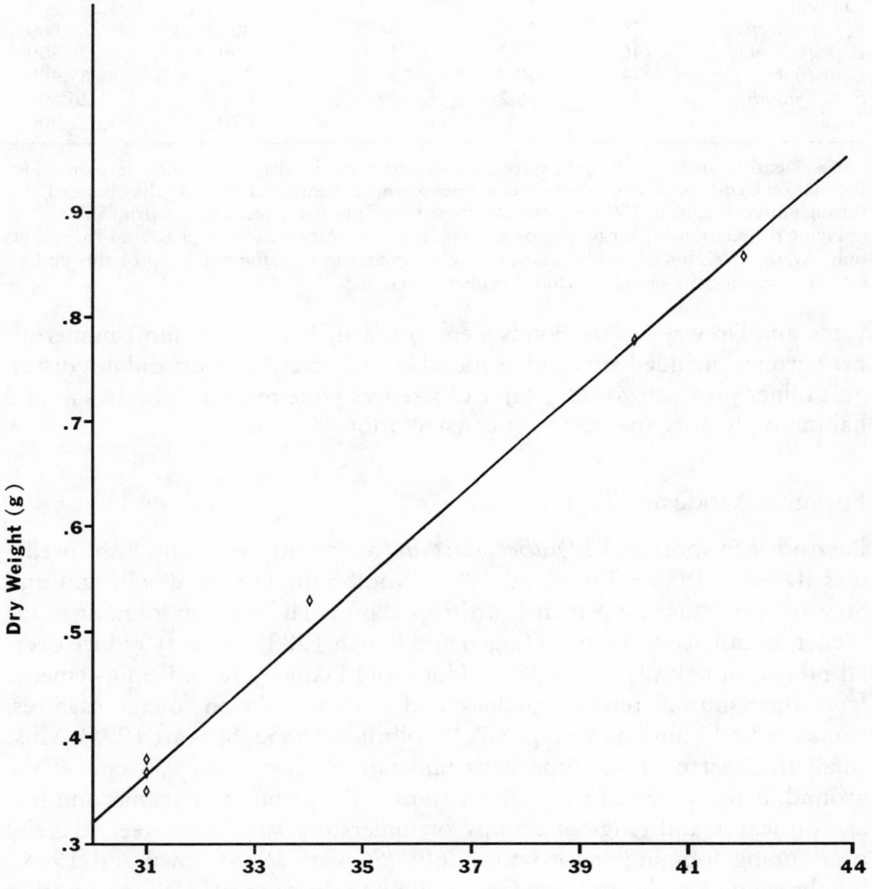

Figure 8.6. The relationship between body length (SVL) and dry mass (grams) for adult *Eleutherodactylus coqui*. The regression equation is as follows: dry mass = 0.00001 $(SVL)^{3.05}$. (Reprinted from Woolbright 1985b).

Table 8.2. Biomass of *Eleutherodactylus* species at El Verde

Species	Sample size	Mean SVL (mm)	Biomass of average individual (mg dry)	Density (N ha $^{-1}$)	Biomass (mg dry ha $^{-1}$)
E. coqui:					
male	79	35.6	540	1,632.5	880,600
female	54	45.9	1,170	1,632.5	1,911,500
preadult	75	13.5	30	17,305.0	484,900
Subtotal					3,276,000
E. portoricensis	54	31.1	360	800	285,800
E. wightmanae	46	20.2	100	400	38,300
E. hedricki	4	30.8	350	200	69,400
E. richmondi	27	28.2	270	100	26,500
Total				22,070	3,697,000

Notes: Densities of *E. coqui* adults were estimated from mark-recapture results (developed in Population Densities). Densities of other species are maximum numbers of adults obtained through direct counts in 100 m^2 plots. Dry mass for all groups was calculated from SVL by applying the relationship measured for *E. coqui* (fig. 8.6). All species except *E. coqui* are adults only. Average SVL for *E. portoricensis, E. wightmanae,* and *E. richmondi* are probably underestimates because the samples included primarily males.

vigne and Drewry 1970). Foods were primarily insects; the most numerous prey groups included ants and beetles. Unfortunately, authors did not distinguish finer prey categories. Major differences were related to body size and habitat of the frog species under consideration.

Foraging Mode and Behavior

Puerto Rican species of *Eleutherodactylus* are sit-and-wait (ambush) predators (Drewry 1970c; Townsend 1985; Woolbright 1985a; Woolbright and Stewart 1987) that eat primarily arthropod prey. The ambush foraging mode is energetically conservative (Taigen and Pough 1983) and may reduce overall predation risk (Andrews 1979; Huey and Pianka 1981). Coquis emerge from their diurnal retreats at dusk and most remain on foliage surfaces, trunks or limbs until dawn (fig. 8.7; Woolbright 1985a; Stewart 1995). Most small frogs sit on low herbaceous understory vegetation <30 cm off the ground. A few preadult frogs, from 10 mm SVL, climb tree trunks and forage on leaves and twigs of canopy or understory shrubs or trees, thereby overlapping foraging space with adults (Stewart 1985; Townsend 1985). Nonbreeding females and some males climb to the canopy to forage, weather permitting. Once individuals reach their perch, they move little or not at all throughout the night depending on weather (Woolbright 1985a). Prey capture movements are usually short distances (<5 cm) and account for about one-half of the total number of moves made by coquis during the night (Woolbright 1985a).

Figure 8.7. An adult male *Eleutherodactylus coqui* sitting in alert posture

Because *E. coqui* forages above the surface of the litter, it plays a very different role in the Luquillo Experimental Forest than does *Plethodon cinereus* in the Hubbard Brook Forest of New Hampshire (Burton and Likens 1975a). The coqui appears to be the most important predator in the nocturnal subweb (Reagan et al., this volume).

Foraging Rates of *E. coqui*

Foraging rates vary according to frog size, time of night, and weather conditions (Woolbright and Stewart 1987). Stomach contents of adult coquis collected at 0600 h indicate that frogs capture an average of 3.2 prey items per night (table 8.3). Assuming 3,265 adults ha^{-1} (see Population Densities), we calculate that adult coquis alone remove more than 10,000 insects ha^{-1} of forest per night. Preadult coquis, sampled between 2200 and 2400 h during the wet season, contained an average of six (smaller) prey per stomach (Townsend 1985). If this figure represents the total nightly intake of these frogs, it indicates that 17,305 preadult coquis eat an additional 104,000 insects ha^{-1} per night. Therefore, this one species alone apparently consumes an estimated 114,000 prey items ha^{-1} per night.

Table 8.3. Common prey of 173 *Eleutherodactylus coqui*

Prey taxon	Number	% total number	Volume (mm^3)	% total volume
PLANTAE: seeds	12	2.2	29.1	<0.1
ANIMALIA				
Mollusca: Gastropoda (shelled)	2	0.4	499.0	1.4
Stylommatomorpha: Sagdidae:				
Platysuccinea portoricensis	1	0.2	904.0	2.5
Arthropoda: Arachnida				
Acarina	13	2.3	1.3	<0.1
Araneida: Sparassidae:				
Stasina portoricensis	9	1.6	1343.1	3.7
other Araneida	15	2.7	499.0	1.4
other Arachnida	8	1.4	318.4	0.8
Anthropoda: Chilopoda and Diplopoda	6	1.1	253.6	0.7
Scolopendromorpha: Scolopendridae:				
Scolopendra alternans	1	0.2	981.7	2.7
Arthropoda: Insecta				
Dictyoptera: Blattellidae:				
Epilampra wheeleri	2	0.4	1060.0	2.9
Neoblattella sp.	4	0.7	437.3	1.2
Pseudosymploce personata	7	1.3	2020.0	5.6
other Blattellidae	20	3.6	2683.7	7.4
Dictyoptera: Cryptocercidae:				
Cryptocercus sp.	1	0.2	805.0	2.2
other Dictyoptera	6	1.1	282.8	0.8
Coleoptera: Curculionidae	15	2.7	227.7	0.6
Scolytidae	17	3.1	14.4	<0.1
other Coleoptera	18	3.2	159.4	0.4
Diptera	17	3.1	32.1	<0.1
Homoptera: Tropiduchidae:				
Ladellodes stahli	6	1.1	270.8	0.7
other Homoptera	16	2.9	466.9	1.3
Hymenoptera: Formicidae:				
Camponotus ustus	36	6.5	279.2	0.8
Iridomyrmex melleus	17	3.1	10.2	<0.1
Myrmelachista ramulorum	19	3.4	14.9	<0.1
Paratrechina microps	30	5.4	235.3	0.7
other *Paratrechina*	61	11.0	102.2	0.3
Pheidole moerens	7	1.3	5.5	<0.1
other Formicidae	30	5.4	198.5	0.5
Apidae: *Apis mellifera*	8	1.4	1364.4	3.8
Lepidoptera: Noctuidae	6	1.1	99.9	0.3
other Lepidoptera	22	4.0	875.2	2.4
Orthoptera: Gryllidae:				
Amphiacusta caraibea	11	2.0	4183.2	11.6
Cyrtoxipha gundlachi	17	3.1	223.5	0.6
Gryllus assimilus	13	2.3	8495.7	23.5
Orocharis sp.	8	1.4	3772.4	10.4
other Gryllidae	17	3.1	1504.4	4.2
Psocoptera	11	2.0	5.4	<0.1
other Insecta	14	2.5	177.9	0.5
Chordata: Amphibia:				
Leptodactylidae:				
Eleutherodactylus sp. eggs	27	4.9	1025.1	2.8

Notes: Taxa are combined to give categories containing >1% by volume or number of prey. For a complete list of prey taxa see appendix 8.A.

Female and noncalling male *E. coqui* forage early in the evening and usually fill their stomachs by 2400 h. Calling males do not feed until the early morning hours after they have finished calling for the night and consequently obtain fewer prey than do noncalling frogs (Woolbright and Stewart 1987). More prey are taken per frog in the wet season (median = 3/adult frog) than in the dry season (2/frog), probably because coquis are less active in dry conditions. On dry, cool nights, fewer frogs are out, and those visible are in water-conserving postures (Drewry 1970c; Pough et al. 1983) and unlikely to catch prey (Woolbright and Stewart 1987).

Diet of *E. coqui*

Coquis prey on invertebrates at all levels of the forest, from ground to canopy. Because 74% of frogs perch on leaves (Townsend 1985), most prey species will be on leaves as well. As with most species of anurans studied (reviewed by Toft 1985; e.g., Loman 1979; Stewart 1979; Emerson 1985; Townsend 1985; Woolbright and Stewart 1987), larger coquis eat larger, and usually fewer food items than small coquis, and their diet is highly diverse. Individuals eat more prey items in wet months than in dry months (Woolbright 1985b; Woolbright and Stewart 1987). Diets of different age groups of *E. coqui* are equally diverse. Prey types are used in different proportions, and some prey types change with age. For example, in a sample of thirty-nine small (7 to 10 mm, SVL), forty-two intermediate-sized (10.5 to 24 mm, SVL), and thirty-two adult coquis (24.5 to 48 mm, SVL) examined during April through July 1979 (K. Townsend pers. comm.), small and intermediate-sized frogs ate primarily ants (74%, 90%, respectively), while only 38% of the adult diet consisted of ants. Larger frogs ate larger prey, including orthopterans up to 22 mm in length (fig. 8.8), moths, and spiders. Whereas no juvenile had an empty stomach, 16% of the adults had empty stomachs; in the adults examined these findings did not differ with gender. Lavigne and Drewry (1970) found Acarina (7.5%) and Hymenoptera (6%) the most numerous prey in stomachs of two juvenile *E. coqui.*

Prey Diversity

The coqui is euryphagous (app. 8.A). The diversity of prey in its diet reflects not only the foraging behavior of the frogs, but the arthropod diversity of the forest (Drewry 1969d, 1970b; Odum et al. 1970c; Reagan et al. 1982; Garrison and Willig, this volume; Pfeiffer, chap. 5, 7, this volume). At least 101 different prey species, representing at least sixty families, twenty-eight orders, nine classes, and five phyla, were identified from the stomachs of 173 adult coquis (mean number of taxa = 3 [range, 0 to 12] with 0 to 17 items/stomach; table 8.3; app. 8.A). Insects constitute the vast majority of prey, and three orders of insects seem particularly important. Hymenopterans (pri-

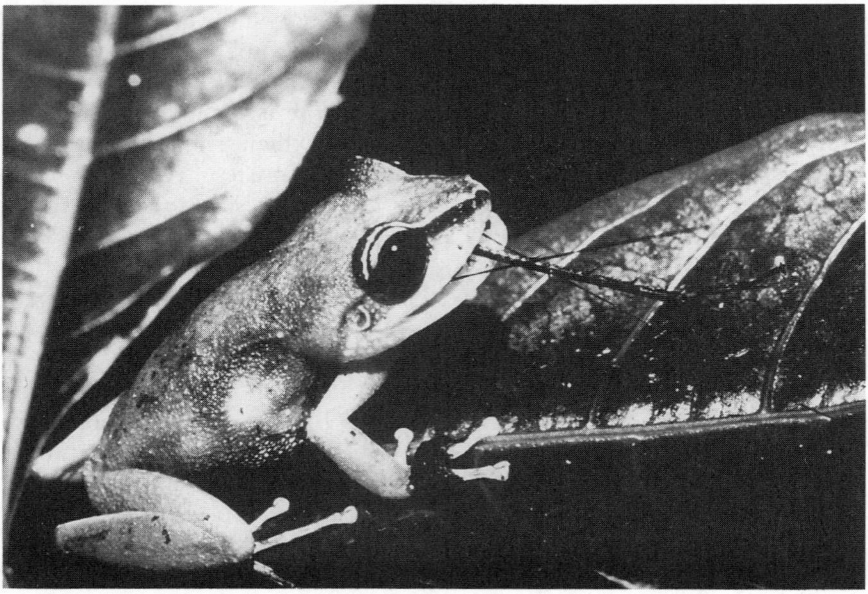

Figure 8.8. An adult *Eleutherodactylus coqui* eating a large insect (orthopteran), too large to fit into the frog's mouth

marily ants) accounted for 38% of the total number of prey found in stomachs but, because of their relatively small size, made up only 6% of the total prey volume. Orthopterans, primarily gryllids (crickets), accounted for 12% of the total number of prey and, because of their large size, 50% of total prey volume. Dictyopterans, primarily blattellids (roaches), accounted for 7% of total prey number and 20% of total prey volume. Therefore, roaches and crickets (large prey) make up 70% of the volume of prey eaten by coquis while ants (small prey) make up 38% of the number of prey eaten (table 8.4).

Because ants are relatively unimportant to adult coquis in terms of volume of prey taken (as well as being high in chitin content), one wonders why coquis eat them in such large numbers. One reason may be the abundance and availability of ants in the forest (Pfeiffer, chap. 5, this volume). Another answer may relate to sexual differences in foraging: 49% of males ate ants, while only 25% of females did ($\chi^2 = 9.33$, $p < .01$, 1 d.f.). We have shown previously that males take fewer prey than expected on nights that they call (Woolbright and Stewart 1987) and that frequently they are on negative energy budgets for those nights (Woolbright 1985b). Therefore, any available food may be consumed. The large number of ants in the coqui diet is doubtless a correlate of the sit-and-wait foraging mode, as it is in other anurans (Toft 1980a, 1981) and in some lizards (Huey and Pianka 1981). The con-

Table 8.4. Major arthropod prey taxa of *Eleutherodactylus* collected near the El Verde Field Station

	E. coqui		*E. portoricensis*		*E. richmondi*		*E. wightmanae*	
Prey taxon	%N	%vol	%N	%vol	%N	%vol	%N	%vol
Arachnida	8.3	8.7	9.8	10.6	23.1	61.0	10.3	9.3
Chilopoda/Diplopoda	1.1	0.7	5.6	1.3	7.7[a]	15.1	11.5[a]	30.7
Dictyoptera	7.2	20.2	7.0	12.1	0	0	1.3	2.4
Coleoptera	9.0	1.1	23.9	34.3	0	0	10.2	4.4
Hymenoptera	37.5	6.1	28.2	3.1	30.8	1.1	38.5	15.8
Lepidoptera larvae	5.0	2.7	8.5	3.8	7.7	13.4	3.8	10.6
Orthoptera	11.9	50.3	5.6	31.9	0	0	0	0

Notes: Only prey taxa representing more than 10%, by prey number or volume, of the diet of any frog species are listed. The numbers of frogs and numbers of prey items found are as follows: *E. coqui* (173, 555), *E. portoricensis* (33, 71), *E. richmondi* (14, 13), *E. wightmanae* (26, 78).

[a]Diplopoda only.

sumption of ants by frogs appears to be determined more by availability than by preference. Jones (1982) found that lowland species of *Eleutherodactylus* ate many more ants in the Lesser Antilles than in Puerto Rico. The numbers of prey eaten by *E. antillensis* consisted of 71% ants on St. John, but only 12% on Puerto Rico. Differences were probably related to prey availability. Ovaska (1991) found ants in 79.6% of 113 *E. johnstonei* stomachs in Barbados.

Other than differences in size of prey and numbers of ants eaten, we found no major sex-related differences in coqui diet. Of prey taxa comprising more than 1% of total number, all were eaten by both males and females except scolytid beetles, the ant *Pheidole moerens,* and coqui eggs, which in our sample were eaten only by males. The beetles and ants are both small prey that were more commonly eaten by males. Conspicuous by their absence are vertebrate prey. Although juvenile *E. coqui* are potentially one of the most abundant prey for adult frogs, we found a juvenile in only one *E. coqui* stomach, and in one *E. wightmanae* stomach. Their consumption is so rare that it appears accidental. Two hatchling *Anolis gundlachi* lizards (15 and 20 mm) are present in coqui stomachs. Leal and Thomas (1992) found a 43 mm adult female coqui containing a 43 mm SVL *Anolis evermanni,* an unusual event. The consumption of coqui eggs probably relates to male competition for territories and nest sites rather than pure energy acquisition (Townsend et al. 1984). Females rarely eat eggs in the field (D. S. Townsend pers. comm.; pers. observation).

A surprising result is the paucity of termites in coqui stomachs. Termites, abundant on tree trunks and in retreats of the frogs, are a potentially important food source. Yet only a few winged stages and no nasutes were found. Evidently the chemicals extruded by the nasutes inhibit consumption by predators (E. A. McMahan pers. comm.). In addition, termites rarely occupy leaves, and they build covered runways (McMahan this volume) that may protect them from frog predation. Termites are frequent prey of anurans in other parts of the world (Stewart 1979; Toft 1982). Four species of *Eleutherodactylus* in lowland Jamaica preyed on ants frequently (44 to 72% of food items), arachnids often (6 to 19%), and termites occasionally (4 to 7%); they rarely took roaches or crickets (0 to 2%) (Stewart 1979; Stewart and Martin 1980). Those frog species forage on or nearer the ground than does *E. coqui.*

Another surprising aspect of coqui diets was the virtual absence, even of smaller individuals, of the extremely abundant walking stick, *Lamponius portoricensis,* that forages commonly on leaves of understory shrubs on which frogs sit. Only one individual was found in coqui stomachs (app. 8.A). Evidently, by moving slowly walking sticks escape predation from coquis, the most numerous ambush predator in the forest. Unusual nonarthropod prey items included three shelled snails, a young coqui, two hatchling lizards mentioned above, and various plant parts. The snails and the young coqui were eaten by female frogs taken in January. It seems likely that the large size of these prey explains their occurrence only in females and that lower feeding success in the dry season (see Prey Diversity; Woolbright and Stewart 1987) explains why such prey are eaten in January. The partially dissolved condition of the snail shells suggests that coquis are capable of digesting shells, but we do not know whether the calcium so obtained has nutritional significance to the frogs. The low frequency of plant parts (primarily seeds, florets) found in stomachs suggests accidental ingestion. Falling seeds may be perceived as animal prey and vegetative matter may be ingested accidentally in the process of feeding. We have seen several coquis gathered in a flowering palm inflorescence feeding on insects concentrated about the florets. Such conditions could easily result in ingestion of florets.

Coqui stomachs contain fewer items than do those of active searchers such as the diurnal and much more actively foraging *Dendrobates pumilio* (Donnelly 1991) or *Anolis* species (Reagan 1986, this volume). Feeding on fewer larger prey is characteristic of ambush predators, while active foragers take many small items. The sit-and-wait mode is most effective in catching large mobile prey such as orthopterans when they are available (Toft 1985). Differences in prey sizes between frogs and anoles may result, in part, from feeding opportunities. The larger prey, orthopterans and dictyopterans, are nocturnal.

Although smaller species of *Eleutherodactylus* do feed in the litter (see Diets of other *Eleutherodactylus* species; Lavigne and Drewry 1970), there is no evidence that *E. coqui* adults forage in the litter. Litter prey species could be captured as frogs move from retreats to elevated perches. Although some litter species are present in coqui stomachs, they are underrepresented. For example, juveniles would be most likely to feed in the litter, but acarines, the most abundant litter arthropod (Pfeiffer, chap. 5, this volume), were rare in juvenile stomachs. Only two soil mites were found in thirty-nine juvenile frogs (Townsend 1985). However, Lavigne and Drewry (1970) found an average of 7.5 acarines in two juvenile coquis. By contrast, Jamaican ground-dwelling species forage heavily on litter species (Stewart 1979; Townsend 1986b).

Prey Size

Small prey size of island lizards (Andrews 1976) and birds (Wilson 1976) is used as evidence of food-limited populations when compared to mainland counterparts. Mean prey sizes of Puerto Rican anurans examined are small, as are prey sizes of Jamaican species (Stewart 1979; Stewart and Martin 1980; Townsend 1986b). However, prey sizes we determined for Puerto Rican species are larger on average than those taken by the *Eleutherodactylus* that Toft (1980a,b, 1981) examined in Panama and Peru. A problem with these comparisons is that the species examined have dissimilar habitats and behavior. The coqui forages at night off-ground while the frogs Toft studied were lowland diurnal litter species.

From all stomachs examined, *E. coqui* eats primarily small prey; the range of prey length was 0.2 to 55 mm. In a sample of thirty-seven males, mean prey length was 6.3 mm; in twenty females it was 10.2 mm. Only six prey exceeded 35 mm, and those were elongate items such as a centipede, a moth, a *Peripatus,* and a thread worm (nematomorphan). In a sample of eighty-six frogs, mean prey length in twenty-nine juveniles was 1.4 (0.5 to 3.0) mm, in thirty subadults 1.8 (0.5 to 8.0) mm, and in twenty-seven adults 4.1 (1.0 to 22.0) mm (Townsend 1985). In sixty adult *E. coqui* from 670 m, where frogs are larger, Cintron (1970) found mean prey length to be 5 mm.

An interesting aspect of these diets is the great degree of overlap in major taxa of prey of the different size classes of frogs. However, because the size of the smallest, as well as the largest, prey eaten increases with body size, there is little overlap in size of prey eaten by frogs that differ by more than 10 mm SVL (fig. 7.9; Woolbright and Stewart 1987). Males take smaller prey than do females. An explanation for sexual differences in prey size is that prey size is positively correlated with frog body size (Woolbright and Stewart 1987; fig. 8.9) and that male coquis are smaller than females (Woolbright 1989). Also, it seems likely that males, on negative energy budgets because they call early in the evening (Woolbright 1985b), have fewer opportunities than females to feed on large prey. Large female coquis typically eat only a few large prey per night (including crickets and roaches) while smaller frogs (males and subadults) tend to eat more but smaller prey (primarily ants). These results parallel observations in Jamaica where the small species, *E. planirostris,* averaged eight items per frog while the three larger species averaged only four, but larger, items per frog stomach (Stewart 1979).

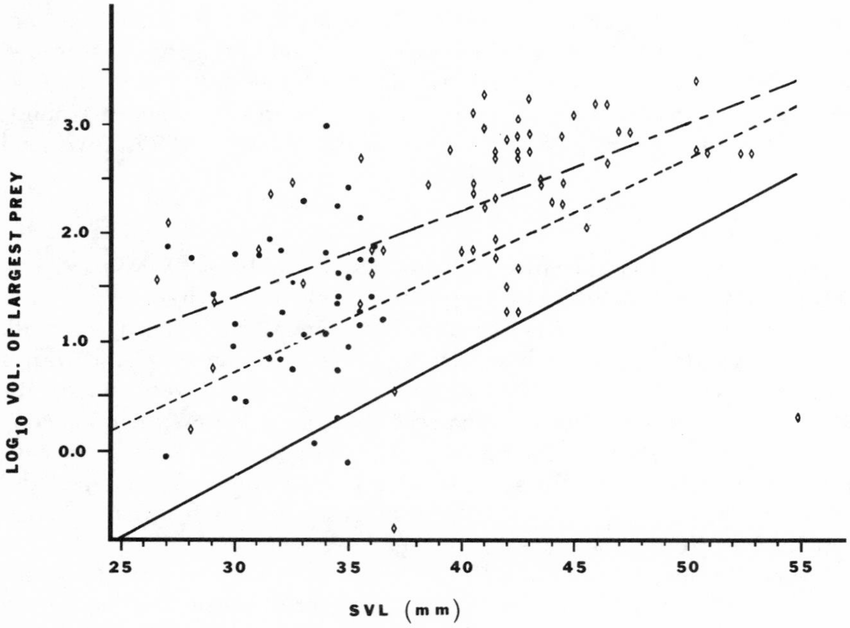

Figure 8.9. The relationship between body size and prey size for noncalling adult *Eleutherodactylus coqui*. Circles are males; diamonds are females. Top line is for volume of largest prey (log vol. = 0.09SVL − 1.50, F = 33.2409, $p < .001$). Middle line is for mean prey volume per stomach (log vol. = 0.10SVL − 2.23, F = 42.9170, $p < 0.001$). Bottom line is for volume of smallest prey (log vol. = 0.11SVL − 3.55, F = 20.8896, $p < 0.001$). Seasons are combined because prey size did not differ seasonally. (Reprinted from Woolbright and Stewart 1987).

Diets of Other *Eleutherodactylus* Species

We collected samples of *E. portoricensis, E. richmondi,* and *E. wightmanae* in order to compare stomach contents of these species to the more extensive data on *E. coqui.* Samples were collected in both wet and dry seasons but were not analyzed separately by season because of small sample sizes. Our results agree with those of Lavigne and Drewry (1970) in that all species feed primarily on arthropods and appear to be generalists. *Eleutherodactylus portoricensis* and *E. wightmanae* samples each contained sixteen orders from five classes, and *E. richmondi* stomachs contained eight orders from four classes (app. 8.A). Although overlap in prey taxa was considerable, the four species appeared to differ in the relative importance of various prey (table 8.4). Differences in numbers of prey per frog are due in part to sample biases. Coqui samples were collected at 0600 h while others were collected in early evening. In addition, coqui samples included many more females than did samples of other species. The lowest number of prey (mean = 0.9) was taken by *E. richmondi,* and the most by *E. coqui* (3.2) and *E. wightmanae* (3.0). The average number of prey per stomach for *E. portoricensis* was 2.2.

The diet of *E. portoricensis* was very similar to that of *E. coqui* (table 8.4), which it resembles both in body size and microhabitat. As with *E. coqui,* ants were the most abundant prey in *E. portoricensis* stomachs, and both cockroaches (dictyopterans) and crickets (orthopterans) were important prey by volume. Unlike the coqui, *E. portoricensis* also fed heavily on beetles (coleopterans) that supplied the largest volume of any prey group. Two coleopteran families, Carabidae and Staphylinidae, were present only in the stomachs of *E. portoricensis.* Also exclusive to *E. portoricensis* stomachs were the spiders, *Leucage regnyi* and *Corrina* sp., and the hymenopteran family, Diapriidae. *Eleutherodactylus portoricensis* forages closer to the ground than does *E. coqui* (Lavigne and Drewry 1970; Drewry 1970c; Stewart unpublished) and shelters in the litter. In the understory, perch heights of the two species are similar (mean = 92 cm for 325 *E. coqui;* 87 cm for twenty-three *E. portoricensis*), but many coquis forage in the canopy as well (table 8.1).

The diets of *E. richmondi* and *E. wightmanae* were more similar to each other than to those of *E. coqui* or *E. portoricensis* (table 8.4). Several components of the diets of *E. richmondi* and *E. wightmanae,* which live on or near the ground, suggest litter foraging. Lavigne and Drewry (1970) also found litter species abundant in their diets. Although both fed heavily on ants, neither contained any orthopterans, and only one dictyopteran was found. Both species ate a comparatively high proportion of homopterans and derived a relatively high volume from lepidopteran larvae and Diplopoda (neither ate Chilopoda). With millipedes making up the predominant standing stock in forest litter (Pfeiffer, chap. 5, this volume), it is not surprising that *E. richmondi* and *E. wightmanae* stomachs contained so many, especially the smaller frogs that live in the litter.

Because of their secretive habits, the number of *E. richmondi* collected for stomach analysis was small and all specimens collected were calling males. Arachnids appeared to be the single most important prey in volume and were second to hymenopterans in number. This is not surprising given 3.5×10^6 spiders ha^{-1} of forest litter (Pfeiffer, chap. 7, this volume). *Eleutherodactylus richmondi* was the only species whose stomach contained an isopod. At 650 m Cintron (1970) found *E. richmondi* eating over 60% beetles and over 20% ants; their mean prey size was 2.5 mm. Lavigne and Drewry (1970) found their diet evenly distributed but with slightly more prey in the orders Coleoptera, Diptera, and Hymenoptera.

Eleutherodactylus wightmanae stomachs contained members of several taxa not found in other species. These included two spiders, *Pseudosparianthus jayuyae* and an unidentified jumping spider (salticid), members of the springtail family Sminthuridae, the fly family Stratiomyidae, the bug family Cicadellidae, and two ant genera, *Solenopsis* and *Ochetomyrmex*. These dietary differences reflect habitat use by *E. wightmanae,* which spends much time in the litter or on low understory vegetation (Lavigne and Drewry 1970). It was unexpected that a species so small as *E. wightmanae* should eat frogs, yet one stomach contained an unidentified juvenile frog.

Puerto Rican data show that frogs of the genus *Eleutherodactylus* are opportunistic feeders. There is much variability between stomach samples collected at different sites and seasons. In all frog stomachs we examined (N = 246; four species; app. 8.A), the greatest prey diversity was found in the following groups: ants (sixteen spp.), roaches (sixteen spp.), beetles (thirteen spp.), spiders (eleven spp.), and gryllids (eight spp.). From fifty-one stomachs of *E. karlschmidti,* Rivero et al. (1963) found that 68% of the number of prey were terrestrial, mostly insects: dipterans 44.5%; hymenopterans, mostly ants, 15%; trichopterans 15%; hemipterans 9%; coleopterans 7%; and a few lepidopterans and ephemeropterans. Cintron (1970) found major food items of three species, *E. coqui* (possibly including *E. portoricensis*), *E. richmondi,* and *E. eneidae,* (N = 10 frogs each species) to be in the taxa Formicidae and Coleoptera. Termites comprised 10% of the diet of *E. eneidae,* 3% of *E. richmondi,* and none in *E. coqui.* Ants comprised nearly 60% of the numbers of prey of *E. coqui* collected at 670 m in the Luquillo Mountains; a few Ichneumonoidea and Chilopoda were also present. Lavigne and Drewry (1970) examined 121 stomachs of eight species of *Eleutherodactylus* taken in the vicinity of El Verde Field Station. The most numerous prey in adults of *E. coqui, E. portoricensis, E. wightmanae, E. eneidae, E. antillensis, E. locustus,* and *Leptodactylus albilabris* were Hymenoptera. Most numerous in *E. richmondi* were Coleoptera, and in *E. brittoni* were Diptera. Second most numerous prey were Diptera in all species except *E. wightmanae* (Acarina), *E. richmondi* and *E. brittoni* (Hymenoptera). Sample sizes were small for most species (mean = 8, range 3 to 18) except *E. coqui* (N = 29). Differences may be due to highly variable local abundances of insects and small sample sizes.

Availability of Prey

Local and Seasonal Differences

Results of stomach content analyses of *E. coqui* indicate that foraging success is higher in the summer than in the winter. Although drier winter conditions undoubtedly explain at least some of this difference (see Foraging Rates), lower food availability could also be a contributing factor. In addition, predictable variation between sites within the forest might influence local densities of frog populations. Therefore we measured temporal and spatial variation in potential prey of *E. coqui* (see app. 8.B.2 for methods).

Table 8.5. Arthropods collected in wet and dry seasons at El Verde

	Numbers of arthropods	
	Wet season	Dry season
Site:[a]		
Sonadora East	51.0 ± 13.80	27.0 ± 8.60
Sonadora West	91.0 ± 25.00	26.0 ± 6.70
Height (m):[b]		
1–5	9.5 ± 1.26	3.1 ± 0.61
6–10	14.0 ± 1.74	2.9 ± 0.56
11–15	15.6 ± 1.43	4.4 ± 0.85
16–20	12.9 ± 1.37	6.9 ± 1.13

[a] Mean (±1 S.E.) number of arthropods collected per string of sticky traps at two sites in the forest in wet and dry seasons (5-m strings with cups every 0.5 m) at El Verde.

[b] Mean (±1 S.E.) number of arthropods collected per sticky trap at four heights above the ground in wet and dry seasons (20 m strings with cups every m).

Arthropod resources appear to be locally and temporally patchy within a season. We measured insect abundance using sticky traps suspended at 0.5 m intervals on 5 m strings. There was an average two-fold difference (range of ratios 1.5 to 2.9) between numbers of arthropods on the string that collected the most arthropods and the string that collected the least within 5 m at the same site on the same night. In addition, the string that was high on one night might be low on another. These results suggest that frogs are unlikely to be able to predict arthropod availability well enough to base territorial defense on food resources. In the wet season there were significant differences between nights within a site (Nested ANOVA: $F = 3.26$, $p < .025$). There were additional differences between sites. The site 50 m west of Quebrada Sonadora averaged many more arthropods than the site 10 m east of Quebrada Sonadora (Nested ANOVA: $F = 16.31$, $p < .01$; table 8.5). Thus it is possible that food availability could contribute to differences in frog population density between sites within the forest, although we have not tested this hypothesis.

More arthropods were collected in the wet season than the dry season both at the Sonadora East site (one-way ANOVA: $F = 74.24$, $p < .001$) and at the Sonadora West site ($F = 253.02$, $p < .001$; table 8.5). The lower dry season numbers did not vary significantly either between nights (nested ANOVA: $F = 2.20$, $p > .10$) or between sites (nested ANOVA: $F = 0.07$, $p > .75$). We suggest that this lower prey availability may explain the broadening of the coqui's dry season diet (see Diet of *E. coqui*) as would be predicted by models of optimal diet choice (Pyke et al. 1977). The relative stability of prey availability between nights suggests that availability of moisture, which influences frog activity levels, is probably responsible for between night variation in the number of empty stomachs in the dry season.

Vertical Differences (Canopy vs. Understory)

We hypothesized that tree-climbing behavior of *E. coqui* (see Horizontal and Vertical Movements) is explained by higher prey availability in the canopy. As an initial test of the hypothesis, we examined arthropod availability along a vertical transect (see app. 8.B.3 for methods). Results showed significant differences in numbers of arthropods collected at various heights above the ground both during the wet season (one-way ANOVA: $F = 3.86$, $p < .025$) and during the dry season ($F = 4.11$, $p < .025$; table 8.5). Numbers of arthropods generally increased with distance off the ground. These results are consistent with findings from stomach contents of ten frogs taken from the canopy at dawn in June 1988. Stomachs contained more prey items, and items were likely to be smaller than those in frogs collected in the understory. Forty-eight percent of the prey items were *Myrmelachista ramulorum,* an arboreal ant. Six species, 15% of the total volume, were not found in stomach contents examined previously (table 8.6; Flynn and Woolbright pers. comm.). Although numbers of possible prey may increase with height off the ground, in an examination of food available in the first 200 cm off the ground, Townsend (1985) found that small prey (0.5 to 3.0 mm) were most abundant at 0 to 50 cm. Therefore, the distribution of small prey mirrors that of small frogs.

Prey identity can indicate the importance of canopy foraging to the frogs. Orthopterans and dictyopterans contribute the greatest volume of food consumed by coquis (see Diet of *E. coqui*). Orthopterans are abundant in the canopy as well as in the understory. Since canopy insects have not been studied thoroughly at El Verde, we do not know whether any species is confined strictly to the canopy (R. Garrison pers. comm.; Reagan, this volume). *Myrmelachista ramulorum,* an arboreal ant, was taken in 5% of coqui stomachs. Seventeen species of homopterans were also present in stomachs. These insects, leaf hoppers and plant hoppers, are most common in the canopy (Garrison and Willig, this volume). Orthopterans comprised 50% by volume of stomach contents. Some very large *Orocharis vaginalis,* a canopy gryllid, comprised 7% of the volume of stomach contents. The canopy spider, *Gasteracantha tetracantha* (Pfeiffer, chap. 5, this volume), was also found in coqui stomachs. The crickets *Gryllus assimilus* (10% of coqui stomachs) and *Amphiacusta caraibea* (7% of stomachs) are strict ground-dwelling prey. Dictyopterans (roaches) are abundant in the understory (Pfeiffer, chap. 5, this volume).

Effects of Frog Density on Prey Availability

Numbers alone suggest that frogs constitute a major component of the forest ecosystem and could potentially exert a top-down control on density of their

Table 8.6. Stomach contents of *Eleutherodactylus coqui* at El Verde

Prey taxon	Numbers	% of total	Volume (mm³)
Arthropoda: Arachnida			
Acarina: Orbatidae	1	0.5	0.05
Araneae: Clubionidae			
Clubiona portoricensis	1	0.5	3.9
Chiracanthium inclusum	1	0.5	70.7
Arthropoda: Insecta			
Dictyoptera: Blattellidae			
Aglaopteryx sp.	2	1	188.0
Cariblatta sp.	1	0.5	7.5
Cariblatta hebardi	2	1	52.5
Coleoptera: Unident. sp.	2	1	16.2
Chrysomelidae	1	0.5	1.3
Curculionidae: *Exophthalmus* sp.	1	0.5	174.5
Homoptera: Tropiduchidae	6	3.1	156.6
Isiidae: *Neocolpoptera portoricensis*	1	0.5	23.8
Cixidae: *Pintalia alta*	1	0.5	3.7
Hymenoptera: Formicidae			
Iridomyrmex melleus	11	5.7	21.9
Myrmelachista ramulorum	93	48.2	44.7
Strumigenys royeri	1	0.5	0.1
Apidae: *Hemisia versicolor*	1	0.5	131.4
Isoptera: *Nasutitermes costalis*	56	29.0	155.1
Lepidoptera			
Unident. larva	1	0.5	6.9
Unident. adult	1	0.5	59.1
Orthoptera: Gryllidae			
Orocharis sp.	7	3.6	738.3
Amphiacusta caraibea	1	0.5	750.1
Chordata: Amphibia			
Leptodactylidae: *Eleutherodactylus* sp. eggs	11	5.4	17.7

Source: From M. Flynn and L. Woolbright (unpublished).

Note: Ten *E. coqui* taken in the canopy at dawn.

prey. If insect prey is limited in the forest understory, then an increase in frog density should cause a decrease in prey abundance. To test the latter prediction, between 2000 to 2200 h in July, Waide and Stewart (unpublished) counted orthopterans and cockroaches on understory vegetation in eight 100 m^2 plots. Four of the plots had increased frog densities following supplementation of frog nest sites with bamboo tubes (total of eighty-three adults in high density plots and twenty adult coquis in control plots). Numbers of insects counted did not differ significantly between plots. In all eight plots, a total of 117 (70%) roaches and fifty-one (30%) orthopterans were counted. Orthopterans comprise 50% of the total volume of coqui stomachs while cockroaches comprise only 15% (table 8.3). While the experiment was preliminary, no effect of differences in frog density on prey density was observed, at least in the understory (but see spider effects in Leaf Surface, Twig, and Trunk Predators).

Is the Forest Food-limited for E. coqui?

Although detection of food limitation in nature may require extensive study, our results to date suggest that the primary factor limiting *E. coqui* populations in the forest is not food shortage. We were not able to demonstrate a reduction in prey insects in high density plots (see above). Food shortages did not prevent significant increases in population density of *E. coqui* following addition of retreat and nest sites (Stewart and Pough 1983; Stewart unpublished). Although during short-term experiments food did not limit population size, long-term effects may result from a lack of foraging opportunity. Coquis may be limited in their ability to forage because of dry weather, or by the calling and brooding behavior of males. During winter months dry conditions often inhibit feeding (Pough et al. 1983), and insect availability is greatly reduced (see Availability of Prey, Local and Seasonal Differences; Foraging Rates; table 8.5). After prolonged dry periods, we find emaciated adults in the forest. Observations in the Activity Transect over several years indicate that temporary dry conditions limit foraging, especially for juveniles (Stewart 1995). In dry periods during 1980 to 1981, up to 60% of coqui stomachs were empty when there had been no rain for three days (Woolbright and Stewart 1987). In a sample of fifty adults 20% of the stomachs were empty in the dry season compared to 2% in the wet season (Woolbright and Stewart 1987). Lack of foraging opportunity due to temporary dry conditions influences salamander populations as well (Jaeger 1978, 1980).

Various data suggest that frog growth is determined both by food availability and the opportunity for frogs to forage. Females produce an egg clutch more frequently in captivity where food is unlimited (Townsend 1984; Woolbright unpublished). Woolbright (unpublished) found that adult coquis reached larger body sizes in experimental plots receiving regular additions of chemical fertilizer than in reference plots. This suggests that "bottom-up" controls may exist over secondary productivity, if not over population density. Evidence also exists for behavioral controls over frog growth. Growth of males in the field is truncated after reproductive activity begins, yet males in captivity continue to grow (Woolbright 1989). Stomachs of calling males contain fewer items and less volume than noncalling males (Woolbright and Stewart 1987). Brooding males take smaller items than nonbrooding males: 56% ($N = 45$) have empty stomachs and 38% have depleted fat bodies (Townsend 1986a). All evidence, therefore, indicates that it is lack of food availability to the frogs, either environmentally or behaviorally determined, rather than food scarcity in the forest, that is influencing growth of individual adult coquis.

FROGS AS PREY

The large amphibian biomass in the forest, composed primarily of *E. coqui,* provides an abundant food source for appropriate predators (Waide, this volume; Pfeiffer, chaps. 5, 7, this volume). Prior studies have found that frogs are preyed upon by various species of vertebrate predators (Wetmore 1927; Pimental 1955; Rolle 1963), primarily birds. Specific behavior of the forest frogs makes them available to different categories of predators. Although calling *E. coqui* and *E. portoricensis* appear especially vulnerable, we have no evidence that they are eaten more frequently than are noncalling frogs (Woolbright 1985b). In our mark-recapture plots, the longest survivors, those living longer than four years, were females (Stewart unpublished). Trunk climbing by coquis, especially preadults and females, makes them vulnerable to trunk-dwelling ambush predators such as tarantulas and amblypygids and possibly to birds (Stewart 1985). Small frogs are especially vulnerable to sparassid spiders; their vulnerability is size dependent (Formanowicz et al. 1981), and their perches are coincident (fig. 8.10).

Mortality Rates of *E. coqui*

Population turnover rates are rapid and preadult populations decline sharply before adulthood (fig. 8.3; Stewart 1995). We do not yet know the maximum age of individual coquis in the forest but we have records of a few frogs that were at least four and six years old. Mortality rates were estimated by applying Jolly's (1965) survivorship formula to data collected during a long term mark-multiple recapture study (Woolbright unpublished). Approximately 94% of adult coquis (or an estimated 3,069 frogs ha^{-1}) do not survive an entire year. Although the higher dry season mortality noted above may be caused in part by physical factors such as water stress, a large proportion of these frogs undoubtedly become prey to other animals. Comparison of preadult and adult population sizes suggests that approximately 81% of preadults (or an estimated 14,017 frogs ha^{-1}) do not survive their first year of life.

Predators

In the trophically complex community at El Verde (fig. 8.11), predation exerts a strong influence on anurans. Although the absence of mammalian predators is often cited (e.g., Waide and Reagan 1983) as a reason for high densities of amphibians and reptiles on islands, Puerto Rico has several bird and invertebrate predators that have a substantial impact on anuran populations (Pfeiffer, chap. 7, this volume; Waide, this volume). Invertebrates apparently eat more juveniles, while birds consume primarily adult anurans. Rolle (1963) found that 8.4% of the diet of red-legged thrushes, *Turdus plumbeus,* consisted of small frogs such as *Leptodactylus albilabris* and *Eleutherodactylus* spp. and a few lizards. Tree frogs (likely the abundant *E. coqui*) comprised most of the diet of nestlings from day five until fledging. Santana C. and Temple (1988) found that up to 42% of nestling diet of *Buteo jamaicensis,* the red-tailed hawk, consisted of *E. coqui.* Up to 40% of the prey items brought to nestling *Otus nudipes,* the screech owl, were *E. coqui* (Snyder et al. 1987).

Documenting predation in the field is difficult. However, during seventeen years of study of *E. coqui* at El Verde, we and others have observed numerous predation events in the field as well as in the laboratory (table 8.7). Predators vary with life stages and as habitat use changes with size of frogs. Overlap in habitat use occurs most often between adults and subadults. Since males guard eggs, habitats of eggs and parental males overlap during brooding, but

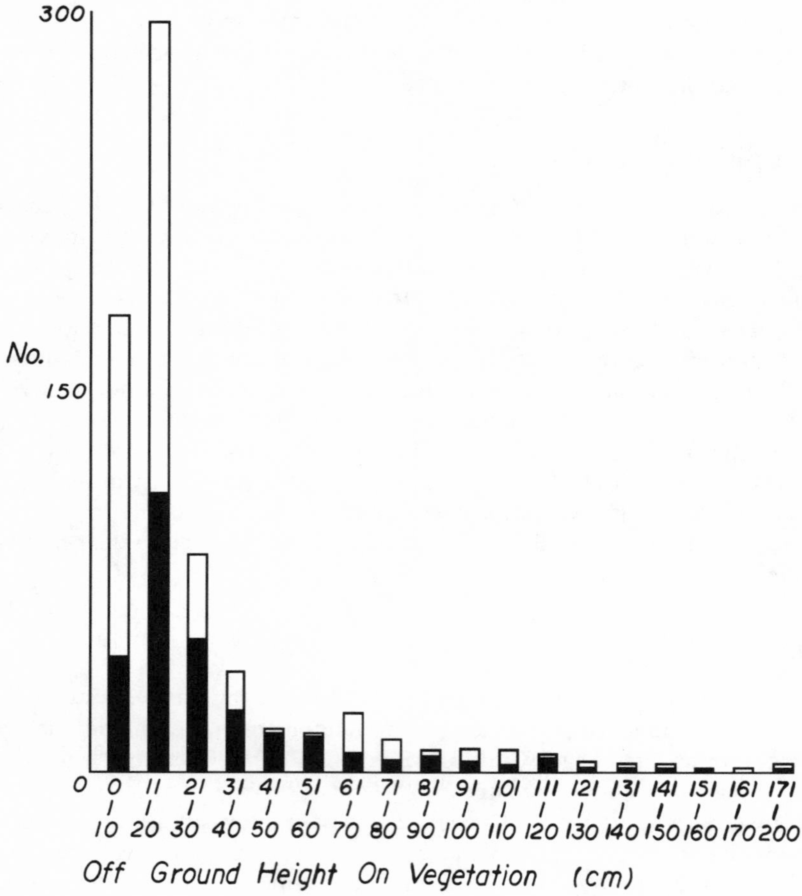

Figure 8.10. The distribution of perch heights of preadult *Eleutherodactylus coqui* (open bars from the ordinate) and crab spiders, *Stasina portoricensis* (solid bars), on the understory vegetation in the Luquillo Forest at El Verde. Bars are superimposed to show overlap in habitat use. (Reprinted from Formanowicz et al. 1981).

predators on eggs and parents differ. We have observations or records of numerous invertebrates and at least nineteen species of vertebrates that feed on *Eleutherodactylus:* two frogs, three lizards, three snakes, eight birds, and three mammals (table 8.7). All mammalian predators are exotics. Of the vertebrate predators, we predict that birds consume the most frogs (Waide, this volume). The generalized diagram in figure 8.10 shows our understanding of directions of major predator-prey interactions, and hence of energy flow, at each life stage of *E. coqui.*

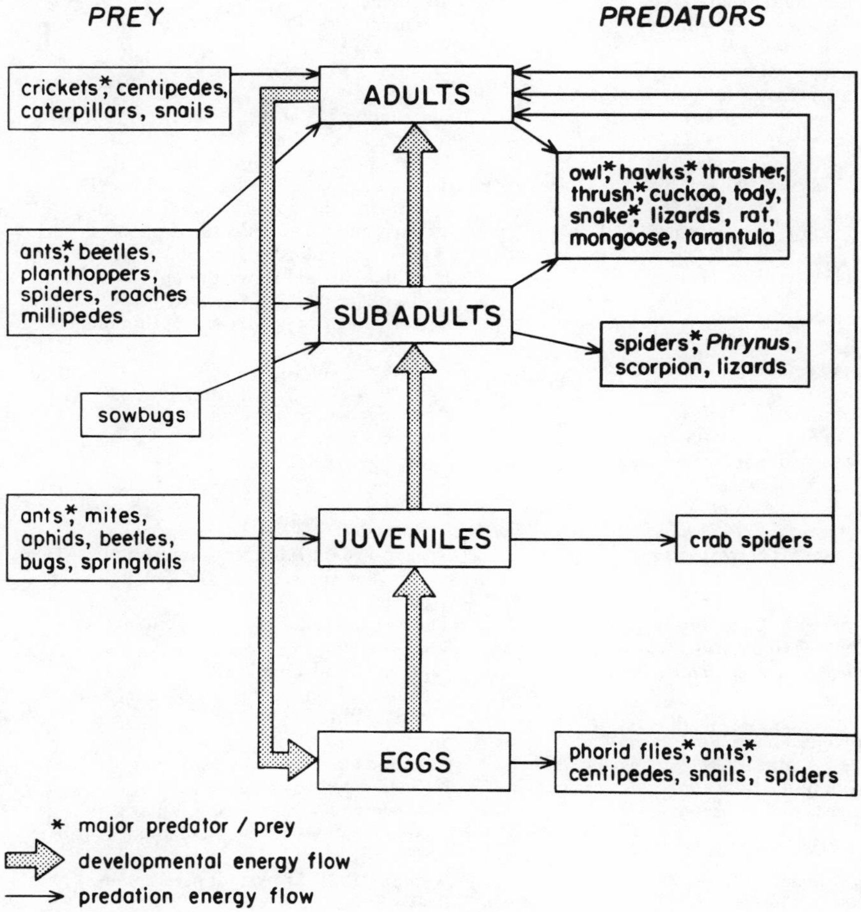

Figure 8.11. Diagram of the food web of *Eleutherodactylus coqui.* Life stages are separate because of different predator-prey relationships at some stages. Compartments enclose major categories of predators or prey. Arrows point to consumers. Asterisks (*) indicate most important groups in the category.

Table 8.7. Predators on *Eleutherodactylus*, primarily *E. coqui*

Predator type	Reference
Predators on *E. coqui* eggs	
Frogs, *E. coqui* adults, primarily males	Townsend 1984; Townsend et al. 1984; pers. obs.
Snake *Arrhyton exiguum*	Schwartz and Henderson 1991
Phorid fly *Megaselia scalaris*	Villa and Townsend 1983; Townsend 1984; pers. obs.
Small red ant *Paratrechina* sp.	Townsend 1984
Ant *Iridomyrmex melleus*	Townsend 1984
Unidentified centipede	Townsend 1984
Millipede *Liomus obscurus*	Townsend 1984
Unidentified litter spider	Townsend 1984
Cricket *Amphiacusta caraibea*	Townsend 1984
Land snail *Subulina* sp.	Rivero 1978
Land snail *Polydontes acutangula*	Townsend 1984; pers. obs.
Invertebrate predators on preadults and adults	
Crab spiders *Stasina portoricensis, Olios*	Formanowicz et al. 1981; Stewart 1995; pers. obs.
Ctenid spiders, wolf spiders	D. Falls pers. comm., pers. obs.
Scorpion *Tityus obtusus*	Thomas and Gaa Kessler, this volume; D. Formanowicz pers. comm.
Amblypygid *Phrynus longipes*	Thomas and Gaa Kessler, this volume; Formanowicz unpublished; pers. obs.
Tarantula *Avicularia laeta*	Formanowicz and Stewart unpublished; pers. obs.
Giant centipedes *Scolopendra alternans*	D. Formanowicz pers. comm.; J. Inmon pers. comm.
Land crab *Epilobocera sinuatifrons*	M. Canals pers. comm.
Vertebrate predators	
Eleutherodactylus coqui	Appendix 8.A, this volume
Eleutherodactylus wightmanae	Appendix 8.A, this volume
Lizard *Ameiva exsul*	Lewis 1986; pers. comm.
Lizard *Anolis gundlachi*	Pers. obs.
Lizard *Anolis cristatellus*	Pers. obs.
Gecko *Sphaerodactylus klauberi*	Thomas and Gaa Kessler, this volume
Snake *Alsophis portoricensis*	Reagan 1984; Rodriguez-Robles and Leal 1993; M. Canals pers. comm.; D. Townsend pers. comm.; pers. obs.
Snake *Arrhyton exiguum*	Thomas and Gaa Kessler, this volume
Boa *Epicrates inornatus*	Reagan 1984
Red-tailed hawk *Buteo jamaicensis*	Santana C. and Temple 1988; J. Wiley pers. comm.; N. Snyder pers. comm.; E. Santana C. pers. comm.
Broad-winged hawk *Buteo platypterus*	Snyder et al. 1987
Pearly-eyed thrasher *Margarops fuscatus*	A. Estrada pers. comm.; W. Arendt pers. comm.
P.R. woodpecker *Melanerpes portoricensis*	N. Snyder pers. comm.
P.R. screech owl *Otus nudipes*	D. Townsend pers. comm.; J. Wiley pers. comm.; W. Arendt pers. comm.; P. Narins pers. comm.
P.R. lizard-cuckoo *Saurothera vieilloti*	Wetmore 1927; W. Arendt pers. comm.
P.R. tody *Todus mexicanus*	R. Waide pers. comm.
Red-legged thrush *Turdus plumbeus*	Wetmore 1916, 1927; Rolle 1963; M. Canals pers. comm.; A. Estrada pers. comm.; A. Arendt pers. comm.; Stewart and Waide unpublished
Mongoose *Herpestes javanicus*	Wolcott 1953; Pimentel 1955
Cat *Felis catus*	Pers. obs.
Black rat *Rattus rattus*	Pers. obs.

Nest Predators

By far, the major predators on coqui eggs are conspecifics of both sexes but especially males (Townsend 1984; Townsend et al. 1984). Parental male *E. coqui* and *E. portoricensis* may eat their own eggs (filial cannibalism) if disturbed early during brooding. Egg-eating by conspecific males (heterocannibalism) is a commonplace behavior of coquis at El Verde. Townsend (1984) found 14.3% of 175 nests cannibalized by conspecifics. Of 100 males captured in our study plots during 1983 to 1984, 15% contained eggs. Coqui eggs provide a substantial food source for coquis (see Energy Budgets of *E. coqui*). Frogs and eggs comprised 5.3% by volume (or 0.2% by number) of the stomach contents of 122 coquis. Males also eat eggs of *E. portoricensis* (D. S. Townsend pers. comm.; pers. observation). Eggs of congeners are too similar in size to distinguish species differences from stomach contents. Stomach contents show that coquis rarely eat young frogs. The few present in stomach contents suggest accidental ingestion.

Several invertebrates have been observed feeding on coqui eggs (table 8.7). The most common are phorid flies, *Megaselia scalaris* (Villa and Townsend 1983; Townsend 1984; pers. observation) that parasitize coqui nests and rapidly consume eggs. At some seasons, phorids are the most abundant insects in the forest (Reagan et al. 1982; Garrison and Willig, this volume). Other invertebrates that have been observed eating eggs are ants, centipedes, millipedes, spiders, and snails.

Leaf Surface, Twig, and Trunk Predators

Most small frogs perch on leaves of low understory herbaceous forbs or on twigs and leaves of low understory shrubs. Major predators of these microhabitats are sparassid spiders that are abundant in the ground cover and understory vegetation (Pfeiffer, chap. 5, this volume). The distribution of juvenile frogs and crab spiders, *Stasina portoricensis,* is coincident (fig. 8.10) and predation is size related: larger frogs can escape even if captured whereas small frogs cannot (Formanowicz et al. 1981). In a field cage experiment designed to see whether the presence of *Stasina* would alter perch heights and sites of *E. coqui* of different sizes, Townsend (1985) found no differences in positions of frogs with or without spiders.

In surveys made at various times throughout the years 1979 to 1985 in the Activity Transect, visible *Stasina portoricensis* occurred in densities of none to 40 per 100 m^2 (Stewart 1995). No other potential predator was found in such high densities. These spiders are also prey for the frogs. In a series of four 100 m^2 experimental plots in which the frog density was increased through addition of retreat sites, *Stasina* populations decreased significantly as frog populations increased (table 8.8). Ctenid spiders, *Oligoctenus ottleyi,* are much less abundant (zero to two per 100 m^2) than crab spiders (one ctenid to ninety-seven *Stasina* seen in the Activity Transect,

Table 8.8. Mean density of crab spiders, *Stasina portoricensis*, and adult *Eleutherodactylus coqui*

Plot	Control			Experimental		
	Frogs	Spiders		Frogs	Spiders	
	mean (S.D.)	mean (range)	S.D.	mean (S.D.)	mean (range)	S.D.
River	8 (3.3)	9 (0–28)	6.9	19 (7.1)	4 (0–12)	4.1
Bridge	7 (3.3)	4 (0–9)	2.9	18 (5.5)	2 (0–7)	2.3
Tower I	4 (2.4)	3 (0–9)	2.3	14 (5.3)	2 (0–7)	2.3
Tower II	6 (2.1)	6 (0–14)	3.6	14 (6.5)	5 (0–17)	4.9
Total	25	22		65	13	

Notes: Data are from 100 m^2 plots without (controls) and with (experimental) 100 bamboo tubes as retreat/nest sites. Results are based on fifteen nocturnal counts March 1980 to January 1983. Numbers of spiders decrease as adult coquies increase ($\chi^2 = 11.8$, d.f. = 1; $p < .001$).

1980 to 1985). We see ctenids (wolf spiders) most frequently in bromeliads and in surface litter. Novo et al. (1985) reported predation by the ctenid spider, *Ctenus vernalis,* on subadult *Eleutherodactylus zugi* in Cuba. Ctenid spiders readily ate *Eleutherodactylus* spp. in Costa Rica (Szelistowski 1985) and we see them eating juvenile coquis. Scorpions, *Tityus obtusus,* are common in the forest. Like coquis, they shelter in the litter in rolled leaves or palm fronds on or off ground during the day, then climb onto leaves, twigs, or tree trunks where they sit in ambush during the night. Scorpions eat *Eleutherodactylus* in the field (Thomas and Gaa Kessler, this volume) and take coquis readily in captivity (D. Formanowicz pers. comm.).

For coquis, trunks are important as calling sites (Townsend 1989), foraging sites, and especially as travel routes for individuals making nightly forays into the canopy to forage. The two most important predators on tree trunks are the amblypygid *Phrynus longipes,* and the tarantula *Avicularia laeta. Phrynus longipes* forages near the ground on trunks and boulders up to at least 2 m, whereas tarantulas are upper trunk or canopy inhabitants. Schwartz (1958) and Thomas (this volume) observed predation on small *Eleutherodactylus* by *Phrynus* in Cuba. Tarantulas build cases on trunks or webs between leaves or in cavities. We have seen both juvenile and adult frogs taken by tarantulas and *Phrynus* in the field. In the laboratory, ten to twenty-three days lapsed between captures of adult coquis by captive tarantulas. Because of the large size of tarantulas, they may be the most important invertebrate predator on adult frogs. Tarantulas are one of the preferred foods of the lizard cuckoo that also eats coquis (Wetmore 1927).

Giant centipedes, *Scolopendra alternans,* eat coquis (D. Formanowicz, J. Inmon pers. comm.) and attack *Bufo marinus* (Carpenter and Gillingham 1984). The forest crab, *Epilobocera sinuatifrons,* which forages widely from streams and burrows on wet nights, eats frogs (M. Canals pers. comm.). We observed a *Bufo marinus,* 110 mm SVL, in the claws of a crab whose carapace was 80 mm long. A crab was seen attempting to eat a *B. marinus* but departed after poison was extruded from the parotoid glands of the toad (R. Garrison pers. comm.). Land crabs, phalangids, and gryllids eat frogs and their eggs in Costa Rica (Hayes 1983).

Because of its large size and omnivorous habits (Rivero 1978), one would expect the introduced *Bufo marinus* to be an important predator on frogs. However, Dexter (1932) examined stomachs of 301 specimens and found no frogs. We found none in one specimen that foraged all night in an area of high frog density. Toads in the forest are small and often appear emaciated as if foraging conditions, for whatever reasons, are less than optimal. Although carnivorous bats prey on calling frogs in Panama (Tuttle and Ryan 1981), no Puerto Rican bats are suitable frog predators (A. Starrett pers. comm.; M. Willig pers. comm.).

ENERGY AND NUTRIENT CYCLING

The importance of amphibians in food chains has been treated in the literature since Kirkland (1904), Surface (1913), and Chandler (1918) estimated the usefulness of the predatory role of toads, frogs, and salamanders relative to pest control. Although diets of amphibians have been studied for many years, only recently have the density and biomass of any amphibian species in a forest community been quantified (Zimka 1974; Burton and Likens 1975a,b; Burton 1976; Pough et al. 1987). Information on biomass alone suggests that amphibians are important in these communities. The most detailed information exists for salamanders, although they are usually much less obvious to the casual observer than are frogs and toads. Biomass of salamanders (1.77 kg wet wt ha^{-1}) in the Hubbard Brook Forest of New Hampshire exceeds that of birds and equals that of mammals (Burton and Likens 1975b). The red-backed salamander, *Plethodon cinereus,* in temperate eastern North America occurs at various estimated densities (900 to 8,900 ha^{-1}, Heatwole 1962; 2,300 to 10,000 ha^{-1}, Jaeger 1979; 1,650 to 27,200 ha^{-1}, Burton and Likens 1975a), depending on the moisture of the habitat. By comparison, we estimate average frog densities at El Verde to be 20,570 ha^{-1} (see Population Densities of *E. coqui*), exceeding not only those of salamanders in some continental temperate hardwood forests but also those in some continental Neotropical forests where salamander densities may exceed 9,000 ha^{-1} (Vial 1968).

Energetically, amphibians are quite efficient. *Plethodon cinereus* expends 60% of its assimilated energy on production (Burton and Likens 1975a) as compared to 0.5 to 3.0% for some birds and mammals (Golley 1968). Assimilation efficiencies are also high (86 to 88% in summer; Fitzpatrick 1973a,b). Since salamanders are absent from the West Indies, our data concern anurans only. Puerto Rican frogs appear to follow these trends with summer assimilation efficiencies of 88.5% (Woolbright 1985b) and production efficiencies of about 59% (see Energy Budgets of *E. coqui*).

Energy Budgets of *E. coqui*

While estimates of biomass and numbers of individuals are important as static indices of community structure, an understanding of ecosystem function requires information on the flow of energy through the system. Ecological efficiency (Ricklefs 1973) depends on the ability of individual links in the food chain to exploit the energy available to them and to convert captured energy to biomass. Because of the coqui's role as an abundant secondary and tertiary consumer, as well as an important prey species for other consumers, we examine the flow of energy through adults of *E. coqui*. Woolbright (1985b) calculated energy intake of adult coquis from measurements of digestive efficiencies in the laboratory and fecal output in the field. Results showed that neither digestive efficiencies nor caloric equivalent of feces varied with sex (although digestive efficiency did vary seasonally). Females produced three times the feces of males in the wet season, and twice as much in the dry season. The study concluded that energy intake for females is approximately 1.6 kJ d^{-1} in the wet season and 0.6 kJ d^{-1} in the dry season, and that of males is 0.6 kJ d^{-1} in the wet season and 0.3 kJ d^{-1} in the dry season. Assuming adult population densities of 3,265 ha^{-1} (see Population Densities of *E. coqui*), a 1:1 sex ratio, and a three-month annual dry season (Odum et al. 1970b), these figures suggest that adult coquis consume about 1.1×10^6 kJ $ha^{-1}y^{-1}$ from the forest. This is much greater than the 1,000 kcal $ha^{-1}y^{-1}$ (= 46,000 kJ $ha^{-1}y^{-1}$) estimated by Burton and Likens (1975a) for temperate zone salamander populations.

Not all energy intake is available for use by animals because some is lost in the form of feces. Woolbright (1985b) measured the caloric equivalent of the daily fecal production of female coquis to be 0.184 kJ in the wet season and 0.138 kJ in the dry season. That of males was 0.069 kJ in the wet season and 0.059 kJ in the dry season. Assuming the same population parameters as above, these figures suggest that coquis excreted 1.6×10^5 kJ $ha^{-1}y^{-1}$ in the form of feces. The remainder of their annual energy intake ($[11 - 1.6] \times 10^5 = 9.4 \times 10^5$ kJ $ha^{-1}y^{-1}$) can be partitioned into metabolic cost and production.

Woolbright (1985b) estimated metabolic expenditures from oxygen consumption rates calculated in the field using Thoday respirometers (Evans 1972). Results suggested considerably higher rates of oxygen consumption than had been reported previously for this species (Taigen and Pough 1983). We suggest that the discrepancy may lie in methods used. Previous studies were performed in the laboratory using long-term captive animals. Our studies were done in the field under ecologically relevant conditions (e.g., at night, exposed to a chorus), using animals captured no more than twenty-four hours before testing. We chose to use our estimates because other studies (e.g., Bucher et al. 1982; Taigen and Wells 1985) have shown that oxygen consumption rates do vary widely with behavioral state. Assuming a conversion factor of 20.5 kJ ml^{-1} oxygen, basic activity budgets from Woolbright (1985b), and population parameters as above, we calculate total metabolic expenditures for adult coquis to be 3.9×10^5 kJ ha^{-1} y^{-1}.

Independent estimates of production should be made only on a size-specific basis because energy expended for growth and probably for egg production varies curvilinearly with body size (Woolbright 1985b). Estimating production as assimilated energy unaccounted for by respiration, we calculate production of adult coquies as 5.5×10^5 kJ $ha^{-1}y^{-1}$. Based on these figures we estimate conversion efficiency (production/assimilation) for the coqui as 59%. This compares favorably with a 60% estimate for the temperate salamander, *Plethodon cinereus* (Burton and Likens 1975a). We show energy flow through the adult population of *E. coqui* in figure 8.12.

Since coqui eggs are so frequently eaten by conspecific males (see Nest

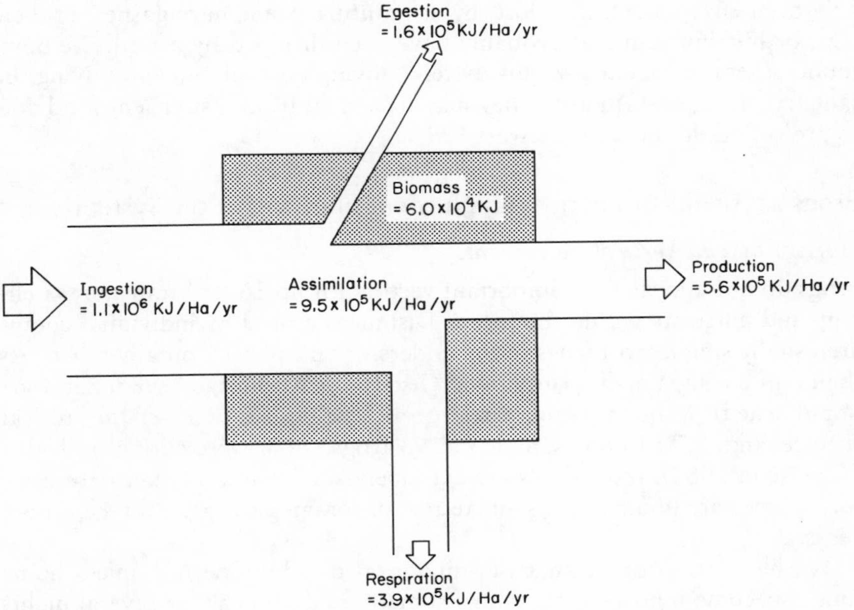

Figure 8.12. Energy flow diagram for the adult *Eleutherodactylus coqui* population at El Verde. Values for egestion, biomass, assimilation, and respiration are independent estimates. Diagram is modelled after Burton and Likens (1975a).

Predators), we calculated the energy source available to the frogs from eggs. One freshly laid coqui egg contains 0.215 kJ energy (Woolbright 1985b). Given a mean clutch size of twenty-eight eggs (Townsend 1984, 1986a) and a daily food intake per adult male of 0.6 kJ d^{-1}, one clutch produces the equivalent of ten days' food for one frog. Males, having consumed a clutch, contained undigested eggs that were still visible in their stomachs after a week. Eggs, then, are an abundant energy source in the forest. Based on numbers of clutches found in plots (up to 6 clutches/100 m^2 in control plots; up to 12 clutches/100 m^2 in high density plots), there are as many as 16,800 eggs ha^{-1} in natural areas, and as many as 33,600 eggs ha^{-1} in high density sites. If 15% of the clutches are cannibalized during their fifteen-day incubation period (Townsend 1984), then egg consumption during the wet season when eggs are plentiful may reach 168 eggs $ha^{-1}d^{-1}$ for a total energy consumption of up to 36 kJ $ha^{-1}d^{-1}$ (a possible energy consumption of five times that amount in high density plots). It is no wonder that males eat eggs and that parental guarding is essential for maintaining high reproductive success.

Decomposers receive an estimated 1.6×10^5 kJ of energy from frog feces. Most frog bodies are consumed directly by predators. Rarely we find dead or dying coquis on the forest floor. Some show no sign of injury and have probably been envenomed, then lost, by a tarantula. Some have gashes, broken legs, or missing skin and probably have been dropped by a bird. We have found several emaciated coquis, barely moving and subsequently dying, in January, April, and August. They may have died from insufficient food due to prolonged dry periods (Stewart 1995).

Frogs as Agents of Energy and Nutrient Flux within the System

Horizontal and Vertical Movements

Frogs do not appear to be important vectors for horizontal transport of energy and nutrients within the forest. Distances moved by individual coquis are usually small; most frogs in the understory move horizontally much less than 6 m during a night (Turner and Gist 1970; Woolbright 1985a; Stewart unpublished). Major foraging sites for all size classes of *E. coqui* are leaf surfaces (fig. 8.7). In one sample of 193 frogs, 74% were sitting on leaves (Townsend 1985). Individuals are extremely site loyal, and males are territorial; some are found in the same retreat or foraging site after three or more years.

We have no evidence that out-migration of adults occurs unless home ranges become inhospitable. Adults sometimes disappear for several nights and reappear days or weeks later (Stewart unpublished; Woolbright 1985a). Presumably they are foraging in the canopy, staying in inaccessible retreats, or guarding eggs. Areas around plots of marked frogs have been surveyed repeatedly. Only when retreats were removed from high-density plots did we find marked frogs more than 1 m from plot boundaries (Stewart unpublished). Adults, when displaced 50 to 100 m, return to their home site (Drewry 1970c; T. Kotliar and B. Milne pers. comm.; Gonser and Woolbright 1995). Since we have no way of marking juveniles, we do not know how far they disperse from hatching sites.

Recent observations show that food is more abundant and larger in the canopy (table 8.6; see Vertical Differences). Benefits should be substantial in order to offset predation and energetic costs to frogs of climbing tree trunks. An increase in food availability would be of greatest importance to rapidly growing subadults and to females forming eggs. As frogs move between the canopy and the forest floor, some energy is transported in both directions via defecation and predation. If canopy frogs are taken by predatory birds, snakes, and large spiders, then transport is from understory to canopy. Probably more energy is gleaned in the canopy by frogs (eating orthopterans, tree crickets, homopterans, ants) and carried back to the understory and forest floor where frogs are eaten by amblypygids, centipedes, and spiders. A frog may forage from leaves of understory shrubs one night, then climb to the canopy the second night. Coquis, therefore, perform a potentially important energy relay between the canopy and the understory and forest floor.

Association with Streams

There is a lesser degree of energy transfer by frogs between terrestrial and aquatic compartments of the forest. Only three species of frogs forage in or around water. *Leptodactylus albilabris* calls and nests in cavities at the edges of streams and temporary rivulets. On wet nights at all elevations subadults and females (to 53 mm SVL) forage widely across the forest floor. Rarely we encountered an adult in one of our plots. We have seen juveniles, 18 mm SVL, in boulder-strewn wet stream beds during the day. The notable dweller of stream and river beds was *E. karlschmidti* (Grant 1931), the largest native anuran. Little is known of its life history. Although it foraged around large boulders in stream beds, numbers of prey included only 32.4% aquatic animals; of those 91% were benthic forms (Rivero et al. 1963). Although formerly in Quebrada Sonadora at El Verde (Drewry 1970c) as well as on El Yunque, *E. karlschmidti* has not been seen in recent years (G. Drewry pers. comm.; R. Joglar pers. comm.; J. A. Rivero pers. comm.). Its disappearance seems widespread throughout the forest, and reasons are unknown. We have seen *E. coqui* sitting on boulders in Quebrada Sonadora where they no doubt forage on streamside invertebrates (Covich and McDowell, this volume). Because of the rarity of such sightings, riparian use appears to be occasional. Apparently when water is low, individuals can cross large rivers without difficulty. Marked coquies do cross the Sonadora in homing experiments (Gonser and Woolbright 1995).

SUMMARY

The only amphibians in Puerto Rico are frogs and toads. Except for *Leptodactylus albilabris,* and the introduced *Bufo marinus,* all frogs in the forest are endemic species of *Eleutherodactylus.* Frogs are broad spectrum carnivores, functioning as second-, third-, and fourth-level consumers in the forest. Of five species of *Eleutherodactylus* in the El Verde region, the most numerous by far is *E. coqui.* It occurs in densities of up to 29 adults/100 m^2 (3,265 adults ha^{-1}), or totals of 20,570 ha^{-1} including all age classes. Only in Jamaica have comparable densities been recorded. Biomass calculations for *E. coqui* indicate 3.28 kg dry mass ha^{-1}. Total numbers of four other species may reach 0 to 8 per 100 m^2, with an estimated total dry mass of 0.42 kg ha^{-1} or 3.70 kg ha^{-1} for all *Eleutherodactylus* species in the forest.

Eleutherodactylus coqui provides the largest component of biomass active in the forest at night, second only to that of anoline lizards, active by day. We estimate that *E. coqui* alone harvests 114,000 prey $night^{-1}ha^{-1}$. Estimates indicate that adults harvest 1.1×10^6 kJ $ha^{-1}y^{-1}$ from the forest, of which 1.6×10^5 kJ is excreted as feces. Metabolic expenditures are estimated at 3.9×10^5 kJ $ha^{-1}y^{-1}$. We estimate production efficiency (production/assimilation) for *E. coqui* as 59%.

Approximately 94% of adult coquis do not survive one year. A few live at least six years. Major predators are spiders (*Stasina portoricensis* on young, *Avicularia laeta* on adults), an amblypygid (*Phrynus longipes*), a snake (*Alsophis portoricensis*), and birds (Puerto Rican screech-owl, pearly-eyed thrasher, red-legged thrush, red-tailed and broad-winged hawks, Puerto Rican lizard-cuckoo). Predator-prey roles switch between *E. coqui* and at least eight invertebrates (notably *S. portoricensis*), and one vertebrate (*Anolis gundlachi*). The role played is size determined.

Eleutherodactylus coqui is a dietary generalist. Stomach contents of 173 adults contained prey from more than 101 species from sixty families, twenty-eight orders, nine classes, and five phyla including conspecific eggs (mean of three taxa per stomach). The greatest volume was provided by invertebrates in the categories Orthoptera, Dictyoptera, Arachnida, and Hymenoptera. Most numerous were Hymenoptera (ants), Orthoptera, Coleoptera, Arachnida, and Dictyoptera. Stomachs with food contained up to seventeen items ranging in size from 0.2 to 55 mm in length. Prey length increases with frog body size. Females, larger than males, consume larger prey. Arthropod prey are locally and temporally patchy within the forest. More prey are available in wet months and at greater heights off the ground. Those conditions may explain canopy foraging by nonbreeding frogs. Males frequently cannibalize conspecific eggs. Each egg contains 0.215 kJ of energy. An average clutch of twenty-eight eggs yields 6.02 kJ, equivalent to ten days of ordinary diet.

Eleutherodactylus coqui, operating mostly in the nocturnal subweb, has a significant impact on the forest ecosystem. That impact is distributed widely through the diverse prey, and also through diverse animal species, both invertebrate and vertebrate, that consume coquis. Frogs play a major role as both secondary consumers (on crickets, ants, some snails, etc.) and as tertiary (or higher) consumers (on spiders, harvestmen, centipedes) in the forest. They also eat several detritivores (ants, cockroaches, some snails, dipteran larvae). Their major foods, by volume, are orthopterans. Although omnivores (ants) constitute the largest number of prey, phytovores constitute the greatest volume. Frogs glean small items and convert them to larger morsels available to a variety of predators. There are many more ectothermic vertebrates than endotherms in the Luquillo forest; although birds are the top level predators, they may exert less control on frog populations than the invertebrate predators such as spiders. The large invertebrate predators may rely heavily on the coqui for food, whereas birds have a greater variety of food sources available.

ACKNOWLEDGMENTS

The following persons have assisted in some way or provided information from personal observations: Wayne Arendt, Phillip Bishop, Jeff Brush, Miguel Canals, Alejo Estrada, Douglas Falls, Michael Flynn, Daniel Formanowicz, Rusty Gonser, Ulmar Grafe, Nancy Humphrey, Jerry Inmon, Rafael Joglar, Tasha Kotliar, John Lasher, Greg Mayer, Bruce Milne, Peter Narins, Jennifer O'Brien, George Preston, George Rapp, Douglas Reagan, Valli Rivera, Eduardo Santana C., Ellen Smith, Noel Snyder, Andrew Starrett, Deborah Bishop, Dan Thero, Richard G. Thomas, Karyn Townsend, Daniel Townsend, Robert Waide, James Wiley, and Randy Zelick. Rosser Garrison kindly identified and measured species in stomach contents of 246 frogs. For that monumental and tedious task we are indeed grateful. William Pfeiffer identified stomach contents of ten canopy frogs. Ryland Loos prepared several of the figures. We thank Robin Andrews, Rosser Garrison, William Pfeiffer, Katheryn Schneider, Catherine Toft, Daniel Townsend, Karyn Townsend, Andrea Worthington, and the editors and reviewers for many helpful comments on the manuscript. Permits to work in the forest were issued by A. Rodriguez Figueroa (Department of Natural Resources, Commonwealth of Puerto Rico) and C. Noble (U.S.D.A., Forest Service). The Center for Energy and Environment Research of the University of Puerto Rico provided facilities and logistical support under a contract with the Office of Health and Environmental Research, U.S. Department of Energy. Funding has been provided by NSF Grant No. DEB-7721349, Oak Ridge Associated Universities Faculty Research Participation, American Philosophical Society Penrose Fund (to MMS); Society of the Sigma Xi, State University of New York at Albany Benevolent Association, the Society for the Study of Amphibians and Reptiles (to LLW), and an NSF LTER grant BSR-881192 to the University of Puerto Rico and the International Institute of Tropical Forestry, U.S.D.A. Forest Service, for the Luquillo Experimental Forest Long-Term Ecological Research Program. Photographs are by the authors.

Appendix 8.A Prey Taxa in Stomachs of Adults of *Eleutherodactylus* spp.

Prey taxon				Predator
PLANTAE		Bursuraceae:	*Dacryodes excelsa* (seeds, flower)	C
		Other seeds		CW
		Thorn		C
		Vegetative fragments		CW
ANIMALIA				
Nematomorpha: (unidentified; possible parasite of arthropod prey)				C
Mollusca:	Class Gastropoda: (unidentified)			CP
	Stylommatophora:	Sagdidae:	*Platysuccinea portoricensis*	C
Arthropoda:	Class Arachnida			
	Scorpionida:	Buthidae:	*Tityus obtusus*	C
	Pseudoscorpionida:	Menthidae:	*Menthus* sp.	CPW
	Amblypygida:	Phrynidae:	*Phrynus longipes*	CR
	Araneae:		(unidentified)	CP
		Barychelidae:	*Psalistops corozalis*	CP
		Pholcidae:	*Modisimus montanus*	C
			M. signatus	C
		Clubionidae:	*Corinna* sp.	P
			Trachelas sp.	C
		Araneidae:	*Gasteracantha cancriformis*	C
			Leucage regnyi	P
		Sparassidae:	(unidentified)	C
			Pseudosparianthus jayuyae	W
			Stasina portoricensis	C
		Salticidae:	(unidentified)	W
	Opiliones:		(unidentified)	CRW
	Acarina:		(unidentified)	CW
	Cryptostigmata			CPRW
Arthropoda:	Class Chilopoda			
	Scutigeromorpha:	Scutigeridae:	(unidentified)	C
	Scolopendromorpha:		(unidentified)	P
		Scolopendridae:	*Scolopendra alternans*	C
	Class Diplopoda			
	Stemmiulida:	Stemmiulidae:	*Prostemmiulus heatwoli*	CPRW
	Polydesmida:		(unidentified)	C
		Cryptodesmidae:	(unidentified)	PW
	Class Crustacea			
	Isopoda:	Oniscidae:	*Philoscia richmondi*	R
	Class Entognatha			
	Collembola:	Entomobryidae:	(unidentified)	CW
			Dicranocentriga sp.	C
		Sminthuridae:	(unidentified)	W
	Class Insecta			
	Blattodea:		(unidentified)	CP
		Blattellidae:	(unidentified)	CPW
			Aglaopteryx facies	C
			Aglaopteryx sp.	C
			Cariblatta sp.	C
			Cariblattoides suave	C
			Epilampra wheeleri	C
			Neoblattella sp.	C
			Plecoptera sp.	C

Prey taxon			Predator
		P. infulata	C
		Pseudosymploce sp.	C
		P. personata	CP
		Symploce sp.	C
	Blattidae:	(unidentified)	C
	Cryptocercidae:	(unidentified)	C
		Cryptocercus sp.	C
Coleoptera:		(unidentified)	CPW
	Carabidae:	(unidentified)	P
	Chrysomelidae:	(unidentified)	C
	Cucujidae:	(unidentified)	CPW
	Curculionidae:	(unidentified)	CPW
	Elateridae:	(unidentified)	CW
	Eucnemidae:	(unidentified)	C
	Lampyridae:	(unidentified)	CP
	Lycidae:	(unidentified)	CP
	Scarabaeidae:	(unidentified)	CP
	Scolytidae:	(unidentified)	CW
	Staphylinidae:	(unidentified)	P
	Trogossitidae:	(unidentified; larva)	C
Diptera:		(unidentified)	CRW
	Anthomyiidae:	(unidentified)	C
	Cecidomyiidae:	(unidentified)	C
	Ceratopogonidae:	(unidentified; larva)	C
	Chironomidae:	(unidentified)	C
	Mycetophilidae:	(unidentified)	C
	Sciaridae:	(unidentified)	C
	Stratiomyidae:	(unidentified)	W
	Tipulidae:	(unidentified)	CP
Ephemeroptera:		(unidentified)	C
Hemiptera:		(unidentified)	CW
	Aradidae:	(unidentified)	P
		Aneuris minutus	C
Homoptera:		(unidentified)	P
	Coccoidea:	(unidentified)	RW
	Cicadellidae:	*Xestacephalus maculatus*	W
	Fulgoroidea:	(unidentified)	CW
	Delphacidae:	*Ugyops occidentalis*	C
	Diaspididae:	(unidentified)	C
	Flatidae:	*Flatoidinus fumatus*	C
		Flatormenis sp.	C
		Pseudormenis virgina	C
	Issidae:	*Neocolpoptera monticulens*	C
		N. puertoricensis	C
		Thionia borinquensis	C
	Tropiduchidae:	*Ladellodes stahli*	C
Hymenoptera:		(unidentified)	CP
	Apidae:	*Apis mellifera*	C
	Braconidae		C
	Diapriidae:	(unidentified)	P
	Formicidae:	(unidentified)	CW
		Anochetus mayri	CW
		Brachymyrmex heeri	CW
		Camponotus ustus	CP

Prey taxon				Predator
			Iridomyrmex melleus	CP
			Mycocepurus smithi	W
			Myrmelachista ramulorum	CP
			Ochetomyrmex auropunctata	W
			Odontomachus brunneus	CPW
			Paratrechina sp.	CP
			P. microps	C
			Pheidole moerens	CPRW
			Solenopsis sp.	W
			Strumigenys sp.	W
			S. eggersi	CW
			S. rogeri	CRW
			Pachycondyla stigma	CPW
	Isoptera:	Termitidae:	*Nasutitermes costalis*	C
	Lepidoptera:		(unidentified; larvae)	CPRW
		Arctiidae:	(unidentified)	C
		Hesperiidae:	(unidentified)	C
		Noctuidae:	(unidentified)	C
		Tineidae:	(unidentified)	C
		"microlepidoptera":	(unidentified)	C
	Microcoryphia:	Machilidae:	(unidentified)	C
	Orthoptera:		(unidentified)	C
		Gryllidae:	(unidentified)	C
			Amphiacusta sp.	C
			A. caraibea	CP
			Anaxipha sp.	C
			Cyrtoxipha gundlachi	C
			Gryllus assimilus	CP
			Laurepa krugii	C
			Orocharis sp.	CP
		Tettigoniidae:	*Analaucomera laticauda*	C
	Psocoptera:		(unidentified)	C
		Epipsocidae:	(unidentified)	C
		Polypsocidae:	(unidentified)	CP
Chordata:	Class Amphibia			
	Anura:	Leptodactylidae:	*Eleutherodactylus* young	CW
			eggs	C
	Class Reptilia			
	Squamata:	Iguanidae:	*Anolis gundlachi*	C

Notes: Complete list of prey taxa found in stomachs of adults of 173 *E. coqui* (C), 33 *E. portoricensis* (P), 14 *E. richmondi* (R), and 26 *E. wightmanae* (W) collected in January, July, and August 1980–1982 in the vicinity of the El Verde Field Station (350 m). All *E. richmondi* were males; other samples included both sexes.

Appendix 8.B Methods

1. Methods for Density Estimates

Some of our density estimates are based on counts in nine 100 m^2 plots (fig. 8.2) that have been surveyed at least twice each year since 1979. Four of eight paired plots were supplemented with sixty retreat-nest sites (small bamboo tubes, "frog houses," attached to tree trunks). Bridge and River Plots (about 350 m altitude) lie in an area dominated by tabonuco and sierra palm; Tower I and Tower II plots lie above the walkup tower (about 450 m altitude) in the transition zone between tabonuco and palo colorado forest where fewer sierra palms occur. Seasonal changes in population density and activity were observed in an Activity Transect, a 2×50 m transect in the forest in which counts of all frogs visible between 2000 and 2100 h were made on three successive nights each fortnight for two years. A 250 m^2 plot (Woolbright plot) in old-growth tabonuco just west of Quebrada Sonadora was used for twenty-five mark-recapture surveys from 1981 to 1983. This plot was expanded to 800 m^2 in 1987 and subsequently used to monitor the impact of Hurricane Hugo.

2. Methods for Local and Seasonal Differences in Prey Availability

We estimated the availability of arthropod prey during *E. coqui*'s nocturnal foraging period using sticky traps. Traps consisted of inverted 266 ml plastic cups covered with Tack Trap® (Animal Repellents, Inc., Griffin, GA). Cups were suspended on strings at 0.5 m intervals from 0.5 to 5.0 m above the ground. Five permanent strings were placed at each of two locations, one 10 m to the east of Quebrada Sonadora, and the other 50 m to the west. Traps were set on the strings at dusk and taken down at dawn. Each site was sampled four times during the wet season and twice during the dry season. This sampling method gave us estimates of variability in numbers of arthropods collected by different strings on the same night, different nights at the same site, different sites within the forest, and different seasons. While no sampling procedure can duplicate the activity of a feeding animal, we feel that our method is a reasonable approximation of a small sit-and-wait forager on an elevated perch in the forest understory. The traps collected a large number of suitable prey, such as coleopterans, orthopterans, and arachnids. They also collected numerous arthropods that are not major items in the coqui's diet (e.g., dipterans). Even so, we expect the data indicate major patterns of availability of potential prey.

3. Sampling for Vertical Differences (Canopy vs. Understory)

We repeated the sampling procedure described above using a 20 m string of sticky traps located at the walk-up tower (fig. 8.2; 450 m). Traps were set at 1 m intervals from 1 m to 20 m above the ground. Samples were taken on four nights during the wet season and two nights during the dry season.

9

Anoline Lizards

Douglas P. Reagan

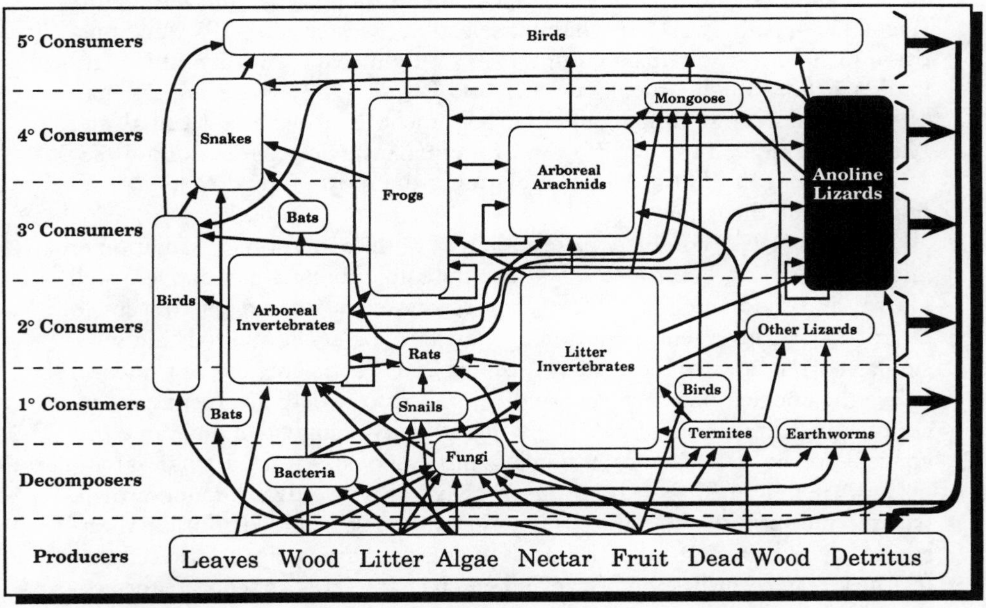

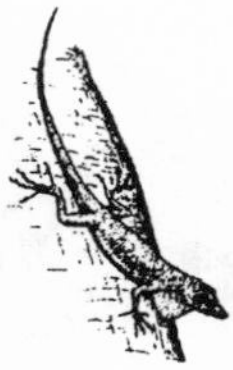

ANOLINE lizards (family Polychridae, genus *Anolis*) are the most conspicuous and abundant vertebrates inhabiting terrestrial ecosystems on Caribbean islands (Williams 1969, 1976; Moermond 1979a). As a group, anoles are diurnal, predominantly insectivorous, and occur throughout all vertical strata of the ecosystems that they inhabit. Anole abundance and high visibility have made them the subject of numerous studies that have contributed to our understanding of island biogeography (Williams 1969, 1972, 1983), resource partitioning (Schoener 1968, 1974), limiting factors (Licht 1974, Andrews 1976), habitat selection (Kiester et al. 1975), competition (Schoener 1969a,b,c; Stamps 1977; Wright 1981; Pacala and Roughgarden 1982; Roughgarden, Rummel, and Pacala 1983; Waide and Reagan 1983; Rummel and Roughgarden 1985), niche relationships (Ruibal and Philibosian 1970; Lister 1976), and foraging behavior (Moermond 1973, 1979a,b; Reagan 1986). Anoles are among the most studied of any vertebrate genus in the tropics.

Williams (1972) conducted a test analysis of the evolutionary radiation of anoline lizards on Puerto Rico because the anole fauna was relatively well known and moderately complex. He introduced the ecomorph concept to describe the convergent evolutionary pattern of a set of animals showing similar correlations of morphology, ecology, and behavior, but not lineage. While the concept has been reasonably applied to widely divergent taxa on continents (Karr and James 1975), Williams convincingly demonstrated this concept in the radiation of a single genus (*Anolis*) within a single archipelago. He described six ecomorphs: crown giant, twig dwarf, trunk-crown, trunk, trunk-ground, and grass-bush. Williams (1983) added additional categories by recognizing subdivisions of the original categories and stated that for the Greater Antilles, body size and perch characteristics separate the ecomorphs. Roughgarden and Pacala (1989) summarize the current understanding of *Anolis* systematics for the eastern Caribbean, including Puerto Rico.

Five anoline species inhabit the tabonuco forests of Puerto Rico (Rand 1964; Turner and Gist 1970; Schoener and Schoener 1971). The Puerto Rican giant anole, *Anolis cuvieri*, a crown giant ecomorph, is the largest species. Three smaller anole species, *A. gundlachi* (trunk-ground ecomorph), *A. evermanni* (trunk-crown ecomorph and generalist), and *A. stratulus* (twig

Table 9.1. Characteristics of the four rain forest anole species

Species	Snout-vent Length[a] (mm)	Weight[b] (g)
Anolis cuvieri	125 (118–140)	49.9 (32.5–56.2)
Anolis gundlachi	58 (42–72)	5.4 (2.2–9.0)
Anolis evermanni	55 (45–64)	4.1 (2.0–7.1)
Anolis stratulus	41 (34–48)	1.8 (1.1–2.8)

[a]Distance from the tip of the snout to the cloacal vent given as mean and range.

[b]Body weight expressed as mean and range.

dwarf), are relatively common within the forest. Table 9.1 provides a brief description of these four forest species. A fifth species, *A. occultus,* occurs near the forest edge or in openings along streams, but is generally rare in tabonuco forest. Two other species, *A. krugi* and *A. cristatellus,* are found at or near ground level at the forest edge and in openings within the forest.

Examining various lines of evidence, Williams (1972) discussed phylogenetic relationships and presented a phylogeny for Puerto Rican anoles based on osteological (Etheridge 1960, 1965), karyotypic (Gorman and Atkins 1969), and electrophoretic (Maldonado and Ortiz 1966) studies. These phylogenies show *A. cuvieri* and *A. occultus* as the most primitive and distinct species. *Anolis gundlachi* diverged from the two closely related species, *A. evermanni* and *A. stratulus* (fig. 9.1) in more recent times. Current evidence indicates that the primary route of the anoline invasions of Puerto Rico was from Hispaniola.

On Caribbean islands where there are no large animals such as those found in mainland ecosystems (e.g., tapirs, jaguars), anoles constitute a substantial portion of the total animal biomass. Their abundance, widespread ecological distribution, and functional role as higher order consumers make them important components of insular animal communities throughout the Caribbean. Recent studies have demonstrated their importance in structuring food webs on Caribbean islands (Schoener and Toft 1983; Schoener and Spiller 1987), and Reagan (1986) described the role of anoles as important consumers in the food web of tabonuco forest at El Verde. This chapter summarizes aspects of anole biology relevant to food web structure and organization in tabonuco forest.

ECOLOGICAL DISTRIBUTION

Investigations of several scientists provide a comprehensive view of the anoles inhabiting a moist tropical forest. Rand (1964) first described the patterns of distribution in Puerto Rican anoles and explained these differences based

Figure 9.1. *Anolis stratulus* is a common anole in tabonuco forest

on variations in structural habitat (perch height and perch diameter) and climatic habitat (exposure to the sun) among anole species. Schoener and Schoener (1971) and Lister (1981) provided additional detail but obtained results in general agreement with Rand. More recent studies added information on the distribution of anoles in the forest canopy (Reagan 1986, 1992, 1995; Dial 1992).

Microhabitat

Vertical Distribution

The forest constitutes a large, diverse three-dimensional habitat that includes sunlit canopy with extensive branches and twigs, subcanopy, and the shaded forest interior where tree trunks and ground litter predominate. This range of structural and climatic variation provides habitat for different anoline ecomorphs within the same areal locations. The vertical stratification of anoline lizards in the tabonuco forest has been described by several researchers (Rand 1964; Turner and Gist 1970; Schoener and Schoener 1971; Moll 1978; Lister 1981). All included study sites in the Luquillo Mountains, and all had the same sampling constraints: lizard sightings were made by ground-based observers. Although lizards could be seen occasionally in the canopy from ground level, reliable quantitative data could not be obtained because of backlighting, distance (20–24 m high canopy), and dense intervening foliage. Ground-level estimates of the relative abundance of anoles during studies in 1980 (table 9.2) indicated that anoles were not reliably detected at distances of more than a few meters (table 9.3). During the 1980s additional studies using vertical transect methods at four tower locations (three near El Verde and one in the Bisley watershed in similar forest) found that anoles were most abundant in the forest canopy (Reagan 1986, 1992).

The vertical distribution patterns of the three common rain forest anoles, *Anolis gundlachi, A. evermanni,* and *A. stratulus,* are presented in figure 9.2. While *A. stratulus* was relatively uncommon near ground level in the forest interior, it was extremely abundant in the forest canopy 8 to 24 m above ground level. Surveys also documented regular use of the canopy by *A. ev-*

Table 9.2. Relative abundance of three common anole species near El Verde

			Anolis gundlachi		*Anolis evermanni*		*Anolis stratulus*	
Transect Number	Wet (N)	Dry (N)	Wet	Dry	Wet	Dry	Wet	Dry
1	84	50	85%	59%	15%	36%	—	5%
2	72	65	84%	71%	16%	25%	—	4%
3	79	60	90%	69%	3%	25%	7%	6%
4	100	84	87%	57%	13%	26%	—	17%
Mean	84	64	86%	64%	12%	27%	2%	9%

Notes: Surveys were taken at four ground-level locations during the wet and dry seasons in tabonuco forest.

Table 9.3. Relative abundance of the three common anole species

	Wet season		Dry season		Mean	
Species	No.	Percent	No.	Percent	No.	Percent
Anolis gundlachi	162	14	136	9	149	11
Anolis evermanni	91	8	86	6	89	7
Anolis stratulus	888	78	1,360	85	1,124	82

Notes: Based on vertical transect data. Numbers are adjusted for average sighting distances among species (*A. gundlachi* = 3.8 m, *A. evermanni* = 2.8 m, *A. stratulus* = 1.2 m) to calculate relative abundance by the Frye strip census method (Overton 1971).

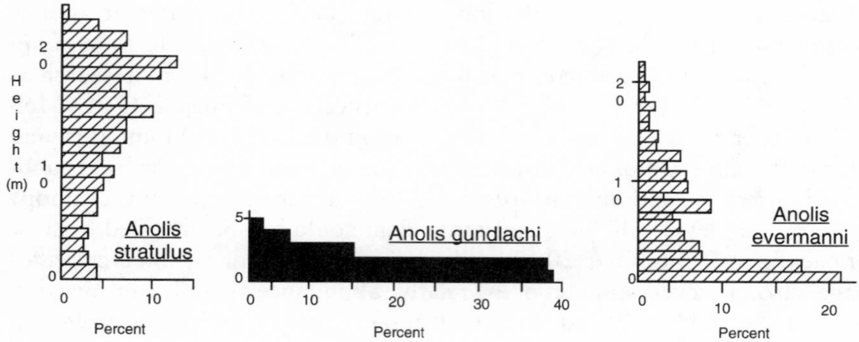

Figure 9.2. The vertical distribution of *Anolis gundlachi, A. evermanni,* and *A. stratulus* in tabonuco rain forest near El Verde, Puerto Rico (Reagan 1992)

ermanni. These results differed from Rand (1964), who found that 65% of the *A. stratulus* that he observed were seen below 10 ft. Differences among study results are attributable to differences in sampling technique (vertical vs. ground level observation). My observations of *A. gundlachi* and *A. evermanni* generally agreed with Rand's findings, except that *A. evermanni* was found distributed throughout the vertical extent of the forest. While the predominance of *A. stratulus* in the forest canopy was a new result, Williams (1983) noted that *A. stratulus* is often found at the outer margins of forests, including the top of the canopy, and Rand (1964) correctly speculated that both *A. stratulus* and *A. evermanni* inhabited the forest canopy.

I rarely observed *Anolis cuvieri* during either ground-level or vertical transect surveys that I conducted in the forest near El Verde, probably because of low population density and cryptic coloration. Williams (1972, 1983) places it in the crown giant ecomorph category, primarily inhabiting the forest canopy. Females are most often observed when they come to ground level to deposit eggs. Juveniles, although they apparently spend much of their time within a few meters of the ground, are slow moving, and cryptically colored (Rand and Andrews 1975; Gorman 1977; Reagan pers. observation). Using radiotelemetry, Losos et al. (1990) monitored the movements of a male *A. cuvieri* in the forest at El Verde and found it most often in the canopy, 15 to 20 m above ground level. Michael Gannon (pers. comm.) indicated that these telemetry readings were performed during midday. The species has been found frequently near ground level at night by Richard Thomas (pers. comm.).

The vertical distribution of anole species within the forest is determined by both microclimatic and microhabitat factors and is generally consistent with the anole ecomorph distribution as described by Williams (1969, 1983). The high abundance of *A. stratulus* in the canopy is noteworthy. The vertical distribution of *A. occultus* in tabonuco forest remains undetermined, but it appears to be limited to the understory.

Perch Size

Of the three common species, only *A. stratulus* appears to prefer perches of a particular size (fig. 9.3). In my observations on perch diameter use by *A. evermanni* based on vertical transect studies, I found a greater preference for perches of small diameter than had been detected in earlier studies. As with *A. stratulus,* observations of *A. evermanni* on small perches during vertical studies were made mostly within the canopy.

The differences among species in the use of perch diameter categories are readily explainable in terms of the vertical distribution of available perches. The majority of available small perches (stems 1–5 cm) were in the forest canopy (Reagan 1992). By comparing the availability of perches in these categories (fig. 9.4) with the observed vertical distributions of the three common anole species (fig. 9.2) we see that most *A. stratulus* were observed in the canopy where small perches are most abundant. The frequency of occur-

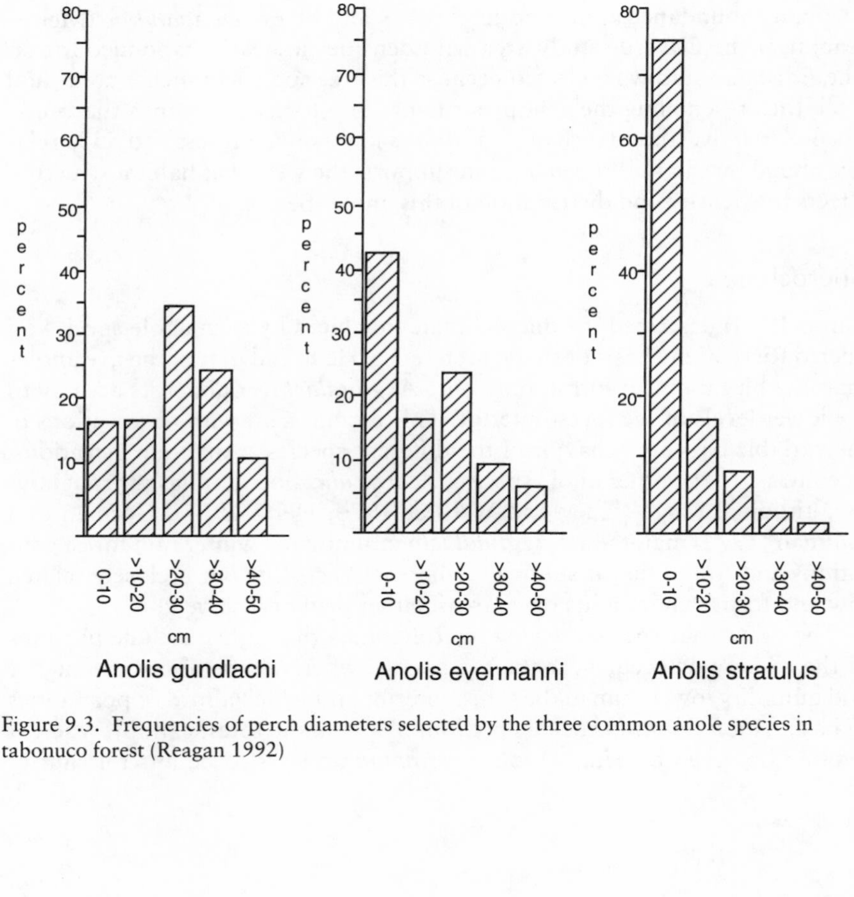

Figure 9.3. Frequencies of perch diameters selected by the three common anole species in tabonuco forest (Reagan 1992)

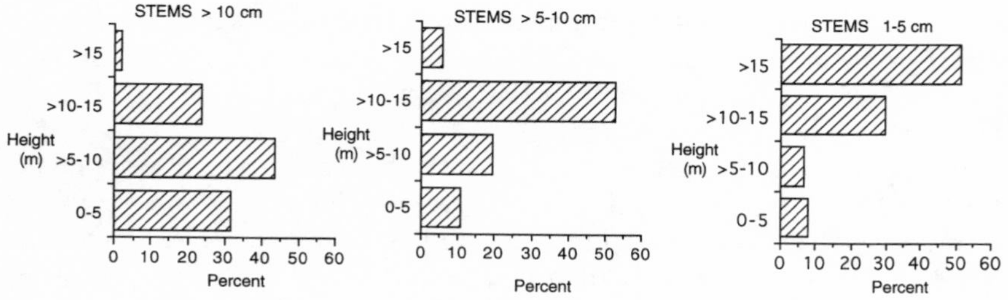

Figure 9.4. The vertical distribution of different size perches in tabonuco forest (Reagan 1992)

rence of *A. gundlachi* on perches of various sizes generally corresponds to the availability of these perches in the lower few meters of the forest (fig. 9.4). *Anolis evermanni* occurs on perches of all sizes, and its vertical distribution encompasses all size categories. It is also the only anole that regularly forages along rocky stream channels throughout the forest and appears to be a genuine generalist with respect to perch size and type.

Reagan (1992) investigated perch size availability by *A. stratulus* by comparing its abundance at ground level in a stand of *Eugenia jambos* (Mertaceae) near the El Verde study area with densities in nearby tabonuco forest. The *Eugenia* stand was selected because this tree species branches at ground level, thus resembling the canopy structure of a forest tree minus the trunk. The 2% relative abundance of *A. stratulus* in tabonuco forest and 53% relative abundance in the *Eugenia* stand supports the view that habitat structure affects the density and distribution of this anole species.

Microclimate

Rand (1964) described the microclimate inhabited by each anole species on Puerto Rico on the basis of shade preference. He found that of the five anoles that inhabit tabonuco forest, only one, *A. gundlachi,* confines its activity to the lower levels of the forest interior. This is consistent with observations of the variable basking behavior of the different species at different elevations. In contrast to the other anole species, *A. stratulus* does not bask throughout its altitudinal range (Huey and Webster 1976; Hertz 1977). Gorman and Hillman (1977) maintained *A. gundlachi* in simulated winter rain forest conditions and found that it survived, whereas *A. cristatellus,* a closely related lowland form that inhabits open areas, either died or lost weight.

The other four species occupy microhabitats that either include portions of the canopy or occur near the forest edge where temperatures are higher and humidity lower than in the forest interior and which afford opportunities to bask. *Anolis stratulus* occurs predominantly in the forest canopy, where it is often observed basking. *Anolis evermanni* appears to be less tolerant of low humidity conditions than is *A. stratulus* and occurs more often in the middle and lower levels of the forest, where conditions are more mesic than in the open upper canopy. Little is known of the microclimatic requirements of *A. cuvieri* and *A. occultus,* although both species inhabit more open habitat and have been observed to bask.

POPULATIONS

In spite of numerous investigations on Puerto Rican anoles, there is little information on population biology of the forest species. Turner and Gist (1970) presented data on the population density of *A. gundlachi* and *A. evermanni* during an investigation of irradiation effects on the forest near El Verde, and I collected additional population data for the three common anole species during recent field studies at El Verde (Reagan 1992). Minimal information is available for other aspects of anole population biology, particularly for the less common anole species, *A. cuvieri* and *A. occultus.*

Relative Abundance

My surveys in mature tabonuco forest at El Verde show that *Anolis gundlachi* was the most abundant species, *A. evermanni* a distant second, and *A. stratulus* the least abundant species at ground level (table 9.2). Less than 1% of all observations were of *A. cuvieri.* These results were generally consistent with the findings of previous studies conducted at ground level (Rand 1964; Schoener and Schoener 1971; Lister 1981). However, the relative abundances for these three species calculated from vertical transect surveys that I conducted at towers in the forest differed substantially from ground level estimates (Reagan 1992). *Anolis stratulus,* the canopy species, was by far the most abundant species followed by *A. gundlachi* and *A. evermanni* (table 9.3). Ground-level observations substantially underestimated canopy use by both *A. evermanni* and *A. stratulus* for reasons already discussed (see vertical distribution section).

Sex and Age Ratios

I calculated sex and age ratios for *A. stratulus* from mark and resight surveys conducted at tower sites in 1981 and 1989. These data were considered more reliable than data from transects where differential activity between sexes and/or age classes could influence estimates. Estimates were made based on the ratios from marked lizards in populations where 86 to 100% of all lizards were marked during multiple mark and resight studies. The adult sex ratio was 1 : 1 (fifty-five males, fifty-four females), and the age ratio was 57% adult and 43% juvenile ($N = 192$).

Adult sex ratios for *A. gundlachi* and *A. evermanni* were approximately 1:1, based on data from ground transect studies conducted at El Verde during 1980–83 (Reagan 1992). Turner and Gist (1970) obtained similar sex ratios (fifty-seven male and fifty-five female *A. gundlachi* at the South Control Center in 1965). Their study involved multiple captures, but no percent total capture data were provided. While Turner and Gist (1970) did not report age ratios, data from tables in their report indicate age ratios for *A. gundlachi* of 66% adults to 34% juveniles and for *A. evermanni* of 67% adults to 33% juveniles. While their results are similar to mine for *A. stratulus,* they cannot be compared rigorously because of differences in time and probable differences in methods.

Of the eighteen *A. cuvieri* incidentally observed at El Verde during field studies from 1980–83, two were unsexed juveniles, five were males, and eleven were females. All were seen within 5 m of ground level. It is likely that females come to ground level to deposit eggs, and are thus observed more frequently than are males.

Population Densities

Multiple mark and resight studies of *A. stratulus* were conducted on vertical transects at tower locations in the forest (Reagan 1992). High resight success (86% in dry season, 100% in wet season) produced relatively accurate population estimates. Because of the vertical orientation of the transect, estimates constituted a point sample with respect to areal extent and produced mean population density estimates of 25,870 ± 7,005 (dry season) and 21,333 ± 6,638 (wet season) *A. stratulus* ha^{-1}, more than 2 individuals m^{-2} of forest habitat (Reagan 1992). Lincoln Index estimates of mark and resight data taken from the north and south walkway towers at El Verde in February-March 1989 were 28,026 ha^{-1} (95% confidence interval of 23,487 to 35,329) and 24,145 ha^{-1} (95% confidence interval of 20,184 to 29,763), respectively, indicating that the high abundances observed in 1980 and 1981 appear to hold for other locations in the forest and may be relatively constant through time.

Turner and Gist (1970) estimated that *A. gundlachi* maintained a population density of 800 per $acre^{-1}$ (approximately 2,000 individuals ha^{-1}) in the forest at El Verde, but they presented no estimates for the other species of forest anoles.

My estimates of anoline relative abundance obtained from vertical transect data were consistent with the estimates of absolute population densities for *A. gundlachi* (Turner and Gist 1970) and *A. stratulus* (Reagan et al. 1982). Multiplying the absolute population density of *A. gundlachi* of 2,000 ha^{-1} (Turner and Gist 1970) by a ratio of the relative abundance of that species and the relative abundance of *A. stratulus* obtained from vertical transects (table 9.3) produced an estimate of 21,500 *A. stratulus*. This estimate approximates the mean population densities determined by multiple mark and recapture/resight methods. Assuming that *A. gundlachi* population densities were generally the same during both studies, the relationship between absolute and relative abundances appears to be consistent.

In the absence of other data, I estimated the population density of *A. evermanni* by the relative abundance approach used to verify the population density of *A. stratulus*. By multiplying the absolute population density of *A. gundlachi* by a ratio of the relative abundances of *A. gundlachi* to *A. evermanni* I obtained a population density estimate of 1,500 *A. evermanni* ha $^{-1}$.

Using my estimates, the combined population density of the three common anole species is approximately 25,000 anoles ha $^{-1}$. This high density is partially attributable to their small size and their ability to partition the habitat vertically. High population densities are well documented for other Caribbean anoles. Gorman and Harwood (1977) estimated densities of *A. pulchellus* up to 20,000 individuals ha $^{-1}$ in grassy lowland areas of Puerto Rico only a few kilometers from El Verde. The seasonal mean estimates of 21,333 to 25,870 for *A. stratulus* in tabonuco forest are the highest population densities reported for any lizard species.

Of the forest anoles, the smallest species, *A. stratulus*, is the most abundant. This contrasts with the situation in forest frogs where the most abundant species, *Eleutherodactylus coqui*, is intermediate in size among the forest species (Stewart and Woolbright, this volume).

Population estimates of *A. stratulus* in different seasons and years indicate the species maintains relatively stable population densities. This is consistent with data for other West Indian anoles (Schoener 1985). Andrews (1991) found year-to-year changes in population density of five- to eight-fold for *A. limifrons* in Panama. Similarly high fluctuations have been found for other mainland anoles (Campbell 1973; Fitch 1975). Reasons for the high fluctuations are variations in rainfall and a short life span for small mainland anoles (Andrews 1991). The longer life span of *A. stratulus* may compensate for annual fluctuations in the environment to maintain more stable population densities.

Hurricane Effects

Hurricanes have a generally negative effect on the overall abundance of forest anoles. Hurricane Hugo struck the Luquillo Mountains of Puerto Rico in September 1989, with maximum sustained winds of 144 km h $^{-1}$ and gusts to 194 km h $^{-1}$ (Scatena and Larsen 1991). These winds removed the leaves from most trees, and in areas exposed to the full force of the winds, knocked down trees and removed tree crowns. The resulting forest structure resembled a forest of telephone poles among 3 to 5 m deep piles of leaf and branch debris. Anole abundance was reduced, probably as a result of the reduction in the three-dimensional habitat structure of the forest which restricted the activity of all anoles to within a few meters of ground level (Reagan 1991).

Increased predation and prolonged dry conditions following the hurricane also appeared to have had detrimental effects on some anole species. One year following the hurricane, *A. stratulus,* had begun reinvading the canopy in areas where branches remained and leaves had reappeared. However, in areas of severe forest damage, populations of this species remained low four years following the hurricane (Reagan unpublished).

Survivorship

Quantitative information on longevity and survivorship rates for Caribbean anoles is scant. Survivorship for *A. stratulus* was calculated from mark and resight data acquired from tower surveys at El Verde (Reagan unpublished). Lizards that had been individually hot-branded during population surveys were easily distinguished from unmarked lizards. A regression line was calculated by the least squares methods, based on the percent marked lizards among the total number of lizards observed during each month in which transects were conducted. The survivorship rate was determined by adjusting the derived regression line for a Y-intercept at 100% and determining the point at which the line intersected the X axis (the time at which, on the average, all lizards in the original population at time zero ceased to exist).

During the ten months of this study the marked population showed a steady rate of decline from 81.5 to 27.0%. A mean population turnover time of 526 days (1.4 years) was calculated for *A. stratulus*. The regression coefficient was -0.19, and the correlation coefficient for the relationship between percent change and time was 0.99, indicating a constant mortality rate throughout life. Survivorship for larger species are probably somewhat higher than the estimate obtained for *Anolis stratulus,* but no estimates are currently available.

Anole Biomass

Seasonal fluctuations in population density and in the age/size class ratios have been documented for *A. stratulus* and are assumed to occur to a similar extent in other anole species. Multiyear population cycles may also exist as they do for vertebrates in temperate ecosystems, and environmental events may alter habitats, food supplies, or directly destroy individuals in lizard populations. Any estimate of anole biomass, therefore, contains an element of uncertainty that is difficult to estimate without long-term studies. Noticeable fluctuations in the population densities of anole species at El Verde have not been detected during the two decades of forest studies at El Verde, and

Table 9.4. Biomass estimates for three common anole species in tabonuco forest near El Verde

Species	Population Density (individuals ha^{-1})	Mean Wt. (g)	Biomass (kg ha^{-1})
Anolis gundlachi	2,000	5.4	10.8
Anolis evermanni	1,500	4.1	6.2
Anolis stratulus	21,500	1.8	38.7
Total	25,000		55.7

the consistency between the few population density estimates which have been made during that period suggest that season-to-season and year-to-year differences probably occur but are not great.

Age and weight data on all three common anole species were collected in conjunction with population and food habit studies. I collected individuals of all size classes of all species to obtain average weight estimates for use in calculating anole biomass (table 9.4). Multiplying average weights by the population density estimates for each species produced a combined biomass estimate of 55.7 kg ha^{-1} wet weight of anoles. This is an underestimate of total anole biomass because it does not include *A. cuvieri,* the giant anole. No population estimates for this species are available, but its larger size (eighteen *A. cuvieri* measured during El Verde field studies had a mean weight of 39.8 g, range: 2.0–56.2 g) indicates that even a low population density of this species could contribute substantially to total anole biomass.

Biomass estimates for *A. gundlachi* (10.8 kg ha^{-1}) are comparable to the results obtained by Lister (1981) who estimated 11.8 kg ha^{-1} during the dry season, the only season for which he estimated the biomass of both adults and juveniles. Lister's estimates for *A. evermanni* (2.944 kg ha^{-1}) and *A. stratulus* (0.736 kg ha^{-1}) were made from ground level surveys and are probably responsible for his substantially lower estimates of anole biomass (table 9.4).

Home Range

Turner and Gist (1970) estimated the home range of *A. gundlachi* as 350 m^2 for males and 250 m^2 for females, based on data for distance between successive captures. They concluded that *A. evermanni* was more vagile than *A. gundlachi* but provided no home range estimate. A home range size of 325.8 m^2 was calculated by the convex polygon method for a single adult male *A. cuvieri* (Losos et al. 1990).

The extremely high *A. stratulus* population density in the forest canopy suggested small individual home ranges and prompted an investigation of its individual home range/territorial relationships. I followed individually branded *Anolis stratulus* for fifty additional surveys over a period of nine months, and intermittently thereafter. Adults and subadults occupied home ranges 6.2 ± 1.2 m in diameter. Male anoles usually occupy larger home ranges than females (Turner and Gist 1970; Schoener 1981). This appears to be true for *A. stratulus*. Of the twenty-six home ranges calculated, only one home range was large enough to occupy the full vertical extent of the 10 to 14 m thick canopy layer. Home range data and field observations revealed that adult male *A. stratulus* occupy three-dimensional territories which they defend in all directions. Adult female and subadult home ranges broadly overlap male territories and each other's home ranges. The resulting configuration of home ranges is best visualized as an assemblage of ellipsoids 6 to 7 m in diameter layered within the canopy stratum of the forest (Reagan 1992).

Three-dimensional home ranges are known for other terrestrial species such as congeneric species of *Peromyscus*, which were found to exhibit considerable arboreal activity (Meserve 1977). Other anoles also occupy three-dimensional home ranges (Rand 1967); however, the home range relationships of *A. stratulus* in tabonuco forest represent the only documented instance of three-dimensional layer of home ranges for a terrestrial vertebrate (Reagan 1992).

Philibosian (1975) studied two allopatric but ecologically similar anole species (*A. acutus* and *A. cristatellus*) and found that they maintained different-sized territories. He suggested that territory/home range requirements limit the density of the breeding population of *A. cristatellus*. Schoener (1981) pointed out that home-range size is related to feeding strategy, and that home-range size is frequently inversely correlated with population density. The small size and three-dimensional shape of *A. stratulus* home ranges appear to provide a suitable configuration for *A. stratulus* foraging and permits the high population densities observed for this species. Males spend considerable time defending their territories and frequently engage in physical combat that occasionally results in tail loss and/or both combatants falling from the canopy (Reagan, pers. observation). A few individuals were also seen returning to the canopy after falls resulting from combat in defense of territorial boundaries in the canopy.

All survival requirements are not met within the canopy home range. Individual *A. stratulus* were also observed at or near ground level, foraging for prey (during the dry season), depositing eggs beneath the litter, or in transit between the canopy and forest floor.

FOOD HABITS

There is a tendency for closely related species that occupy the same habitat to consume different foods (Zaret and Rand 1971; Hespenheide 1975; Fraser 1976; Mares and Williams 1977). This trend has also been documented for insular anole communities (Schoener and Gorman 1968), but results usually have been based on prey size (Sexton et al. 1972; Moll 1978; Pacala and Roughgarden 1982). In anole studies where prey taxa were considered, taxonomic identification was usually to the ordinal level, except for ants which were identified to species. Microhabitat differences exist for predators (anoles) and prey, but identification to ordinal levels may be too gross to discern these differences.

I conducted an investigation of food habits during 1980 and 1981, concurrent with investigations of their population and distribution. The principal objective of this study was to determine the types and quantities of prey consumed by the four forest anole species in as much detail as possible. I then used this information to investigate species, sex, and seasonal differences in food habits.

I collected adults of the three common species (*A. stratulus, A. evermanni,* and *A. gundlachi*) during wet and dry seasons. Additional *A. evermanni* were collected from the rocky channel of the Quebrada Sonadora after I noted that this was the only anole species seen foraging regularly in this habitat in addition to the forest interior. Because of its relative scarcity, specimens of *A. cuvieri* had their stomachs pumped (Sexton et al. 1972) and were subsequently released near capture locations.

At El Verde, Garrison and Reagan (unpublished) found a total of 1,848 prey items in 145 anole stomachs, of which 55.3% were identified to species or genus, 89% to family, and 99.5% to order. Thirty-four animal orders were represented in gut contents, twenty of which were insects.

The giant anole, *A. cuvieri,* fed primarily on snails (Gastropoda), butterfly and moth larvae (Lepidoptera), beetles (Coleoptera), walking sticks (Phasmodea), and plant material (fig. 9.5). A portion of the tail of an adult male

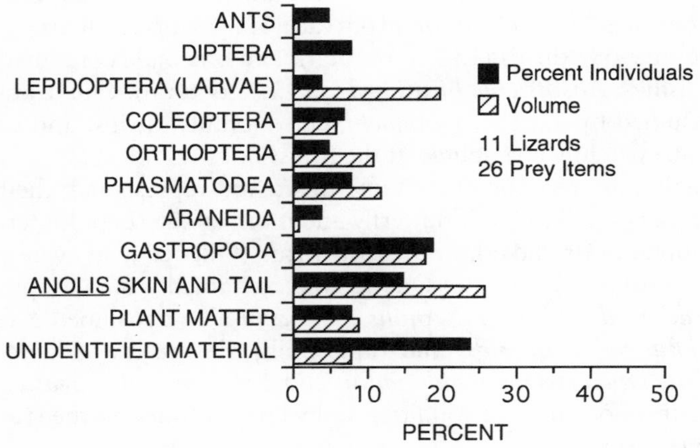

Figure 9.5. Prey of *Anolis cuvieri:* percentage of individuals and volume consumed (Garrison and Reagan unpublished)

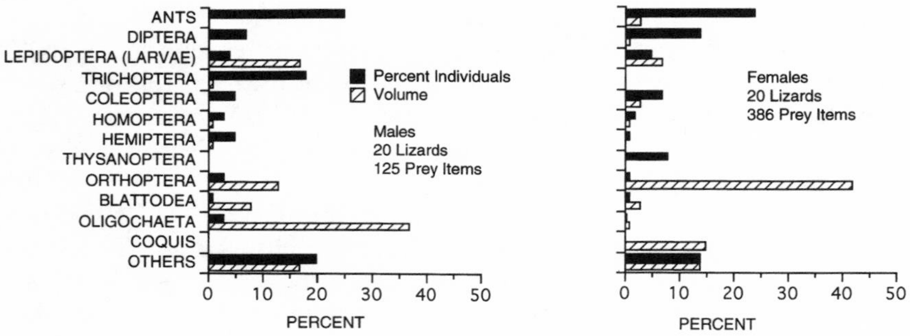

Figure 9.6. Prey of *Anolis gundlachi:* percentage of individuals and volume consumed (Garrison and Reagan unpublished)

A. gundlachi was found in one stomach. Only one ant species (*Odontomachus brunneus,* 8 mm) was consumed, presumably because *A. cuvieri* tends to cue on prey items larger than most ant species. The mean number of taxa, (1.6), animal taxon richness (1.5), and mean number of animals per lizard (1.6) were almost identical because almost every prey item found was unique.

Wolcott (1924) documents snails as part of the diet of *A. cuvieri* collected in the lowlands of central Puerto Rico. Large animal items and vegetable matter are also recorded as food for another giant anole, the Cuban knight anole, *Anolis equestris* (Branch 1976; Dalrymple 1980). This species is similar to *A. cuvieri* in size, general habitat, and mode of foraging.

Ants were found in nearly all stomachs of *A. gundlachi* and comprised about 30% of the total number of prey consumed by both sexes (fig. 9.6). The numbers of Diptera, Thysanoptera, and Trichoptera are high because many items of these groups were often consumed by individual lizards. One female consumed fifty chironomid larvae (12.9% of total prey), another contained fifty-one thrips (13.2% of total), and one male contained twenty-four caddisflies, *Alisotrichia hirudopsis,* (19.2% of total). Unlike ants, which were consumed by most *A. gundlachi,* chironomids, thrips, and caddisflies appear to be patchily distributed at the forest floor.

Although ants were the most numerous taxon represented, their volume contribution is small (<8%). Butterfly and moth larvae (Lepidoptera), crickets (Orthoptera: Gryllidae), and earthworms (Oligochaeta) were the most important groups by volume. One female contained a small leptodactylid frog (*Eleutherodactylus* sp.). *Anolis gundlachi* also consumed a terrestrial isopod (*Philocia richmondi*) and three millipede species (*Glomeridesmus marmoreus, Docodesmus maldonadoi,* and *Prostemmiulus heatwolus*). All of these arthropods inhabit soil litter and were not found in the stomachs of other anole species.

Anolis evermanni is an opportunistic feeder able to exploit a variety of

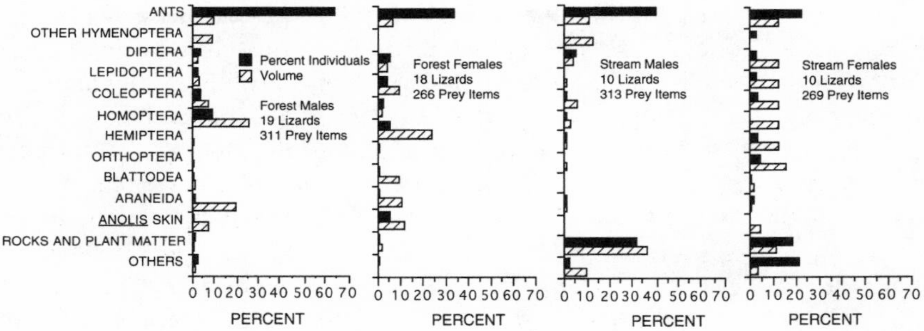

Figure 9.7. Prey of *Anolis evermanni:* percentage of individuals and volume consumed (Garrison and Reagan unpublished)

prey. Ants were numerically the most important items consumed by *A. evermanni* (high of 65% in forest males to a low of 25% in creek females from the Quebrada Sonadora); (fig. 9.7). Homoptera, primarily Fulgoroidea, numerically constituted more than 10% of the prey of forest *A. evermanni* but less than 5% of the creek *A. evermanni*. Vegetable matter was the second (for males) and third (for females) most numerous food item in creek *A. evermanni*. Seeds of *Caesaria arborea, Cecropia schreberiana,* and another undetermined species were consumed. Spiders contributed more than 15% of the total prey volume for forest anoles, but less than 5% for creek specimens. Insects that characteristically dwell on the water surface (Hemiptera: Velidae), one emerging damselfly larva (*Telebasi vulnerata*), and riffle beetles (Elmidae) were found in creek specimens, indicating that these lizards feed at the water's edge. These taxa were not found in the stomachs of other anoles.

Wolcott (1924) also found a variety of prey items in a sample of ten *A. evermanni,* nine of which were collected from a coffee grove on El Yunque in the Luquillo Mountains. The species composition of the diet differed substantially from the results of our studies at El Verde. Wolcott determined that beetles (Coleoptera) and bugs (Homoptera) constituted 80% of the lizards' diet, while ants, moths, and caterpillars made up most the remainder. Fulgorid bugs were 27% of the total diet. No vegetable material had been eaten by any of the lizards examined. The dietary differences are probably due in part to differences in available food in the various microhabitats used by this species rather than to shifts in dietary preference. The high taxon richness values for *A. evermanni* from streamside habitat (one adult female had seventy-seven different prey items representing twenty-one species, twenty invertebrates, and one seed) reflect the higher diversity of prey found in this microhabitat.

Ants were the most commonly consumed items by adult *A. stratulus,* followed by fulgoroid Homoptera (planthoppers) and Diptera (fig. 9.8). By vol-

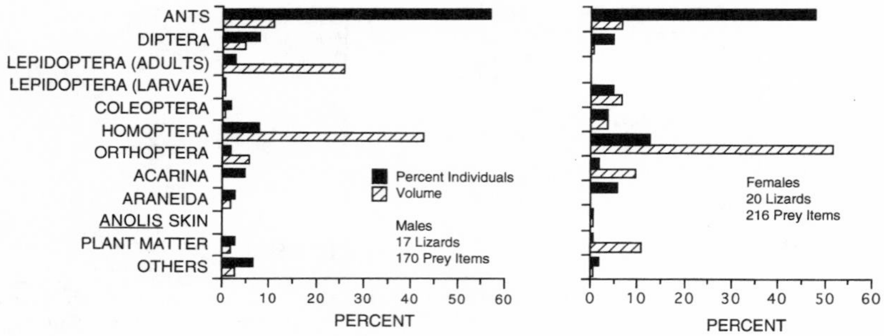

Figure 9.8. Prey of *Anolis stratulus:* percentage of individuals and volume consumed (Garrison and Reagan unpublished)

ume, fulgoroids were clearly the most important group consumed (46% in males, 50% in females). Two males each ate a relatively large noctuid moth (18 mm long) which accounted for the high (27%) volume component for this group in male lizards.

The findings of Lister (1981), who also investigated the food habits of the three common anole species in tabonuco forests in the Luquillo Mountains, differed from ours at El Verde. He found that ants constituted the greatest amount of biomass in diet of *A. stratulus,* while our data indicated that planthoppers contributed more appreciably to its diet. He also stated that homopterans and ants were not among the dominant order collected in sticky trap samples. Fulgorids display a great diversity of form, and many species resemble other insect orders. Some dietary differences may be due to differences in available prey between the locations and time of sample collection. However, based on observations at El Verde, it is possible that many of his beetle identifications may really have been flatid or issid planthoppers. As for the ants in sticky traps, ants are social insects that lay odor trails for other ants to follow. If one avoids a trap others will do the same, thus leading to an underrepresentation of ants in the traps.

Like Lister, Wolcott (1924) found that ants composed a substantial portion of the diet of *A. stratulus.* He analyzed the gut contents of fifty lizards from several locations in the lowlands and uplands of Puerto Rico. It is probable that differences between our results and his are due to differences in prey availability in these diverse habitats.

Analysis of Prey by Taxa

Orthoptera: Gryllidae (crickets). The most common Orthopteran in tabonuco forest at El Verde are crickets (Gryllidae). Their occurrence in anole stomachs is presented in table 9.5. *Cyrtoxipha* and *Orocharis* were

Table 9.5. Distribution of three common ant species among stomach samples of three species of *Anolis*

	A. gundlachi		*A. evermanni* Forest		*A. evermanni* Stream		*A. stratulus*	
Taxon	Male	Female	Male	Female	Male	Female	Male	Female
Myrmelachista ramulorum[a]	1	1	96	28	8	7	61	45
Iridomyrmex melleus[a]	9	5	104	40	45	2	19	33
Pheidole moerens[b]	11	53	—	9	1	22	2	2

Source: From Garrison and Reagan (unpublished).

[a]Species found in sticky trap samples at Tower (May-June 1981): *Myrmelachista ramulorum* (1, 3, 11, 13 m), *Iridomyrmex melleus* (4, 14, 15, 18 m).

[b]Species found in litter samples: *Pheidole moerens*.

found at various heights in the forest during sticky trap studies of insects conducted concurrently with anole food habit studies. *Cyrtoxipha gundlachi,* a small (5 mm long) green bush cricket, and at least five species of *Orocharis* commonly occur among the foliage. No specimen was ever observed on the ground or in litter traps. *Aurogryllus* and *Gryllus* are large (over 7 mm long) common ground species. Both were found only in the stomachs of *A. gundlachi,* and never during vertical sticky trap sampling. Nemobiine crickets are also ground dwellers (Borror et al. 1981).

HOMOPTERA: FULGOROIDEA (PLANTHOPPERS). Planthoppers are the most common homopterans in tabonuco forest (Pfeiffer, chap. 5 this volume). Drewry (1970b) lists forty-six species from the forest at El Verde. Fourteen of these were found among prey consumed by anoles. Vertical distribution studies of insects using sticky traps at the tower showed that 76% of planthoppers occur in the upper 6 m of the canopy compared to 12% at both the middle and lower 6 m of the forest (Reagan et al. 1982). The concentration of planthoppers in the canopy correlates with the vertical distribution pattern of *A. stratulus,* supporting the observation (discussed in the following section) that this species forages primarily within the forest canopy while *A. evermanni* and *A. gundlachi* do not.

HYMENOPTERA: FORMICIDAE (ANTS). Ants were numerically the predominant dietary constituents of the three common species of *Anolis.* One *A. cuvieri* ate one ant and is not further considered here. Lavigne and Drewry (1970) list thirty-three species of ants from the forest at El Verde. Three of these, *Myrmelachista ramulorum, Iridomyrmex melleus,* and *Pheidole moerens* are common, and all three are consumed by anoles. *Myrmelachista ramulorum* is primarily arboreal while *I. melleus* occurs from ground level to the canopy. It appears to be the most abundant ant species in the forest.

Pheidole moerens colonies are present under rocks on the forest floor (Lavigne and Drewry 1970); this species was the only ant found in litter samples.

Pheidole moerens was the most common ant in stomachs of *A. gundlachi* and creek specimens of *A. evermanni,* but it was poorly represented in the more arboreal *A. stratulus* and forest specimens of *A. evermanni* (table 9.5). *Myrmelachista ramulorum* was the predominant ant in the diets of both sexes of the most arboreal anole, *A. stratulus,* common in forest *A. evermanni,* and rare in *A. gundlachi* and creek *A. evermanni.* Not surprisingly, *I. melleus* was found in all three lizards species, but was most commonly represented in forest specimens of *A. evermanni,* with which it shares the same general pattern of vertical distribution.

Milstead (1957) indicated that it was generally unnecessary to proceed beyond the ordinal level of prey identification in analyzing the differences in food habits among species of whiptail lizards. Analysis of data on crickets and ants in the diet of anoles from El Verde, however, support Lavigne and Drewry (1970), who stated that prey identification beyond the family level provides additional insights into the microgeographic distribution of Puerto Rican forest anoles.

Foraging Rates

The importance of a species in a community food web depends on its abundance, what it eats, what it is consumed by, and the rate at which these transfers of energy and nutrients occur. Quantitative information on any of these subjects is difficult to obtain. This situation is particularly true of foraging rate data for anoline lizards, the most abundant diurnal predators in tabonuco forest habitat. The data presented here on foraging rates for anoles is based partially on detailed field studies, on extrapolations from gut content studies conducted at El Verde, and from correlations with food habit investigations of other researchers in the tabonuco forest on Puerto Rico.

Several studies have investigated seasonal differences in food habits and foraging by anoles. Sexton et al. (1972) studied the seasonal food habits of *Anolis limifrons* in Panama and detected no differences in the frequency of food found in stomachs between wet and dry seasons. Licht (1974) experimented with food supplementation for anoles in Puerto Rico and suggested that island anoles were probably food limited, based on his observations of fat body condition, the amount of food present in anole stomachs, and the consumption rate of food supplements. Andrews (1976) also concluded that anoles on Caribbean islands were probably food limited.

The studies of Lister (1981) and Reagan et al. (1982) on anoles in the tabonuco forest in the Luquillo Mountains of Puerto Rico provide some sup-

Table 9.6. Movement and foraging data for adult *Anolis stratulus* near El Verde

Time of Day	Total No. of Moves		No. of Prey Captures	
	Wet	Dry	Wet	Dry
Morning	165	313	4	6
Midday	304	420	17	9
Afternoon	178	306	7	12
Total	647	1,039	28	27

Source: Data from Reagan (1986).

Note: Data from wet- and dry-season tower surveys in tabonuco forest.

port for the concept of food limitation in island anoles. Few of the anoles analyzed for gut contents had empty stomachs, but some differences between wet and dry seasons were apparent; *A. stratulus* showed decreased numbers of prey items and reduced prey biomass during the dry season. Reagan (1986) calculated foraging rates of 20.5 prey items d^{-1} during the wet season and 18.0 items d^{-1} during the dry season. These estimates differed in the same direction as those of Lister (1981), who found 25.6 prey items per lizard during wet season and 15.4 prey items per lizard during dry season for *A. stratulus*. Based on observations of movements and prey capture, substantially more activity was required to obtain prey during the dry season than during the wet season (table 9.6). Nearly twice as many moves were required to obtain the same number of prey during the dry season compared to the wet season.

An additional estimate of foraging rate was obtained from analyses of *A. stratulus* gut contents studies in the forest near El Verde (Reagan et al. 1982). Stomachs contained an average of 11.5 prey items each during wet season and 8.7 items each during the dry season. All lizards were collected at between 1100 and 1400 hours, approximately halfway through the available foraging time. Doubling these estimates produces consumption rate estimates of 23 items d^{-1} in wet season and 17.4 items d^{-1} during the dry season. A mean of 9.5 items d^{-1} was calculated for juvenile *A. stratulus* from stomach content data (Reagan et al. 1982). An estimated mean daily consumption rate of 3.4×10^5 prey items ha^{-1} was obtained by combining the average daily consumption rate for adults and juveniles, population structure, and mean population density.

Foraging rate observations were made only on *A. stratulus*. Stomach content data available for *A. gundlachi* and *A. evermanni* were used to provide estimates of foraging rates for these species. Estimates of 25.4 prey per lizard for *A. gundlachi* and 34.6 prey per lizard for *A. evermanni* produced mean daily consumption rate estimates of 50,800 prey ha^{-1} for *A. gundlachi* and

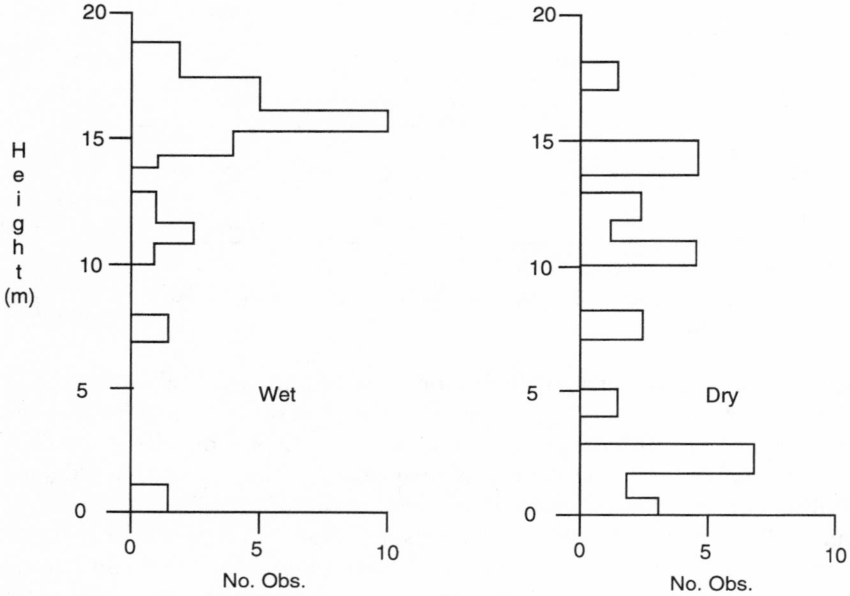

Figure 9.9. Prey capture heights for *Anolis stratulus* during wet season (September 1982) and dry season (March 1983) at the El Verde Tower (Reagan 1986)

5.19×10^4 prey ha $^{-1}$ for *A. evermanni* (Garrison and Reagan unpublished). Although these numbers are considerably smaller than estimates for *A. stratulus,* their contribution to overall consumption of prey biomass is substantial. The mean prey volume per lizard consumed by *A. gundlachi* is more than eight times greater than that consumed by *A. stratulus,* while *A. evermanni* consumes nearly twice as much prey volume per lizard as *A. stratulus.*

Statistical analysis of data on foraging movements and prey capture location for *A. stratulus* indicated seasonal and time-of-day differences, but no differences between sexes (Reagan 1986). Individuals moved 60% more often during the dry season than during the wet to obtain the same number of prey items (table 9.6). Individuals generally foraged in the canopy above 10 m during the wet season, but expanded their foraging area and captured most of their prey at ground level or below 10 m on trees during the dry season (fig. 9.9).

Pacala and Roughgarden (1982) demonstrated a basic assumption of competition theory: for Caribbean anoles in the Lesser Antilles, the extent of competition between species is inversely related to the amount of resource partitioning. A similar situation may occur at El Verde. The downward shift of *A. stratulus* in distribution and foraging location, from the canopy to the lower forest, during the dry season reduces the vertical stratification of the three common anole species and increases the potential for competition among them.

PREDATORS

Of the several animal species that regularly consume anoles, the most important predators are birds. Major avian predators include the pearly-eyed thrasher and the Puerto Rican lizard cuckoo, but other species such as red-tailed and broad-winged hawks also feed on anoles. Other than the Indian mongoose and feral cats, both introduced species, there are no mammalian predators likely to eat anoles. The role of anoles in the diet of birds and mammals are addressed in detail in chapters 11 and 12, respectively.

Anoles are major prey in the diet of the snake *Alsophis portoricensis*. The Puerto Rican boa probably also feeds to some extent on anoles, although this feeding relationship has not been documented (Reagan 1984). Anole species also regularly feed on each other. *Anolis evermanni* feeds on small *A. gundlachi* and *A. stratulus*. A tail fragment from a male *A. gundlachi* was recovered from the stomach of an *A. cuvieri*. While cannibalism within a species has not yet been observed, it may well occur.

Invertebrate predation on anoles also occurs and is a distinctive feature of the overall community food web. The tailless whip scorpion *Phrynus longipes* (Amblypygida), tarantula *Avicularia laeta*, and centipede *Scolopendra alternans* have been observed feeding on anoles in the forest. These species are also known predators of leptodactylid frogs, *Eleutherodactylus* spp. (Steward and Woolbright, this volume). Chapter 14 provides a discussion of the role of invertebrate predation and food loops in the community food web.

SUMMARY

Anoline lizards are abundant and conspicuous members of the animal community in the forest at El Verde. They are diurnal, primarily insectivorous, and forage throughout the vertical extent of tabonuco forest. While these lizards occupy the same general trophic level as frogs, there is little overlap in the species of prey consumed because frogs and their prey are nocturnal (Stewart and Woolbright, this volume).

The four anole species studied at El Verde adopted different strategies for partitioning the available food resources within the forest, essentially following the ecomorph patterns described by Williams (1972). *Anolis cuvieri* is a canopy omnivore which consumes a variety of plant and animal food items, most of which are too large to be consumed by the three common but smaller anoline species. *Anolis gundlachi* is a typical sit-and-wait trunk-ground species (Moermond 1973), which forages in the surface litter or on the lower trunks of trees. The smallest species, *A. stratulus,* actively forages for small insects in the forest canopy. The fourth species, *A. evermanni,* is a true generalist, which forages from ground level near streams to the upper forest canopy.

Population densities of *A. stratulus* are estimated at between 21,333 and 28,026 individuals ha $^{-1}$, the highest reported density for any lizard. The small size and three-dimensional shape of *A. stratulus* home ranges layered in the forest canopy permit these high densities. A mean population turnover time of 526 days (1.4 years) was estimated for this species. Population estimates for *A. gundlachi* and *A. evermanni* were 2,000 and 1,500 individuals ha $^{-1}$, respectively.

The sex ratio for the three common forest anoles, *A. stratulus, A. gundlachi,* and *A. evermanni* is approximately 1 : 1. The combined biomass estimate for the three common forest anoles is 55.7 kg ha $^{-1}$.

The diversity of prey items consumed by forest anoles indicates that all species are largely opportunistic predators. Prey species consumed represent those available in the particular microhabitat and foraging habits of the predator species. The large, slow-moving *A. cuvieri* eats predominantly large, slow prey including snails, lepidopteran larvae, beetles, walking sticks, and plant material. *Anolis gundlachi* consumes mostly ants, lepidopteran larvae, and earthworms by volume. Flies, thrips, and caddisflies are also represented in the diet. The small canopy dwelling *A. stratulus* consumes tiny ants and planthoppers that are abundant in the canopy. The high taxon richness values for *A. evermanni* from streamside habitat reflect the higher diversity of prey found in that microhabitat and the opportunistic food habits of *A. evermanni* in the forest.

Schoener (1968) noted that larger lizards generally take larger prey than do smaller species. This trend is apparent for the four species of anoles in tabonuco forest on Puerto Rico, but no statistical differences were detected when comparisons were made among species, sexes, and seasons. Data were characterized by large variances in prey size (length, volume, and weight; Garrison and Reagan unpublished).

A number of vertebrate and large invertebrate predators eat anoles. Principal anole predators include the Puerto Rican lizard cuckoo; the colubrid snake, *Alsophis portoricensis;* and the red-tailed hawk. Tarantulas (*Avicularia*), centipedes (*Scolopendra*), and tailless whip scorpions (*Phrynus*) also feed on anoles. Anoles are also consumed by other anoles (Reagan et al., this volume).

Considerable information has been obtained on the role of anoles in the structure and function of insular terrestrial ecosystems of Puerto Rico, beginning with the research of Wolcott (1924) on the food of Puerto Rican lizards and continuing through the ongoing investigations of the Luquillo Experimental Forest Long-Term Ecological Research (LTER) Program. Anoline lizards exert a substantial impact on the structure and function of the forest food web at El Verde. They are important insectivores in the diurnal subweb and, together with frogs which play a similar role in the nocturnal subweb, they are important in the transfer of energy and nutrients to top predators in the food web of the forest community. As a result of recent studies, we now know that in terms of biomass, numbers, and nutrient and energy cycling, anoles constitute a dominant component of the animal community within the forest.

ACKNOWLEDGMENTS

I thank Rosser Garrison for analyses of anole stomach contents used in this chapter. The data were developed for a different purpose and have not yet been submitted for publication in their completed form. The textual clarity and content of the chapter were improved by reviews of earlier versions of this chapter by Meg Stewart, Jonathan Losos, Larry Woolbright, and Bob Waide, my most merciless and helpful critic. Initial research was supported by the U.S. Department of Energy under contract No. HA 0203010 to the Center for Energy and Environment Research of the University of Puerto Rico. Portions of the research were performed under grant BSR-8702336 from the National Science Foundation to the Center for Energy and Environment Research of the University of Puerto Rico and the Institute of Tropical Forestry as part of the Long-Term Ecological Research Program in the Luquillo Experimental Forest. Additional support was provided by the Forest Service (U.S. Department of Agriculture) and the University of Puerto Rico.

10

Nonanoline Reptiles

Richard Thomas and Ava Gaa Kessler

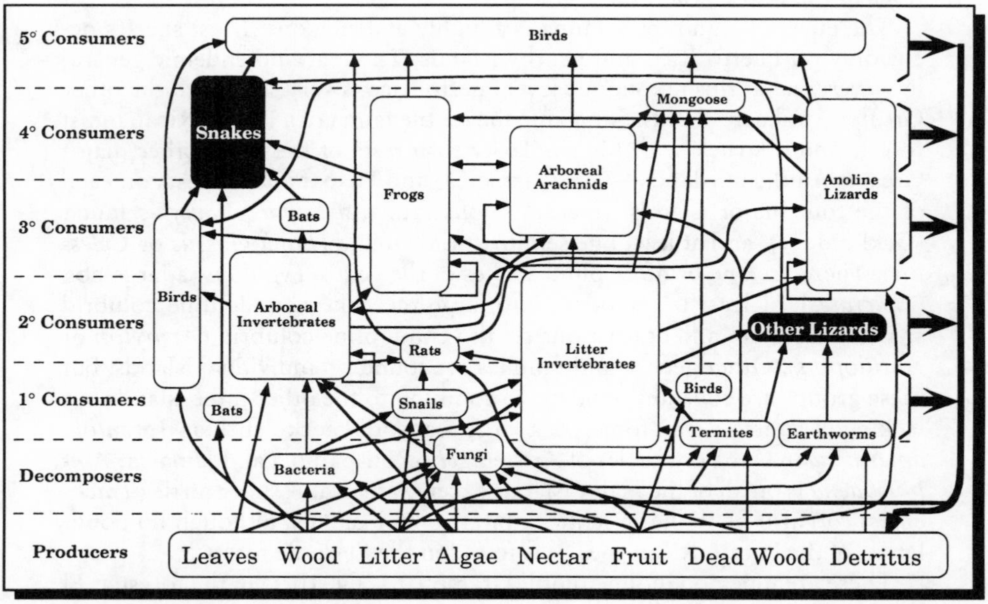

NONANOLINE reptiles span almost all consumer trophic levels, with the exception of the first, in the El Verde food web. Nonanoline lizards feed on invertebrates that are primary consumers and decomposers; large snakes feed mostly on vertebrates. Due to the small diversity of the group and relatively low densities of many of the species, their contributions to the overall energy flows in the web may be minimal. Nonetheless, snakes and nonanoline lizards contribute a large number of connections to the web's structure.

The Puerto Rican reptile fauna is a highly endemic one. Most species occur only on Puerto Rico and nearby islands. There are no endemic genera, however, and at this taxonomic level Puerto Rico is very similar to the other Greater Antillean islands. Surveying the reptile fauna of a Puerto Rican moist forest, one is struck with the similarity to forests of the three other major Greater Antillean islands—Cuba, Jamaica, and Hispaniola. Forests on each of the four major islands have an *Anolis* fauna, a *Sphaerodactylus* fauna (Gekkonidae), and at least one semifossorial anguid (*Diploglossus* or *Celestus*). There are one or more blind snakes of the genus *Typhlops,* a large boa (*Epicrates*), at least one moderate to large racerlike xenodontine colubrid snake (*Alsophis*), and at least one small xenodontine colubrid (*Arrhyton* or *Antillophis*). Other genera and families are found on individual islands, but these groups are constant. One lizard common to the other three islands but lacking in Puerto Rican forests is a large scansorial gecko such as *Tarentola americana* in Cuba or species of *Aristelliger* on Hispaniola and Jamaica. *Amphisbaena* is on all of the major islands except Jamaica. A slider turtle (*Trachemys*) can also be found in lentic habitats on all islands, although no population of the Puerto Rican species is found in the El Verde forest.

There are nine nonanoline reptiles (table 10.1, fig. 10.1) in the forest at El Verde. (The genus *Anolis* is discussed in chapter 9.) There are two typhlopid blind snakes (*Typhlops platycephalus* and *T. rostellatus*), one amphisbaenian (*Amphisbaena caeca*), an anguid lizard (*Diploglossus plei*; formerly *D. pleei*), two geckos (*Sphaerodactylus macrolepis* and *S. klauberi*), a boa (*Epicrates inornatus*), and two colubrids (*Alsophis portoricensis* and *Arrhyton exiguum*). The same assemblage of nine species is found throughout Puerto

Table 10.1. Sizes of nonanoline reptiles at El Verde

Species	Snout-Vent Length (mm)	Weight (g)
LACERTILIA: GEKKONIDAE		
Sphaerodactylus klauberi	24–32, X = 27.47 SE = .72 (N = 19)	.3–.8, X = .48 SE = .04 (N = 19)*
Sphaerodactylus macrolepis	24–32, X = 27.59 SE = .71 (N = 29)	.1–.9, X = .54 SE = .04 (N = 30)*
LACERTILIA: ANGUIDAE		
Diploglossus plei	48–112, X = 81.78 SE = 1.83 (N = 79)	2.7–20.63, X = 8.0 SE = .73 (N = 89)*
AMPHISBAENIA: AMPHISBAENIDAE		
Amphisbaena caeca	143–270, X = 177.35 SE = .70 (N = 28)	2.2–8.2, X = 6.27 SE = .82 (N = 7)
SERPENTES: TYPHLOPIDAE		
Typhlops platycephalus	112–337, X = 225.06 SE = 5.08 (N = 36)	.9–15.0, X = 5.2 SE = .35 (N = 79)*
Typhlops rostellatus	139–202, X = 177.35 SE = 5.14 (N = 14)	1.1–3.5, X = 1.88 SE = .34 (N = 6)
SERPENTES: BOIDAE		
Epicrates inornatus	430–1630, X = 1164.8 SE = 126.55 (N = 10)	20.35–1417.5, X = 528.31 SE = 126.60 (N = 10)*
SERPENTES: COLUBRIDAE		
Alsophis portoricensis	210–810, X = 554.07 SE = 41.96 (N = 13)	2.7–226.8, X = 87.33 SE = 10.65 (N = 29)*
Arrhyton exiguum	134–325, X = 265.18 SE = 10.06 (N = 28)	1.7–24.8, X = 13.54 SE = 1.02 (N = 29)*

Notes: Range, mean (X), standard error of mean (SE), sample size (N) is given for each species. All samples comprise only adults; SVLs and weights are sometimes from different specimens; asterisks indicate weights at least in part from preserved specimens.

Rico in moist forest habitats up to about 600 m. In some of these forests, one or two other species of *Amphisbaena* occur. The heliothermic teiid lizard, *Ameiva exsul,* reaches the periphery of the forest in open or edge situations but does not occur under continuous canopy; we do not consider it here.

All of the reptiles of the forest at El Verde are endemic to the Puerto Rican bank islands, an archipelago united by a shallow submarine platform and extending from Puerto Rico in the west throughout the Virgin Islands (except St. Croix) to Anegada (Heatwole and MacKenzie 1967). *Alsophis portoricensis* and *Sphaerodactylus macrolepis* extend beyond the limits of the bank to some adjacent small islands. Of the nine nonanoline species, five (*Alsophis portoricensis, Amphisbaena caeca, Arrhyton exiguum, Typhlops platycephalus,* and *Sphaerodactylus macrolepis*) are eurytopic and occur on other islands of the bank. *Diplogossus plei, Sphaerodactylus klauberi, Epicrates inornatus,* and *Typhlops rostellatus* are confined to Puerto Rico proper and, with the exception of *E. inornatus,* are restricted to mesic, interior, forested habitats (see Schwartz and Henderson 1991 for distributions).

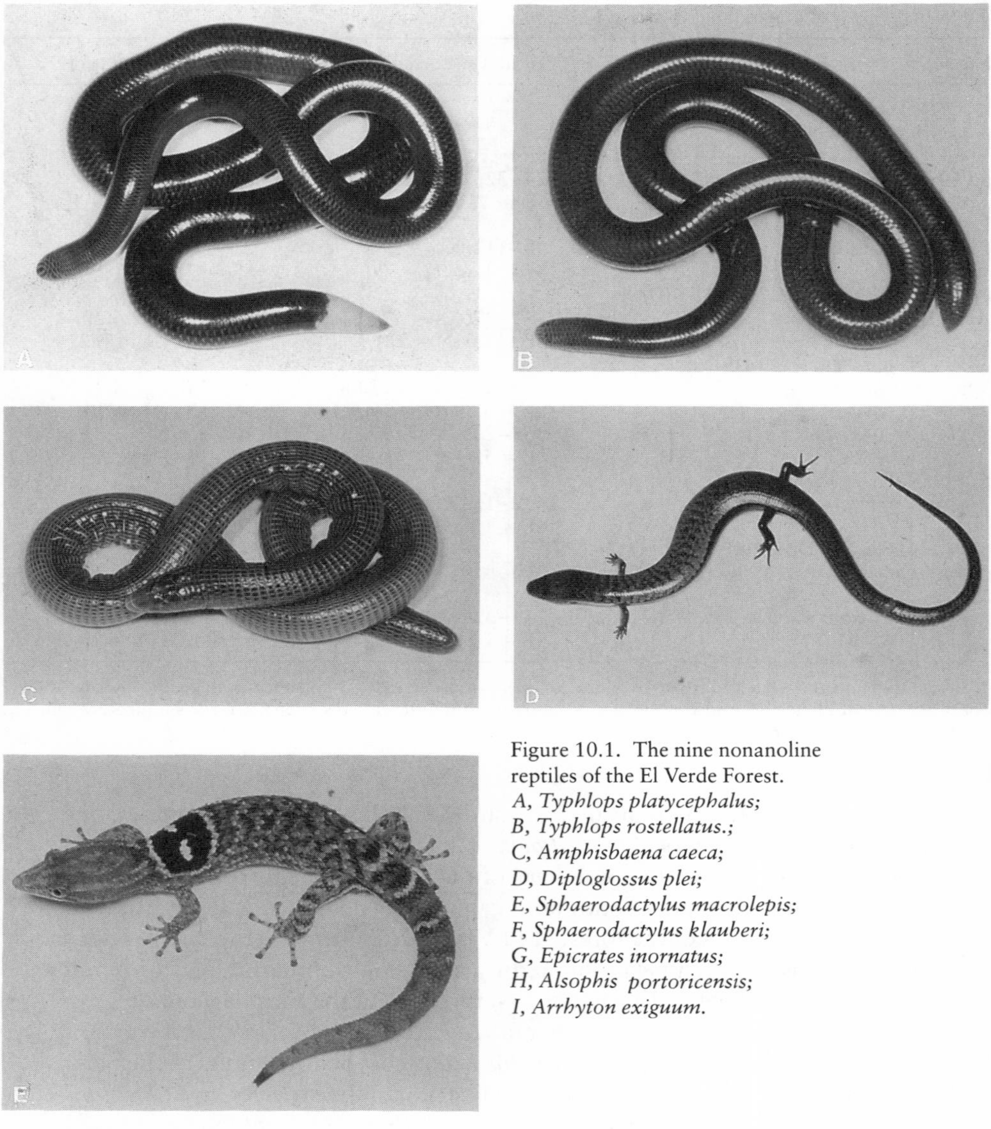

Figure 10.1. The nine nonanoline reptiles of the El Verde Forest. *A, Typhlops platycephalus; B, Typhlops rostellatus.; C, Amphisbaena caeca; D, Diploglossus plei; E, Sphaerodactylus macrolepis; F, Sphaerodactylus klauberi; G, Epicrates inornatus; H, Alsophis portoricensis; I, Arrhyton exiguum.*

MICROHABITATS

Waide and Reagan (this volume) have described the major features of the forest habitat. We here focus on some aspects that are of importance to small reptiles, particularly fossorial and cryptic ones. In the vicinity of El Verde the terrain is characterized by moderate to steep slopes. The soil is predominantly clayey (lateritic); granitic boulders and cobbles are found irregularly

on the surface throughout the region. On ridges the clayey soil may be exposed at the surface, barely covered with a thin layer of leaf-litter. In areas of thicker litter there is a layer of dark humic soil of variable depth, often 3 to 4 cm. A dense, planar network of fine roots often occurs in the lower part of the humic layer and the upper part of the inorganic soil. The thickness of vegetation debris and humic soil depends on local conditions.

Thick litter drifts accumulate behind rocks on slopes, in depressions, and around buttresses; palms may accumulate deposits of dead fronds around their bases. The result is that the forest floor is a complex mosaic of vegetative litter associated with areas of humic soil, thinner leaf-litter, nearly bare clay, root networks, and rocky areas that are not only accumulate litter and humus but add further complexity to the subsurface microhabitats used by a variety of reptile species.

Feeding Modes and Habitat Occurrence

To facilitate visualizing size differences among the nonanoline reptiles of the forest we present data on size and weight in table 10.1. The mean weight of

Epicrates inornatus is over 1000 times greater than that of *Sphaerodactylus klauberi*. We discuss these reptiles according to predominant microhabitat occurrence: fossorial (*Amphisbaena caeca* and the two species of *Typhlops*), semifossorial (*Diploglossus plei*), cryptic (*Sphaerodactylus* species, *Arrhyton exiguum*), and emergent (*Alsophis* and *Epicrates*). Emergent species forage to a large extent in exposed situations both on the ground and above ground in vines, bushes, and trees.

Fossorial Species

Like other amphisbaenians, *Amphisbaena caeca* is a specialized burrower with reduced eyes, a peculiar annular scale arrangement that increases traction, along with other integumental modifications for underground locomotion, and a compact, wedgelike skull that enables the head to function as a tunneling device. The generalized amphisbaenians, among which are included all four of the Puerto Rican species, burrow by compacting the soil with ramming movements of the head (Gans 1969).

Despite their burrowing specializations, amphisbaenians are generalized predators, capable of eating very small insects such as termites, or, with their sharp teeth and powerful jaws, of biting pieces from larger prey such as earthworms, large beetle larvae, or even small vertebrates. The muscular tongue aids ingestion (Thomas 1965). Amphisbaenids appear to locate their prey by sensing low-frequency sounds made by soil animals in movement; each columella is attached to the corresponding mandibular ramus and serves to transmit ground-borne sounds to the inner ear (Wever and Gans 1973). Olfaction, however, is also important, and in captivity, *Amphisbaena caeca* will locate carrion by this means. *Amphisbaena caeca* is the most widespread species on the island but is found principally in shaded situations (Thomas 1966), where it burrows in the upper soil and humus layers beneath rocks, accumulations of dead vegetation, abandoned termite nests, and rotten logs. It does not appear to burrow in clay.

Typhlopid blind snakes are also highly modified for subsurface existence. Like amphisbaenians, they have reduced eyes (but also capable of light/dark perception). Their scales are smooth and overlapping, and the skull, although rigid in comparison to those of other snakes, is not as reinforced as are those of amphisbaenians. The rostral portion of the skull is expanded and bulbous to accommodate the enlarged olfactory organs. Their mouths are small and ventral, and their reduced jaws are highly modified for the ingestion of very small prey such as ants and termites. The modifications used by normal snakes for swallowing large prey are absent, and they have no such biting ability as do the amphisbaenians.

Blind snakes move along paths of least resistance through yielding soil or along root channels and crevices; they do not tunnel. They are found beneath rocks and logs and in or under mounds of dead vegetation, but the exact range of depths and microsites that the Puerto Rican species occupy has not been adequately characterized. The principal means of locating prey appears to be olfactory, although touch is the proximate cue for seizing the prey (Thomas unpublished). *Typhlops rostellatus* is the most commonly encountered species within the forest. *Typhlops platycephalus,* which is more eurytopic, occurs in the El Verde forest but is less common and possibly restricted to more open situations.

Semifossorial Species

Diploglossus plei (fig. 10.1*D*) is a long-bodied, short-limbed, shiny lizard of moist, wooded habitats, from 50 m to at least 670 m. Even though it is occasionally seen in surface leaf litter, it is most readily found deep in the lower layers of thick drifts of leaf litter or other decomposing vegetation. It is seldom seen on the surface. We do not know what relative contributions to prey capture are made by sight, olfaction, or touch. In the deepest layers in which *D. plei* is found, sight must play a minor role.

Cryptic Species

Arrhyton exiguum (fig. 10.1*I*) is a small snake, usually found beneath rocks, logs, or vegetation piles but occasionally seen moving on the surface during the day. It occurs up to at least 600 m elevation. Like most other snakes its expansible, kinetic jaw structure allows it to engulf prey of greater diameter than its head and body. Recent observations indicate that *Arrhyton exiguum* has venom effective in subduing prey (Thomas and Leal 1993).

The geckos of the genus *Sphaerodactylus* include some of the smallest lizards in the world (Russell and McWhirter 1987). In being largely diurnal they are atypical gekkonids. The eyes are mobile, have temporal foveas, and are capable of forward vision over the slender snout, permitting some degree of binocular overlap of the visual fields (Underwood 1951). Binocular vision appears to be important for their feeding, since they characteristically stalk their prey with the eyes rotated forward (pers. observation).

Sphaerodactylus macrolepis (fig. 10.1*E*) is widespread in Puerto Rico and on other islands of the bank and occurs in habitats ranging from distinctly xeric to wet forest. It is a leaf-litter inhabitant in some lowland areas, undergoes a habitat shift in forested areas, where it becomes more exclusively an inhabitant of thick deposits of vegetation debris and aboveground situations. *Sphaerodactylus macrolepis* has been found 2 to 10 m off the ground associated with bromeliads or under the bark of trees, where its eggs are also frequently found. In the forest it is less commonly found than *S. klauberi;* because of its somewhat different microhabitat, we know nothing of its relative abundance.

Sphaerodactylus klauberi (fig. 10.1*F*) is an inhabitant of interior, upland, moist habitats, where it is a leaf-litter inhabitant and not particularly scansorial. Density estimates of this species in the tabonuco forest are 1 individual m^{-2}. Individuals of *S. klauberi* are seen moving in leaf-litter during the day, but they can also be found, sometimes in high concentrations, in thick litter piles. Both species of this genus choose relatively dry microsites, even in very moist forest. The distribution of *S. klauberi* extends to the highest peaks of Puerto Rico, at elevations of over 1000 m.

Emergent Species

The two large, emergent species, the boa *Epicrates inornatus* (fig. 10.1*G*) and the racer *Alsophis portoricensis* (fig. 10.1*H*), possess the usual ophidian mechanisms for engulfing large prey. Like *A. exiguum, Alsophis portoricensis* is a rear-fanged xenodontine colubrid. The venom of *A. portoricensis* functions to subdue large prey (Thomas and Prieto Hernandez 1985) and to facilitate digestion (Rodríguez-Robles and Thomas 1992). It is an active diurnal snake, abundant in the lower parts of the forest, but it also forages in trees. Individuals of *A. portoricensis* have been observed resting and foraging in the forest canopy 18 to 20 m above ground level (Gillingham and Reagan pers. observation). *Epicrates inornatus,* like other boas, subdues its prey by constriction. It is a nocturnal forager, often foraging above the ground in trees or on rocks.

PREDATION

We have little direct data on what species prey on the nine reptiles. Over the years, we have received reports of unspecified snakes and anoles being fed upon by the red-tailed hawk (*Buteo jamaicensis*) and of small lizards (anoles or *Sphaerodactylus*) being preyed upon by the pearly-eyed thrasher (*Margarops fuscatus*). These and other birds are very likely significant predators on small reptiles.

Introduced mammals—the rats, *Rattus rattus* and *R. norvegicus,* and the feral cat, *Felis cattus*—are found in the forest and will prey on small vertebrates (Stewart and Woolbright, this volume); whether they prey substantially on small, cryptic reptiles is unknown. The mongoose, *Herpestes auropunctatus,* which seems to occur at low population densities in the forest, is infamous for its contribution to the extinction of moderate-sized terrestrial reptiles on some Lesser Antillean islands (Baskin and Williams 1966) and no doubt preys to some extent on reptiles in the tabonuco forest (Willig and Gannon, this volume).

The smaller species of these reptiles are, to an unknown but probably significant extent, preyed upon by other reptiles. Javier Rodríguez observed an adult *Alsophis portoricensis* capture and eat a *Sphaerodactylus nicholsi* (ca. 30 mm total length), which shows that *A. portoricensis* does not disdain very small prey. Larger anoles will eat smaller anoles of their own and other species, and we have observed *Anolis cristatellus* feed on *Sphaerodactylus* on different occasions (Gaa and Leal unpublished). Alfonso Silva observed an *Anolis cristatellus* attempting to eat a juvenile *Alsophis portoricensis*. Amphisbaenids are formidable small predators. Two species, including *A. caeca*, were observed to feed on juveniles of the Hispaniolan colubrid snake, *Antillophis parvifrons*, in captivity (Thomas 1966). *Alsophis portoricensis* is reported to have eaten *Arrhyton exiguum* and *Typhlops* sp. (Schwartz and Henderson 1991).

The forest is rife with predaceous arthropods, which regularly consume small vertebrates (Reagan et al., this volume). The large centipede *Scolopendra alternans*, which reaches a length of over 15 cm, is commonly encountered. Centipedes, including *S. alternans*, are noted for taking prey larger than themselves (Cloudsley-Thompson 1958; Carpenter and Gillingham 1984; pers. observation). The arachnid (Amblypygi), *Phrynus longipes*, which is found in crevices during the day, has been observed feeding on anoles, including an adult male *A. stratulus* (Reagan, this volume). The common scorpion, *Tityus obtusus*, which is mostly found above the ground, has been observed feeding on an *Eleutherodactylus* (Thomas pers. observation) and might well capture *Sphaerodactylus*. Mygalomorph spiders, such as *Cyrtopholis portoricae* and *Avicularia laeta*, are likely predators on small reptiles. The land crab *Epilobocera sinuatifrons*, which is very common in vegetation debris and rock piles, is another potential predator. These land crabs will feed on carrion and may capture live prey (see Covich and McDowell, this volume). Thomas observed a similar land crab in Hispaniola carrying a freshly killed *Amphisbaena innocens*.

FOOD HABITS

Food habit information on nonanoline reptiles of Puerto Rico is limited. Rivero (1978) provides general descriptions of food for various species. Since we have small sample sizes of the nonanoline reptiles from the forest at El Verde, we have supplemented our stomach content data with specimens in systematic collections from other areas of the island outside the forest at El Verde. We collected specimens by searching in the most appropriate microhabitats, principally by turning rocks, logs, and abandoned termite nests and by raking in vegetation debris. Specimens were killed by immersion in ice shortly after being collected and were later fixed in formalin and stomach contents removed for examination. Our data on *Sphaerodactylus macrolepis* are from specimens collected at Cabezas de San Juan, the northeastern extreme of Puerto Rico (Gaa-Ojeda 1983). Specimens of *Diploglossus plei*

Table 10.2. Stomach contents of *Amphisbaena caeca*

Food	Individual Prey	Percentage[a]
ANNELIDA		
Oligochaeta	1	1.4
MOLLUSCA		
Gastropoda		
adults (slugs)	2	2.7
eggs	12	16.2
ARTHROPODA		
Acari	4	5.4
Opiliones	1	1.4
Araneida	1	1.4
Diplopoda	3	4.1
Collembola	1?	1.4
Isoptera		
Nasutitermes	6	8.1
Thysanura	1	1.4
Coleoptera		
adults	5	6.7
larvae	17	22.9
Hymenoptera		
Formicidae		
adults	3	4.1
pupae	12	16.2
Diptera	3	4.1
CHORDATA		
Reptilia		
Anolis	2	2.7
TOTAL	74	

Note: Twenty *Amphisbaena caeca* individuals examined.

[a]Based on numbers of individuals.

came from the forest at El Verde and from other areas of the island, principally the moist limestone forests from the vicinity of Arecibo to the Cambalache Forest Reserve. The prey data for the typhlopids is based principally on examination of scats of specimens from a locality south of Arecibo and other moist lowland forest sites in north-central Puerto Rico.

The specimens that we examined produced information consistent with the sparse literature. In the prey of *Amphisbaena caeca* (table 10.2), beetle larvae constitute the largest single prey category, which is not surprising since most of the *A. caeca* came from humic soil around rotting vegetation. Ants constitute the second largest category. Ten unidentified pupae were in the stomach of one *A. caeca,* and remains of the ants *Cyphomyrmex* sp. and *Pheidole* sp. were found in two others. The gastropod eggs were all from one animal.

The major prey of *Typhlops platycephalus* from two sites in northern

Puerto Rico are ants (68%) of seven species; termites (mostly *Nasutitermes*) are second (17%), with oribatid mites, a few beetle larvae, and a single snail making up the rest of the prey (Thomas unpublished). Those snakes came from pasture and pasture-woods ecotone, and from preliminary data it seems that *T. platycephalus* from forest consumes more rotten-wood insects, such as termites and minute beetle larvae.

Typhlops rostellatus is an inhabitant of forested or upland habitats. Of the three specimens so far examined, 52% of the prey (156 total items) is *Parvitermes*, 28% ants, and 19% minute beetle larvae, of one of the families (Micromalthidae?) eaten by *Amphisbaena caeca*. Some potential prey, possibly those with good defenses, are avoided. For example, *T. platycephalus* eats only a limited subset (six out of twenty-two) of the species of ants available in the pasture habitats where we have studied it.

The most important prey (by number) in the stomachs of the thirty-six (of eighty examined) specimens of *Diploglossus pleei* were stemmiulid millipedes, earwigs, and beetles (table 10.3). All the gastropods were slugs. It may be that *Diploglossus* is to some degree a millipede specialist, since many millipedes exude repugnatorial fluids, some of them very potent quinone- or hydrogen cyanide-based mixtures (Woodring and Blum 1965). Feeding ex-

Table 10.3. Prey of *Diploglossus plei*

Food	Individual Prey	Percentage[a]
MOLLUSCA		
Gastropoda	7	9.8
ARTHROPODA		
Araneida	4	5.6
Chelonethida	1?	1.4
Opiliones	1	1.4
Chilopoda		
Scolopendra	2	2.8
Diplopoda		
Stemmiulidae	17	23.9
Polydesmida	1	1.4
Other	2	2.8
Isoptera	5	7.0
Dermaptera	13	18.3
Coleoptera		
adult	14	19.7
larvae	1	1.4
Diptera (pupa)	1	1.4
Hymenoptera		
Formicidae	1	1.4
Lepidoptera	1	1.4
TOTAL	71	

Note: Fifty *Diploglossus plei* specimens examined.

[a] Based on numbers of individuals.

tensively on millipedes may involve countervailing adaptations. Inchaustegui et al. (1985) also found a high incidence of millipedes among the prey of the Hispaniolan *Diploglossus warreni.* All but one of the beetles consumed were adults, not larvae, in contrast to the higher proportion of beetle larvae in the prey of *Amphisbaena caeca. Diploglossus plei* ate staphylinids, carabids, curculionids, and some that are not identified to family. The snails eaten were minute juveniles. Four specimens of *D. plei* had plant fibers in their stomachs, probably accidentally ingested.

White et al. (1992) examined stomach contents of *Typhlops syntherus, Amphisbaena gonavensis,* and a semifossorial anguid, *Sauresia agasepsoides,* all from a xeric region of Hispaniola. Despite the differences in island and habitat, the similarities are notable. Like *A. caeca, A. gonavensis* consumed a large proportion of beetle larvae. Termites were the major prey of the *Typhlops;* no ants were recorded. The anguid, a much smaller species than *D. plei,* ate a high proportion of pupal, larval, and adult beetles and no millipedes.

We examined stomach contents of thirty-four specimens of *Sphaerodactylus klauberi* and thirty-nine *S. macrolepis* (table 10.4). The *S. klauberi* are from the forest at El Verde and other forested areas of the island; but since we obtained very few *S. macrolepis* in the forest, we present data for a sample of that species from a lowland semixeric site in northeastern Puerto Rico (Gaa-Ojeda 1983). The site for *S. macrolepis* was a more disturbed site (coastal coconut grove) than those for the *S. klauberi.* The prey of both species is what one would expect for small litter inhabitants. Mites are important prey for *S. klauberi* but not for *S. macrolepis.* Collembolans, on the other hand, predominate in the *S. macrolepis* sample. Probably because of the coastal habitat, the prey of *S. macrolepis* includes fewer taxa than *S. klauberi* (thirteen of twenty-four vs. twenty-two of twenty-four). Interestingly, ants were not important prey for either species.

Although most of the specimens of *Arrhyton exiguum* from our collections had no stomach contents, two contained *Anolis pulchellus* and two *Anolis cristatellus.* G. Rogowitz (pers. comm.) observed an *Arrhyton exiguum* in the field eating an *Anolis gundlachi.* Captive specimens have eaten *Anolis, Sphaerodactylus,* and *Eleutherodactylus* (Thomas and Leal 1993; Thomas pers. observation). Schwartz and Henderson (1991) reported finding *Eleutherodactylus* eggs (47%), *Sphaerodactylus* (11%), and *Anolis* (42%) among nineteen prey items in stomachs of *A. exiguum.* Since the prey items are few, and no indication is given of the number of snakes that had eaten *Eleutherodactylus* eggs (one snake might have eaten a clutch of eggs), we cannot conclude that *A. exiguum* is to any extent a dietary specialist on this kind of prey.

The boa, *Epicrates inornatus,* probably feeds mainly on endothermal prey (Rivero 1978), at least the adults. Unfortunately, actual data on the natural

Table 10.4. Prey taxa of *Sphaerodactylus*

Food	*S. klauberi*	*S. macrolepis*
MOLLUSCA		
Gastropoda	3 (1.9)	—
ARTHROPODA		
Isopoda	21 (13.4)	15 (10.3)
Acari	25 (15.9)	1 (0.7)
Opiliones	3 (1.9)	—
Chelonethida	—	4 (2.7)
Araneida	24 (15.3)	11 (7.5)
Diplopoda	3 (1.9)	—
Chilopoda	1 (0.6)	—
Collembola	22 (14.0)	94 (64.4)
Orthoptera	1 (0.6)	5 (3.4)
Hemiptera	—	1 (0.7)
Homoptera	5 (3.2)	—
Psocoptera	1 (0.6)	2 (1.4)
Isoptera	6 (3.8)	2 (1.4)
Thysanoptera	1 (0.6)	2 (1.4)
Coleoptera		
adults	9 (5.7)	—
larvae	10 (6.4)	—
Hymenoptera		
Formicidae	2 (1.3)	3 (2.1)
Diptera	9 (5.7)	6 (4.1)
Trichoptera?	2 (1.3)	—
Lepidoptera		
adult	1 (0.6)	—
larvae (geometrid)	7 (4.5)	—
CHORDATA		
Amphibia		
Eleutherodactylus	1 (0.6)	—
TOTAL	157	146

Source: S. macrolepis data from Gaa-Ojeda (1983), based on specimens from a lowland locality.

Notes: Data from thirty-four *S. klauberi* and thirty-nine *S. macrolepis* individuals. Percentages (in parenthesis) are based on numbers of individual prey.

prey of this not uncommon species is very spotty. It has been reported to feed on bats (Rodríguez and Reagan 1984) and on the introduced *Rattus rattus* (Reagan 1984). The young of *E. inornatus,* at least, will eat *Anolis* and *Eleutherodactylus* in captivity (Reagan 1984), and one ate a *Bufo marinus,* which incidentally proved fatal to the snake (J. Wunderle pers. comm.). A captive adult constricted and ate a large *Ameiva exsul* (M. Leal pers. comm.). The very similar *Epicrates striatus* of Hispaniola feeds on mammals (*Rattus rattus, Mus musculus*), birds, and lizards (*Anolis*) with a dietary transition from lizards to endotherms occurring from 60 to 80 cm snout-vent length (Henderson et al. 1987). As Henderson and Crother (1989) observed, the large *Epicrates* have almost certainly undergone dietary change with the extinction of the pre-Columbian, Greater Antillean, nonflying mammal fauna. (Puerto Rico had at least five native rodents and a large insectivore; Woods 1989a,b; also Willig and Gannon, this volume).

Alsophis portoricensis feeds readily on lizards (*Anolis* and *Ameiva*) in captivity; it will also accept mice, although its venom seems to be less effective on mammals (Thomas and Prieto Hernandez 1985). M. Leal (pers. comm.) found a mouse in the stomach of a wild-caught individual. Woolbright and Stewart (this volume) note an instance of predation on *Eleutherodactylus*, and Javier Rodríguez (pers. comm.) observed one to capture and eat a *Sphaerodactylus nicholsi* in the field. *Alsophis portoricensis* thus appears to be an opportunistic predator. Species of *Anolis* and *Eleutherodactylus* probably form the major food of this snake in the El Verde area. Schwartz and Henderson (1991) reported that of thirty-two prey items from stomach contents of *A. portoricensis*, 9.4% was *Eleutherodactylus*, 9.4% *Sphaerodactylus*, 62.5% *Anolis*, and 15.6% *Ameiva*. Other prey consumed included *Arrhyton exiguum*, *Typhlops* sp., *Hemidactylus* sp., and *Iguana iguana* (from a Virgin Islands snake).

POPULATION AND BIOMASS

We did not estimate the numbers or biomass of any of the species. The more cryptic and fossorial species (*Diploglossus plei*, *Amphisbaena caeca*, and the two species of *Typhlops*) appear to be abundant in scattered patches of suitable habitat: areas of thick vegetation debris, around tree bases, in root networks, and rock jumbles—situations extremely difficult to sample. We tried some random sampling for these species and conclude that to get any usable results will require many workers, a long time, or both and would also have to involve destructive sampling.

Although often perceived as rare because of their fossorial habits, species of *Amphisbaena* and *Typhlops* are very common in some situations. In doline pasture-edge habitats, it is possible for one person to find as many as twenty specimens of *Typhlops platycephalus* in an hour by turning rocks. In some areas *Amphisbaena caeca* is much more commonly found than in tabonuco forest at El Verde. Species of *Typhlops* and *Amphisbaena* have been found in large numbers (hundreds in a few days' time) by collectors in Hispaniola and are often extremely abundant in cultivated fields and in semixeric woods (Thomas pers. observation), although there are no data that can be converted to individuals per unit of area. That we do not find the fossorial and semifossorial reptiles in such apparent abundance in the El Verde forest (and generally in mesic forests throughout the West Indies) may be deceptive, since in open habitats rocks and logs concentrate these kinds of animals, apparently by offering protection from high temperatures.

Sphaerodactylus klauberi appears fairly uniformly distributed in the leaf litter. Pfeiffer (chap. 5, this volume) found that *S. klauberi* was the most common reptile in leaf litter samples, with an estimated density of 1 individual m^{-2}. The other species that we discuss here are much less regularly encountered. Since *Epicrates inornatus* is both partly arboreal and nocturnal, it is difficult to evaluate its abundance, except that it is probably more abundant than its classification as an endangered species would indicate.

SUMMARY

The nine nonanoline reptiles of the tabonuco forest include three lizards, two small diurnal geckos (*Sphaerodactylus klauberi* and *S. macrolepis*), a semi-fossorial anguid (*Diploglossus plei*), a limbless burrower (*Amphisbaena caeca*), two subsurface blind snakes (*Typhlops platycephalus* and *T. rostellatus*), two colubrid snakes (*Alsophis portoricensis* and *Arrhyton exiguum*), and a boa (*Epicrates inornatus*). In general the feeding of these species ranges from the subsurface humus and soil (the anguid, the amphisbaenian, and the two blind snakes) to the leaf litter (the two geckos and one colubrid snake), the ground surface, and the arboreal zone (the other colubrid and the boa). Of course there is some overlap in the foraging area of some of these species, but considering their relative abundances, their greatest feeding impact is on the soil and leaf-litter arthropods. Secondarily, their impact is on the *Anolis* lizards and thirdly on other vertebrates.

All of these nine reptiles found in the tabonuco forest are to some extent opportunistic predators within the limits of their sizes, jaw structures, activity periods, and microhabitats. Nevertheless, the higher frequency of certain prey indicates some degree of dietary specialization. The two blind snakes (*Typhlops*) are the most specialized feeders among the species we discuss. Ants and termites strongly predominate in their diets, but they also eat a variety of small invertebrates from the soil and rotting vegetation. *Amphisbaena caeca* also preys principally on species from soil and decaying vegetation but is able to eat larger prey than the blind snakes. Millipedes are the most prevalent prey of *Diploglossus plei,* but a variety of litter invertebrates is also eaten. The two species of *Sphaerodactylus* feed principally on small litter arthropods. The two snakes *Alsophis portoricensis* and *Arrhyton exiguum* are probably generalized predators that feed on both cryptic and surface species; the former also forages arboreally. The boa *Epicrates inornatus* appears also to be a generalist that, as it attains large size, restricts its prey more to birds and mammals.

These nine reptiles can be roughly ranked by apparent prey specialization: *Typhlops rostellatus* (principally termites), *Typhlops platycephalus* (ants and termites), *Diploglossus plei* (millipedes), *Alsophis portoricensis* (lizards), *Arrhyton exiguum* (lizards), *Epicrates inornatus* (ectothermic vertebrates), *Amphisbaena caeca* (larval coleoptera) *Sphaerodactylus macrolepis* (collembollans), and *S. klauberi* (no strongly modal prey). True feeding specialization should involve specific adaptations for certain prey. Thus, myrmecophages such as *Typhlops* have evolved not only extreme jaw modifications but behavioral specializations and countermeasures against the prey defenses. A species adapted to eat prey of a limited size range may appear specialized, if that size range happens to include a common species. The abundance of collembollans in the diet of *Sphaerodactylus macrolepis* may be due to this factor. The same is probably true for the abundance of coleopteran larvae in the diet of *Amphisbaena caeca*. In order to determine feeding specificity, experiments in prey choice should be done.

In conclusion, the actual impact of the nine nonanoline reptiles is unknown, since we do not have estimates of their biomasses. However, six of the species (*Amphisbaena caeca, Diploglossus plei, Sphaerodactylus klauberi, Sphaerodactylus macrolepis, Typhlops platycephalus,* and *Typhlops rostellatus*) we estimate to be important in the subsurface component of the food web. The other three (*Alsophis portoricensis, Arrhyton exiguum,* and *Epicrates inornatus*) prey on small vertebrates, which are at higher tropic levels. As Reagan has noted (this volume), the anoles constitute an important component of the aboveground food web, and these three snakes are predators largely within that component and largely on the anoles.

ACKNOWLEDGMENTS

We thank John Calderón, Ariel Díaz, Luis Olmedo, Javier Rodríguez, and Manuel Vélez for help in collecting and identifying invertebrates. Manuel Leal offered criticisms of the manuscript. The photographs in figure 10.1 were taken by S. Blair Hedges (*A, B, C, D, E, F, I*), Javier Rodríguez (*H*) and Richard Thomas (*G*). We thank the Department of Natural Resources of Puerto Rico for permits to collect specimens.

11

Birds

Robert B. Waide

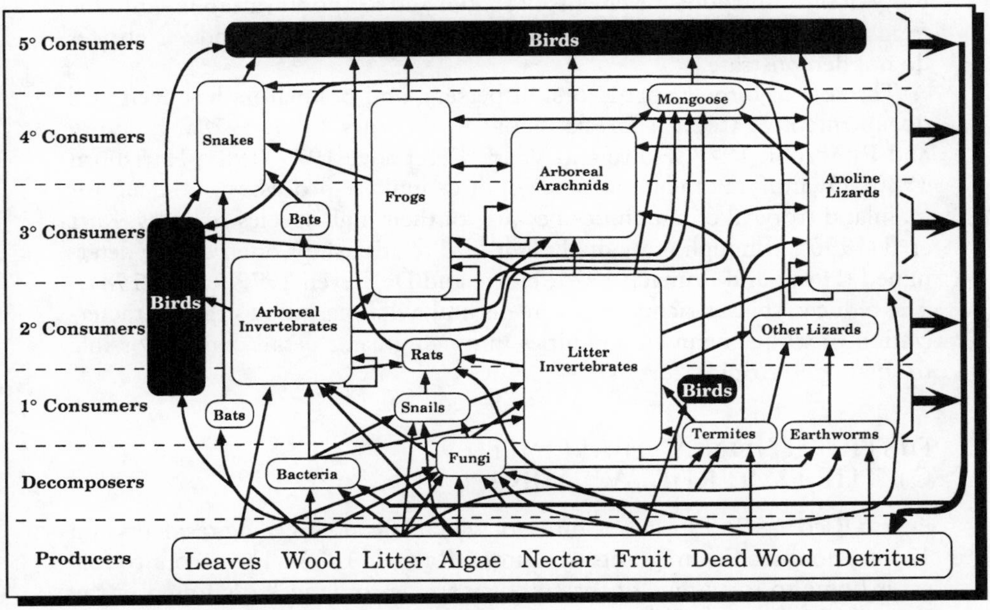

BIRDS exert a strong influence at multiple points within the El Verde food web (see diagram above). Birds are top predators in many of the food chains in the web and have the potential to influence population dynamics of their prey, including frogs and lizards. Along with bats, birds are the only native dispersers of fruit. Because some birds feed from many trophic levels, they have the potential to alter the structure of the food web by seasonal diet switching. Along with the flexibility to vary diet, the mobility of birds introduces elements of spatial and temporal variability into the food web that more specialized and sedentary invertebrates and vertebrates do not demonstrate.

The role of birds as predators, dispersers, and pollinators has been well documented for tropical forests (Snow 1965; Stiles 1975, 1978a, b; Howe and Estabrook 1977; Howe and Vande Kerckhove 1979, 1981; Hartshorn 1980). Quantifying the importance of birds in these processes is difficult for mainland tropical communities because of their high species richness (Karr et al. 1990), although individual plant-bird relationships often can be determined (Howe and Primack 1975; Howe and De Steven 1979; Howe 1981). The reduced species richness of some island avifaunas allows the characterization of whole avian communities in much greater detail than is possible for mainland tropical forests.

DISTRIBUTION AND AFFINITIES OF THE EL VERDE AVIFAUNA

Puerto Rico has 269 species of birds, of which 106 are breeding residents and 126 are nonbreeding migrants or visitors (Raffaele 1983). The tabonuco forest is home to forty-nine of the sixty-six species of land birds found in the Luquillo Mountains (Wiley and Bauer 1985), forty (thirty-one resident, nine migrant) of which have been recorded at El Verde (table 11.1). Of these, thirty are found regularly in the forest surrounding the El Verde Field Station and are considered to constitute the avian component of the tabonuco forest food web.

These thirty common species exclude species rarely seen at El Verde (e.g., elfin woods warbler, *Dendroica angelae;* sharp-shinned hawk, *Accipiter striatus;* broad-winged hawk, *Buteo platypterus;* ovenbird, *Seiurus aurocapillus;* black-cowled oriole, *Icterus dominicensis*) and species that are confined to water (spotted sandpiper, *Actitis macularia;* belted kingfisher, *Ceryle alcyon*), the air (black swift, *Cypseloides niger*), or man-made clearings or roadsides (loggerhead kingbird, *Tyrannus caudifasciatus;* gray kingbird, *T. dominicensis*). Likewise the list includes none of the thirty-six bird species introduced to Puerto Rico because few of these exotics are found in the Luquillo Experimental Forest (LEF) and then in low numbers.

The thirty common species at El Verde occur in eight orders, each represented by a single family, and the Passeriformes, which comprises four families (one of which, the Emberizidae, has five subfamilies). Most families or subfamilies are represented by only one or two species, the exceptions being the Thraupinae (tanagers) and Columbidae (pigeons and doves) with three species each and the Parulinae (warblers) with eight species (all migrants).

Of the thirty common species constituting the El Verde avifauna, eight (27%) are endemic to Puerto Rico, and two others (7%) are found only in Puerto Rico and the Virgin Islands (Raffaele 1983). One species (3%) has a Lesser Antillean distribution, three (10%) reach the other islands of the Greater Antilles and the Bahamas, and one is widespread throughout the West Indies. Seven species (23%) have distributions that include the North, Central, or South American mainland, and eight species are winter migrants from North America. Thus, half (15/30 = 50%) of the avifauna at El Verde and two-thirds (15/22 = 68%) of the breeding avifauna are West Indian endemics.

Two additional species that probably occupied tabonuco forest have become extinct within the last hundred years. An endemic parakeet (*Aratinga maugei*) became extinct in the late nineteenth century, and the last observation of the white-necked crow (*Corvus leucognaphalus*) occurred at El Yunque in the LEF in 1963 (Raffaele 1983). The endangered Puerto Rican parrot, once reduced to only thirteen birds in the wild, reached a population of about forty-nine individuals before Hurricane Hugo, but it has not been seen in the El Verde area since 1981. Three fossil species (an endemic barn-owl, *Tyto cavatica;* a quail-dove, *Geotrygon larva;* and the Cuban crow, *Corvus nasicus*) may once have occurred at El Verde (Raffaele 1983).

Comparison of Avifaunas on Tropical Islands and Mainlands

The most striking difference between island and mainland avifaunas is the great reduction in species richness on islands (Lack 1976). Many Neotropical bird families are poor over-water colonizers and hence are absent from the islands of the West Indies. To the extent that studies have focused on the

Table 11.1. Diets of birds at El Verde

Taxon	Trophic Level[d]	Principal Food[e]	Foraging Location
Order: Falconiformes			
Family: Accipitridae			
red-tailed hawk[a] (*Buteo jamaicensis*)	3	rats, birds	canopy
broad-winged hawk (*Buteo platypterus*)	2, 3	reptiles, birds, invertebrates	canopy
sharp-shinned hawk (*Accipiter striatus*)	3	birds	canopy
Order: Charadriiformes			
Family: Scolopacidae			
spotted sandpiper (*Actitis macularia*)	2	crustaceans, insects	streams
Order: Columbiformes			
Family: Columbidae			
scaly-naped pigeon[a] (*Columba squamosa*)	1	fruit	canopy
zenaida dove[a] (*Zenaida aurita*)	1	seeds	litter
ruddy quail-dove[a] (*Geotrygon montana*)	1	seeds	litter
Order: Psittaciformes			
Family: Psittacidae			
Puerto Rican parrot[a] (*Amazona vittata*)	1	fruit	canopy
Order: Cuculiformes			
Family: Cuculidae			
Puerto Rican lizard-cuckoo[a] (*Saurothera vieilloti*)	2, 3	lizards, spiders, Lepidoptera, insects	understory
Order: Strigiformes			
Family: Strigidae			
Puerto Rican screech-owl[a] (*Otus nudipes*)	2, 3	insects, lizards, birds	understory
Order: Apodiformes			
Family: Apodidae			
black swift (*Cypseloides niger*)	2	flies, beetles	open air
Family: Trochilidae			
Puerto Rican emerald[a] (*Chlorostilbon maugaeus*)	1, 2	nectar, spiders, insects	understory
green mango[a] (*Anthracothorax viridis*)	1, 2	nectar, Diptera, spiders	understory

Order: Coraciiformes			
Family: Alcedinidae			
belted kingfisher (*Ceryle alcyon*)[b]	3	fish, shrimp	streams
Family: Todidae			
Puerto Rican tody[a] (*Todus mexicanus*)	2, 3	Diptera, Coleoptera	understory
Order: Piciformes			
Family: Picidae			
Puerto Rican woodpecker[a] (*Melanerpes portoricensis*)	1, 2	insects, fruits	canopy
Order: Passeriformes			
Family: Tyrannidae			
Puerto Rican flycatcher[a] (*Myiarchus antillarum*)	1, 2	Hemiptera, coleoptera, caterpillars, fruit	canopy
loggerhead kingbird (*Tyrannus caudifasciatus*)	1, 2, 3	fruit, insects, reptiles, amphibians	clearings
gray kingbird (*Tyrannus dominicensis*)	1, 2	fruit, insects	clearings
Family: Muscicapidae			
red-legged thrush[a] (*Turdus plumbeus*)	1, 2, 3	fruit, caterpillars, lizards	litter
Family: Mimidae			
pearly-eyed thrasher[a] (*Margarops fuscatus*)	1	fruit, vertebrates, invertebrates	canopy
Family: Vireonidae			
Puerto Rican vireo[a,c] (*Vireo latimeri*)	1, 2	Orthoptera, Homoptera, Lepidoptera, fruit	canopy
black-whiskered vireo[a] (*V. altiloquus*)	1, 2	fruit, insects	canopy
Family: Emberizidae			
Sub family: Parulinae			
black-and-white warbler[a,b] (*Mniotilta varia*)	2	Coleoptera, Orthoptera	understory
parula warbler[a,b] (*Parula americana*)	2, 3	spiders, Homoptera, Coleoptera	understory
Cape May warbler[a,b] (*Dendroica tigrina*)	1, 2	nectar, insects	understory
black-throated blue warbler[a,b] (*D. caerulescens*)	1, 2, 3	insects, fruit, spiders	understory
black-throated green warbler[a,b] (*D. virens*)	2	insects	canopy
prairie warbler[a,b] (*D. discolor*)	2, 3	fulgoroids, spiders	understory
elfin woods warbler (*D. angelae*)	2	insects	canopy
ovenbird[b] (*Seiurus aurocapillus*)	1, 2	snails, insects, seeds	litter
Louisiana waterthrush[a,b] (*S. motacilla*)	2	insects	litter
American redstart[a,b] (*Setophaga ruticilla*)	2	fulgoroids, Diptera	understory
Sub family: Coerebinae			
bananaquit[a] (*Coereba flaveola*)	1, 2	nectar, caterpillars, spiders	understory

(*continues*)

Table 11.1. (*continued*)

Taxon	Trophic Level[d]	Principal Food[e]	Foraging Location
Sub family: Thraupinae			
Antillean euphonia[a] (*Euphonia musica*)	1	mistletoe	canopy
stripe-headed tanager[a] (*Spindalis zena*)	1	fruit	canopy
Puerto Rican tanager[a] (*Nesospingus speculiferus*)	1, 2	caterpillars, fruit	understory, canopy
Sub family: Icterinae			
black-cowled oriole (*Icterus dominicensis*)	2, 3	Coleoptera, spiders, earwigs	canopy
Sub family: Emberizinae			
Puerto Rican bullfinch[a] (*Loxigilla portoricensis*)	1, 2	fruit, insects	understory, canopy
black-faced grassquit[a] (*Tiaris bicolor*)	1	seeds	openings

Sources: Names correspond to nomenclature used in the Checklist of American Birds (AOU 1983). Data on principal foods from Wetmore (1916) modified by results from Waide (unpublished), Wiley (pers. comm.), and Arendt (pers. comm.). Foraging locations are from Raffaele (1983), Wunderle et al. (1987), Waide (unpublished), Wiley (pers. comm.), and Arendt (pers. comm.).

[a] Species regularly found at El Verde.

[b] Breeds in North America, winters in Puerto Rico.

[c] Breeds in Puerto Rico, winters in South America.

[d] 1, 2, 3 = primary, secondary, and tertiary consumers, respectively. Species are included in a trophic level if at least 10% of their diet comes from the next lowest trophic level.

[e] In order of importance. Major invertebrate groups in the diet are listed when possible.

reduced number of species on islands, functional similarities between island and mainland avifaunas have not received the attention they deserve. For example, island and mainland communities support similar avian biomass, but the distribution of biomass among guilds is quite different (see below). Comparison of the complexity of mainland communities with the relative simplicity of island communities provides insight into the variety of functional ecosystem configurations that is possible.

Factors Associated with Reduced Species Numbers on Islands

As MacArthur et al. (1972) point out, islands are colonized by a nonrandom subset of mainland avifaunas. Many widespread and diverse Neotropical families, including the Momotidae (motmots), Galbulidae (jacamars), Bucconidae (puff birds), Capitonidae (barbets), Ramphastidae (toucans), Dendrocolaptidae (woodcreepers), Furnariidae (ovenbirds), and Pipridae (manakins) are absent not only from Puerto Rico but from the entire West Indies. The absence of these groups from oceanic islands is apparently because of their poor colonizing ability; the effect is to reduce greatly the size of the potential pool of colonists for insular bird communities.

Despite the reduced species pool for the West Indies, Terborgh and Faaborg (1980) found that insular bird communities in humid lower montane forest and coastal sclerophyll scrub were saturated; that is, as the available species pool increased on larger islands, the number of breeding species found in these habitats reached an asymptote. An upper limit of about twenty coexisting species was found in forests on islands ranging in size from Hispaniola (76,000 km^2) to Montserrat (85 km^2). This generalization also applies to El Verde, where twenty-two resident species occur within tabonuco forest. Increased numbers of species on larger islands were a result of increased habitat diversity rather than a product of enhanced within-habitat species richness.

In one sense, therefore, complex mainland and simple island forest communities are functionally equivalent in that they are resistant to colonization. Terborgh and Faaborg (1980) suggest that the difference in the number of species required to saturate island and mainland habitats is the result of a difference in the upper limits of competitive ability in the two kinds of communities. They postulate that in relatively competitor-free insular environments, birds become adapted to exploit a wider range of habitats at the expense of competitive ability within any one habitat. Species packing is less on islands, but at least on larger islands, insular bird communities are as resistant to invasion as are continental communities.

The reduction of species number in West Indian avifaunas is thus attributable to two factors, one extrinsic and one intrinsic to the region. The poor colonizing abilities of many continental bird groups reduces the available pool from which insular communities are drawn. Those species that do become established on islands become habitat generalists and use the resources that might support several species under more competitive conditions. The high rate of natural disturbance in Caribbean islands (Walker et al. 1991) probably makes a generalist strategy even more adaptive.

ANNUAL AND SEASONAL CHANGES IN AVIAN POPULATIONS

The avifauna of El Verde experiences both annual and seasonal fluctuations in population density that are related to patterns of reproduction and mortality and to long-distance migration. Annual changes in density estimates may reflect long-term trends, but few tropical sites have sufficiently lengthy records to detect such tendencies (but see Faaborg 1982; Faaborg et al. 1984; Karr et al. 1990). Seasonal changes in density estimates may reflect fluctuations in actual numbers or seasonal cycles in behavior such as breeding, dispersal, or migration (fig. 11.1). A thorough understanding of temporal changes in populations is necessary to appreciate the dynamic nature of the food web.

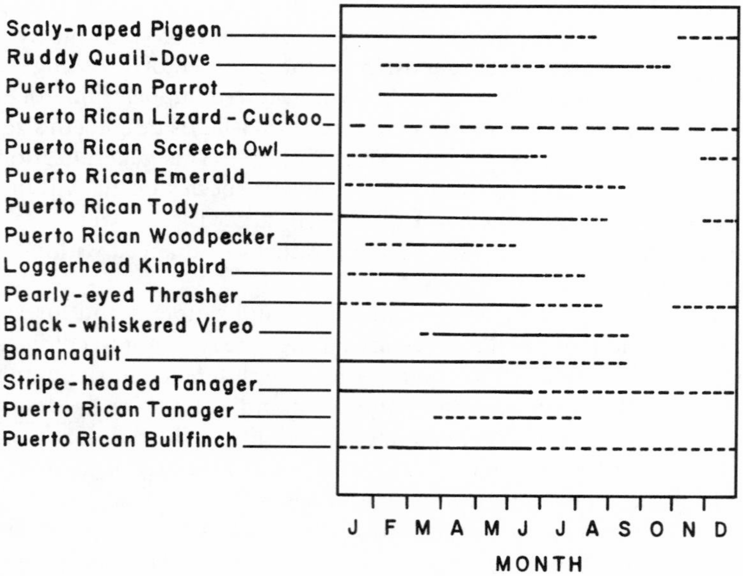

Figure 11.1. Breeding seasons of birds in the Luquillo Experimental Forest. Principal nesting periods are indicated by a solid line, sporadic nesting by a dotted line, and species for which there are no data by a dashed line. Data are from Recher (1970) modified with information from J. Wunderle (pers. comm.) and W. Arendt (pers. comm.).

Figure 11.2. Bananaquit (*Coereba flaveola*). Photo by José Colon.

Relative Abundance

Knowledge of relative abundance is influenced by the survey methodology employed, for which reason multiple methods are often used in community studies. Spot maps, transect counts, and mist-netting are frequently used techniques. Specific details of methods used to determine relative and absolute abundances are discussed in appendix 11. Abundance is expressed in individuals ha^{-1}, which is the norm for birds; for comparison with other groups, see appendix 14.B.

Spot Maps

The bananaquit (fig. 11.2) is the most widespread and abundant of Puerto Rican birds, constituting over 50% of the individuals in some surveys in the LEF (table 11.2; Recher 1970; Kepler and Kepler 1970; Wunderle et al. 1987). Black-whiskered vireos, scaly-naped pigeons, and Puerto Rican todies were the next most abundant species in descending order in spot-map counts taken during the breeding season.

Mist Nets

The ruddy quail-dove was the most abundant species in ground-level mist net samples from closed-canopy forest at El Verde (table 11.3). Ruddy quail-

Table 11.2. Population densities at El Verde

	Spot Counts					Transects	
	1964	1965	1966	1981	1982	1981	1982
red-tailed hawk	—	—	—	0.2	—	—	—
Puerto Rican emerald	0.4	0.4	0.4	0.4	0.2	0.4	1.9
ruddy quail-dove	0.6	0.5	0.4	1.1	1.1	2.0	0.3
scaly-naped pigeon	0.4	0.3	0.3	2.0	2.9	4.2	5.2
Puerto Rican lizard-cuckoo	—	—	—	—	0.2	0.2	—
Puerto Rican tody	1.9	2.0	2.2	0.6	2.2	2.2	2.7
Puerto Rican woodpecker	—	—	—	0.4	0.4	0.4	0.4
Puerto Rican flycatcher	—	—	—	0.2	—	0.2	—
pearly-eyed thrasher	1.2	1.6	0.6	0.2	0.2	0.6	0.8
red-legged thrush	—	—	—	—	0.2	0.1	—
bananaquit	13.3	12.0	8.8	3.9	5.0	6.6	13.2
black-whiskered vireo	3.3	3.0	1.8	2.3	4.8	5.2	11.6
Puerto Rican tanager	0.7	1.0	0.8	0.4	1.3	2.2	4.7
stripe-headed tanager	0.2	0.2	0.2	0.1	0.9	0.3	0.9
Puerto Rican bullfinch	0.6	0.8	0.2	0.3	0.7	0.2	0.4

Sources: Data from 1964–1966 are from Recher (1970). Data from 1981–1982 are from Reagan et al. (1982).

Notes: Densities expressed as individuals ha^{-1}. Spot counts in 1964–1966 were carried out in April and May. Data from 1981 were obtained in June and July and from 1982 in April and May.

Table 11.3. Bird captures at El Verde

	Plot 1	Plot 2	Plot 3	Plot 4	Total
ruddy quail-dove	9	16	16	3	44
Puerto Rican tody	0	2	7	1	10
Puerto Rican emerald	2	2	2	2	8
Puerto Rican tanager	0	6	0	0	6
bananaquit	3	2	0	1	6
pearly-eyed thrasher	1	2	2	0	5
black-throated blue warbler	0	0	0	2	2
ovenbird	0	0	1	1	2
stripe-headed tanager	0	1	0	0	1
red-legged thrush	0	0	0	1	1
Total captures	15	31	28	11	85
Number of nets	11	10	10	10	
Number of days	3	3	4	3	
Captures/net day	0.45	1.03	0.70	0.37	0.64

Source: Data are from Reagan et al. (1982).

Note: Each plot 1 ha (see chap. 1).

doves were also predominant in mist net samples from rain forest in Dominica but not in Guadeloupe (Terborgh et al. 1978).

The rank abundance distribution of mist net captures from El Verde (fig. 11.3) shows clearly the dominance of the quail-dove. The quail-dove is followed in rank by the bananaquit, Puerto Rican emerald, Puerto Rican tody, Puerto Rican tanager, and pearly-eyed thrasher. The remaining eight

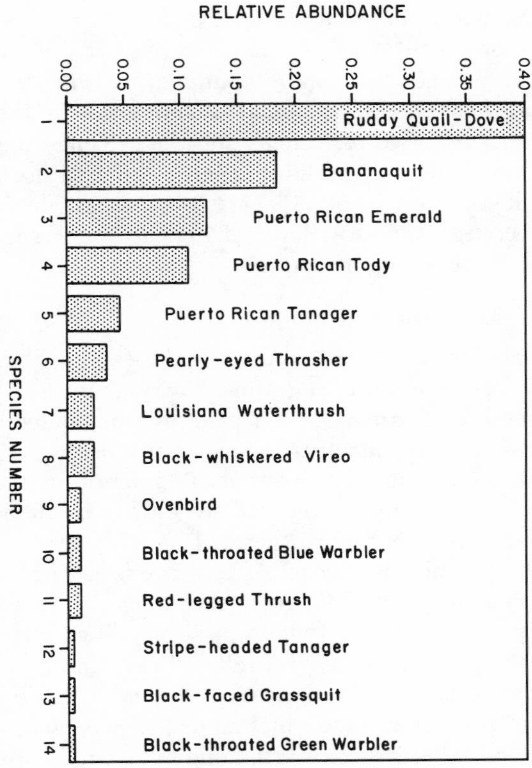

Figure 11.3. Relative abundance of species caught in mist nets at El Verde. Data are from table 11.3 and Wunderle et al. (1987).

species in the sample (57%) were rare (less than 2% of the sample; Karr 1971). Mist net samples from mainland tropical forests have a much larger proportion of rare species (85% from Pipe Line road in Panama, Karr 1990; 90% from the central Amazon, Bierregaard 1990). The relatively small sample size from El Verde (168 individuals) may partially explain the relatively low proportion of rare species; as sample size increases, more rare species are likely to be added. If all the species at El Verde that have the potential to be captured by mist nets are included, 82% of the species would be considered rare.

Annual Changes in Abundance

A comparison of spot-map counts from 1964 to 1966 (Recher 1970) with spot-map counts and transect censuses from 1981 shows that four (Puerto Rican emerald, Puerto Rican tody, bananaquit, Puerto Rican bullfinch) of ten species did not change in abundance. Five species did not appear in

Recher's censuses but occurred in my 1981 counts. Five other species (scaly-naped pigeon, ruddy quail-dove, black-whiskered vireo, stripe-headed tanager, Puerto Rican tanager) were more abundant in 1981 censuses, and one (pearly-eyed thrasher) was less abundant. Differences in abundance of the Puerto Rican tanager between Recher's and the present study are probably due to the species' habit of travelling in large foraging flocks, resulting in extreme local fluctuations in abundance. Increases or decreases in calculated densities between 1964 to 1966 and 1981 for other species reflect real changes in abundance during this period.

Seasonal Changes in Abundance

Densities calculated from transect counts (fig. 11.4) demonstrate a variety of patterns that reflect seasonal changes in activity as well as fluctuations in number. In general, birds are more conspicuous when defending territories, feeding fledglings, or foraging in post-reproductive flocks and less conspicuous during periods of incubation and molt. Populations are lowest immediately before breeding and highest immediately after. Combinations of these factors result in the patterns shown in fig. 11.4. Differences among species are the result of differences in activity pattern and breeding behavior.

During July to September 1980, eight of thirteen species had their lowest densities of the year. This period follows peak breeding for the avifauna as a whole (March to June; fig. 11.1) and reflects a diminution of singing and territorial defense and reduced activity related to molt (Waide unpublished). Seven of thirteen species reach their highest densities between December and March. This period reflects increased singing associated with territorial defense and the onset of reproductive activities. Exceptions to this general pattern usually result from differences in breeding behavior. Ruddy quail-dove density is highest in June to August (Recher 1970) during the main breeding period for this species. Bananaquits and Puerto Rican todies show less pronounced peaks extending over much of the year resulting from year-round singing and nest-building in the former (although breeding itself is seasonal; J. Wunderle pers. comm.) and year-round territoriality in the latter (Kepler 1977). The scaly-naped pigeon shows two peaks, the first (November to December) associated with aggregations of foraging birds in the plots and the second (March to June) with singing males. The density of pearly-eyed thrashers shows a small peak in October, probably as a result of breeding activity in older pairs, which commence nesting from October to December (W. Arendt pers. comm.). The bulk of the population begins to breed in March, as suggested by the rise in observations in that month.

Response to Disturbance

The pattern of chronic disturbance resulting from frequent hurricanes has a strong effect on the composition of the avian community at El Verde. Both direct mortality from high winds and rainfall and post-hurricane conditions

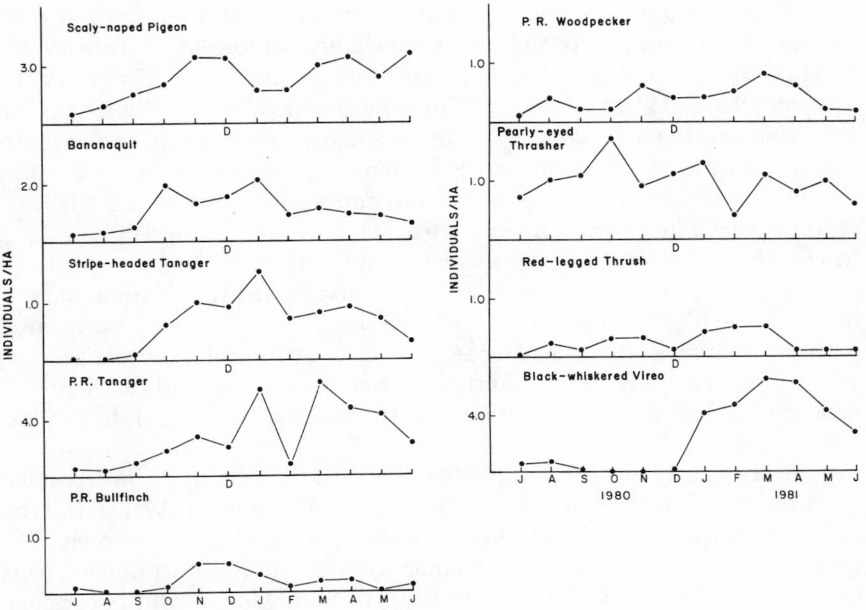

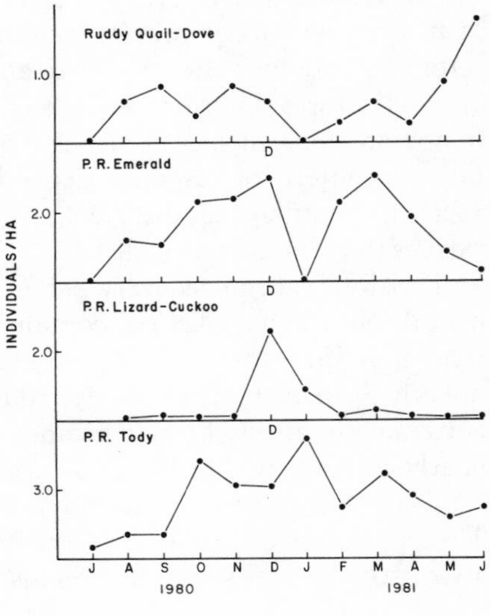

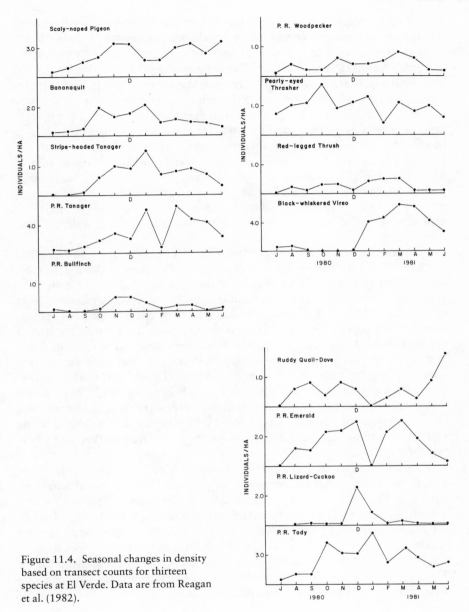

Figure 11.4. Seasonal changes in density based on transect counts for thirteen species at El Verde. Data are from Reagan et al. (1982).

contribute to changes in community structure. Hurricanes cause short-term disruptions in food supply and long-term changes in forest structure (Waide 1991a). Studies from St. John, U.S. Virgin Islands (Askins and Ewert 1991), El Verde (Waide 1991b), Jamaica (Wunderle et al. 1992), and Mexico (Lynch 1991) indicate that frugivores, granivores, and nectarivores suffer more short-term effects and recover more slowly from hurricanes than do insectivores or omnivores (Waide 1991a). Flowers, fruit, and seeds are stripped from vegetation during a hurricane, and new stocks take weeks or months to produce. In contrast, some insects may increase rapidly in response to a hurricane (Torres 1992), and small vertebrate prey survive with populations largely intact (Reagan 1991, Woolbright 1991). Avian primary consumers are faced with a temporary loss of resources, resulting in mortality or emigration, while predators may find conditions better because of outbreaks or improved visibility of prey. Major shifts in food web structure can thus follow a hurricane.

Longer-term effects are less predictable. At El Verde, most bird populations recovered to pre-hurricane levels within nine months (Waide 1991b), although some species (i.e., ruddy quail-dove) responded more slowly. The avian community approached its pre-hurricane composition within five years of Hurricane Hugo (Waide unpublished). In both Mexico and Nicaragua, species richness and bird abundance recovered within seventeen months of the hurricanes. However, both sites exhibited marked and potentially long-lasting changes in community composition. Frugivores and nectarivores were reduced in number after the hurricane in Mexico (Lynch 1991), whereas species of the forest interior were less common in Nicaragua (Will 1991). North temperate migrant species also responded to secondary effects of hurricanes, but the pattern of response depended on the species and habitat (Waide 1991a). Overall, generalists (e.g., American redstart) are less affected by hurricanes than habitat specialists.

The difference in recovery patterns between island (El Verde) and continental (Nicaragua, Mexico) communities parallels the gradient of hurricane frequency from east to west in the Caribbean. Continental sites are less frequently disturbed by hurricanes, and their avifaunas recover slowly. El Verde is frequently affected by hurricanes, and the avifauna recovers quickly and nearly completely. The broad diets and habitat requirements of species at El Verde (see below) may be adaptive under conditions of frequent disturbance. The pattern of disturbance in the West Indies may reinforce the tendency away from habitat specialization discussed above.

COMMUNITY STRUCTURE

Biomass

The reduced species richness of insular bird communities often is accompanied by increased densities for particular species (Crowell 1962; Grant

1966a,b). As a result, species-poor islands may have an equivalent or greater density of individuals than nearby continental sites (Terborgh and Faaborg 1980). This phenomenon, termed density compensation, counteracts two trends (the tendency for species numbers to be reduced on islands and the tendency for heavier species to be poor colonists; MacArthur et al. 1972) that otherwise might act to reduce the total biomass of insular bird communities.

No consistent difference in standing crop biomass of resident birds between islands and continents exists for the few tropical and subtropical sites for which data are available (table 11.4). Islands tend to have higher densities of birds and lower mean weights per individual than do continents. Puerto Rico has a relatively high biomass compared to both islands and continental sites, principally because of the predominance of pigeons and doves in the community.

The weakness of using biomass to compare different communities is that most estimates of biomass are made during the breeding season. In some areas, the composition of breeding and wintering communities is quite different as a result of the presence of migrant species. In some tropical continental (Waide 1981) and subtropical island (Emlen 1977) sites, the large increase in particular guilds during the winter period without an accompanying increase in food supply suggests that breeding density is considerably below any limit that might be imposed by food. Hence, comparison of communities that differ in the importance of winter migrants is risky unless biomass estimates are available for the nonbreeding season as well. Unfortunately, great difficulties are inherent in estimating the density and community biomass of nonterritorial species.

There is a strong negative correlation ($r = -.94$; $p = .02$) between the biomass of resident communities and the proportion of migrants in samples from the same communities (table 11.4). The sample size is small (only five of the studies give data on the proportion of migrants), and interpretation of the correlation is confounded by differences in latitude and habitat. If the correlation represents a causal relationship between migrant numbers and breeding season biomass, it suggests that the biomass of wintering communities may be more similar among sites than the breeding season biomass.

Guild Signatures

The distribution of avian species or biomass among feeding guilds produces a signature characteristic of each avian community. In complex communities, species distribution among guilds is commonly used to determine the guild signature for the community (table 11.5). Because data on population density and diet are not generally available for species-rich tropical communities, guild signatures based on biomass are less commonly presented. Emlen (1978) apportioned fractions of each species' biomass in Florida and Bahama pine communities among seventeen foraging guilds that combined

Table 11.4. Biomass of tropical and subtropical island and continental bird communities

Location	Habitat	No. Species Resident Land Birds	Number (ind ha^{-1})	Biomass (kg ha^{-1})	Mean wt (g ind^{-1})	Source
Islands						
Grand Bahama	pine forest	31	—	0.13	—	Emlen 1978
Bermuda	scrub	6	18.7	0.59	31.6	Crowell 1962
Puercos	deciduous forest	16	43.2	1.23	28.4	MacArthur et al. 1972
Tres Marías	deciduous forest	34	54.3	1.66	30.6	Grant 1966a
Puerto Rico	subtropical wet forest	31	33.0	1.68	50.9	Waide 1987
New Guinea	lowland rain forest	>100	69.1	4.96	71.8	Bell 1982
Continents						
Florida	pine forest	24	—	0.09	—	Emlen 1978
Panama	shrub	50	28.5	1.01	35.4	Karr 1971
Liberia	subtropical wet forest	50	—	1.24	34.9	Karr 1975
Panama	moist forest	56	36.4	1.32	36.2	Karr 1971
Panama (Barro Colorado)	moist forest	131	22.7	1.41	62.1	Willis 1980
Peru (Cocha Cashu)	moist forest	240	—	2.00	—	Robinson and Terborgh 1990

Notes: ind = individuals. Dashes indicate data not available.

Table 11.5. Avian species in feeding guilds

Locality	Feeding guild					
	Insectivores	Omnivores	Frugivores	Nectarivores	Raptors	Insect/Others
Costa Rica[a]	45.2	27.4	11.0	7.8	6.4	2.1
Panama[a]	49.0	26.3	12.4	4.8	7.2	0.3
Brazil[a]	52.3	22.0	13.3	4.0	7.0	1.3
Peru[a]	49.1	20.8	17.5	4.2	7.8	0.6
Costa Rica[b]						
La Selva	46.9	31.2	10.9	6.3	4.7	0.0
Osa-Ridgetop	46.2	26.9	11.5	7.7	7.7	0.0
Osa-*Mora*	31.5	27.8	18.5	13.0	9.3	0.0
Barranca	39.7	31.7	9.5	7.9	11.1	0.0
Taboga-*Brosimum*	35.8	32.8	7.5	9.0	14.9	0.0
Barba	42.3	42.3	3.8	11.5	0.0	0.0
La Chonta	43.8	31.3	18.8	6.3	0.0	0.0
Australia[c]	52.1	22.2	13.7	—	12.0	—
Malaya[c]	56.9	23.9	13.7	—	5.6	—
Puerto Rico[d]	16.7	25.0	33.3	8.3	12.5	4.2

Sources:

[a]Karr (1990).

[b]Orians (1969).

[c]Harrison (1962).

[d]Wetmore (1916) and Waide (unpublished) from El Verde.

Notes: Figures are percentages of avian species in each feeding guild. Omnivores are species that feed on two or more trophic levels. Dashes indicate that the category was not used in that study. Insect/others reflects species that take both insects and other invertebrate or vertebrate prey, but not necessarily from different trophic levels.

both diet and substrate. A similar method is used in this study to assign biomass of Puerto Rican birds to seven food categories (table 11.6).

Guild signatures based on species are remarkably similar for a wide range of continental avian communities (table 11.5). Neotropical lowland forests ranging from Costa Rica to Brazil are all dominated by insectivores, omnivores, and frugivores (in that order), with smaller but roughly equal proportions of insect/nectarivores and raptors. Rain forests in Australia and Malaya have a higher proportion of insectivores than the Neotropical sites.

The distribution of species among guilds is quite different for Puerto Rico compared to mainland tropical sites, reflecting the greater proportion of frugivores and the absence of insectivores on West Indian islands (Terborgh and Faaborg 1980). The abundance of lizards and frogs as potential prey at El Verde leads to the relatively high proportion of raptors and the existence of a specialist predator on lizards listed in table 11.5 under "other." The difference between the guild signatures of continental and island communities is even more pronounced when biomass is considered (table 11.6). Insects support 69 to 91% of avian biomass in tropical continental communities and

Table 11.6. Avian biomass supported from seven food categories

Locality	Vertebrates	Insects	Fruit	Seeds	Nectar	Carrion	Other
Brazil[a]	—	69.4	22.3	0.0	0.9	—	7.4
Peru[b]	—	91.2	8.3	0.0	0.5	—	—
Peru[b]	—	76.0	17.4	4.2	2.4	—	—
Barro Colorado[c]							
Old forest	4.7	38.5	51.9	0.0	0.5	4.7	—
Young forest	3.6	34.5	53.7	0.0	0.8	5.6	0.1
Hispaniola[b]	—	20.2	74.0	4.6	1.2	—	—
Puerto Rico							
El Verde[d]	2.6 (0.04)	12.9 (0.22)	70.9 (1.19)	10.0 (0.17)	3.5 (0.06)	—	—
Guanica[e]	—	27.1	57.3	9.6	6.0	—	—
Dominica[b]	—	7.8	82.9	6.4	2.9	—	—
St. Kitts[b]	—	9.2	78.4	10.9	1.4	—	—
Mona[e]	—	0.1	63.6	36.1	0.1	—	—

Sources:

[a]Bierregaard (1990).

[b]Terborgh and Faaborg (1980).

[c]Willis (1980).

[d]Reagan et al. (1982) and Waide (unpublished).

[e]Terborgh and Faaborg (1973).

Notes: Figures are percentages of avian biomass supported from each food category and are calculated from mist net samples except for Barro Colorado (Panama) and El Verde (Puerto Rico). At these sites, percentages are based on census data, observations of foraging behavior, and stomach contents. Numbers in parentheses are biomass in kg ha^{-1} for each trophic group. Dashes indicate where data are unavailable.

0.1 to 27% in the West Indies. Fruits, seeds, and nectar support higher proportions of the biomass in West Indian communities than on continents.

On Barro Colorado Island, which has a relict mainland fauna, the proportions of biomass supported by insects and fruit are intermediate between values from continents and islands. This suggests that as extinctions accumulate on Barro Colorado, the guild signature is shifting from domination by insectivores to domination by frugivores. Faaborg (1979) showed that extinction rates differed among avian families on land bridge islands, leading to a shift in guild signature that depends on island size. However, since the cause of extinctions on Barro Colorado Island has not yet been completely resolved (Karr 1982a,b; Loiselle and Hoppes 1983; Sieving 1992), interpretation of the present situation there solely in terms of island biogeographic theory is unwarranted.

Species Rank Abundance

The distribution of individuals among species is similar for samples collected with mist nets in Puerto Rico and Central and South America. The ten species most commonly captured in mist nets at El Verde (fig. 11.3) compose

97% of the individuals captured. In mist net samples from Central and South America, the ten most common species make up 40 to 64% of the total captures (Karr et al. 1990) but contribute only 9 to 14% of the captured species. At El Verde, 14% of the species captured (i.e., the top two species) compose 58% of the total captures, a value similar to that from mainland sites.

The major difference between the sample from El Verde and those from continental sites is the proportion of rare species. At El Verde, eight of fourteen species each contained more than 2% of the total sample. Comparable figures are eleven of seventy-six for Limbo Hunt Club in Panama (Karr 1990) and twenty of 143 for *terra firme* forest in Brazil (Bierregaard 1990). Thus, although the number of species captured is an order of magnitude higher in Brazil, the difference in the number of common species is much less pronounced.

In both mainland and island sites, detailed knowledge of fewer than twenty species provides a good picture of the magnitude of the contribution of the avifauna to ecosystem processes such as consumption, nutrient cycling, and energy flow, at least in the understory. Birds are important participants in other ecosystem processes such as pollination and seed dispersal, and the key species in these processes are probably also common. Therefore, even in diverse communities useful food webs can be constructed using only a subset of the total diversity. Keystone mutualists (Gilbert 1980), which have a large influence on ecosystem structure that may be unrelated to abundance, need to be included in these webs. Identification of these keystone species is critically important in understanding and comparing tropical forest ecosystems. Knowledge of the organization of insular ecosystems should provide a focus for continental studies and should pinpoint species whose existence is disproportionately important to the structure of tropical forest ecosystems.

DIET

The diet of birds in the LEF and other forested areas of Puerto Rico was examined in detail by Wetmore (1916). I also undertook analysis of stomach contents as part of the present study. In addition, a large number of observations of foraging are available for individual species (Kepler 1977; Snyder et al. 1987; Waide unpublished; Kepler and Kepler unpublished). The principal food of each species at El Verde, the trophic level from which foods are taken, and the general foraging location appear in table 11.1. A complete list of published observations of feeding is included in the general food web (appendix 14.A).

The diets of the sixteen most common species at El Verde are separated into categories in table 11.7. The categories reflect the most commonly mentioned food types in Wetmore's (1916) analysis of the diet of Puerto Rican birds. The proportion of the diet attributed to each feeding category is based

Table 11.7. Diets of common bird species at El Verde

	Fruit	Seeds	Nectar	Total Vegetable Matter	Insects	Spiders	Other Invert.	Lizards	Frogs	Birds	Mammals	Vertebrates/ Animal Matter
red-tailed hawk					+		22.9	12.5	+	33.3	31.3	77.1/100.0
ruddy quail-dove		100.0		100.0								0.0/ 0.0
scaly-naped pigeon	100.0			100.0								0.0/ 0.0
Puerto Rican lizard-cuckoo					14.6	6.9		78.6	+			78.6/100.0
Puerto Rican screech-owl					57.6	14.5	+	10.8	+++	17.0		27.8/100.0
Puerto Rican emerald			82.1	82.1	10.1	7.8						0.0/ 17.9
Puerto Rican tody	+	2.4		2.4	85.9	8.2		3.5	+			3.5/ 97.6
Puerto Rican woodpecker	24.9	+		34.2	58.3	1.4	+	—6.2—				6.2/ 65.8
Puerto Rican flycatcher	13.9			15.8	83.1			—1.1—				1.1/ 84.2
pearly-eyed thrasher	87.2			87.2	4.1	2.6	4.6	—1.5—				1.5/ 12.8
red-legged thrush	63.5			63.5	25.5		2.6	—8.4—				8.4/ 36.5
bananaquit	24.0		46.0	70.0	21.3	7.7						0.0/ 29.0
black-whiskered vireo	57.8			57.8	33.8	7.7		—0.6—				0.6/ 42.2
Puerto Rican tanager	25.7	8.3		40.1	52.0	1.5	0.7	—6.1—				6.1/ 59.9
stripe-headed tanager	100.0			100.0								0.0/ 0.0
Puerto Rican bullfinch	70.1			70.1	29.7	0.2						0.0/ 29.9

Sources: Data are from stomach analyses by Wetmore (1916) except for nectarivores, whose diet is distributed among the classes according to foraging observations by Snow and Snow (1971), Lack (1976), and the author.

Note: Each figure is a percentage of the diet contributed by the food category.

on stomach analyses (Wetmore 1916; Waide unpublished) modified by foraging observations for the two nectarivores (Puerto Rican emerald, bananaquit), whose principal food is not easily detected in stomach samples.

Carnivores

The largest avian predator in the forest is the red-tailed hawk, which feeds on vertebrates and large invertebrates. The most commonly taken prey items are rats and small birds, although centipedes of the genus *Scolopendra* are also prevalent in the diet (Santana and Temple 1988). In Puerto Rico, red-tailed hawks reach their highest densities (and higher densities than in most North American continental sites) on the eastern slopes of the Luquillo Mountains, where they are an important predator of the endangered Puerto Rican parrot (Santana and Temple 1988). These hawks prey on the introduced small Indian mongoose (*Herpestes auropunctatus*), which makes them the top predator in the forest.

At the time of the arrival of Europeans in the West Indies, red-tailed hawks probably fed at least partly on native mammals, many of which are now extinct or rare on the islands. Introduced mammals (house mouse, *Mus musculus;* rats *Rattus rattus, R. norvegicus;* and mongoose) now compose the greatest part of the diet of red-tailed hawks in lowland and mountainous sites in Puerto Rico (Santana and Temple 1988). At five nests in tabonuco and colorado forest zones in the LEF, nestlings were fed mostly mammals (41% of prey items) and birds (27%). Nestlings in the cloud forest were fed reptiles and amphibians (47%) more often than mammals (10%), with birds (19%) still as a major part of the diet (Santana and Temple 1988).

Red-tailed hawks achieve high densities in the LEF despite the fact that continuous forest is reported to be poor habitat for this species (Brown and Amadon 1968). Santana and Temple (1988) suggest that high numbers of hawks in the LEF are supported by the high biomass of reptiles (Reagan, this volume) and amphibians (Stewart and Woolbright, this volume) in the forest canopy. Tropical forest canopies in general tend to support a higher diversity and biomass of vertebrate prey than do temperate forest canopies (Bourliere 1983; Owen 1983).

There are two small populations of the broad-winged hawk in Puerto Rico, one in the LEF and one in Río Abajo. Although the Luquillo population is confined principally to the eastern slopes, a few individuals occasionally appear at El Verde. In the LEF, the diet of broad-winged hawks is similar to that of red-tailed hawks, but the former takes more large invertebrates, frogs, and lizards and fewer rats than the latter (Snyder et al. 1987).

The sharp-shinned hawk has a more restricted diet than the other Puerto Rican raptors, feeding principally on small birds (Wetmore 1916; Snyder and Wiley 1976; Snyder et al. 1987). The species is rare and local throughout

Puerto Rico (Delannoy 1992) and has been observed infrequently at El Verde (Recher and Recher 1966; Waide pers. observation).

The Puerto Rican screech-owl and the Puerto Rican lizard-cuckoo are nocturnal and diurnal predators, respectively, that include appreciable proportions of vertebrates in their diets. The screech-owl supplements its principally insectivorous diet with birds, lizards, and amphibians. Captive individuals can localize *E. coqui* in the dark (Narins unpublished), and they also take live frogs in captivity. As their name implies, lizard-cuckoos are specialists on *Anolis* lizards, but they also include insects and amphibians in their diet. Frogs were not found in the five stomachs examined by Wetmore (1916), but three times as many frogs as lizards (49:18) were brought to young in two nests observed by Snyder et al. (1987), suggesting that lizard-cuckoos exhibit less selectivity in prey under the demands of feeding young.

Insectivores

The four resident birds at El Verde that are principally insectivorous are the Puerto Rican flycatcher, Puerto Rican tody, Puerto Rican vireo, and elfin woods warbler. The flycatcher is extremely rare at El Verde, as it is throughout the LEF (Snyder et al. 1987). The tody, on the other hand, is one of the more common birds in the forest. Adult todies are almost completely insectivorous and have a broad diet, feeding on a minimum of fourteen insect orders and forty-nine families (Kepler 1977). Diptera (31%) and Coleoptera (23%) are the most common groups in the adult diet; nestlings are fed primarily Homoptera (30%), Coleoptera (25%), and Lepidoptera (16%; Kepler 1977). Small *Anolis* are taken by adults (Wetmore 1916, Kepler 1977), but frogs are fed more often to nestlings than are lizards (Waide unpublished).

The elfin woods warbler is a recently discovered (Kepler and Parkes 1972) Puerto Rican endemic whose range is restricted principally to remnant montane forests in the Luquillo Mountains and the Central Cordillera (Cruz and Delannoy 1984a). Although the species reaches its greatest density at higher elevations (900 to 1000 m; Cruz and Delannoy 1984a), it has also been observed at lower elevations (450 m; Waide unpublished) in the LEF. The few observations of the elfin woods warbler at El Verde suggest that a breeding population does not exist at this location.

The foraging ecology of the elfin woods warbler was studied in the Maricao Forest Reserve in western Puerto Rico (Cruz and Delannoy 1984b). At this site, the elfin woods warbler is entirely insectivorous, gleaning insects from leaves and branches in the forest canopy. The Luquillo population also is known to be completely insectivorous (Snyder et al. 1987).

The Puerto Rican vireo is found more commonly in the western part of the island in dense tangles on the limestone hills, but a few individuals have

Figure 11.5. Puerto Rican tanager (*Nesospingus speculiferus*). Photo by José Colon.

been seen at El Verde. This species is largely insectivorous and consumes mostly Orthoptera, Homoptera, and Lepidoptera. Small berries (mostly *Miconia prasina* and *Xanthoxylum* sp.) composed 14% of the diet and were particularly common in the diet during December in birds collected by Wetmore (1916) outside of the LEF.

Nine species of migrant wood warblers have been recorded at El Verde, and eight of these (black-and-white warbler, parula warbler, black-throated blue warbler, black-throated green warbler, prairie warbler, Cape May warbler, Louisiana waterthrush, and American redstart) are regular visitors. The diet of each of these species is composed mostly of insects, but black-throated blue warblers, parula warblers, and Cape May warblers also consume fruit and nectar (Kepler and Kepler unpublished). None of the migrant species is very common at El Verde, but taken together they represent an important component of the avifauna. Except for the four resident species mentioned above, no other leaf- or trunk-gleaning insectivorous birds occur in the El Verde avifauna.

Omnivores

The Puerto Rican tanager (fig. 11.5) and Puerto Rican woodpecker are omnivorous, each taking insects, fruits and seeds, lizards and frogs, arachnids, and other invertebrates in that order of importance (Wetmore 1916). The Puerto Rican tanager searches dead leaves and vine tangles for insects, whereas the woodpecker extracts insects from dead wood. Both species consume many fruits in the family Rubiaceae. The woodpecker particularly eats palm fruits as well as fruits of the genera *Miconia, Ficus,* and *Psychotria.*

Red-legged thrushes, black-whiskered vireos, pearly-eyed thrashers, and Puerto Rican bullfinches also eat plants and animals, but in contrast to the woodpecker and Puerto Rican tanager, these species emphasize fruit rather than insects. Each of the four species consumes a wide range of fruits (Wetmore 1916). For example, pearly-eyed thrashers have been observed taking nineteen different species of fruit. Thrashers have been seen feeding seven species of fruit to nestlings. The fruits of the sierra palm (*P. montana*) make up 21.9% and 21.8% of the diets of adults and nestlings, respectively (Snyder et al. 1987).

Of the four species, only the thrush consumes an appreciable amount of vertebrate prey, mostly *Anolis*. Captive red-legged thrushes readily searched through litter and consumed any frogs discovered (Stewart and Waide unpublished). Free-living birds consume frogs, lizards, and nestling birds (A. Arendt pers. comm.).

Frugivores and Granivores

Frugivores and granivores compose the largest group of species in the El Verde avifauna. Scaly-naped pigeons, stripe-headed tanagers, Puerto Rican parrots, and Antillean euphonias are completely frugivorous. The former three species take a wide variety of fruits. The pigeon prefers large fruits, especially those of palms, wild figs (*Ficus* sp.), *Cordia* sp., and *Genipa americana* (Wetmore 1916), which it swallows whole. Stripe-headed tanagers consume a wide variety of fruits and occasionally leaves. Most of the fruits are small and swallowed whole, but Wetmore (1916) observed individuals opening and consuming the pulp from oranges and soursop (*Anona muricata*).

More information is available on the diet of the Puerto Rican parrot than for any other frugivorous species because of the long-term effort to protect the last remaining wild population of this species (Snyder et al. 1987). Parrots have been observed to feed on fifty-eight different species of plant, generally taking fruit but also feeding on flowers, flower bracts, pericarp, bark, leaves, buds, and twigs. The most commonly consumed fruits are those of the sierra palm (167 observations) and the tabonuco (forty-nine observations), but fruits of other trees, shrubs, and vines are taken as well (Snyder et al. 1987). Parrots do not eat fruits in proportion to their availability. In the tabonuco zone, the fruits of a number of common tree species (e.g., *Cecropia schreberiana, Micropholis garciniaefolia, M. chrysophylloides, Schefflera morototoni*) are not eaten. Parrots were formerly common throughout the LEF, but declined drastically in the 1950s and 1960s.

In contrast to the preceding three species, the Antillean euphonia is a specialist on mistletoe (*Phoradendron* spp.). Fifty-one stomachs of this species examined by Wetmore (1916) had only mistletoe seeds. More recent observations (Wunderle pers. comm.) suggest that euphonias occasionally consume fruits of the families Melastomaceae and Rubiaceae.

Ruddy quail-doves, zenaida doves, and black-faced grassquits have mostly granivorous diets. Stomachs of quail-doves captured at El Verde contained seeds of *Dacryodes excelsa, Schefflera morototoni, Laetia procera, Palicourea riparia, Clusia rosea, Cecropia schreberiana, Marcgravia racemiflora,* and *Ficus sintenisii,* as well as fruit, rocks, feathers, insect parts, and a few ants (Waide unpublished). Quail-doves are terrestrial foragers that often open decaying fruit to extract the seeds (Wetmore 1916). They often aggregate under fruiting trees to feed on fallen fruit during the nonbreeding season. At El Verde, quail-doves are more commonly found in areas where *Dacryodes* and *Sloanea* are the dominant trees; they avoid areas where *Prestoea* predominates (Recher 1970).

Zenaida doves and black-faced grassquits occur along edges of the forest or where large man-made gaps result in the growth of grasses. Neither species is found in the forest interior. Seeds of grasses and sedges are the principal food of the black-faced grassquit (Wetmore 1916). Zenaida doves have been recorded feeding on the fruits and seeds of 105 plant species in Puerto Rico (Wiley 1991). Small numbers of invertebrates are also taken.

Nectarivores

Three largely nectarivorous species occur in the LEF, two hummingbirds (Puerto Rican emerald, green mango) and the bananaquit. Stomach analyses do not adequately describe the diet of these species, because the amount and origin of nectar consumed cannot be determined. Wetmore (1916) describes the stomach contents of both hummingbird species as being composed nearly completely of insects, but field observations indicate that 82% of the emerald's foraging effort is directed toward flowers, mostly of the shrub *Palicourea riparia* (Waide unpublished). Since small insects may be ingested along with nectar at flowers, the relative importance of insects and nectar in the diet of these species cannot be determined.

Wetmore (1916) found that 149 bananaquit stomachs collected in the islands of Vieques and Culebra, as well as Puerto Rico, contained 98% animal matter. Field observations from the LEF and other Caribbean islands (e.g., Trinidad, Snow and Snow 1971) suggest that fruit, nectar, and invertebrates are all important elements in the bananaquit's diet (table 11.7).

FORAGING DISTRIBUTION

Habitat Selection

Habitat selection in resident Puerto Rican birds occurs at large (between forest types; Wiley and Bauer 1985) and small (between gaps and undisturbed

Figure 11.6. Forest edge at the boundary between tabonuco forest and a man-made gap. Canopy birds often follow the edge down into gaps where they are caught in mist nets.

forest; Wunderle et al. 1987) scales. At an intermediate scale, bird assemblages were nearly identical between four 1 ha plots within tabonuco forest (see chap. 1 for map and description of plots). Green mango hummingbirds, Puerto Rican vireos, and migrant warblers were seen more often in the most recently disturbed plot (Zou et al. 1995). Migrant warblers prefer disturbed or early successional areas in the wet tropics (Willis 1966; Karr 1976; Hutto 1980; Blake et al. 1990).

Wunderle et al. (1987) examined assemblages of birds caught in mist nets in mature forest and large and small gaps within the forest at El Verde. No species were restricted to small gaps (treefalls) or the intact forest, but green mangos were caught only in large gaps. Bananaquits, Puerto Rican todies, black-throated blue warblers, and Puerto Rican bullfinches were all captured less frequently in forest than in small gaps, where they apparently descend from the canopy to lower levels (fig. 11.6). Bananaquits, Puerto Rican todies, and ruddy quail-doves also were captured more frequently in large gaps than in mature forest. Migrant species were captured more commonly in gaps than in forest.

Nets placed across streams captured significantly more individuals in small and large gaps than nets away from streams, but a similar trend was not found in mature forest (Wunderle et al. 1987). These higher capture rates were apparent even in the drier months, when water was not present in the temporary streams. Species that are particularly attracted to streams include the bananaquit, Puerto Rican tody, Puerto Rican emerald, green mango, and Louisiana waterthrush.

Fewer species are found in heavily managed forest stands within the LEF than at El Verde. Cruz (1988) found eighteen resident bird species in a stand of Caribbean pine (*Pinus caribaea*), eleven of which were regularly observed. In a mahogany (*Swietenia macrophylla*) plantation, Cruz (1987) found thirteen resident species including only eight that were observed regularly. In contrast, El Verde has thirty-one resident species, twenty-two of which are commonly found. The most striking difference between native forest at El Verde and plantations is the scarcity or absence of some frugivorous species (Antillean euphonia, scaly-naped pigeon, ruddy quail-dove) in the plantations.

Vertical Stratification

MacArthur et al. (1966) demonstrated that birds in seven habitats in Puerto Rico (including tabonuco forest at El Verde) recognized fewer vertical layers of vegetation than do birds in temperate North American or in mainland tropical habitats. They further suggested that this phenomenon was not based on intrinsic differences in the structure of the forests, but rather on different behavioral responses by the bird species. Thus, in a gradient from structurally simple to more complex vegetation types, the change in bird species among habitats and the subdivision of vertical space within habitats were less in Puerto Rico than in mainland temperate and tropical sites.

Data on foraging heights of particular species at El Verde support this generalization (fig. 11.7). Most species forage over a wide range of heights (Reagan et al. 1982). Of the ten species for which there are more than ten observations of perch height, only three fail to use all of the available vertical space. The ruddy quail-dove generally was observed picking its way through the ground litter, whereas scaly-naped pigeons and black-whiskered vireos foraged on fruits within the canopy.

Species that use the greatest diversity of foraging heights also tend to have a broad diet (fig. 11.8). Species that are limited to a single kind of food (such as the ruddy quail-dove) are limited in their foraging behavior by the distribution of that food. Those with a more diverse diet have fewer limitations on their foraging behavior. The prominent exception to the above generalizations, the Puerto Rican emerald (fig. 11.8), uses a comparatively narrow range of foraging heights but consumes nectar, insects, and arachnids. However, the vertical foraging range of the emerald may be underestimated because of the difficulty in seeing such a small bird in the canopy. In addition, many of the arthropods consumed by hummingbirds are associated with nectar sources. Hence, the three sources of food for the emerald may not be independent, resulting in an unexpectedly restricted range of foraging heights.

The most common species of migrant wood warblers in the LEF generally are observed in the mid-to-upper canopy (Kepler and Kepler unpublished). Black-throated blue warblers forage more in the understory than do other migrant species. In addition, males of this species forage significantly higher than females (mean = 5.56 vs. 3.36 m; $p < .01$; Kepler and Kepler unpublished). This difference in foraging height may be related to habitat segregation of the sexes in the LEF. Male black-throated blue warblers are more commonly found in tall stature mature forests, and females occur in shrubby second growth in Puerto Rico (Wunderle pers. comm.). Redstarts (mean foraging height = 5.31 m; Kepler and Kepler unpublished), Cape May warblers (6.52 m), black-and-white warblers (6.81 m), and parula warblers (7.42 m) are the most common arboreal warblers at El Verde.

BIRDS IN THE EL VERDE FOOD WEB

Maintenance Energy

The year-round maintenance of a relatively large amount of biomass at El Verde results in an estimate of respiratory energy six times that found at Hubbard Brook, New Hampshire, and twice as great as any temperate site yet

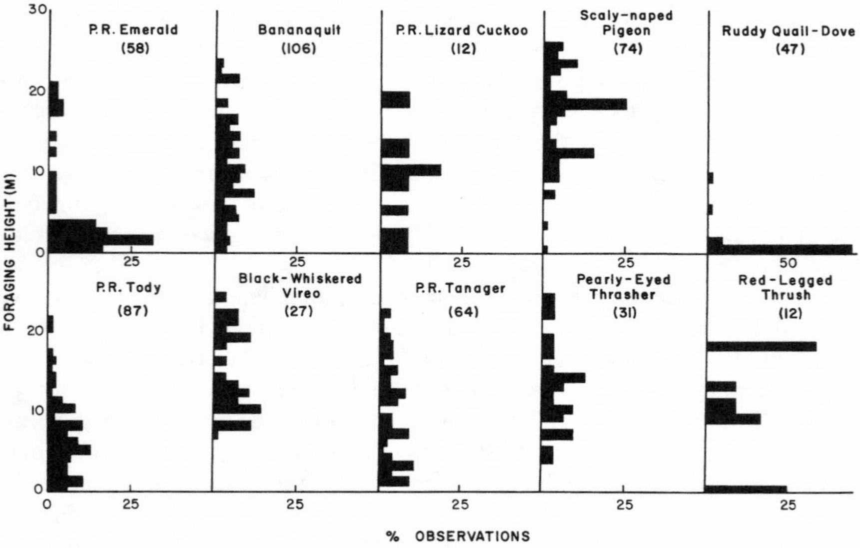

Figure 11.7. Distribution of foraging heights in meters for ten bird species at El Verde. Sample sizes are in parentheses under each species' name. Data are from Reagan et al. (1982).

known (table 11.8; Wiens 1977). Production energy was not included in the analysis because data are lacking on reproductive and turnover rates for birds at El Verde. Holmes and Sturges (1975) estimated net production in the avifauna at Hubbard Brook as only 0.1% of energy assimilated. The figure is likely to be even lower in tropical birds, which have small clutch sizes and fewer broods than their temperate counterparts. Because of seasonal changes, temperate avifaunas in general probably expend less energy than tropical avifaunas. Given that net primary productivity is higher in the tropics, the proportion of productivity consumed by birds may be quite similar in the two areas.

Feeding Relationships

Birds have their largest impact at the top and the bottom of the food web at El Verde. Frugivorous and granivorous birds compose 81% of the avian biomass at El Verde and, among vertebrates, share the fruit- and seed-eating guild only with bats and the introduced *Rattus*. Because of the scarcity of mammalian carnivores, the top predators at El Verde are also mostly birds. However, insectivorous birds are relatively unimportant at El Verde,

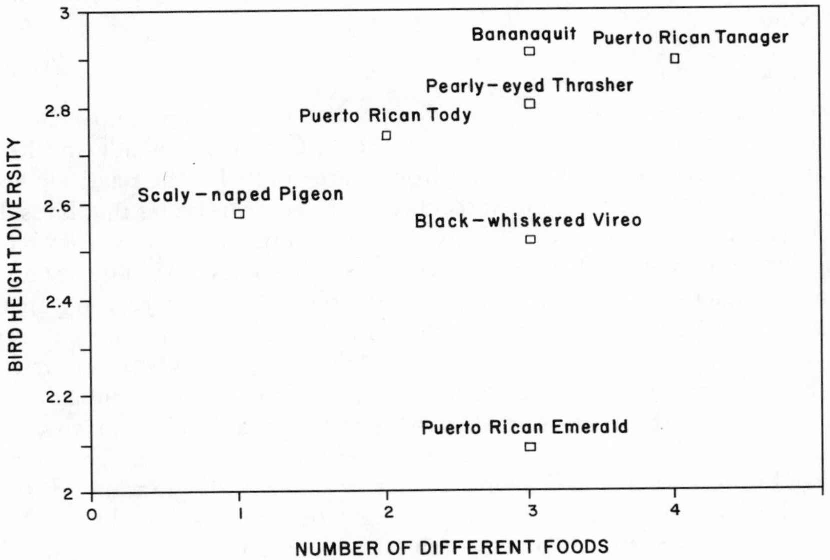

Figure 11.8. Plot of bird height diversity calculated from the data in figure 11.7 versus the number of different food categories (fruit, seeds, nectar, invertebrates, vertebrates) consumed. Species were scored for a category if more than 5% of their diet as shown in table 11.7 came from that category.

Table 11.8. Estimated energy expended for standard metabolism by El Verde avifauna

	Mean wet weight (g)	kcal $ind^{-1}d^{-1}$	kcal $ha^{-1}d^{-1}$	kcal $ha^{-1}y^{-1}$
red-tailed hawk	1,036.0	132.3	26.5	9,661.2
Puerto Rican emerald	3.1	2.0	0.4	143.8
ruddy quail-dove	132.0	29.8	32.8	11,955.6
scaly-naped pigeon	305.0	54.6	158.4	57,798.1
Puerto Rican lizard-cuckoo	77.1	20.2	4.0	1,472.8
Puerto Rican tody	6.8	3.5	7.7	2,792.8
Puerto Rican woodpecker	55.6	15.9	6.4	2,324.8
Puerto Rican flycatcher	22.9	8.4	1.7	611.6
pearly-eyed thrasher	96.3	23.7	4.7	1,730.0
red-legged thrush	84.2	21.5	4.3	1,569.8
bananaquit	10.2	4.7	23.3	8,512.9
black-whiskered vireo	20.0	7.6	36.5	13,306.7
Puerto Rican tanager	33.4	11.0	14.3	5,224.2
stripe-headed tanager	29.7	10.1	9.1	3,322.0
Puerto Rican bullfinch	32.1	10.7	7.5	2,733.3
Total			337.6	123,159.6
Maintenance energy (× 2.5)			844.0	307,899.0
Hubbard Brook				52,344.5

Source: Hubbard Brook data from Holmes and Sturges (1973).

where the intermediate trophic levels are dominated by lizards, frogs, and arachnids.

Birds as Predators

Numerous authors have suggested that predation is an important factor in structuring island communities (Grant 1966a; Case 1975; Waide and Reagan 1983; Faeth 1984; Pacala and Roughgarden 1984). The relatively low biomass of birds supported by insect prey at El Verde indicates that birds do not greatly affect insect populations. The most common insectivorous bird, the Puerto Rican tody, feeds principally on small insects at the rate of about 0.8 individuals min^{-1} (Kepler 1977). Over the course of a year, a pair of todies foraging on an average-sized territory (0.7 ha) will consume 60 insects m^{-2}. This value, while impressive for a bird the size of the tody (mean weight at El Verde = 6.8 g), is insignificant when compared with the annual consumption of insects by frogs and lizards (Stewart and Woolbright, this volume; Reagan, this volume). Certain insect groups that are more commonly taken by todies (Diptera, Coleoptera) may be exposed to significant pressure from bird predation, but no direct evidence of this effect is available.

In contrast, pressure on other vertebrate populations from avian predators may be important. The top predators in the food web at El Verde are principally birds, and observations indicate that predation on anoles and frogs is widespread and frequent. Predation pressure may vary considerably in space and time. The normal diet of adult pearly-eyed thrashers on Puerto Rico, Mona Island, and the U.S. Virgin Islands is 1.1% anoles by number (Snyder et al. 1987). During the breeding season, however, 9.6% of the items delivered to nests were anoles (Snyder et al. 1987). Anoles composed 12% of the diet of nestling red-tailed hawks (fig. 11.9) in rain forest in the LEF, but 47% in cloud forest (Santana and Temple 1988). The high proportion of raptors in the avifauna at El Verde (table 11.5) further suggests that avian predation on vertebrates may be important in this ecosystem.

Evidence from other West Indian islands suggests an important role for avian predation in the control of anole populations (McLaughlin and Roughgarden 1989). The main potential anole predators on Anguilla, St. Martin, and St. Eustatius are the pearly-eyed thrasher and the American kestrel (*Falco sparverius*). On Anguilla, predation pressure on lizards is low because kestrels are uncommon and pearly-eyed thrashers there do not eat vertebrates. On St. Martin and St. Eustatius, where a frugivorous congener occurs, pearly-eyed thrashers are important anole predators, and McLaughlin and Roughgarden (1989) conclude that anole populations are at least partially regulated by avian predation on these latter islands.

Dispersal and Pollination

Information on the importance of birds as agents of seed dispersal and pollination at El Verde is indirect. No studies have addressed the relationships between individual plant species and their avian dispersal agents, but infor-

Figure 11.9. A nest of the Red-tailed hawk (*Buteo jamaicensis*) in the LEF. Photo by José Colon.

mation from a variety of sources suggests that bird dispersal is common and potentially important. For example, experimental germination of seeds in soil samples from forest plots identified plants of eight light-requiring species that had no nearby sources of seed (R. F. Smith 1970b).

A large proportion (74%) of species of trees and shrubs in Puerto Rico and the Virgin Islands have fruits with a fleshy exocarp, a fleshy pulp, or an aril; seeds of these species are adapted to dispersal by vertebrates (Flores and Schemske 1984). The only native frugivorous vertebrates are birds and bats. The abundance of frugivores in the avifauna at El Verde and the paucity of frugivorous bats (Willig and Gannon, this volume) indicates that birds are the principal dispersers of seeds. Introduced *R. rattus* also consume fruit and may disperse seeds (Layton 1986).

Information on avian pollination at El Verde is far from complete. In a very wet area such as the LEF, wind-borne pollen is ineffective, and Ogle (1970) was not able to recover any pollen from traps at El Verde. In the dwarf forest in the LEF, the number of plant species with an insect/bird pollination syndrome is much higher than the number with a wind pollination syndrome (Nevling 1971), but pollination efficiency and fruit set is low for both kinds of plants. Hernández-Prieto (1986) found that the green mango hummingbird was the only avian pollinator of *Tabebuia rigida* (Bignoniaceae) in the dwarf forest and that the Puerto Rican emerald and the bananaquit robbed nectar from this species.

At El Verde, the Puerto Rican emerald and the green mango are the only avian pollinators. The Puerto Rican emerald forages heavily on *Palicourea riparia,* the dominant understory shrub (Waide unpublished), which has a short-tubed flower. Wunderle et al. (1987) note a positive relationship between green mango captures and *Heliconia* flowering at El Verde. No other information on avian pollination is available.

Analysis of pollinator communities at other sites in Puerto Rico suggest that flowers are adapted to either long-billed (green and Antillean [*Anthracothorax dominicus*] mangos) or short-billed (Puerto Rican emerald) hummingbirds (Kodric-Brown et al. 1984). Specificity for different flower morphologies is maintained because nectar in flowers with long tubes is not accessible to short-billed hummingbirds and flowers with short tubes do not produce enough nectar for efficient foraging by long-billed hummingbirds. Within groups of species with the same flower morphology, the blooming periods are displaced so that some species are in flower at all times (Kodric-Brown et al. 1984).

SUMMARY

Birds have a strong effect on food web structure because they take prey from all trophic levels and because their mobility introduces spatial and temporal variability into the web. The avifauna of tabonuco forest at El Verde is characteristic of Caribbean islands in having a lower number of resident species (thirty-one) than comparable mainland sites, and thus an analysis of the food web can include all species in the community. Behavioral and ecological correlates of reduced species richness are the most important factors structuring the avian community at El Verde.

The most common bird at El Verde is the bananaquit, which contributes about 50% of the individuals detected in point counts. Other common species include the black-whiskered vireo, scaly-naped pigeon, and Puerto Rican tody. Comparisons of density between 1964 to 1966 and 1981 indicate that most species had the same or higher densities in 1981.

The main difference in mist net samples from El Verde and continental sites lies in the large number of very rare species that occur in continental sites but not at El Verde. The guild signature of the El Verde community is quite different from Central and South American communities. The avifauna at El Verde has fewer insectivores and more frugivores than a wide range of continental communities. However, in both mainland and island communities, knowledge of the biology of the twenty most common species provides sufficient information to estimate the roles of bird communities in many ecosystem processes.

Preferences for large gaps, stream courses, and secondary stands were shown by some species, but in general most bird species were found throughout tabonuco forest regardless of disturbance history. A positive relationship was found between vertical foraging range and diet breadth. Caribbean birds live in a situation in which periodic natural disturbances (hurricanes) result in both major modifications of the habitat and fluctuations in specific dietary resources. Flexibility in both foraging behavior and diet could be adaptive under these circumstances.

The total biomass of birds per hectare at El Verde (1.68 kg) is relatively high compared to continental sites because of higher density per species on the island (33 individuals ha^{-1}) and larger individual body weight (mean = 50.9 g). Preliminary evidence suggests that winter communities are more similar in total avian biomass than summer communities, further emphasizing the important role of North American migrants in some tropical communities. The flow of energy through the avifauna at El Verde (307,899 kcal $ha^{-1}y^{-1}$) is at least double that found in temperate sites.

The most important trophic function performed by birds at El Verde is that of predator. Most of the upper-level predators in the food web are birds. In contrast, birds play a relatively small role as predators on lower-level insect consumers compared to other vertebrates. Thus, top-down control is most likely to express itself through predation on frogs and lizards.

Most birds at El Verde consume fruit, seeds, or nectar. Along with bats, birds are thus important agents of dispersal and pollination in tabonuco forest. The absence of terrestrial and arboreal mammals at El Verde indicates that bird dispersal is potentially much more important in the tabonuco forest than in mainland forests. Few data are available to examine the relationship between rates of avian dispersal and plant success in the tabonuco forest. A similar situation exists with regard to the importance of birds as pollinators.

ACKNOWLEDGMENTS

I thank the many friends and colleagues who made useful comments on this chapter and contributed in other ways to its completion. Fellow ornithologists Wayne and Angela Arendt, Cameron Kepler, Jim Wiley, and Joe Wunderle shared their observations with me and commented on several versions of the manuscript. Fellow authors Gerardo Camilo, Meg Stewart, and Mike Willig were generous with their time and contributed substantially toward improving the chapter. Jose Colon permitted the use of several of his excellent photographs. Eva Cortes, Albert Muñis, and Nilda Sosa were always available to assist me with the many details of preparing the manuscript and figures. Alejo Estrada Pinto shared with me an understanding of tabonuco forest developed over thirty years as manager of the El Verde Field Station. Doug Reagan's unwavering enthusiasm for this project helped keep us both on track. Finallly, I thank my wife Valli and my daughters Valiangelic and Dayanara for their support and patience during the long process of bringing this work to completion.

Appendix 11 Methods

Spot (territory) maps, transects, circular plots, and mist nets all have been used to study bird populations at El Verde (Reagan et al. 1982; Wunderle et al. 1987; Waide and Narins 1988; Waide 1991; Pardieck and Waide 1992). Although each method has strengths, spot maps probably provide the best comparison of annual variability, whereas a combination of the other methods can be used to estimate seasonal variability.

The best features of each technique can be exploited by calibrating transect surveys to a spot-map count performed in the same location (Holmes and Sturges 1975), the approach used in this study. Estimates of population density were made by mapping territories (Int. Bird Census Comm. 1970) in a 9 ha grid (see chap. 1) during the breeding season. Mapping was performed toward the end of the main breeding period in 1981 and at the beginning of breeding in 1982 (Reagan et al. 1982). Also, monthly counts were made along four transects during 1980 to 1981 using methods described by Emlen (1971). One of these transects (Plot 3) was within the 9 ha grid.

In both 1981 and 1982 densities calculated from transect counts performed by the author tended to be higher than densities calculated from spot maps, especially for the more common species (table 11.2). One possible explanation for this difference is that moving birds were counted more than once along the transect line. Observers may have moved too slowly along the 300 m transect lines, allowing active birds to change position and be counted again during the course of the census. This type of error is more likely when several individuals of the same species are singing at once; hence, as measured by the two methods, rare species have more similar densities than common species. The spot maps probably reflect population densities more accurately and therefore were used in calculations. However, spot maps are not accurate for rare species or for any species outside the breeding season. Transect counts were used to calculate densities of species that were not detected in spot maps and to compare seasons.

Energy calculations

Until practical methods of measuring metabolic rates in free-living vertebrates are devised, the calculation of energy flow in communities will remain crude. Different investigators have adopted varying methods of calculating energy flow (Holmes and Sturges 1973; Weiner and Glowacinski 1975; Wiens 1977). For comparative purposes, we have used the method of Holmes and Sturges (1973), described below.

Metabolic energy expenditures for resting birds in thermoneutral condition can be predicted from the Lasiewski-Dawson (1967) equation: $M = 129\ W^{0.724}$, in which $M = \text{kcal d}^{-1}$ and $W =$ the weight of each species in kg. Multiplying the metabolic expenditure for each species by density and summing gives a minimum estimate of daily energy expenditure for the avifauna. Multiplication by 365 converts this figure to a minimum annual estimate. Populations are assumed to be relatively constant from year to year. Data from Recher (1970) and the present study show this latter assumption to be reasonable for most species. The annual estimate is multiplied by 2.5 to adjust for the energetic costs of normal movement, thermoregulation, molt, and fat deposition (see Holmes and Sturges 1973). Only those species with densities greater than 0.1 ha^{-1} are included in the analysis.

12

Mammals

Michael R. Willig and Michael R. Gannon

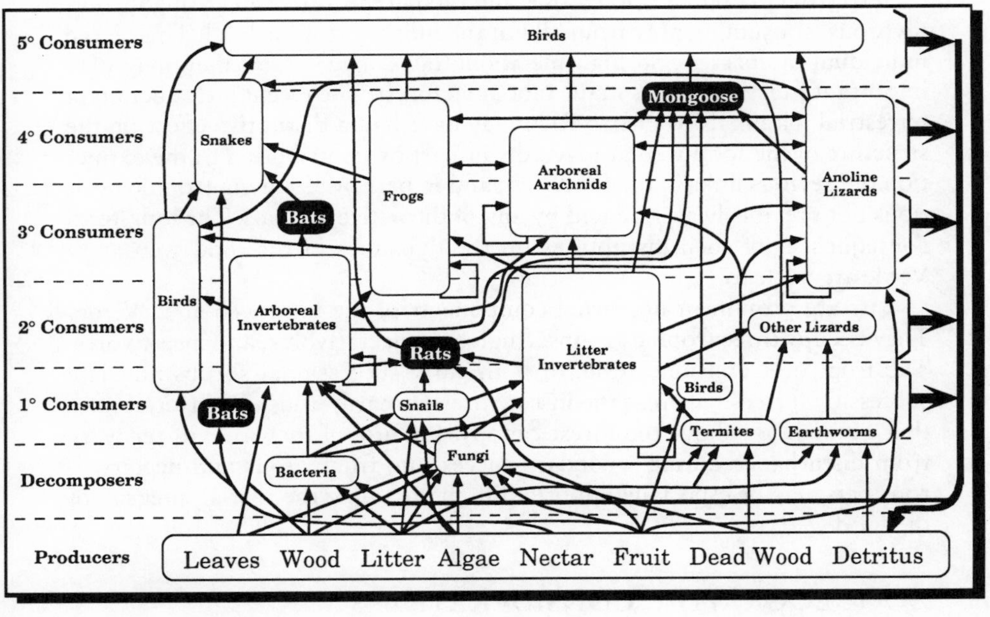

MAMMALS make important contributions to the food web at El Verde. They do so in the usual manner, by affecting pathways of energy flow and nutrient cycling. Perhaps equally important, they affect the spatial heterogeneity of nutrients, as well as the spatial distribution and genetic structure of plant populations. Indeed, because of their role in pollination, and fruit or seed dispersal, a number of mammal species may be keystone mutualists in the tabonuco forest.

Terrestrial mammals (mongooses and rats) in the food web are highly omnivorous, consuming prey from all but the quinary trophic level. Their large individual biomasses and high metabolic rates suggest that they may play important functional roles in the tabonuco forest. The recent introduction of terrestrial mammals to Puerto Rico may have had a disruptive effect on the structure of the food web at El Verde, in part by contributing to the extinction of previous links in the food web, and in part by assuming trophic positions not previously represented by any of the native animals. The long-term consequences of such disruptions on the dynamics of the food web at El Verde are unclear.

Bats are prominent nocturnal components of the food web at El Verde. They occupy three trophic groups: frugivores, nectarivores, and insectivores. The frugivores consume fruits of both early successional shrubs and late successional trees, whereas the insectivores primarily consume invertebrates that occur in or above the forest canopy. Because of their role in recovery from disturbance during secondary succession, frugivorous and nectarivorous bats may be especially important in maintaining the spatial integrity of the food web.

ZOOGEOGRAPHIC CONSIDERATIONS

The composition of the mammalian community of the tabonuco forest at El Verde is the product of a variety of biogeographic and human-mediated events. In general, the mammal fauna of Puerto Rico is depauperate compared to tropical mainland areas of similar size and habitat diversity (table 12.1). Low species richness is affected in part by well-documented biogeographic processes related to the island's size and distance from potential sources of colonization. Moreover, Griffiths and Klingener (1988) observed that, compared to other islands in the Greater Antilles, the bat fauna of Puerto Rico is well below its equilibrial species richness based upon the MacArthur-Wilson model of island biogeography (MacArthur and Wilson 1967). This observation is consistent with the hypothesis that colonization from Central and North America was differentially directed to Cuba and Jamaica because of extensive land bridges (e.g., the Grand Bahama, Rosilind, Seranillas and Pedro banks in the Caribbean) and an enlarged mainland (e.g., the Nicaraguan Plateau and the southern Florida peninsula) during Pleistocene glacial maxima. Puerto Rico was relatively isolated even during periods of the Pleistocene when sea level was lower than it is presently, and insufficient time has elapsed since the last glacial maximum to allow dispersal of North and Central American–derived stock to Puerto Rico. In addition, Puerto Rico has experienced a number of mammalian extinctions since the end of the Pleistocene (Anthony 1918, 1926; Baker and Genoways 1978; Best and Castro 1981; Choate and Birney 1968; Morgan and Woods 1986) that include one insectivore (*Nesophontes edithae*), three bats (*Macrotus waterhousii waterhousii, Monophyllus plethodon frater,* and *Phylloncyteris major*), two edentates (*Acratocnus odontrigonus* and *A. major*), and six rodents (the heptaxodontids, *Heptaxodon bidens* and *Elasmodontomys obliquus;* the capromyid, *Isolobodon portoricensis,* possibly introduced to Puerto Rico from Hispaniola by Amerindians; and the echimyids, *Heteropsomys insulans, Homopsomys* [= *Heteropsomys*] *antillensis,* and *Proechimys corozalos*). Moreover, the fossil mammalian fauna of Puerto Rico is depauperate for the island's present size, compared to other fossil faunas on islands in the West Indies, a discrepancy further magnified if the island's larger size during the last glacial interval is considered (Morgan and Woods 1986). Unfortunately, the fossil record does not indicate the range of habitats that these extinct mammals occupied, making it impossible to determine which, if any, species occurred in the tabonuco forest.

Other than domesticated animals, three rodents (*Rattus rattus,* the black rat, *R. norvegicus,* the Norway rat, and *Mus musculus,* the house mouse) and one carnivore (*Herpestes auropunctatus,* the Indian mongoose), have been introduced to Puerto Rico by Europeans in post-Columbian times. Of these, two (*R. rattus* and *H. auropunctatus*) are found at El Verde and represent the entire nonvolant mammal fauna. Regardless of the causes of nonvolant mammal extinctions, the composition of the mammal community of the tabonuco forest is clearly of recent origin in evolutionary and perhaps ecological time.

The only extant native mammals in Puerto Rico are bats. Thirteen species representing five feeding guilds presently are found on the island. The piscivore guild is represented by only one species, *Noctilio leporinus mastivus,* the greater bulldog bat. The aerial insectivore guild comprises five species:

Table 12.1. Species richness of various mammalian assemblages

Taxon	Puerto Rico	Costa Rica							Panama		Colombia			Brazil	
		1	2	3	4	5	6	7	8	9	10	11	12	13	14
Marsupialia	0	6	8	7	5	5	0	6	5	6	—	—	—	4	—
Insectivora	0	0	0	0	2	0	2	2	0	0	—	—	—	0	—
Chiroptera	13	86	81	82	60	45	16	34	29	32	14	16	29	34	25
Emballonuridae	0	9	9	7	4	0	0	0	2	2	0	0	2	1	1
Noctilionidae	1	2	2	2	0	0	0	0	1	1	0	0	0	1	1
Mormoopidae	3	4	4	4	3	1	0	2	1	1	0	0	0	1	1
Phyllostomidae	5	48	44	47	35	34	11	25	22	24	11	12	25	21	14
Natalidae	0	0	0	1	0	0	0	0	0	0	0	0	0	0	1
Furipteridae	0	2	2	0	0	0	0	0	0	0	0	0	0	1	0
Thyropteridae	0	1	1	1	0	1	0	1	0	0	1	0	0	0	0
Vespertilionidae	2	11	9	10	10	8	5	6	2	2	1	2	2	2	4
Molossidae	2	9	10	10	8	1	0	0	1	2	1	2	0	7	3
Primates	0	3	4	3	2	3	0	3	2	3	—	—	—	2	—
Edentata	0	6	7	5	6	3	0	4	5	4	—	—	—	3	—
Lagomorpha	0	1	1	2	1	1	1	1	1	1	—	—	—	1	—
Rodentia	0	17	20	15	16	10	10	16	14	13	—	—	—	10	—
Carnivora	0	13	17	19	12	10	7	14	10	10	—	—	—	3	—
Artiodactyla	0	3	4	3	0	0	1	3	2	2	—	—	—	1	—
Perissodactyla	0	1	1	1	0	0	1	1	0	0	—	—	—	0	—

Sources: Costa Rica (Wilson 1983), Panama (Fleming 1973), Colombia (Thomas 1972), and Brazil (Mares et al. 1981, Willig 1983, Willig and Mares 1989).

Notes: Introduced and domesticated mammals are not included in the species richness of any site. Site code: 1 = La Selva, 2 = Osa, 3 = Guanacaste, 4 = San Jose, 5 = San Vito, 6 = Cerro de la Muerte, 7 = Monteverde, 8 = Balboa, 9 = Cristobal, 10 = Hormiguero, 11 = Pance, 12 = Zabaletas, 13 = Caatinga, 14 = edaphic Cerrado.

Table 12.2. Relative importance of bat species at El Verde

Taxon	Mean wet weight (g)	Guild	Numerical dominance	Biomass dominance
Phyllostomidae				
A. jamaicensis	46.8	Frugivore	0.40	0.62
S. rufum	22.7	Frugivore	0.37	0.28
M. redmani	9.3	Nectarivore	0.15	0.05
B. cavernarum	49.6	Frugivore	0.02	0.03
E. sezekorni	19.1	Frugivore	0.02	0.01
Vespertilionidae				
E. fuscus	15.8	Aerial Insectivore	0.02	0.01
P. parnellii	12.5	Aerial Insectivore	<0.01	<0.01
P. quadridens	5.6	Aerial Insectivore	<0.01	<0.01
L. borealis	10.0	Aerial Insectivore	<0.01	<0.01

Note: Estimates of relative importance are based upon numerical dominance and biomass dominance. See text for details and discussion of dominance indexes.

Pteronotus quadridens fuliginosus, the sooty mustached bat; *P. parnellii portoricensis,* Parnell's mustached bat; *Mormoops blainvillii cuvieri,* Blainville's ghost-faced bat; *Eptesicus fuscus wetmorei,* the big brown bat; and *Lasiurus borealis minor* (= *L. minor*), the red bat. The nectarivore guild contains one species, *Monophyllus redmani portoricensis,* the Puerto Rican long-tongued bat. The frugivore guild contains four species: *Brachyphylla cavernarum intermedia,* the Antillean fruit-eating bat; *Artibeus jamaicensis jamaicensis,* the Jamaican fruit-eating bat; *Stenoderma rufum darioi,* the red fig-eating bat; and *Erophylla sezekorni sezekorni* (reported as *E. bombifrons* by some authors), the buffy flower bat. The molossid insectivore guild contains two species, *Tadarida brasiliensis antillularum,* the Brazilian free-tailed bat; and *Molossus molossus debilis,* Pallas' free-tailed bat. Only nine species (four frugivores, one nectarivore, four aerial insectivores; table 12.2) have been recorded from the tabonuco forest at El Verde (Tamsitt and Valdivieso 1970; Jones et al. 1971; Genoways and Baker 1975; and this report for *L. b. minor* and *P. q. fuliginosus*).

BATS

Foraging Ecology

In general, the population biology and foraging ecology of bats are poorly known; this is especially true of tropical species. The paucity of data is attributable in part to nocturnal activity patterns, as well as to the rapid rate at which bats process food, making identification of dietary constituents difficult (Willig et al. 1993). More specifically, little ecological information of any sort has been published concerning bats in the tabonuco forest of Puerto Rico prior to 1990. Density estimates are not known for any bat species in the forest; however, the relative importance of each species (compared to other bats in the community) can be assessed from netting records in two ways. Numerical dominance (ND) for each species can be measured as n_i/N, where n_i is the number of captured specimens of species i and N ($= \Sigma\, n_i$) is the total number of captured specimens regardless of taxonomic identity. Biomass dominance (BD) for each species is given by

$$n_i Y_i / \sum_{i=1}^{s} n_i Y_i$$

where Y_i is the mean biomass of species i and s is the number of species in the community. Both ND and BD for each species are listed in table 12.2. As a consequence of faunal depauperization associated with low origination (immigration and speciation) and high extinction rates, island species may frequently experience competitive release and enjoy elevated local densities compared to mainland taxa (MacArthur and Wilson 1967). Bats at El Verde do not demonstrate this phenomenon, although other vertebrates (e.g., *Eleutherodactylus coqui* and anoline lizards) clearly maintain high densities. In fact, netting success for bats in the tabonuco forest is considerably less (5 to 20%) than that of tropical or subtropical mainland sites.

Frugivores

Four species of frugivorous bat (*A. jamaicensis, S. rufum, E. sezekorni,* and *B. cavernarum*) have been captured at El Verde. Together, they constitute the bulk of the bat fauna in terms of numbers and biomass (total ND = 0.81; total BD = 0.94; see table 12.2).

The Jamaican fruit-eating bat, *A. jamaicensis* (fig. 12.1), has a wide distribution in tropical and subtropical habitats of the New World. Moreover, it is the predominant bat species at El Verde (ND = 0.40; BD = 0.62). Its population biology has been studied intensively in moist tropical forest of Panama and dry tropical forest of Mexico (Morrison 1978a,b,c,d, 1979, 1980). A summary of its foraging ecology is provided by Fleming (1982) and Handley et al. (1991). Although figs are a major dietary item elsewhere, these bats primarily consume *Cecropia schreberiana* (table 12.3) at El Verde. Preliminary data from the wet season, based upon telemetrically monitored bats, suggest that *A. jamaicensis* does not exhibit foraging site fidelity at El Verde (Gannon and Willig unpublished).

The reproductive biology of *A. jamaicensis* is geographically variable (Wilson 1979; Willig 1985a,b) throughout its range and is poorly documented on Puerto Rico in general and at El Verde in particular. The proportion of reproductively active females during the wet season is statistically variable from year to year (G-Test of Independence: $G = 6.16$, d.f. = 20,

Figure 12.1. Photograph (courtesy of R. J. Baker) of an adult male *Artibeus jamaicensis*, the Jamaican fruit-eating bat, from Puerto Rico

$.025 > p > .01$): 68% were pregnant or lactating ($N = 45$ adult females) in 1982 (Willig and Bauman 1984), whereas 36.6% ($N = 30$) were reproductively active in 1988 and 1989 (Gannon 1991; Gannon and Willig 1992). In part, this difference may be a methodological bias because the 1988–1989 data are based solely upon field palpation of live females (underestimates of true reproductive activity) whereas the 1982 data are based upon necropsy. Data for the dry season are few (33.3% of six adult females were reproductively active in 1988 and 1989) but do not suggest seasonal differences in reproductive activity beyond the variation characteristic of the wet season alone.

The red fig-eating bat, *S. rufum* (fig. 12.2), is endemic to Puerto Rico and the nearby islands of St. John and St. Thomas. Although generally reported to be rare, it was almost as abundant as *A. jamaicensis* at El Verde prior to Hurricane Hugo (table 12.2) and during the wet season was more abundant at some locations. Genoways and Baker (1972) summarized published information concerning the biology of this species. Although its common name implies a diet of figs, neither published (Scogin 1982; Willig and Bauman 1984) nor current data (table 12.3) indicate that this species eats figs. The

Table 12.3. Diet of frugivorous bats at El Verde

	Bat Species		
Dietary Item	*A. jamaicensis* (*N* = 40)	*S. rufum* (*N* = 36)	*E. sezekorni* (*N* = 5)
Cecropia schreberiana	0.65	0.38	—
Piper aduncum	—	—	0.80
Piper hispidum	—	0.04	—
Piper glabra	0.05	—	—
Manilkara bidentata	0.03	0.23	—
Prestoea montana[a]	0.05	0.27	0.20
Unknown plant material	0.23	0.08	—

Notes: Consumption of each dietary item is reported as the percentage of digestive tracts that contain that item. *N* = number of bats with stomach contents.

[a]Tentative identification of fruit pulp.

bulk of the diet of *S. rufum* at El Verde (88% of the stomachs containing food) comprises *Cecropia schreberiana, Manilkara bidentata,* or *Prestoea montana* (table 12.3). All of the adult females collected in June and July (1983) at El Verde ($N = 12$) were pregnant or lactating (Willig and Bauman 1984). When all data from that study are combined with similar reproductive records from 1989–1990 (Gannon 1991; Gannon and Willig 1992; museum records for El Verde prior to 1989 [specimens deposited in Carnegie Museum of Natural History or The Museum, Texas Tech University]), it appears that reproductive activity during the wet season (72.3% of sixty-five adult females pregnant or lactating) is statistically indistinguishable (G-Test of Independence: $G = 0.58$, d.f. $= 1$, $.5 > p > .1$) from that during the dry season (80% of twenty-five adult females pregnant or lactating). Although the presence of simultaneously pregnant and lactating specimens indicates that *S. rufum* is polyestrous, the statistical comparison suggests that the population is asynchronous, with an invariant and high proportion of adults reproductively active throughout the year. Such asynchronous polyestry is characteristic of species whose food does not vary greatly in abundance on a seasonal basis (Wilson 1979).

Detailed consideration of space use by *S. rufum* prior to Hurricane Hugo (Gannon 1991; Gannon and Willig unpublished) is available only for the wet season (June-August). Neither total home range nor foraging home range size differ between sexes, but consistent differences for each exist between adults and subadults (table 12.4). More specifically, adults have consistently smaller total home ranges than do subadults (fig. 12.3). These differences are consistent with a number of behavioral hypotheses. Subadults are less experienced than adults and as a consequence may traverse larger areas in search for food, or may at times be excluded by adults from feeding on trees with

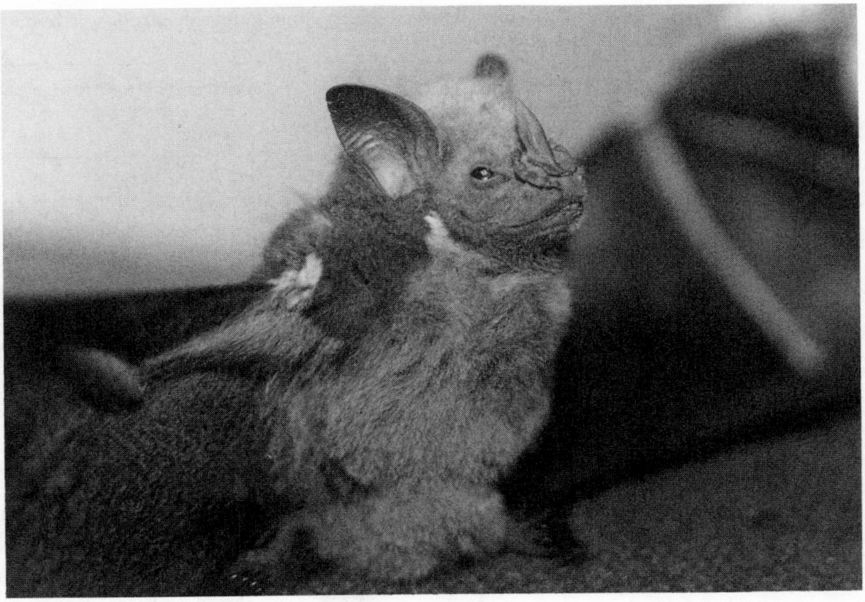

Figure 12.2. Photograph (courtesy of M. R. Gannon) of an adult male (*A*) and juvenile female (*B*) *Stenoderma rufum*, the red fig-eating bat, from Puerto Rico

Table 12.4. Two-way analysis of variance (sex versus age) of minimum convex polygon home range size for captures of *Stenoderma rufum*

		Total Home Range[a]			Foraging Home Range[b]		
Source	d.f.	MS	F	P	MS	F	P
Age (A)	1	44815.2	7.36	0.01	4638.4	11.23	0.03
Sex (S)	1	9.5	0.02	0.90	1080.3	2.62	0.12
A × S	1	69.9	0.12	0.74	543.4	1.32	0.26
Within	22	609.0			412.1		

Note: All captures during wet season.

[a] All captures.

[b] Night captures.

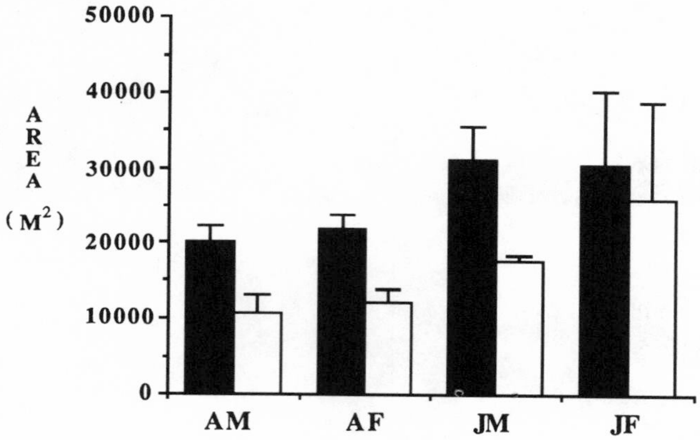

Figure 12.3. Bar diagram of the minimum convex polygon estimates of total home range (solid bars) and foraging home range based upon activity only during the night (open bars) for *Stenoderma rufum* at El Verde (AM = adult males, AF = adult females, SM = subadult males, and SF = subadult females). Vertical lines represent standard errors.

high fruit set. Both scenarios result in larger total home range size in subadults than in adults. The social system of *S. rufum,* as revealed from telemetry, is one in which both males and females are solitary in roosting behavior with no evidence of polygeny or harem formation. Moreover, considerable overlap in space use by all age-sex categories exists (fig. 12.4). These two characteristics suggest the absence of factors promoting sex-specific differences in foraging, especially in light of the relatively even year-round abundance of *Cecropia* and other bat fruits (most fruit falls to the forest floor uneaten) in the tabonuco forest (Devoe 1990).

A log of nightly movement patterns of focal bats, with each individual's

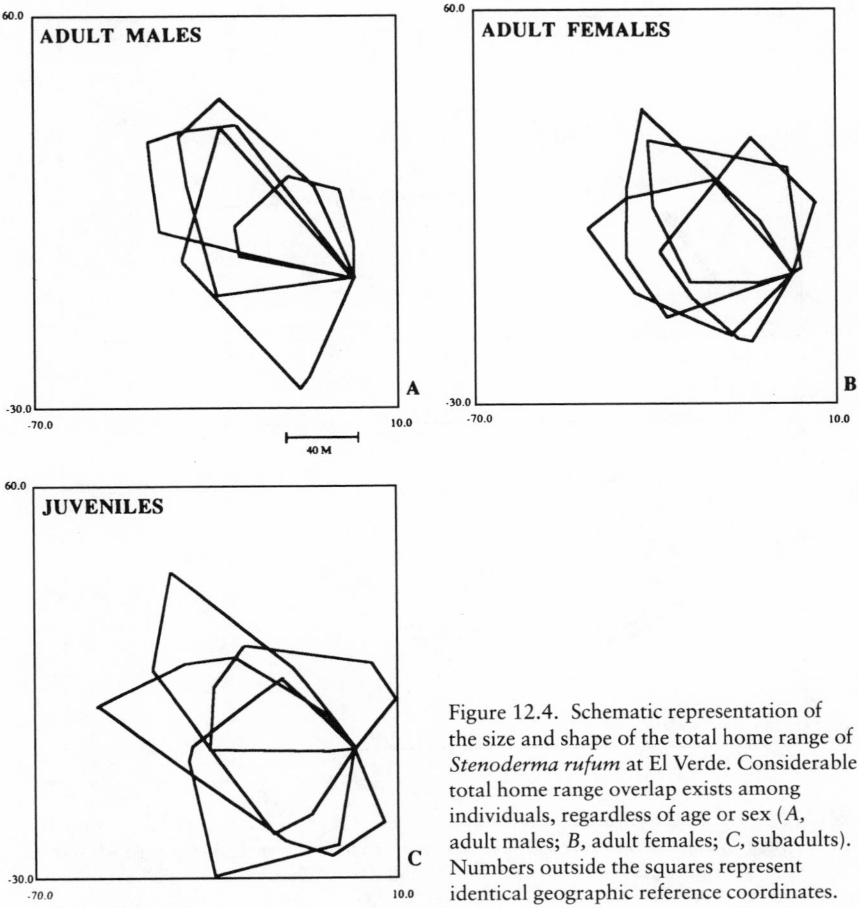

Figure 12.4. Schematic representation of the size and shape of the total home range of *Stenoderma rufum* at El Verde. Considerable total home range overlap exists among individuals, regardless of age or sex (*A*, adult males; *B*, adult females; *C*, subadults). Numbers outside the squares represent identical geographic reference coordinates.

position recorded at five-minute intervals over a four- to eight-hour period, can be superimposed on long-term home range data (fig. 12.5). Consideration of the results suggests that day roosts are scattered throughout the total home range with only a few located outside the foraging area. Thus, commuting time is negligible. Although a major night roost may exist in which a bat repeatedly returns to a single tree after foraging, most sites are visited only once. The area used on a particular night rarely encompasses more than 75% of the entire foraging home range, but occasional forays outside the total home range suggest that *S. rufum* exhibits a foraging strategy sensitive to the detection of fruiting trees in surrounding habitat. The emerging view of the total home range of an individual is one in which space is used hetero-

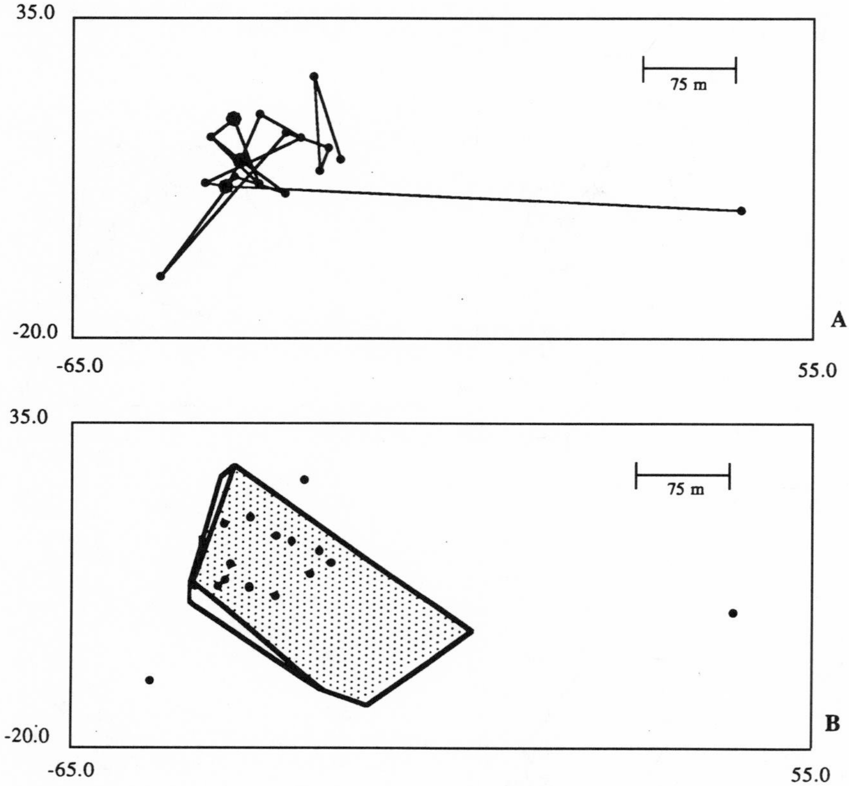

Figure 12.5. Detailed records of movements (lines) every 5 m (over an eight-hour period) for an adult male *Stenoderma rufum* (*A*) indicates repeated use of three focal night roosts (larger dots) and many singly used minor night roosts (small dots). When these detailed data are superimposed (*B*) upon the total home range and foraging home range (shaded area) of the same individual over a longer time period, it is clear that most minor and all major roosts occur within the estimated ranges, but that occasional long-distance sorties outside the total home range occur.

geneously for particular behaviors. A few areas of intense use (peaks) may be contrasted with areas of infrequent use (valleys), giving rise to a varied behavioral topography (fig. 12.6) even during short time periods.

Comparison of the details of nightly movement parameters as they relate to age or sex characteristics, as well as to lunar illumination (table 12.5) substantiates that the way in which total home ranges are used does not differ among adult males, adult females, or subadults (age-sex is not significant for any parameter) even though the terrain over which subadults forage is greater than that of adults (table 12.4). Mainland bat populations may suffer

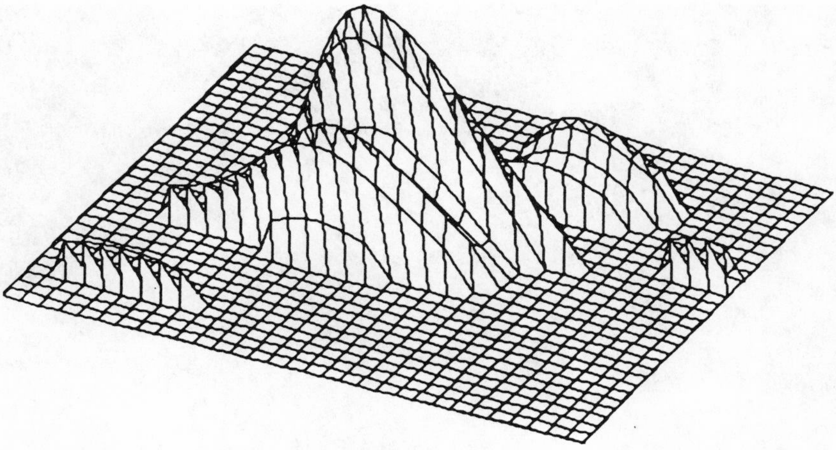

Figure 12.6. A minimum area probability plot or space utilization distribution (Anderson 1982) representing the total home range (in the shaded x-y plane) with its frequency of use plotted on the vertical z-axis. This gives rise to a topographic representation of space use for an adult female *Stenoderma rufum*. Areas of intense space-use in the background (peaks) are distinct from the isolated areas of minor use in the foreground.

Table 12.5. Statistical comparison (analysis of variance) of the effects of lunar phase and age-sex categories on movement parameters of the red fig-eating bat at El Verde

		Lunar		Age-Sex		Lunar × Age-Sex	
Movement Parameter	d.f.	F	P	F	P	F	P
Mean Distance Per Hour	10	0.004	0.95	0.341	0.72	0.163	0.85
Maximum Distance	10	0.096	0.76	0.128	0.88	0.390	0.69
Minimum Distance	10	2.501	0.15	0.177	0.84	0.051	0.95
Mean Distance Per Move	10	0.545	0.48	0.508	0.62	0.022	0.98
Number of Moves Per Hour	10	0.550	0.48	0.102	0.90	0.173	0.84
Percent Time Roosting	10	0.061	0.81	0.031	0.97	1.007	0.40

appreciable predation pressure, resulting in modification of foraging behavior during periods of high lunar illumination (i.e., lunar phobia). The tabonuco forest harbors few nocturnal predators capable of subduing bats. As a result, *S. rufum* should not experience severe selection to reduce foraging activity during high lunar illumination even though closely related taxa exhibit such behavior on the mainland (Morrison 1978b). The absence of a significant lunar effect for each foraging parameter (table 12.5) corroborates this prediction and further indicates that the absence of lunar phobia is consistently exhibited by subadults as well as by adult males and females.

Figure 12.7. Photograph (courtesy of M. R. Gannon) of an adult female *Brachyphyllum cavernarum*, the Antillean fruit-eating bat, from Puerto Rico

The Antillean fruit-eating bat, *Brachyphylla cavernarum* (fig. 12.7), occurs on Puerto Rico, the Virgin Islands, and throughout the Lesser Antilles to St. Vincent and Barbados. Netting records suggest that it is uncommon in the tabonuco forest (table 12.2). Swanepoel and Genoways (1983) summarized published information on the biology of *B. cavernarum;* Nellis (1971) and Nellis and Ehle (1977) provided the most detailed information concerning its behavior and natural history (see also Silva-Taboada and Pine 1969). None of the digestive tracts of *B. cavernarum* that were captured at El Verde ($N = 8$) contained food; however, a number of reports summarized in Swanepoel and Genoways (1983) have indicated that it consumes fruit, flowers, pollen, and insects. *Manilkara* fruit and the flowers of palms (*Prestoea montana*) are likely dietary constituents at El Verde, as they are consumed in other tropical and subtropical habitats.

The buffy flower bat, *E. sezekorni* (fig. 12.8), is endemic to the Greater Antilles and associated islands. It is an uncommon species at El Verde (table 12.2). Its general biological features are summarized by Baker et al. (1980); few ecological data are available. The diet appears to consist mostly

Figure 12.8. Photograph (courtesy of M. R. Gannon) of an adult female *Erophylla sezekorni*, the buffy flower bat, from Puerto Rico

of *Piper aduncum* fruits at El Verde (table 12.3); however, this observation is based on analysis of the stomach contents of only five specimens.

Nectarivores

The nectarivore guild at El Verde is represented by only one species, *M. redmani* (fig. 12.9), the Puerto Rican long-tongued bat. The distribution of *M. redmani* includes the Greater Antilles and some of the southern Bahama Islands. It is common in the tabonuco forest, but never was caught in large numbers at any one site (table 12.2). Homan and Jones (1975) summarized published biological information on this species; ecological data are few, but suggest a diet of nectar, soft fruit, and possibly insects. None of the digestive tracts from specimens collected at El Verde contained seeds or insect remains, although a viscous liquid (= nectar or fruit pulp?) was found in the stomach of some specimens. Of the four adult females collected in the dry season (March), one was pregnant (Gannon 1991; Gannon and Willig 1992).

Figure 12.9. Photograph (courtesy of M. R. Gannon) of an adult male *Monophyllus redmani*, the Puerto Rican long-tongued bat, from Puerto Rico

Aerial Insectivores

Although four bat species (*E. fuscus, P. parnellii, P. quadridens,* and *L. borealis*) constitute the insectivore guild at El Verde, they represent a minor portion of the bat fauna as assessed by ground netting (table 12.2). However, insectivorous bats were frequently seen foraging in open areas (7 to 12 m above the ground) near El Verde Field Station throughout the night, but especially at dusk. Netting records probably provide underestimates of the guild's importance in the tabonuco forest. Frugivores and nectarivores would more frequently encounter ground nets while foraging in the understory of the forest, whereas aerial insectivores would infrequently encounter ground nets while foraging.

The big brown bat, *E. fuscus* (fig. 12.10), has a cosmopolitan distribution in the New World and although rare at El Verde compared to species in other guilds was more frequently caught than other insectivores (table 12.2). Based upon analyses of four digestive tracts which contained food, a variety of insect taxa constitute the diet (Blattodea [a blattelid, *Cariblatta* sp.]; Homoptera [a cixiid, *Pintalia* sp.; a cicadellid leafhopper; a flatid, *Colpoptera*

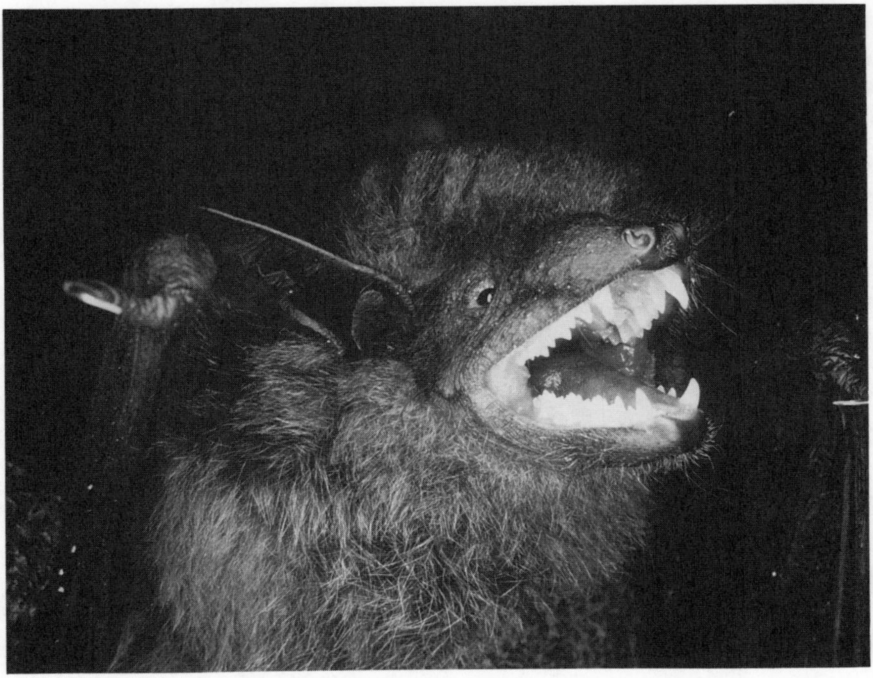

Figure 12.10. Photograph (courtesy of M. R. Gannon) of an adult female *Eptesicus fuscus,* the big brown bat, from Puerto Rico

sp. or *Neocolpoptera* sp.]; Hemiptera [a lygaeid, probably *Ozophora* sp.]; Coleoptera [a scarabaeid, probably *Phyllophaga* sp.]; Hymenoptera [a formicid, probably *Iridomyrmex melleus*]; and Isoptera [a termitid, *Nasutitermes* sp.]).

Parnell's mustached bat, *P. parnellii* (fig. 12.11), is distributed throughout much of the tropical and subtropical New World, including the Greater Antilles. The most recent systematic analysis of the relationship of this species to other mormoopids (ghost-faced bats) is provided by Smith (1972). It is rare in the tabonuco forest at El Verde (table 12.2). A general summary of the biology of *P. parnellii* is provided by Herd (1983); its diet includes Lepidoptera (moths), but mainly comprises Coleoptera (beetles).

The sooty mustached bat, *P. quadridens* (fig. 12.12), is endemic to the Greater Antilles, with *P. q. fuliginosus* occurring on Puerto Rico, Hispaniola, and Jamaica (Silva-Taboada 1976). It is rare at El Verde (table 12.2), and most information concerning the biology of the species is based upon the study of specimens associated with caves outside of the tabonuco forest (Rodriguez-Duran and Kunz 1992). In Puerto Rico, *P. quadridens* are known to forage

Figure 12.11. Photograph (courtesy of M. R. Gannon) of an adult female *Pteronotus parnellii*, the Parnell's moustached bat, from Puerto Rico

as far as 9 km from their day roost (Rodriguez-Duran 1984), and in Cuba it evinces the ability to home from distances of up to 30 km (Silva-Taboada 1979). The species feeds primarily on insects in the forest understory, with Coleoptera, Lepidoptera, Orthoptera, Hymenoptera, Diptera, Hemiptera, and Homoptera composing the bulk of the diet (Rodriguez-Duran 1984; Silva Taboada 1979)

The red bat, *L. borealis,* is migratory and occurs from temperate North America to temperate South America. Shump and Shump (1982) provide a summary of its biology, but most data are derived from temperate North America. Specimens roost in trees and shrubs, at times on or near the ground. Based upon records from ground netting, it is rare at El Verde (table 12.2). The contents of a single stomach included many flying male formicids, probably carpenter ants, *Camponotus ustus.*

Bat Ectoparasites

It is no more clear for bats in the tabonuco forest than for most animals in general whether parasites have an effect on population-level processes of

Figure 12.12. Photograph (courtesy of M. R. Gannon) of an adult male *Pteronotus quadridens,* the sooty moustached bat, from Puerto Rico

their hosts (Anderson 1982; Holmes 1983a; May 1983). Bat ectoparasites (Webb and Loomis 1977) and endoparasites (Ubelaker et al. 1977) have received varied attention, but mostly for New World leaf-nosed bats from a systematic perspective. Even so, prior to 1990, documentation of the ectoparasite fauna at El Verde was limited, based on data for only a few bat specimens, fewer than ten individuals in the case of most bat species. Herein, we provide an updated list of ectoparasites from bats captured at El Verde. We augment published records of ectoparasite infestation with those recently (1989–90) obtained from moderate samples (*A. jamaicensis,* $N = 86$; *S. rufum,* $N = 42$; *M. redmani,* $N = 22$; *E. sezekorni,* $N = 4$; and *Eptesicus fuscus,* $N = 3$) by Gannon (1991) and Gannon and Willig (1994b, unpublished) in the tabonuco forest. Flies (Streblidae) and mites (Labidpocarpidae, Macronyssidae, Spinturnicidae, Spelaeororhynchidae) are the predominant ectoparasites of bats from El Verde, whereas ticks (Argasidae) are rare and only known from *E. sezekorni* (table 12.6).

Incidence (percentage of bats with parasites), prevalence (mean number of parasites per bat), and density (mean number of parasites per infested bat) of ectoparasites of *S. rufum, A. jamaicensis,* and *M. redmani* at El Verde have

Table 12.6. Host-parasite associations for bats at El Verde

Host	Ectoparasite	Family
A. jamaicensis	*Megistopoda aranea* [a,c,g,*]	Streblidae
	Paralabidocarpus foxi [a]	Labidocarpidae
	Paralabidocarpus artibei [d,*]	Labidocarpidae
	Periglischrus iheringi [a,b,c,e,g,*]	Spinturnicidae
	Periglischrus vargasi [a,b,*]	Spinturnicidae
	Sphelaeorhynchus praecursor [a,c,f,g,*]	Spelaeorhynchidae
	Aspidoptera phyllostomatus [g,*]	Streblidae
	Trichobius intermedius [a,g,*]	Streblidae
	Trichobius robynae [a]	Streblidae
B. cavernarum	*Lawrenceocarpus micropilus* [a,d]	Labidocarpidae
	Lawrenceocarpus puertoricensis [a]	Labidocarpidae
	Radfordiella oudemansi [a,c]	Macronyssidae
	Trichobius truncatus [a,c,g,*]	Streblidae
	Periglischrus cubanus [g,*]	Spinturnicidae
E. fuscus	*Spinturnix bakeri* [g,*]	Spinturnicidae
E. sezekorni	*Trichobius robynae* [a]	Streblidae
	Trichobius truncatus [c]	Streblidae
	Ornithodoros viguerasi [c]	Argasidae
	Ornithodoros sp. [g,*]	Argasidae
	Periglischrus cubanus [g,*]	Spinturnicidae
M. redmani	*Spelaeorhynchus monophylli* [a,b,c,f,g,*]	Spelaeorhynchidae
	Trichobius cernyi [g,*]	Streblidae
	Trichobius robynae [a,g,*]	Streblidae
	Trichobius truncatus [a,c]	Streblidae
	Trichobius sp. (near *sparsus* Kessel) [b,*]	Streblidae
	Nycterophilia parnelli [g,*]	Streblidae
	Periglischrus vargasi [g,*]	Spinturnicidae
S. rufum	*Paralabidocarpus artibei* [a,d,*]	Labidocarpidae
	Paralabidocarpus foxi [a,*]	Labidocarpidae
	Paralabidocarpus stenodermi [a,*]	Labidocarpidae
	Periglischrus iheringi [a,b,c,g,*]	Spinturnicidae
P. Parnellii	*Cameronieta thomasi* [b,*]	Spinturnicidae

Sources: After Gannon and Willig 1994b.

[a] Webb and Loomis 1977.

[b] Tamsitt and Valdivieso 1970.

[c] Tamsitt and Fox 1970a.

[d] Tamsitt and Fox 1970b.

[e] Rudnick 1960.

[f] Fain et al. 1967.

[g] Gannon and Willig 1994b.

Note: Some ectoparasite records are from Puerto Rico in general, without documented occurrence on specimens from El Verde in particular; only parasites documented from El Verde are indicated with an asterisk.

been recorded by Gannon and Willig (unpublished). Prevalence differed as a consequence of host age, but not sex, with juveniles harboring higher numbers of ectoparasites in the cases of *S. rufum* and *A. jamaicensis.* In contrast, prevalence did not differ between seasons for any bat species. Significantly different parasite assemblages occurred on adult male, adult female, and juvenile *A. jamaicensis.* Moreover, *S. rufum, A. jamaicensis,* and *M. redmani* each had statistically distinctive ectoparasite assemblages.

TERRESTRIAL CONSUMERS

Excluding domesticated or feral mammals (dogs and cats), two terrestrial consumers (*R. rattus* and *H. auropunctatus*) occur at El Verde. Although two additional commensal species, *Mus musculus* and *Rattus norvegicus,* occur on Puerto Rico, no museum specimens or published records document their occurrence at El Verde. Moreover, no long-term studies of the ecology of *R. rattus* or *H. auropunctatus* have been undertaken in the tabonuco forest. Nonetheless, six vertebrate species (Puerto Rican short-eared owl, *Asio flammeus portoricensis;* the Puerto Rican boa, *Epicrates inornatus;* Puerto Rican parrot, *Amazona vittata;* the snake *Alsophis portoricensis;* and possibly both the Key West quail-dove, *Geotrygon chysia,* and the Puerto Rican whip-poor-will, *Caprimulgus noctitherus*) are considered to be endangered partly because of the introduction of rats or mongooses to Puerto Rico (Raffaele et al. 1973). The endangered birds construct ground or exposed nests which are vulnerable to predation by terrestrial mammals. Moreover, the decline or extinction of the mammalian insectivore, *Nesophontes edithae,* is circumstantially associated with competitive or predatory interactions with *Rattus* during post-Columbian times (Morgan and Woods 1986).

The black rat, *R. rattus* (fig. 12.13), is a common omnivore at El Verde; it was most likely inadvertently introduced to the island by Ponce de Leon in 1508 (Snyder et al. 1987). Population studies suggest a density of approximately 40 individuals ha^{-1} in tabonuco forest (see Weinbren et al. 1970; Brown et al. 1983), whereas up to 281 individuals ha^{-1} have been estimated in Puerto Rican parrot habitat in the palo colorado forest (Layton 1986). Immatures compose up to 35% of the population at El Verde, with December through February representing the months with the smallest proportion of immature individuals; recapture records indicate low survival rates even though few predators occur at El Verde (Weinbren et al. 1970).

Massive infestations of the liver parasite, *Capillaria hepatica,* characterized the specimens obtained by Weinbren et al. (1970). However, it is unclear how such parasitism is related to demographic properties of tabonuco forest populations of black rats. In a later study, 50% of the adults and 22% of the juveniles from a tabonuco forest population of black rats ($N = 25$) were infected with parasites, a rate of parasitism similar to that of populations

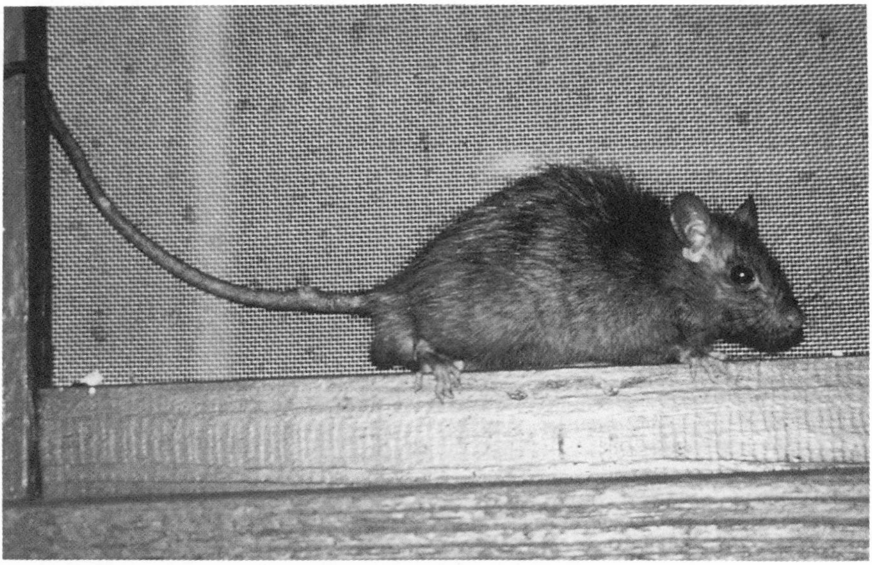

Figure 12.13. Photograph (courtesy of M. R. Gannon) of *Rattus rattus*, the black rat, from Puerto Rico

in palo colorado forest, but different from that within elfin forest (Layton 1986). These data certainly imply a possible regulatory effect by endoparasites on rat populations in the tabonuco forest. Regardless of the identity of the regulatory agent, maximum survival of rats is about two years, but mean survival is usually one year or less (Weinbren et al. 1970).

The black rat is arboreal in the tabonuco forest; it is frequently observed foraging in vines and trees throughout the night as well as in the day. Similarly, *Rattus* activity in palo colorado forest appears to be primarily arboreal, with diurnal denning sites located in the canopy (Layton 1986). Laboratory experiments showed that *R. rattus* will consume a variety of fruits from the tabonuco forest, including *Byrsonima coriacea, Cecropia schreberiana, Drypetes glauca, Ormosia krugii, Allophylus occidentalis, Inga vera, Dacryodes excelsa, Prestoea montana, Palicourea riparia, Tetragastris balsamifera,* and *Cordia borinquensis* (Weinbren et al. 1970). Snyder et al. (1987) suggest that sierra palm (*Prestoea montana*) is an important food source for rats and that their nests are associated with the vine, *Marcgravia sintenesii.* A statistical analysis (reported herein) of the consumption data from the laboratory (table 6 in Weinbren et al. 1970) indicates that all foods are not equally preferred (analysis of variance; $V_1 = 10$, $V_2 = 33$; $F = 5.18$; $p < .001$). An a posteriori Student-Neuman-Keuls Test (experiment-wise error rate = 0.05) revealed that *D. excelsa* and *T. balsamifera* were least preferred fruits; consumption of *C. schreberiana* and *C. borinquensis* was indistinguishable from that of all fruits; and the six other species constituted a group of most preferred fruits. Sastre-De Jesus (1979) reported that rats consume and disperse seeds of the late successional tree, *Buchenavia capitata,* in the tabonuco forest. Similarly, both immature and mature fruits of *Inga vera* were consumed by black rats at El Verde (Muniz-Melendez 1978). Additional studies of black rat foraging ecology under natural conditions are needed to confirm the existence of preferences when resource abundance reflects the ecological conditions in the tabonuco forest.

In addition to mongooses, which are almost exclusively diurnal, terrestrial predators of black rats (Pimentel 1955), red-tailed hawks (*Buteo jamaicensis*) and broad-winged hawks (*B. platypterus*) prey upon rats during the day (Layton 1986). Although aggregations of the Puerto Rican boa (*Epicrates inornatus*) have been observed feeding on bats as they leave caves at dusk (*B. cavernarum* and *M. redmani,* Rodriguez and Reagan 1984; *E. sezekorni,* Armando Rodriguez-Duran pers. comm., July 1994), the geomorphology of the Luquillo Mountains makes this phenomenon unlikely in the vicinity of El Verde. Nonetheless, the Puerto Rican boa has been found to consume adult rats in the tabonuco forest at El Verde (Reagan 1984; Thomas and Gaa Kessler, this volume).

Indian mongoose, *H. auropunctatus,* was introduced into Puerto Rico in 1877 in an attempt to control the rat population (Wadsworth 1949). Its density in the tabonuco forest is low, consistent with the observation that mongooses are not tree-climbers and avoid forested areas because they offer little chance to obtain food or adequate shelter (Pimentel 1955). Mongooses forage during the day and are relatively inactive during the night. They occupy burrows located near boulders, logs, or roots.

Mongooses may be among the most omnivorous species at El Verde. Pimentel (1955) reported that the diet of mongooses contained approximately 11% plant material, 56% insects, 17% reptiles, 12% myriapods, 8% arachnids, 3% mammals, 1% asteroids, and 1% amphibians by volume; in some areas, *Anolis* lizards constituted a major portion of the diet. Vilella (unpublished) provides the only dietary information concerning mongooses at El Verde; subsequent analyses are entirely based on that data. Stomach content analysis of thirteen male and six female mongooses from the tabonuco forest revealed exceptionally euryphagic diets, including both terrestrial and aquatic components. Males consumed a greater variety of food types (mean = 3.2; standard deviation = 1.5; range, 1–5) than did females (mean = 2.2; standard deviation = 1.0; range, 1–3). Nonetheless, most individuals that contained three or more food items in the stomach had vertebrates, invertebrates, and plant material as dietary constituents. Based on the percent of examined mongooses (N = 18) with food items in the stomach, dietary constituents included birds (11%); lizards in the genus *Anolis* (50%);

frogs in the genus *Eleutherodactylus* (11%); crabs in the genus *Epilobocera* (17%); freshwater shrimp in the genus *Attia* (17%); centipedes in the genus *Scolopendra* (33%); coleopterans (28%); orthopterans (33%); spiders (6%); and plant (fruits, seeds) material (61%), especially fruits from taxa in the Flacourtiaceae and Melastomataceae.

RESPONSE TO DISTURBANCE

Natural disturbances can have large effects on ecosystem structure and function depending on their scale, intensity, and frequency. On 18 September 1989, the eye of Hurricane Hugo passed within 10 km of El Verde, providing a rare opportunity to evaluate the effects of an infrequent but large-scale and high-intensity disturbance (Walker et al. 1991) on tropical bat species. Gannon and Willig (1994a) compare demographic parameters of the three most common phyllostomid bats (*A. jamaicensis, S. rufum,* and *M. redmani*) before and after the hurricane; their work forms the basis for the discussion that follows. In general, population levels, as estimated by numerical dominance (ND), biomass dominance (BD), and captures per net hour for the three species were affected by Hurricane Hugo in a species-specific fashion.

Artibeus jamaicensis was negatively affected by Hurricane Hugo, with a severe decrease in numbers, as well as in relative importance (fig. 12.14). Its numbers remained low for two years after the hurricane. Because it is a strong flier that moves large distances (Handley et al. 1991), immediate reduction in numbers may reflect movement of individuals from severely affected areas to less affected areas of the forest, or alteration of foraging patterns by cave-dwelling individuals. Data collected since the rainy season of 1991 indicate that *A. jamaicensis* has returned to the level of dominance it evinced prior to Hurricane Hugo.

Relative to other species, *M. redmani* was affected positively by Hurricane Hugo. Both its biomass dominance and numerical dominance at El Verde increased compared to pre-hurricane levels (fig. 12.14). However, only a slight increase occurred in actual numbers of captured individuals. This relative increase is due mostly to the decrease of the other two, previously dominant species. Nonetheless, the small increase in actual numbers may be attributed, in part, to the rapid and sizable increase in the presence of flowering plants in the open forest understory after Hurricane Hugo.

Numbers of *S. rufum* declined to about 30% of pre-hurricane levels and did not recover after three years (fig. 12.14). Moreover, analyses of telemetry data (two years before and two years after Hurricane Hugo) indicate that foraging and total home range size of individuals expanded to encompass an area approximately five times larger than its pre-hurricane size (fig. 12.15). The cost of foraging, in terms of time and energy, may be considerably elevated over pre-hurricane scenarios. The protracted decline in numbers of *S. rufum*

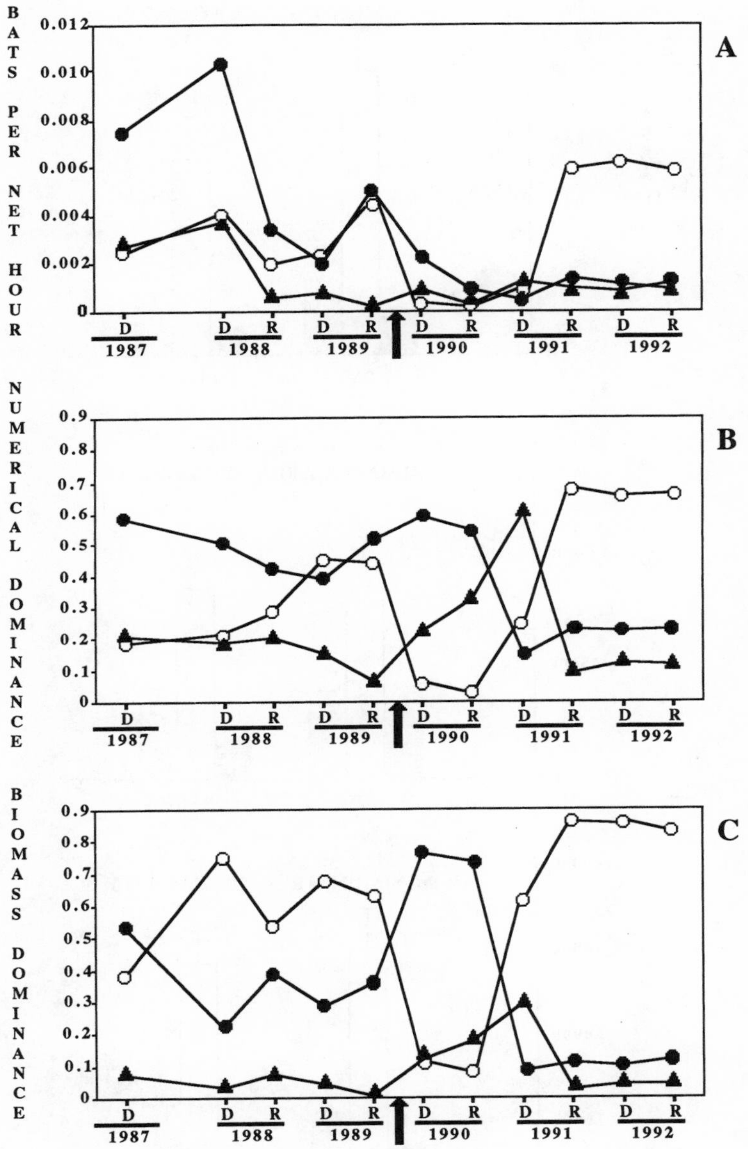

Figure 12.14. Long-term population trends of three common phyllostomid bats (*Artibeus jamaicensis,* open circles; *Stenoderma rufum,* closed circles; *Monophyllus redmani,* closed triangles) from a single netting locality in the tabonuco forest at El Verde based on number of specimens per net hour of sampling effort (*A*), numerical dominance (*B*), and biomass dominance (*C*). An arrow indicates the occurrence of Hurricane Hugo. Modified from Gannon and Willig (1994a).

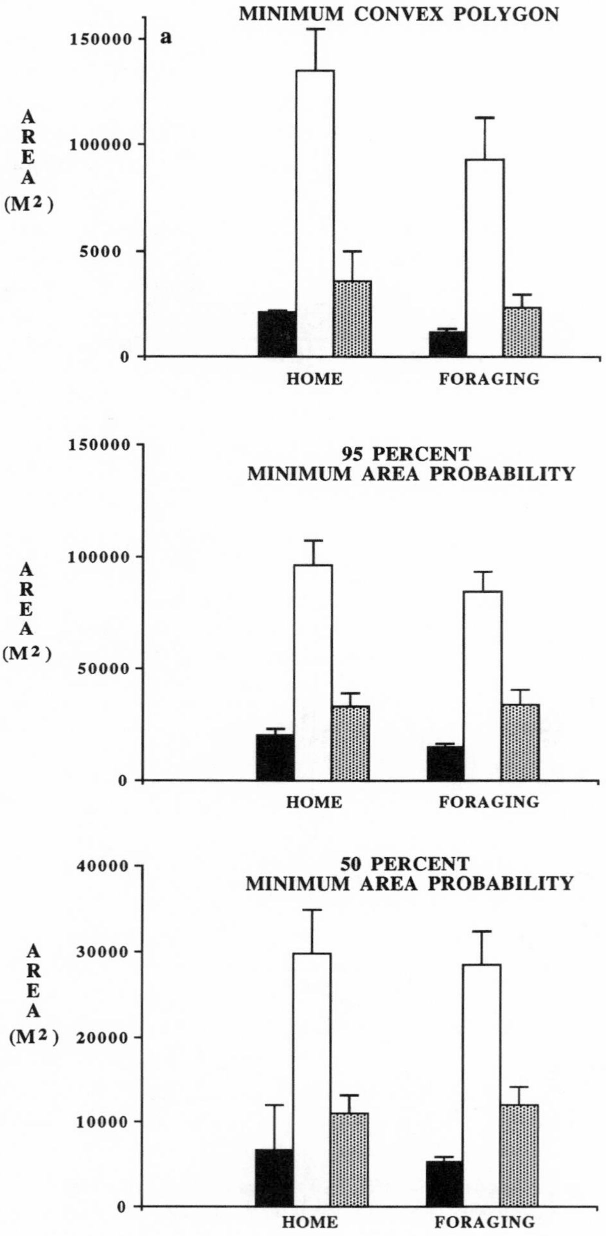

Figure 12.15. A comparison of mean total home range and foraging home range of *Stenoderma rufum*, before (solid bars) and after Hurricane Hugo (one year post-Hugo, unshaded bars; two years post-Hugo, stippled bars). Modified from Gannon and Willig (1994a).

may be related to four factors. First, increased exposure to high temperature, precipitation, and wind at roost sites (tree canopy) during and immediately after Hurricane Hugo likely caused immediate reductions in density. Second, the inability to disperse out of the tabonuco forest subsequent to the hurricane, as indicated by small foraging and total home range sizes, may have resulted in continued exposure to severe or detrimental microclimatic conditions, resulting in additional mortality and reduced fecundity. Third, decreased availability of fruit, especially that of *Cecropia,* for an extended period after the hurricane may have exacerbated microclimatic effects. Finally, the local population at El Verde could not be rescued by immigration from other demes because of the fragmented geographic distribution and low population levels of the species on Puerto Rico (Willig, Stevens, and Gannon unpublished).

The impact of Hurricane Hugo on reproduction of *S. rufum* occurred in two stages. The first stage reflects the greater susceptibility of juveniles to hurricane-induced alteration of resource and refuge characteristics (on average, proportions of juveniles decreased from 40% before to 17% immediately after Hurricane Hugo). The second stage reflects reduced fertility of adult females, as well as reduced survivorship of pregnant females and their offspring; few of the post-hurricane females were reproductively active (on average, proportions decreased from 93% before to 29% after the hurricane).

TROPHIC COMPARISONS

A recent study by Stevens and Willig (unpublished) has examined the composition of seven bat guilds from a variety of mainland communities throughout the New World; each has been netted on a monthly basis for at least twelve consecutive months (fig. 12.16). The bat community at El Verde is sufficiently well studied that one can be confident of its composition as well, making comparison with mainland tropical sites meaningful. Moreover, the clearly delimited species pool of potential bat colonists from Puerto Rico makes the distinction between island and tabonuco forest effects possible.

Most tropical bat communities from the mainland contain representatives of all (six of eleven sites) or at least six (ten of eleven sites) of the seven bat guilds, whereas El Verde harbors representatives of only three guilds (fig. 12.16). The absence of sanguinivores is an island-specific consequence of their inability to disperse to or persist on oceanic islands rather than a unique attribute of the tabonuco forest. Most of the islands of the Caribbean do not support blood-feeding bats. It is unlikely that any of the vampire bats could maintain sufficiently large populations to persist in the tabonuco forest given the absence of suitable nonvolant mammalian prey or suitable caves for roost sites. Similarly, the island of Puerto Rico does not harbor any species of gleaning carnivore (e.g., bat species in the Phyllostominae); their

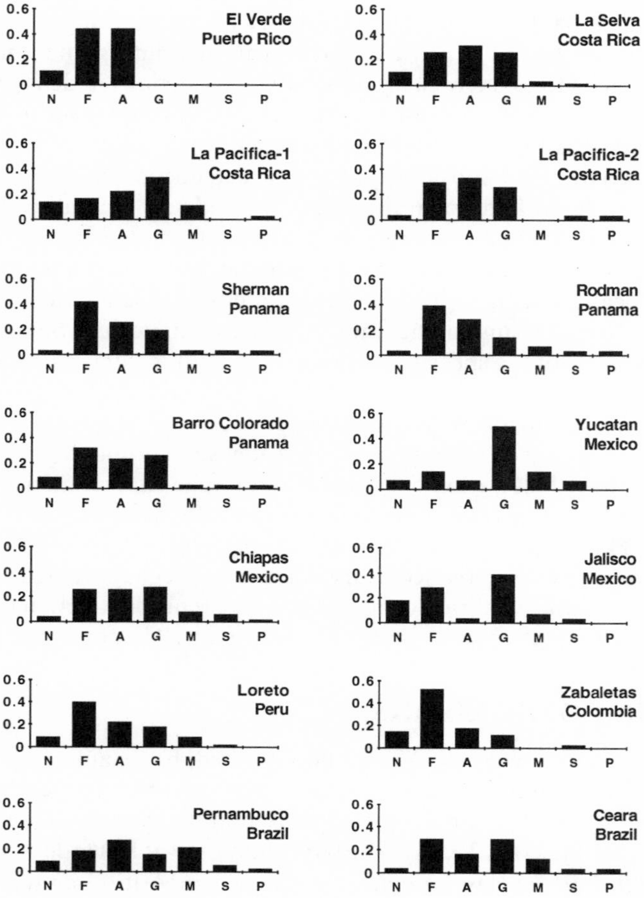

Figure 12.16. Guild structure (on the ordinate, proportion of species represented by a particular guild) of the bat community in the tabonuco forest of El Verde, as well as at eleven other tropical sites throughout Central and South America (La Selva, Costa Rica, La Val and Fitch 1977; La Pacifica-1, Costa Rica, Fleming et al. 1972; La Pacifica-2, Costa Rica, La Val and Fitch 1977; Sherman, Panama, Fleming et al. 1972; Rodman, Panama, Fleming et al. 1972; Barro Colorado, Panama, Bonaccorso 1975; Zabalatas, Colombia, Thomas 1972; Yucatan, Mexico, Bowles et al. 1990; Chiapas, Mexico, Medellin 1993; Jalisco, Mexico, Iniguez Davalos 1993; Loreto, Peru, Ascorra unpublished.; Pernambuco, Brazil, Willig 1982; Ceara, Brazil, Willig 1982). Guild codes (on the abscissa) are N, nectarivore; F, frugivore; A, aerial insectivore; G, gleaning carnivore; M, molossid insectivore; S, sanguinivore; P, piscivore.

absence from the community is likely an island biogeographic artifact. The absence of piscivores and molossid insectivores is not primarily an island attribute because the species pool of bats on Puerto Rico does comprise species in these guilds. Their absence from the tabonuco forest is more likely a consequence of the ecological characteristics of the forest. Streams are steep, fast running, and contain few fish (Covich and McDowell, this volume); as a consequence, *Noctilio,* the fish-eating bat, would have to rely almost exclusively on insects to persist at El Verde. Molossid insectivores (e.g., *Tadarida* or *Molossus*) have never been recorded from El Verde or the Luquillo Experimental Forest (Willig, Stevens, and Gannon unpublished); the density of insect prey above the forest canopy may be insufficient to support a local population of these species.

Compared to mainland tropical communities, the three feeding guilds present at El Verde (nectarivore, frugivore, and aerial insectivore) comprise larger percentages of the faunas primarily as a consequence of the absence of species in each of the other four guilds. Nonetheless, mainland communities are commonly dominated by species in the frugivore or aerial insectivore guild. The aerial insectivore guild at El Verde may be limited because of the abundant diurnal and nocturnal predators (frogs and lizards) that reduce available prey (Reagan, this volume; Stewart and Woolbright, this volume).

DISCUSSION

A discussion of the impact and role of mammals at El Verde is difficult to develop in the absence of detailed information from throughout the year on the population biology and foraging ecology of the common mammal species. In particular, three species (*A. jamaicensis, S. rufum,* and *R. rattus*) deserve future study because their relatively large size, high metabolic rate, probable high biomass, and mobility suggest an important role in affecting energy and nutrient budgets of the tabonuco forest ecosystem, or in shaping landscape patterns. All three species may be important seed-dispersal agents. For example, *Manilkara bidentata* is a dominant tree in the tabonuco forest and appears to be exclusively dispersed by bats (You 1991), most likely *S. rufum* (table 12.3; Gannon and Willig 1994a). In addition, both *A. jamaicensis* and *S. rufum* consume fruit of *Cecropia schreberiana,* a dominant tree in the tabonuco forest that produces fruit throughout the year. In general, *Cecropia* plays a major role in secondary succession and forest regeneration after disturbance (e.g., tree falls, landslides, or hurricanes). Considerable evidence suggests that as a result of coevolution, bats and *Cecropia* are mutualists (for a review, see Heithaus 1982). Fruits obtained from a tree may be dispersed several hundred meters to several kilometers (Fleming and Heithaus 1981). Moreover, food transit time in the digestive system of bats is short (usually less than twenty minutes) and defecation usually occurs in flight. The many tiny seeds contained within *Cecropia* fruit lose their cohesiveness after passing through the digestive system, and because of air turbulence, the seeds fall to the ground in a trail up to 400 × 30 cm (Charles-Dominique 1986). Seed germination of *Cecropia* is enhanced by passage through the digestive tract of bats (Fleming and Heithaus 1981).

Adult *A. jamaicensis* require 43.9 kJ d^{-1} to support basal metabolic needs and minimum maintenance flight (Morrison 1978a) whereas *C. schreberiana* fruit contains 4,675 calories (and 32.6 mg nitrogen) g^{-1} dry weight (Scogin 1982). Discounting costs of foraging (which cannot be estimated because search and travel time are not known for *A. jamaicensis* in the tabonuco forest) and assuming an assimilation efficiency of 20% (see Morrison 1978a, 1980), then an adult *A. jamaicensis* must consume at least 11.2 g dry weight of *Cecropia* a night to maintain a positive energy balance, and perhaps substantially more, because much of the caloric value of *Cecropia* is contained within the seeds, which for the most part pass intact through bat digestive systems.

Like *Cecropia,* a number of species of *Piper* contribute to forest regeneration as early successional species in light gap areas. Moreover, bats are important seed-dispersal agents for *Piper* and have been considered keystone mutualists in some tropical forests (Fleming 1985). Although the data are few from the tabonuco forest, they suggest a one-to-one correspondence between bat consumers and *Piper* species in that *P. glabra* is consumed only by *A. jamaicensis, P. hispidum* is consumed only by *S. rufum,* and *P. aduncum* is consumed only by *E. sezekorni.* These bat species also consume fruit from other species, and the common bats (*A. jamaicensis* and *S. rufum*) consume appreciable quantities of *Cecropia* as well. This may facilitate the deposition of later successional seeds (*Cecropia*) in light gap areas that are in early stages of secondary succession (dominated by *Piper* species).

The two common frugivores at El Verde appear to have similar diets. Each consumes five food types, four of which are shared dietary constituents. Nonetheless, the proportional contribution of the various components (after combining the three *Piper* species with *M. bidentata* to obtain sufficiently large sample sizes) are significantly different (G-Test of Independence; $G = 13.78$; d.f. $= 3$; $p = 0.003$), indicating real dietary differences between *A. jamaicensis* and *S. rufum* at El Verde.

Because *R. rattus* is a large abundant consumer in the tabonuco forest, it potentially has a significant impact on food web structure and function. Moreover, its euryphagic habits suggest a broad range of primary producers upon which it may have an effect via seed dissemination or predation. Unlike the scenario for bats, time may not have permitted coevolutionary adjustment between the black rat and its prey. The manner in which the black rat affects the abundance and distribution of its prey could have disruptive effects on any equilibrial conditions which may have existed among components of the food web in pre-Columbian times.

SUMMARY

The mammal fauna of Puerto Rico is depauperate compared to equivalent areas of the mainland or to other islands in the Caribbean as a consequence of biogeographic and anthropogenic factors. In the same manner, the tabonuco forest at El Verde contains fewer mammal species than it might otherwise support because the pool of colonist species on the island is small and functionally different than that on the mainland. With the exception of feral animals (i.e., cats and dogs), the mammal community at El Verde comprises two introduced terrestrial mammal species (Indian mongooses and black rats) and nine bat species, the only native mammals.

Mongooses and rats are relatively large, mobile omnivores; they have been implicated in the extinction of a number of bird species in the food web at El Verde. Black rats in particular may occupy a dominant role in the food web because they forage on the ground as well as in the canopy, they have high population densities (281 individuals ha^{-1}), and they consume a variety of fruits, invertebrates, and vertebrates. Although the density of mongooses is relatively low, they are among the most omnivorous species in the food web, feeding on all but the quinary consumer trophic level. The high mobility of both of those terrestrial mammals suggests that they may transport seeds a considerable distance from parent trees, thereby affecting survivorship and spatial distribution of seedlings.

Bats are highly mobile, nocturnal consumers and occupy frugivore (four species), nectarivore (one species), and insectivore (four species) feeding guilds at El Verde. The absence of some bat guilds (e.g., sanguinivores) from El Verde is a consequence of island biogeographic phenomena, whereas the absence of others (e.g., foliage gleaning carnivores) may be a consequence of the elevated density of arboreal frogs and lizards. Because of bat behavior, accurate estimates of population density are not possible using conventional multiple mark and recapture techniques. Nonetheless, numerical or biomass dominance, based on netting records prior to Hurricane Hugo, indicate that frugivorous species (ND = 0.81; BD = 0.94) are far more abundant than insectivorous (ND = 0.03, BD = 0.02) or nectarivorous (ND = 0.15, BD = 0.05) counterparts. In particular, two frugivores, *A. jamaicensis* (ND = 0.40, BD = 0.62) and *S. rufum* (ND = 0.37, BD = 0.28), are the most dominant bat species in the forest. Unfortunately, estimates of biomass and numerical dominance for insectivorous bats are biased. Insectivorous species consume prey that occur in or above the forest canopy; as a consequence, these species are less likely to be caught in ground nets than are frugivorous or nectarivorous bats.

Many bats are keystone mutualists in tropical ecosystems, primarily as a consequence of their role in seed dispersal and pollination. In fact, a number of early successional shrubs (e.g., *Piper* spp.) and late successional trees (e.g., *Manilkara*) in the tabonuco forest rely on bats for pollination or seed dispersal. These activities may be particularly important because recurrent hurricanes and tropical storms continually create patches (e.g., landslides and treefalls) within the forest that undergo secondary succession.

The effect of Hurricane Hugo on bat populations was species specific. *Artibeus jamaicensis* exhibited an immediate and drastic decline in captures, followed by relatively rapid recovery. In contrast, *S. rufum* declined more gradually and has not yet (1992) recovered to pre-hurricane levels. The considerable demographic data available on *S. rufum* suggests that philopatric and canopy-roosting species are particularly susceptible to large-scale and intense disturbances such as hurricanes and tropical storms.

In conclusion, mammals consume prey from a variety of trophic levels in the food web at El Verde and are represented by herbivorous (frugivores, granivores, nectarivores) and carnivorous species. Although primarily nocturnal, they forage at a diversity of vertical strata within the forest, including the litter, the shrub, subcanopy, and canopy levels, and even above the canopy. Mongooses are highly omnivorous and consume prey from four trophic levels in the grazing circuit. Rats are similarly omnivorous, prey on a variety of decomposers, producers, and consumers, and consequently occupy trophic positions in grazing and detrital circuits. Bats are more stenophagic, with particular species being either herbivores, consuming fruit and nectar, or insectivores, consuming arboreal invertebrates.

ACKNOWLEDGMENTS

We would like to express our appreciation to Doug Reagan and Bob Waide for their creative execution of duties as editors of this volume. Moreover, Doug as responsible for MRW's introduction to the Luquillo Mountains, and Bob consistently provided support and encouragement since our initiation at El Verde. Along with Alan Covich, Rosser Garrison, Meg Stewart, and Larry Woolbright, they have made ecological research at El Verde productive and enjoyable. Many students aided in gathering the data used in this chapter; G. Camilo, J. Cary, R. Colbert, S. Cox, D. Ficklin, M. Krissinger, K. Lyons, M. Mays, D. Paulk, E. Sandlin, D. Smith, R. Stevens, A. Towland, and R. Van Den Bussche deserve particular thanks. Not least among those worthy of acknowledgment is Alejo Estrada-Pinto—botanist, field assistant, and logistician. Financial support was primarily provided by the Department of Energy through programs (faculty participation and travel contracts) administered by Oak Ridge Associated Universities, as well as through a grant from the International Institute of Tropical Forestry (Forest Service, U.S. Department of Agriculture). Supplemental support in later years was provided through grant BSR-8811902 from the National Science Foundation to the Terrestrial Ecology Division, University of Puerto Rico, and the International Institute of Tropical Forestry as part of the Long-Term Ecological Research Program in the Luquillo Experimental Forest. Additional funds were provided by Texas Tech University (State of Texas Organized Research Funds administered by the College of Arts and Sciences; summer grants from The Graduate School; and research support from the Department of Biological Sciences), the American Museum of Natural History (Theodore Roosevelt Fund), Bat Conservation International, the American Society of Mammalogists (Grants-in-aid of Research), and the Terrestrial Ecology Division of the University of Puerto Rico. The clarity and content of the manuscript was improved by reviews of earlier versions of this chapter by H. H. Genoways, M. A. Mares, O. J. Reichman, M. M. Stewart, and L. L. Woolbright, along with D. P. Reagan and R. B. Waide.

13

The Stream Community

Alan P. Covich and William H. McDowell

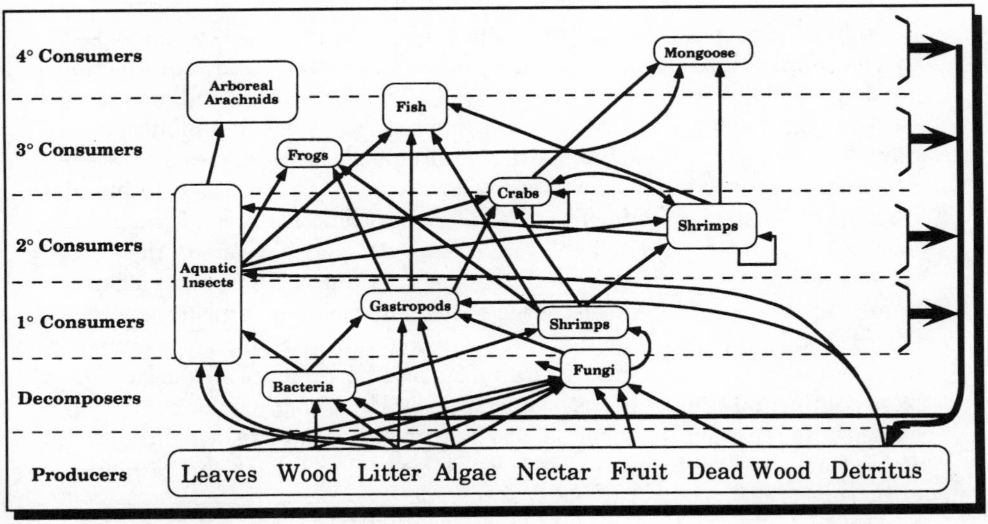

IN THIS chapter we summarize information regarding species which live in the streams of the Luquillo Experimental Forest and describe the general nature of their trophic connections among both aquatic and terrestrial components. Differences in access to certain sections of headwater streams by top-level carnivores as well as different degrees of spatial and temporal variability among detrital inputs apparently control the main structure of food webs in El Verde streams. The location and height of waterfalls along the stream channels greatly restrict which fishes, decapods, and gastropods reach headwater pools as they migrate upstream from coastal waters. Once species migrate into these headwaters, individual growth and reproduction are determined by food availability, as well as by competition and predation.

We suggest that there are only a few physical variables that influence the flow of energy in the stream network that drains the steep, forested slopes of El Verde. Food web interactions among terrestrial and aquatic organisms are thought to be strongly influenced by (1) geomorphic features of the catchment, such as the rapid rise in elevation from the coastal rivers to the headwater streams; (2) the intensity and frequency of rainfall events and high temporal variability of stream discharge that regulate inputs and transport of detrital food supplies and silt; and (3) biogeographic parameters, especially the low diversity of consumers resulting from the isolation of Puerto Rico from continental sources of freshwater species. These characteristics are similar to those reported for other small Neotropical streams (Allee and Torvik 1927; Darnell 1956; Brinson 1976; Dietrich et al. 1982; Barnish 1984; Boon et al. 1986).

The Luquillo headwater streams, as those of other Caribbean islands, have a low number of species relative to mainland streams. The representative food web in small, mid-elevational, montane streams within the Luquillo Experimental Forest consists of several size classes of predaceous eels and omnivorous mullets, together with four species of gobiid fishes, two species of grazing gastropods, eleven species of decapod crustaceans, and more than sixty species of aquatic insects. Depending upon their sizes and feeding adaptations, these shrimps, crabs, and insects have varied feeding niches including those of herbivores, detritivores, omnivores, and carnivores. Most of these invertebrates can play major trophic roles in high-elevational, headwater streams where fishes and other vertebrate predators are rarely found. Atyid shrimp are very abundant herbivores/detritivores in these headwater streams, where they constitute most of the biomass. Densities and rates of secondary production, however, are probably highest among the smaller, fast-growing aquatic insects. Data on primary productivity in these streams are generally lacking, although recent and on-going studies are beginning to fill this gap. Long-term data on stream flow, nutrient concentrations, and sediment transport are often limited for small Neotropical streams. However, as discussed below, data on these important parameters have been directly measured (or can be estimated) from hydrologic records collected over a period of twenty-five years in the Luquillo forest.

PHYSICAL CONTROLS ON STREAM FOOD WEBS

Streams and rivers drain relatively narrow watersheds as they flow through several types of forest at El Verde. The variations in stream width as well as the height and type of trees along the stream banks determine the extent of shading and the quantity and quality of detrital inputs to the stream consumers. The biotic community is strongly influenced by the steep slopes (approximately 25% on average) in the stream channels and by the frequency of high waterfalls flowing over exposed bedrock and large boulders. The height and steepness of waterfalls influence access of the biota to upstream habitats as well as provide important microhabitats for some specialized species.

The research discussed in this chapter focuses on three streams in the vicinity of the El Verde field station that pass through or are adjacent to many of the terrestrial sites discussed in previous chapters. Quebrada Sonadora is the main tributary of the Rio Espiritu Santo at elevations above 500 m. The Sonadora originates at approximately 975 m elevation in the cloud forest and flows past the site of the El Verde field station (at 350 m) in the tabonuco rain forest before joining the Espiritu Santo at 180 m elevation. The general features of this basin are reviewed by Lugo (1986) and McDowell and Asbury (1994), and a description of geomorphologically similar streams is provided by Scatena (1989).

The Prieta and Toronja are both similarly sized tributaries of the Sonadora and are two of our study sites. These are small, steep-gradient, boulder-lined streams. Boulders in the channels are typically well rounded and up to several meters in diameter. Catchment basins of these tributaries are amphitheater shaped. These "swallows" provide storage areas for leaf accumulation during dry periods. Similarly, leaf litter accumulates on the tops of large boulders within the active channel and on flood plains and wetlands adjacent to larger streams, where elevational profiles of channels level off for short distances.

The rate of lateral and downslope movement of leaf litter across the landscape is influenced by several biotic and abiotic factors, such as fungal growth, slope, and soil wetness (Lodge and Asbury 1988).

Energy Sources

Direct solar inputs to the montane stream ecosystem are restricted to those pools located in gaps along the riparian corridor. These openings occur following tree falls along the steep stream banks, especially after intense storms. Bank erosion and windthrows occur frequently in some locations, particularly in unstable steep terrain dominated by early successional trees (such as *Cecropia scheberiana* and *Prestoea montana,* sierra palms). In contrast, those pools in riparian zones that are dominated by late-successional trees (such as *Dacryodes excelsa,* tabonuco) rarely are exposed to direct inputs of solar energy, and thus primary productivity is generally low (Karen Buzby and Cathy Pringle pers. comm. 1991).

Terrestrially produced leaf litter, small twigs, and fruits constitute a relatively continuous input of detritus-based energy to these small streams (Bloomfield et al. 1993; Zou et al. 1995; Vogt et al. in review). Two peaks of leaf fall occur each year (March through May and September through October) in the tabonuco forests at El Verde (Wiegert 1970a; LaCaro and Rudd 1985). During the dry season (January through April) accumulations of leaf litter and fruit fall from the vegetation near the stream occur as a result of increased input and decreased washout. In preliminary studies we found that newly fallen leaves of trees such as the dominant tabonuco decayed rapidly relative to other species of late successional trees such as ausubo, *Manilkara bidentata* (fig. 13.1). To determine the quality of available foods, we have also begun to characterize leaves in terms of their carbon/nitrogen (C/N) ratios for representative tree species within the riparian zone (table 13.1). The time course of nitrogen enrichment and resultant changes in C/N ratios in decomposing leaves is initially very rapid (e.g., *Buchenavia capitata,* fig. 13.2).

Because most headwater streams in tabonuco forests receive relatively continuous amounts of leaf detritus as a common source of energy, the retention and storage of organic matter are key variables in providing food and cover for stream consumers. Coarse woody debris (branches and tree trunks) and organic debris dams along the Sonadora are uncommon, apparently because of the very high peak discharges. Leaf accumulations in riffles and pools of the Sonadora are also limited because of the very high peak discharges. The hydrologic power of this large stream is frequently sufficient to carry large, uprooted trees downstream many meters before boulders impede their movement. Debris dams (up to 1 m in height) can form on the smaller

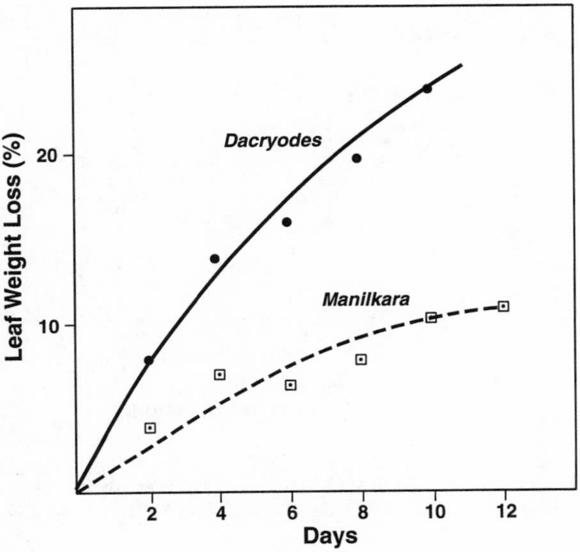

Figure 13.1. Percentage of weight loss over time for two species of late-successional trees in Quebrada Prieta (Covich and McDowell unpublished).

Table 13.1. Carbon/nitrogen ratios (mass/mass) of riparian species' leaf litter

Low C/N (<50)	High C/N (>50)
Drypetes glauca	*Buchenavia capitata*
Inga fagifolia	*Clusia gundlachia*
Trichilia pallida	*Manilkara bidentata*
Tabebuia heterophylla	*Homalium racemosum*
Sapium laurscerasus	*Dacryodes excelsa*
	Croton poecilanthus
	Sloanea berteriana

streams with complex boulder substrata, especially after storms with high winds and rain.

The slow breakdown of some debris dams (those created by hardwood species such as tabonuco and ausubo) evens out the temporal distribution of high-quality leaf litter to detritivores in headwater streams (Covich et al. 1991; Covich et al. in press; Johnson et al. in press). Changes in supplies of high-quality detritus also result because some leaves decompose relatively quickly (Stout 1980; Webster and Benfield 1986; Covich 1988b) and may be synchronized in when they fall. There are also delays in decomposition of

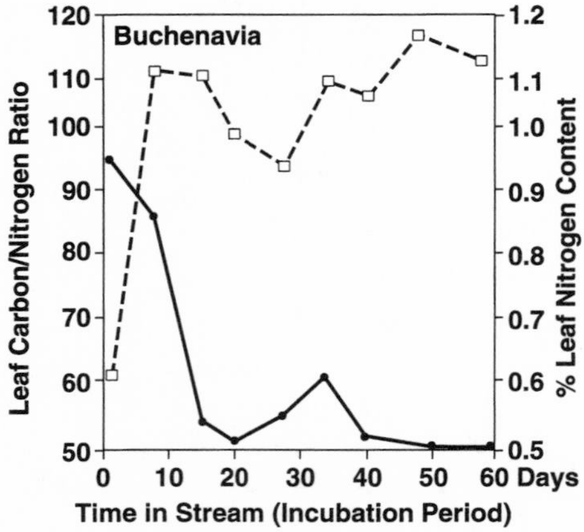

Figure 13.2. Changes in carbon/nitrogen ratios (mass/mass) over time in *Buchenavia capitata* leaves incubated in a laboratory stream at El Verde. C/N ratios are solid line; percentage leaf nitrogen content are indicated with dashed line. (Covich and McDowell unpublished).

refractory organic detritus in the form of roots, sticks, and blades of palm fronds that are retained in complex substrata (Covich and Crowl 1990) but break down relatively slowly in streams (Vogt et al. in review).

Rates of decomposition of many woody species that characterize the riparian vegetation (such as sierra palms) can be rapid, especially if insect attacks increase surface penetration by fungi and bacteria. The relatively constant warm temperatures also result in high rates of decomposition by bacteria and fungi, but different tree species have distinct patterns of decay depending on the chemical contents of their leaves and wood (Padgett 1976; Stout 1980; Bloomfield et al. 1993; Vogt et al. in review; Buzby in prep.). Temperature shows little seasonal variation, ranging from 18° to 23° C annually in the Sonadora. Throughout the year an increase of 3° C is generally observed along the Sonadora from 975 to 180 m elevation. Variations in water temperatures from day to night are restricted to less than 3° C (Covich unpublished).

Variable Rainfall, Discharge, and Nutrient Limitations

Variability of rainfall and stream discharge is a major factor determining food web structure through its influence on the availability of detrital resources (e.g., nitrogen content of suspended particles, fig. 13.3). During prolonged periods of below-average rainfall and runoff there may be extreme low flows that also alter food web composition and function through the accumulation of organic detritus and increases in the population densities of some consumers. Although larger streams generally do not dry completely, the uppermost headwater tributaries vary markedly in their permanence.

Without flushing events caused by high storm flows, first- and second-order streams may accumulate relatively large amounts of organic detritus and inorganic sediments that decrease pool depths. Reduced pool depths and volumes may also expose some prey to increased risks of predation as refuge sites are left high and dry. In shallow pools, some prey species are no longer protected by their ability to swim up into surface waters that are out of reach of bottom-dwelling predators. On the other hand, damage and mortality due to physical scour are minimized during prolonged dry periods. Well-lit pools in light gaps may also accumulate attached algae (periphyton) during these dry periods if nutrients from organic decomposition are available. In contrast, rapid increases in stream flow may (1) dilute dissolved nutrients required for sustained algal growth; (2) wash out algal and detrital foods in scoured channels; (3) fill crevices (refugia from predation) with silt in depositional pools; and (4) dislodge some life-history stages of many bottom-dwelling or open-water species.

Because of the steep catchment slopes, stream discharge can increase very rapidly following storm events: ten-fold increases in discharge in less than an hour have been recorded by the U.S. Geological Survey. Between storm events, clarity of large streams rapidly returns to pre-storm levels. These channels are also sufficiently wide (up to 15 m) to allow direct sunlight to penetrate the water. In these large streams growth of periphyton may be an important potential source of energy for grazers. Although algal biomass is relatively low, some grazing does occur on periphyton-covered rocks (Pringle et al. 1993). Nitrogen concentrations in these montane streams are generally

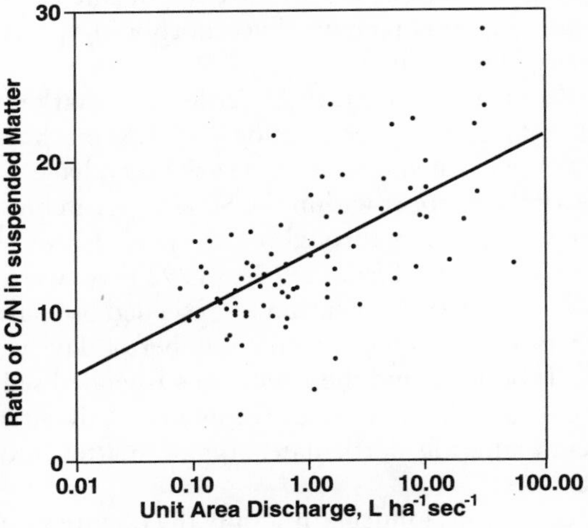

Figure 13.3. Relationship between carbon/nitrogen ratios (mass/mass) of suspended sediments and unit area discharge for Quebrada Sonadora (data plotted from McDowell and Asbury 1994).

Table 13.2. Dissolved and particulate matter in selected Luquillo streams

Parameter	Quebrada Sonadora	Rio Icacos	Quebrada Toronja
NH_4-N	0.011	0.014	0.011
NO_3-N	0.056	0.066	0.062
Particulate organic N	0.024	0.025	0.015
Dissolved organic N	0.13	0.11	0.15
Particulate organic C	0.41	0.35	0.19
Dissolved organic C	2.13	1.58	1.36
Dissolved inorganic C	2.93	11.9	10.20
Na	4.36	5.07	7.52
Ca	2.22	3.33	6.27
Mg	1.40	1.20	4.42
K	0.22	0.51	0.28
Cl	6.80	6.24	8.82
SO_4	2.71	2.24	2.67
SiO_2	9.72	18.3	22.9
Suspended sediments	5.0	11.8	7.0

Notes: Data are from March 1983 to February 1986. All values are expressed in mg l^{-1}. Dissolved phosphorus concentrations were below detection limits of 0.001 mg l^{-1} (McDowell and Asbury 1994).

low (table 13.2) and dissolved phosphorus is often below detection limits (0.001 mg l^{-1}) so that even in streams with full sunlight the algal growth is limited. In small first- and second-order streams such as Quebrada Prieta, the riparian vegetation completely covers the narrow (3 to 5 m) channel and prevents bright sunlight from entering the streams. In these small streams retention and storage of organic detritus are important variables, which are directly associated with stream channel geomorphology, complexity of substrata, and intensity of storm flows.

Rainfall records for sixteen years at El Verde show marked seasonal variation in maximal and minimal precipitation (fig. 1.5, this volume). This seasonality, as well as the marked interannual variability, have a direct influence on runoff and detrital transport within the Sonadora catchment. The annual distribution of rainfall intensity throughout the year also varies considerably. For example, during the period from 1967 to 1992 there were only four years (1970, 1979, 1981, and 1987) when runoff exceeded 300 cm (fig. 13.4). Yet these years differ markedly in terms of the number of days with average discharge exceeding 1000 l s^{-1} and the export of suspended sediment and particulate organic carbon. These intense storms indirectly alter food webs by exporting nutrients, altering particulate organic matter, eroding channels, and causing landslides.

Variable flow conditions influence not only the quantity but also the quality of suspended matter. Most suspended organic particles are leaf fragments that have been broken down by both microbial degradation and physical fragmentation. The ratio of carbon to nitrogen increases as unit area of discharge increases (fig. 13.3). Because nitrogen-enriched food particles provide filter feeders with high-quality food resources, the time for optimal filtering would appear to be during relatively low flows (<1.0 l $ha^{-1}s^{-1}$) but when turbulence is sufficiently high to keep organic particles in suspension.

Concentrations of many dissolved substances show much less variability with discharge than suspended sediments or particulate carbon (McDowell

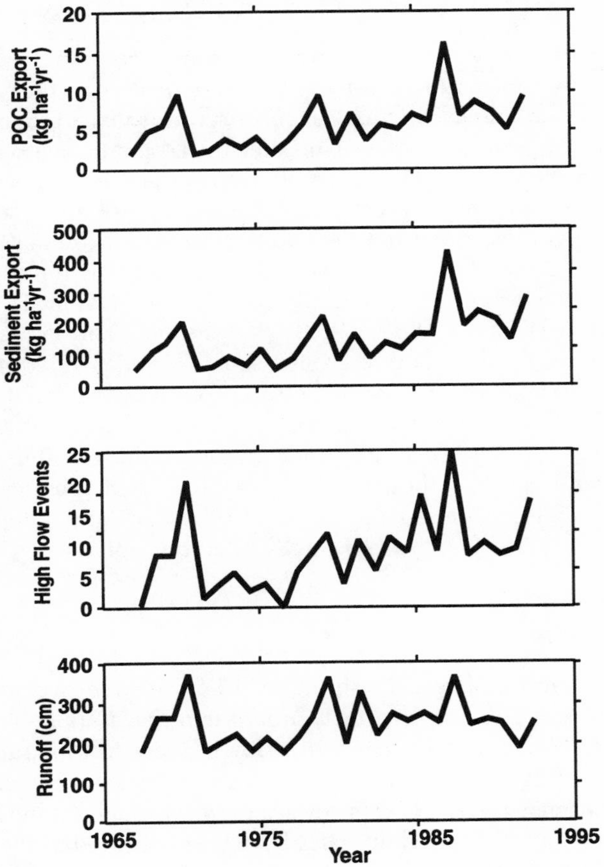

Figure 13.4. Long-term data (1963–1992) on runoff, high-flow events (when runoff exceeded 300 cm), and export of particulate organic carbon (POC) and suspended sediments from Quebrada Sonadora. Data on stream discharge prior to 1983 are estimated from discharge at a gaging station downstream from the Sonadora (Rio Espiritu Santo). Exports of POC and suspended sediments are calculated from regression equations found in McDowell and Asbury (1994).

and Asbury 1994). Concentrations of most of the major cations and anions are inversely related to discharge because of considerable dilution. However, dissolved organic carbon is an important and consistent exception, showing pronounced increases in concentration with increasing discharge and runoff. Apparently, some organic debris is washed into the stream during major rain storms and is a source for dissolved organic carbon; aggregates of this material may be important food resources for filter-feeding insects and shrimp.

BIOTIC CONTROLS ON STREAM FOOD WEBS

One advantage for food-web analyses resulting from the remoteness of Puerto Rico's easternmost location in the Greater Antilles is the relatively low number of freshwater species that can constitute any given food web. As discussed in previous chapters, the island's terrestrial fauna is generally less diverse than on the mainland or on some of the larger islands of the Greater Antilles. In comparison with terrestrial habitats, there are relatively fewer aquatic species in the streams of the Luquillo Experimental Forest. The oceanic barrier to dispersal of freshwater organisms severely limits the number of species in these food webs, especially among upper trophic levels of vertebrate predators (Erdman 1961, 1972, 1984; Burgess and Franz 1989) and among some major invertebrate groups such as predatory stoneflies (Baumann 1982), some herbivorous and detritivorous mayflies (Edmunds 1982), and caddisflies (Flint 1978). Freshwater mammals such as otters are completely lacking. Others, such as manatees, are no longer abundant in the shallow coastal rivers, although they may have been present historically (Powell et al. 1981). Caimans and turtles also are currently unrepresented in the coastal rivers (see Thomas and Gaa Kessler, this volume).

Fish Distributions

The fish fauna above 600 m elevation (fig. 13.5) apparently consists of only a single species of gobiid fish, *Sicydium plumeri,* that is an active grazer on periphyton (table 13.3 and discussion below). These fish migrate upstream from the sea and grow to 15 cm in length in the fast-flowing waters of the Sonadora (Erdman 1961, 1972). They are found throughout the headwaters of major rivers, and larger adult males have been reported at the higher elevations (Erdman 1986; L. Nieves pers. comm., 1993). From May until October the females lay eggs in crevices of rocks along the stream bottom. Apparently many eggs are dislodged and washed out to sea during peak stream flows. One month later, these dislodged eggs develop into marine postlarvae (25 to 27 mm in length) and are thought to migrate along the coastline and swarm into the mouths of large rivers (Erdman 1986). The juveniles can climb up and around very steep waterfalls by using their specialized mouthparts and pelvic fin sucking discs to ascend to elevations more than 600 m above sea level. The right and left pelvic fins are fused to form a distinctive cup-shaped sucking disc that is used to hold onto boulders while grazing in very fast flow. In ancestral marine habitats these "suction cups" held gobies in place in the surge zone where waves dislocated other fishes (Ford and Kinzie 1982).

At low and intermediate elevations three species of gobies (*Awaous taiasica, Bathygobius soporator,* and *Gobiomorus dormitor*) recently were collected in the Bisley catchment of the Rio Mameyes (L. Nieves pers. comm., 1991, 1994). Unlike *Sicydium plumeri,* these gobies ascend rivers as juveniles and spawn in brackish waters (McKaye 1980). This catadromous life history is typical of several of the fish species that have invaded insular freshwater stream habitats. As bottom-dwelling fish they consume prey such as freshwater shrimp and aquatic insects (table 13.3).

A fifth species of fish, the mountain mullet (*Agonostomus monticola*), is found at low to intermediate elevations (fig. 13.5) and in less steeply sloped channels. It is very well adapted for life in clear, turbulent mountainous streams with deep pools. The mountain mullet moves downstream to spawn in brackish waters but spends most of its adult life in fresh water (Winemiller

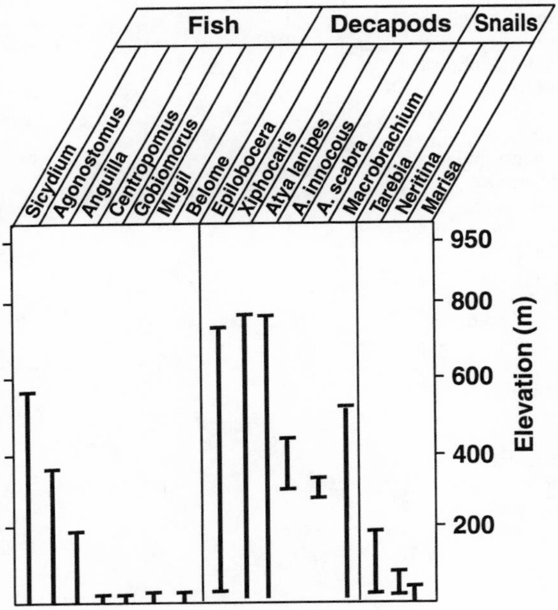

Figure 13.5. Elevational limits of distributions of major faunal elements along the Rio Espiritu Santo drainage (from unpublished report by Bhajan et al. 1978).

Table 13.3. Feeding relationships of the main consumer species in Luquillo streams

Taxon	Herbivore	Carnivore	Omnivore	Detritivore
Phylum Chordata				
Class Pisces				
Family Gobiidae				
Sicydium plumieri	J-A			
Awaous taiasica			J-A	
Bathygobius soporator			J-A	
Gobiomorus dormitor			J-A	
Family Mugilidae				
Agonostomus monticola			J-A	
Family Anguillidae				
Anguilla rostrata		A		
Phylum Arthropoda				
Class Crustacea				
Order Decapoda				
Family Atyidae				
Atya lanipes	J-A			A
A. innocous	J-A			A
A. scrabra	J-A			A
Micratya poeyi	J-A			A
Xiphocaris elongata	J-A			A
Family Palaemonidae				
Macrobrachium carcinus	J		A	
M. heterochirus	J		A	
M. crenulatum	J		A	
M. faustinum	J		A	
M. acanthurus	J		A	
Family Pseudothelphusidae				
Epilobocera sinuatifrons	J		A	
Class Insecta				
Order Odonata				
Family Coenagrionidae				
Enallagma coecum		A-J		
Telebasis vulnerata		A-J		
Family Aeshnidae				
Aeshna psilus		A-J		
Order Hemiptera				
Family Belostomatidae				
Belostoma subspinosum		A-J		
Family Veliidae				
Microvelia spp.		A-J		
Family Saldidae				
One sp.				
Order Coleoptera				
Family Dytiscidae				
Copelatus posticatus		J-A		
Family Hydrophilidae				
Enochrus debilis		J-A		

Table 13.3. (*continued*)

Taxon	Herbivore	Carnivore	Omnivore	Detritivore
Order Trichoptera				
Family Hydrobiosidae				
Atopsche trifida				J
Family Glossosomatidae				
Cariboptila orophila	J			
Family Philopotamidae				
Chimarra albomaculta	J			
C. maldonadoi	J			
Family Polycentropodidae				
Polycentropus zeneta	J			
Antillopsyche tubico	J			
Cernotina n. sp.	J			
Family Helicopsychidae				
Helicopsyche minima	J			
H. singulare	J			
H. n. sp.	J			
Family Xiphocentronidae				
Xiphocentron haitiensis	J			
X. borinquensis	J			
Family Hydroptilidae				
Alisotrichia hirudopsis	J			
Alisotrichia spp.	J			
Hydroptila martorelli	J			
Mayatrichia sp.	J			
Ochrotrichia spinosissima	J			
O. marica	J			
O. juana	J			
O. ceer	J			
Oxyethira puertoricensis	J			
O. janella	J			
Family Calamoceratidae				
Phylloicus pulchrus	J			J
Family Hydropschidae				
Smicridea alticola	J			
S. protera	J			
Macronema matthewsi	J			
Order Diptera				
Family Chironomidae				
Xestochironomus sp.				J
Stenochironomus innocuus				J
Rheotanytarus sp.				J
Cricotopus sp.	J			
Larsia sp.				
Labrundinia sp.	J			
Pentaneura sp.	J			

Note: J, juvenile; A, adult.

1983; G. A. Cruz 1987). In Puerto Rico it reaches up to 30 cm in length and a weight of 0.2 kg (Erdman 1972). Like some of the gobies these fish consume small shrimp and insects as well as some algae and detritus (see table 13.3 and discussion of omnivores below).

Another low-elevation (fig. 13.5) predator is the eel, *Anguilla rostrata.* Although eels swallow their prey whole and consume any species within the range appropriate to their mouth size (Moriarty 1978), their small teeth are especially well suited for predation on shrimp and aquatic insects (chironomids are reported in the stomachs of a wide size range of eels), as well as small fishes. Their highly elastic stomach allows them to rapidly consume large numbers of prey; they can be extremely opportunistic whenever encountering an appropriate group of prey. When prey are scarce they can also go for several days without feeding. A female caught in Puerto Rico was over 91 cm in length (Erdman 1972). Worldwide, the fifteen species of eels all share the same basic life cycle of reproducing at sea and, after long-distance migrations, returning upstream to specific areas for feeding; males and females have different patterns of migrations.

Eels and many other species of fishes are found in coastal habitats (Corujo Flores 1980); species richness is highest where large rivers reach the sea because of the high diversity of marine fishes that are well adapted to this estuarine mixing zone. Of the nine large rivers that drain the Luquillo Experimental Forest, only one estuarine lagoon is well studied from the standpoint of fish distributions and feeding relationships. At the mouth of the Rio Espiritu Santo there is a complex assemblage of sixty species in thirty families of fishes (Bhajan et al. 1978; Corujo Flores 1980). As distance from the Atlantic Ocean and the river's mouth increases, the diversity of fishes decreases; the number and height of waterfalls further limits fish diversity.

Although a large number of fishes have been introduced to Puerto Rican streams and rivers (Ortiz Carrasquillo 1980; Erdman 1984) there are very few native species. Puerto Rico is not unusual in its low diversity of indigenous freshwater fishes (Burgess and Franz 1989). Indeed, there are no native species of fish restricted to freshwater (primary species) in the West Indies (Miller 1966). As Miller (1982) points out, this total absence has led some biogeographers to conclude that "there have been no continuous land connections to either Central or South America during a time frame (late Cretaceous–early Cenozoic) when such fishes could have invaded the region. To others . . . cypriniforms entered the area via past continental connections but have become extinct in the West Indies." Miller further notes that in Puerto Rico "only a single, secondary fish, *Poecilia vivipara,* is known from the island and its natural occurrence there has been questioned." Secondary freshwater species are those that can tolerate brackish waters. In both the West Indies and the mainland of Central America there is a paucity of primary freshwater species of fishes, and a high percentage of the total fish fauna is derived either from marine species or salt-tolerant, secondary species. Several reviews (e.g., Miller 1982; Briggs 1984, 1987; Miller and Smith 1986) of the evolutionary history of the region discuss how the presence of numerous low-elevation rivers allows marine-derived species to dominate freshwater fish communities.

Decapod Distributions

The most common consumers in the steeply sloped streams are decapods, with reports of at least ten species of freshwater shrimp and one species of freshwater, "semi-aquatic" crab (fig. 13.5, table 13.3). Decapods generally have a relatively thick protective exoskeleton with heavily calcified chelipeds (or pincers). Their weight is supported in part by the water where they spend most of their time feeding, finding mates, and in some cases defending territories. However, some species can move on land along the stream banks while migrating or foraging. This adaptation for short periods of amphibious travel allows them to occupy headwater habitats and to exploit terrestrial resources and refugia as long as the relative humidity is high. Most overland activity occurs at night, although some adult and larger juvenile freshwater crabs are active during the day along stream banks and on top of boulders in pools.

The rigid exoskeleton serves to reduce desiccation and to protect decapods from many types of predators, but these crustaceans must periodically shed their covering as they grow. For several days before and after molting (ecdysis) their muscles are not well attached or functional, so that movement, foraging and defense are temporarily suspended. During ecdysis the exoskeleton is soft until it again becomes highly calcified (from resorbed internal storage and from ingestion of calcium in food when feeding does resume). While in this soft phase the decapods are highly vulnerable to predation and cannibalism. As a consequence the availability of refugia in tight-fitting crevices or burrows is an essential attribute to suitable habitat for these crustaceans (Covich and Thorp 1991).

The amphibious potamonid crab, *Epilobocera sinuatifrons,* completes its life cycle entirely in fresh water (Chace and Hobbs 1969; Villalobos-Figueroa 1982). Their life history is unusual for crabs because they lack free-swimming larvae and the female broods the eggs and young (as do freshwater crayfish). Once the juveniles become free-ranging, they are usually found in gravel beds and leaf packs in the stream. Adults can move up and down the stream from pool to pool and throughout the leaf litter on land in search of fallen fruits and seeds as well as terrestrial gastropods. The adult's inflated carapace allows its gills to remain moist even when outside its bank burrows (that extend to the groundwater level) or away from stream pools. These crabs have a highly vascularized lining in the branchial chamber which functions as a lung for aerial respiration (Diaz and Rodriquez 1977). The genus is widespread in the Greater Antilles at altitudes between 100 and 300 m.

Macrobrachium carcinus, a widespread species throughout the Caribbean and Gulf coasts of the Neotropics, is one of the region's largest palaemonid shrimp (22 cm carapace length, 27 cm chelae length, collected in April 1982 by J. C. Luvall in Quebrada Prieta). Villalobos-Figueroa (1982) reports they are found generally in the upper portions of rivers to 1500 m above sea level where currents are rapid and dissolved oxygen concentrations are above 6 mg l^{-1}. They are very aggressive predators on several species of small shrimp and are cannibalistic at high densities. All the palaemonid and atyid shrimps that live in these streams produce larvae that must reach brackish water to complete their development before returning upstream to grow and reproduce (Chace and Hobbs 1969; Villamil and Clements 1976; Hunte 1978; Villalobos-Figueroa 1982).

A second species of palaemonid shrimp, *M. crenulatum,* is also widespread in the Sonadora and its tributaries (Canals 1979). These shrimp have dimorphic chelae with distinctive spines on the inflated dactyl of the larger chela. At least three additional species (*M. acanthurus, M. heterochirus* and *M. faustinum*) have been reported at various elevations in the Espiritu Santo drainage (Bhajan et al. 1978; J. C. Montero-Oliver and A. Estrada-Pinto pers. comm., 1989). *M. acanthurus* is generally reported to occur in lower reaches with slow currents, dissolved oxygen above 4 mg l^{-1}, and water temperatures between 23° and 26° C (Villalobos-Figueroa 1982).

Atyid shrimp are the most abundant large consumers in these streams (Covich 1988b; Covich et al. 1991; Covich et al. in press; Johnson et al. in press) with five species, *Atya lanipes, A. innocous, A. scabra, Xiphocaris elongata,* and *Micratya poeyi* occurring in different microhabitats (Villamil and Clements 1976). Their biogeographic distributions are well known (Hunte 1978; Hobbs and Hart 1982; Villalobos-Figueroa 1982).

Shrimp and other crustaceans, as well as insects, penetrate crevices in stream channels where they feed and escape larger predators (Crowl and Covich 1994). Groundwater-fed streams throughout the world are characterized by various taxa of crustaceans that live primarily in deep crevices in gravel and cobble substrata below the surface of stream channels; some are restricted to these "stygobiotic" habitats such as in caves or groundwaters, while others (termed "phreatobites") move in and out of deep, subterranean, interstitial waters (Covich and Thorp 1991). The distributions and trophic dynamics of these crustaceans in Puerto Rico are not well understood, but some cave habitats have been analyzed in western portions of the island. Stock (1986) reports that such subsurface dwellers in the West Indies apparently persisted throughout prolonged dry periods during the Pleistocene when surface streams completely disappeared.

Aquatic Insect Distributions

Relatively little is known about the distributions of aquatic insects in Puerto Rican streams (Garcia-Diaz 1938; Flint 1964; Drewry 1970b; Ferrington et al. 1993; Masteller and Buzby 1993). Compared to temperate zone mainland streams and Neotropical continental streams, there are fewer species of some major aquatic insect groups in Puerto Rico; e.g., stoneflies are completely absent from the West Indies (Illies 1969; Baumann 1982). Other groups such as caddisflies (Trichoptera) are more diverse but still relatively low in species richness in comparison to mainland sites (Flint 1964, 1978; Flint and Masteller 1993). The caddisflies are represented by at least 165 species in the West Indies, and the dragonflies and damselflies (Odonata) by approximately 102 species (Flint 1978; Paulson 1982). The larger islands generally have more species. Of considerable interest are the differences in degree of endemism and regional distribution. Differences in adult size, flying ability, habitat specificity, and geologic history are thought to generally influence dispersal among the Caribbean islands (see Covich 1988a for a review). Work in Puerto Rico is focusing on caddisfly distribution because the taxonomy of the more than thirty-five species is sufficiently well studied to permit accurate identification and temporal comparison (Flint 1978). Several new species have recently been collected (table 13.3) and are discussed by Flint and Masteller (1993).

Recent collections of aquatic insects at Quebrada Prieta exceeded sixty species (Flint and Masteller 1993; Masteller and Buzby 1993; Pescator et al. 1993; E. Masteller pers. comm.). For example, during February 1989 Masteller collected twenty taxa of caddisflies using emergence traps, kick nets and black-light traps in and around Quebrada Prieta. These included representatives of ten different families. He also collected leptophledid mayflies, two families of Heteroptera (predatory Veliidae and Saldidae), four families of aquatic beetles (predatory Dytiscidae, herbivorous Hydrophilidae, Elmidae, and Limnichidae) and nine families of aquatic dipteran flies. Masteller found that one species of caddisfly, *Cariboptila orophila* dominated the streams. A very rare caddisfly species, *Macronema matthewsii,* was found in one stream and is known to be associated with root mats (O. Flint pers. comm.).

Among the net-winged midges (Blephariceridae) that are characteristic of fast-flowing streams, C. L. Hogue (pers. comm., 1987) collected three species in the Sonadora near El Verde (*Paltostoma argyrocinta, Paltostoma* sp., and *Maruina* sp.). Multiple species of *Paltostoma* are unusual; Puerto Rico is the only Caribbean island known to have more than one species. These larvae cling to rocks in swift waters with a single row of specialized ventral suckers. They are thought to feed on microscopic aquatic plants (table 13.3). Some twenty-seven taxa of chironomids were collected from a wide range of habitats (Ferrington et al. 1993).

Gastropod Distributions

Aquatic snails are primarily restricted to lower elevations (fig. 13.5), where streams are less steeply sloped and flows are relatively slow. The freshwater neritid limpet *Neritina punctulata* is similar to those snails found in other coastal streams and may also be characterized by seasonal upstream migrations as reported in Costa Rica (Schneider and Frost 1986) and Hawaii (Ford and Kinzie 1982). Recent studies by Pyron and Covich (in review) indicate that *Neritina* migrate along the Rio Espiritu Santo but do not enter the Sonadora tributary. *Thiara granifera* is a widely distributed prosobranch snail that has recently been inadvertently introduced into Puerto Rico and other Caribbean islands; it has not yet been collected in the Sonadora but does occur downstream in the Espiritu Santo. *Marisa cornuaretis,* an ampullarid snail, was introduced into many low-elevation streams and lakes as a predator to control the schistosome-vector snail (*Biomphalaria glabrata*) through consumption of the vector's eggs and possibly as a competitor for similar foods. Neither of these species has been found in the Sonadora.

FEEDING GROUPS

Herbivores and Detritivores

Some fish, such as *S. plumeri,* are grazers and have specialized teeth for scraping periphyton off rock surfaces. Their feeding is clearly indicated by large, distinct grazing scars left on boulders. They can feed on vertical rock faces in the Sonadora, where there are microhabitats free of atyid shrimp grazing and which support a well-developed periphyton community. Similar feeding sites are reported for mainland species in this genus (Winemiller 1983).

The shrimp *Atya lanipes* is one of the most abundant grazers in headwater streams. These shrimp are widespread and quickly recover after major storm events. Following Hurricane Hugo, when decomposing leaves and algal periphyton were exceptionally abundant, there were rapid increases in populations along the Prieta that exceeded maximal densities observed prior to the hurricane (Covich et al. 1991). These increases leveled off and five years after Hurricane Hugo the relative abundances of *Atya* were similar to those observed before the hurricane (Covich et al. in review).

A. lanipes, as well as *A. innocous,* have modified chelae that are fan-like brushes with sharp setules. They scrape these cheliped fans across periphyton (Pringle et al. 1993) on the larger cobbles and boulders (which are stable and not rolled or scoured during peak flows). The fans of toothed bristles have microscopic hairs that hold the removed algae and bacteria while the food is transferred by the cheliped fans to brushy appendages that clean the cheliped fans and transfer a bolus of food to the mandibles (Fryer 1977; Felgenhauer and Abele 1983, 1985). Linear "scars" or "trails" are left behind on rock surfaces after the shrimp have scraped off the periphyton or layers of silt deposited by storm-generated, turbid waters. These scars are larger than those left by snails and smaller than those of gobies.

These same shrimp switch to a filtering feeding mode at flow rates above 5 cm s^{-1} (Covich 1988b). The cheliped fans are extended into the flow and intercept suspended particles over a relatively large size range (from bacteria to leaf fragments and fecal pellets from other shrimp). Both species of *Atya* are known to form rows of up to fifteen or more shrimp in a line facing into the flow. They occur primarily in pools of both the larger and smaller streams but also filter-feed in fast-flowing riffle habitats (Covich 1988b). They switch to scraping microbes from decomposing leaves that accumulate during dry periods in slow-flowing pools. A third species, *A. scabra,* is also an effective filter-feeder and is well adapted to very fast flow. It is more specialized and lacks the fine-toothed scrapers on its filtering setae (Fryer 1977).

A fourth species of atyid shrimp, *Xiphocaris elongata,* is endemic to the West Indies (Villalobos-Figueroa 1982). It still retains its pincers and lacks the brushes and cheliped fans typical of other atyid species. *X. elongata* grasp onto and consume a very wide range of larger particles at the water's surface (e.g., leaf fragments, dead insects, small flowers, and fruits that have fallen into the pools). In Quebrada Prieta they were found to grasp the numerous small flowers (with droplets of nectar) on the sierra palm (*Prestoea montana* = *Euterpe globosa*) as soon as these florets dropped into the pools (A. Covich pers. observation). *Xiphocaris* was also seen to actively swim upstream while carrying fallen fruits of *Rourea glabra* or even small, dead animals. They are very abundant and reach densities of 20 individuals m^{-2} in shallow stream pools where predatory fishes are absent. *Xiphocaris* swim among the other atyid shrimp; no aggressive interactions among these consumers are evident.

Aquatic insects also are important detritivores in first- and second-order streams. *Phylloicus pulchrus* (Calamoceratidae) is a widespread caddisfly, the larvae of which build protective cases from circular leaf discs (Flint 1964). Wiggins (1978) lists the genus *Phylloicus* as a detrital "shredder." Wiggins (1977, p. 70) notes that the gut contents of larvae from another species of this genus consist of "primarily filamentous algae and vascular plant tissue—probably from detritus." Caddisflies grind their food in gastric mills that make identification of particular food items very difficult. Several other species of caddisflies are reported from Luquillo (Drewry 1970b). The microdistribution and feeding ecology of several species are currently under study (K. Buzby, O. Flint, and E. Masteller pers. comm., 1990). Two species in each of two genera of chironomids, *Xestochironomus* and *Stenochironomous,* whose larvae are typically xylophagic, were common in woody debris (L. Ferrington pers. comm., 1990).

Another common insect detritivore is *Microvelia* sp. (Veliidae). These hemipterans occur in slow-flowing pools in first- and second-order streams. They apparently consume dead terrestrial insects that fall onto the quiet surface waters of the pools. Many other aquatic insects are known to feed on decaying leaves and sticks or graze on periphyton; a partial list is available (table 13.3).

Omnivores

Among the palaemonid shrimps the adults are generally omnivorous and active scavengers. For example, *Macrobrachium carcinus* consumes smaller shrimp, molluscs, small fish, insects, algae, macrophytes, and decomposing leaf litter. The long, thin chelae of *Macrobrachium carcinus* are well adapted to probe into crevices for prey as well as for aggressive interactions with other shrimp. Some species (*M. heterochirus*) have dimorphic chelae, one chela being heavier and more powerful in crushing prey or "disarming" conspecifics. They typically hide in crevices or bank burrows during the day and forage in the deeper pools at night. Juveniles and intermediate-aged shrimp are often seen perched vertically on steep-sided rocks just below the water's surface. They forage as "sit-and-wait" predators on damselflies and other insects associated with the air-water surface.

Above the water, from the terrestrial perspective of the food web, this same boulder habitat is used by *Anolis evermanni.* These small lizards perch on top of the boulders waiting for emerging aquatic insects or aerial flying adults that land near the water's edge (see below).

The most widely omnivorous decapod is the freshwater crab, *Epilobocera sinuatifrons.* They have large, powerful chelae capable of breaking shells of the thickest aquatic or terrestrial snails or cracking open fruits from the sierra palm and other hard-covered seeds. Adult crabs can feed on the largest shrimps, if they are quick enough to catch them. Adults will carry food out of the water and onto the bank at night. This land-based feeding lowers chances of losing choice food items to other crabs or shrimps that are highly chemoreceptive in tracking food odors carried downstream in currents. The first individuals to locate prime food items will also "monopolize" resources by transporting them into their burrows along the stream banks and in the flood plains of larger streams. Although some adult crabs can cache certain foods such as nuts and fruits, most individuals seem to consume the more perishable foods immediately (A. Covich pers. observation). Adult female crabs carry large numbers of juveniles ventrally after they hatch (as do other freshwater crabs and crayfish). This parental care protects the young from predation, cannibalism, and desiccation (when the adult female moves across the forest floor in search of food). Later, free-roaming juvenile crabs remain in gravel-bottomed riffles, under large rocks, or in leaf packs where they apparently consume aquatic insects and fine detritus. Adults are known hosts to copepods that live on gill filaments (G. M. Sponholtz pers. comm., 1989).

Carnivores

The only fish species in the headwater streams that are known to be mainly carnivorous (Moriarty 1978; Cruz 1987) are mountain mullets, *Agonostomus monticola,* and eels, *Anguilla rostrata,* which are rarely encountered around El Verde. The diet of *A. monticola* in the Rio Espiritu Santo estuary, downstream from El Verde, is reported by Corujo Flores (1980) to include shrimp (33%), adult insects (50%), and plant material such as seeds (17%). At higher elevations in the Bisley catchment mullets are known to consume aquatic insect larvae and small crabs (L. Nieves pers. comm., 1990). Winemiller (1983) observed that *A. monticola* fed on both terrestrial and aquatic insects in the Rio Claro, on the Costa Rican Oso Peninsula, although filamentous algae and terrestrial vegetative detritus were also found in some stomachs.

Eels and mountain mullets are less well adapted for moving upstream than gobies. However, these two predatory species likely consume juvenile shrimp and gobies migrating upstream. These prey quickly seek shelter to avoid larger predators such as sight-feeding fishes and birds during daylight periods. Upstream migrations by shrimp and gobies from lower elevations occur at night. During the day shrimp often seek refuge in rock crevices and under organic debris.

Other fishes known to feed on invertebrates are larger gobies such as *Awaous taiasica, Bathygobius soporator,* and *Gobiomorus dormitor* (L. Nieves pers. comm., 1990). Analyses of stomach contents are now underway for these species in the Bisley catchment of the Luquillo Experimental Forest (L. Nieves and M. Pyron pers. comm., 1994).

Carnivorous insects such as odonate larvae are reported from El Verde (Drewry 1970b), but little is known about their densities or feeding habits. As either stalking or sit-and-wait predators, the odonate larvae are found among submerged leaf packs and within the sand/gravel substrata where they can feed on a wide range of small aquatic insects and crustaceans. They use their extensible labrium (a highly modified pair of pincer-like lobes, palpi, that are held below the head and front thorax) to grasp their motile prey (Corbet 1980; Paulson 1982). This extended jaw-like trap is shot forward within 1/100 second and can grab even rapidly swimming prey that are then brought back and chewed by the mandibles. Free-flying adults forage widely and consume many aerial insects, especially small dipterans. On the Central American mainland some species form feeding aggregations (Young 1980). The damselfly *Telebasis vulnerata* (Coenagrionidae) was collected in pools of Quebrada Prieta where it could feed on mayfly and chironomid larvae. Other odonates (such as dragonflies, *Argia, Aeshna,* and the damselfly, *Hetaerina*) reported from El Verde are generally considered to be active predators on small insect prey; rarely are they cannibalistic (Corbet 1980).

Other insect predators are hemipterans (*Belostoma* spp., *Buenoa* spp., and *Notonecta indica*), caddisfly larvae (*Atopsyche trifida*) and dytiscid beetle larvae. The dytiscids inject their prey with proteolytic enzymes and can consume species much larger than themselves, such as small fish. The reported predatory interactions between ants and fish (or other aquatic vertebrates) reported for mainland streams (Hoenicke 1983) have not yet been documented.

CONNECTIONS WITH TERRESTRIAL FOOD WEBS

Important linkages between aquatic and terrestrial food webs are found along the stream riparian zone and are discussed below. In some instances the same species shifts from consuming aquatic food resources to terrestrial foods. For example, damselflies and dragonflies feed initially as larvae on aquatic prey and then shift to terrestrial prey after becoming territorial, free-flying adults along the stream banks or at considerable distances from pools where their stream-based larval development occurred. Freshwater crabs also forage at some distance from streams as adults but are generally restricted to aquatic habitats during the early stages of development.

The most abundant terrestrial vertebrate predators are lizards and frogs. As Reagan (this volume) notes, *Anolis evermanni* feeds on a wide range of emerging adult aquatic insects from its boulder perch sites on the Sonadora. These lizards eat dipterans, mayflies, caddisflies, damselflies, and hemipterans. *A. evermanni* is unique among the four species of *Anolis* in taking advantage of the stream habitat in its horizontal spatial distribution. This species is extremely opportunistic. For example, one female had seventy-seven food items in her stomach representing twenty species of invertebrates. *A. gundlachi* also includes some aquatic insects in its diet but to a much lesser extent than *A. evermanni.*

Frogs such as *Eleutherodactylus karlschmidti* were reported to be relatively common along the boulders of the Sonadora and other stream beds

(Drewry 1970b; Rivero 1978). This semiaquatic frog is the only native species known to feed on aquatic species of prey. Rivero et al. (1963) report that aquatic benthic invertebrates constitute nearly one-third of the diet of *E. karlschmidti.* Unfortunately, this frog is now very rare and may be locally extinct around El Verde and other previously collected sites (G. Drewry pers. comm., 1985; L. Woolbright, M. Stewart, and R. Thomas pers. comm., 1988). *Leptodactylus albilabris* is known to be associated with streams but its food habits are unknown.

Spiders also consume emerging adult aquatic insects. Spiders of the genus *Tetragnatha,* the long-jawed orb-weavers, are among the most abundant spiders worldwide. They are characterized by their elongate abdomens and long jaws, and they are typically found in stream-side habitats. *Tetragnatha tenuissima* is abundant at El Verde. R. Gillespie and L. Bishop (pers. comm., 1989) report that this species is nocturnal and builds fragile orb webs in twigs or grass stems, often directly over permanent streams. The webs are commonly positioned 3 to 6 m above the water and entrap small dipteran insects which emerge from the stream. The small red spider, *Wendilgarda clara,* spins its web very close to the water's edge. Two other species, *Alcimosphenus borinquenae* and *Leucauge regnyi,* are also commonly found near streams and temporary stream beds (Pfeiffer, chap. 7, this volume). Some spider webs are "robbed" of prey by odonates (O. Fincke pers. comm. 1990). Odonates are another potential connection for moving nutrients and energy from stream pools into the forest food web as they not only consume aquatic and terrestrial insects, but are often eaten by bats and birds (O. Fincke pers. comm., 1990).

Relatively few consumers of decapods are known. Juvenile crabs are known to be attacked and consumed by ants and one stomach of an *Eleutherodactylus coqui* was found to contain a juvenile crab (M. Stewart and L. Woolbright pers. comm., 1987). Centipedes are another predator reported to feed on juvenile crabs (K. Buzby pers. comm., 1991). Although birds such as herons, hawks, and kingfishers are reported to include species of aquatic prey (crabs, shrimps, and insects) in their diets, these birds only rarely are found within the forest (Waide, this volume). A great blue heron and a night heron were observed feeding along the Quebrada Prieta on various sizes of freshwater crabs and shrimps (T. Crowl and A. Covich pers. observation, 1989). Remains of atyid shrimp were found in fecal matter thought to be from the Puerto Rican boa, *Epicrates inortatus,* on top of a boulder in the middle of a pool in the upper zones of Quebrada Prieta (A. Covich pers. observation, 1989). Distributions of the Puerto Rican boa are known (Reagan 1984), but their feeding habits in and around streams are not yet documented. The introduced Indian mongoose is known to feed around small streams (Pimental 1955) but direct feeding relationships with stream-based food webs are not reported. Recent observations of shrimp remains in mongoose feces are unexplained but may result from ingestion of feces from a wide range of stream-based consumers (G. R. Camilo pers. comm., 1994).

PATTERNS OF FEEDING RELATIONSHIPS

Food Web Structure and Feeding Specificity

In general, headwater food webs are "short" and composed of relatively few links within the maximal length of any food chain. The webs are also highly interconnected and similar in composition throughout the El Verde drainage. As elevation increases along any stream channel there are some distinctive longitudinal changes, as have been observed in other Caribbean islands (Hynes 1971; Harrison and Rankin 1976). The larger, coastal tributaries typically have more fish while the smaller, headwater streams are dominated by crabs and are repeatedly recolonized by a restricted number of decapods and insects following washout of bottom-dwelling species during intense floods (Covich et al. 1991).

The very high densities of several types of shrimp in some headwater streams apparently result from a lack of fish or other vertebrate predators and an abundant food supply of leaf detritus; high algal growth rates may also be associated with high grazer/detritivore densities. We find no evidence of highly specific feeding relationships among the relatively small numbers of top carnivores or among the lower trophic level consumers that can generally switch from algal grazing to detritivory on a wide range of leaf litter.

In several studies of other tropical regions there are reports of considerable overlap in food use among fishes (e.g., Bishop 1973; Winemiller 1983, 1989a,b, 1990; Dudgeon 1987), especially in those instances where fish feed on drifting insects or detritus. However, in some other tropical rain forest streams where fishes dominate food webs, there are documented feeding specializations with relatively little overlap among several co-occurring species (e.g., Zaret and Rand 1971; Moyle and Senanayake 1984). The roles of various vertebrate and invertebrate consumers vary considerably in different locations where food-web structure can be influenced by biogeographic and local distributions. As previously described, we find that local restrictions in species distributions within El Verde are often influenced by steepness of the topography and differences in ability to migrate upstream and over or around waterfalls.

Productivity and Food Web Structure

Detailed studies of stream food webs are scarce even for temperate-zone headwaters but there are indications that energy flow does limit the number of trophic levels. For example, Hildrew et al. (1985) examined midges and other aquatic insects in a small, fishless stream in southern England. They suggested that the observed highly generalized feeding relationships could be a result of very low stream productivity. They conclude that "all possible items are included in the diet of component species" because of meager food supplies.

Briand (1983, 1990) contends that the physical environment is a major determinant of food-web structure. In his analysis of twenty-one freshwater food webs, Briand (1985) concludes that stream food webs were relatively more complex, wider, and shorter than lake and large-river webs. Food-web height is measured by the maximal number of trophic species linked together within the food web. Width is measured by the maximal number of trophic species within any trophic level. Briand (1985) suggests that because food webs in small, shallow, forested streams generally are detritus based, their webs are comparatively shorter than more efficient webs in lakes and deeper rivers that are based on algae and submerged macrophytes. He states that "there may not be enough energy left at the upper trophic levels of streams to support higher carnivores."

Energy flow in detrital pathways will partially depend on detrital quality. Bacteria and fungi living on decomposing leaves can be very nutritious (Cummins and Klug 1979; Cummins 1988), and detritus-based food webs can be highly productive (Howard-Williams and Junk 1977; Bowen 1983; Payne 1986; Covich 1988a). Not all phytoplankton and macrophyte productivity is equally nutritious, and some species are of relatively low quality (e.g., Porter 1977; Leibold 1989; Newman et al. 1990).

Environmental Variability and Food Web Structure

Briand (1990) and Briand and Cohen (1987, 1990) have recently reviewed effects of environmental variability on food-web organization. In his analysis of forty published food webs Briand (1990) notes that connectance (a ratio of observed to possible feeding interactions that is used as a measure of the degree of shared prey among predators) is lower in fluctuating environments. He suggests that the differences in connectance patterns between food webs in fluctuating and constant environments may relate to differences in optimization of feeding: "in fluctuating systems, environmental perturbations do limit time available for feeding. Therefore, it would appear advantageous for the consumer species to rely on briefer but more intense periods of predation" and specialize on particular prey, whereas in constant environments "weaker interactions may be tolerated since they can be exploited on a more continuous and reliable basis" (Briand 1990, p. 53). Periods of high stream flow could decrease connectance in aquatic food webs. In addition to interfering with locomotion and chemoreception through a combination of high

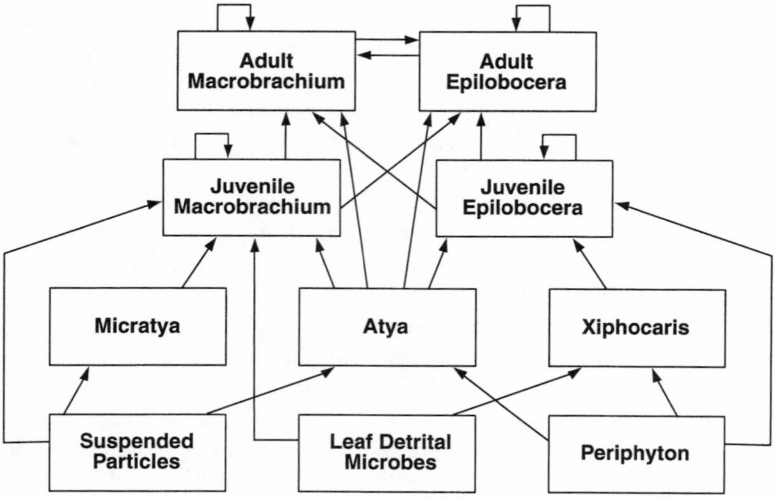

Figure 13.6. Decapod crustacean subweb of the aquatic food web. Connections among major genera and age classes of consumers denotes energy flows from detrital and algal food resources.

velocity and increased suspended sediments, very high stream discharge can wash organisms from the stream, thereby physically limiting the numbers of encounters among predators and prey.

Cycles, Reversals, and Food Loops

Cohen (1990) defines "cycles" within a food web as a directed sequence of one or more links starting from, and ending with, the same species. In the published food webs he studied there have not been more than two links in any cycle and these are apparently rarely observed (or reported). Similarly, Pimm (1982) concludes from his review of temperate-zone terrestrial food webs that predatory reversals, or food loops, are scarce.

At El Verde, if some juvenile crabs that are vulnerable to predation by frogs were able to grow into large adults (in perhaps two to three years), they could conceivably consume those frogs encountered along the riparian zone. Other examples of predatory reversals within the decapods are between crabs and species of *Macrobrachium*. The young crabs are potential prey for adult *Macrobrachium* and vice versa. Odonate larvae can readily consume post-larvae (stages I and II) of *Macrobrachium* as they drift downstream after hatching in the upper reaches of these streams. Later, however, the adult shrimp can be observed during the day clinging to vertical rock surfaces where they wait for adult free-flying odonates and terrestrial insects to land near the water's edge. The shrimp grab these prey with their chelae and carry them into the water. At night the shrimp also perch on top of the rocks and move along the stream banks where they may encounter other insect prey.

SUMMARY

First- and second-order streams draining the tabonuco watershed at El Verde are distinctively different from most temperate zone streams because (1) they receive a continuous supply of forest-produced leaf litter; (2) daily discharge is unpredictable over annual time scales; (3) extremely variable frequencies and magnitudes of high rainfall events ("floods") wash out leaf detritus and scour the substrata thereby disrupting life history patterns and predator-prey dynamics; and (4) freshwater vertebrates are low in abundance relative to decapods.

These headwater streams differ from their mainland counterparts in having a predominance of "coastal" species that move upstream from brackish water lagoons rather than true freshwater species. Food webs of these insular tropical streams are relatively simple in terms of numbers of interacting species and are comparatively "short" in terms of numbers of trophic levels because they lack the diversity of higher trophic-level consumers (fig. 13.6), especially vertebrate carnivores. The upper trophic levels are dominated by invertebrate omnivores such as freshwater crabs and shrimps; carnivorous fishes and large bird predators such as herons and kingfishers are infrequently associated with stream-based food webs at these elevations.

14

The Community Food Web: Major Properties and Patterns of Organization

Douglas P. Reagan, Gerardo R. Camilo, and Robert B. Waide

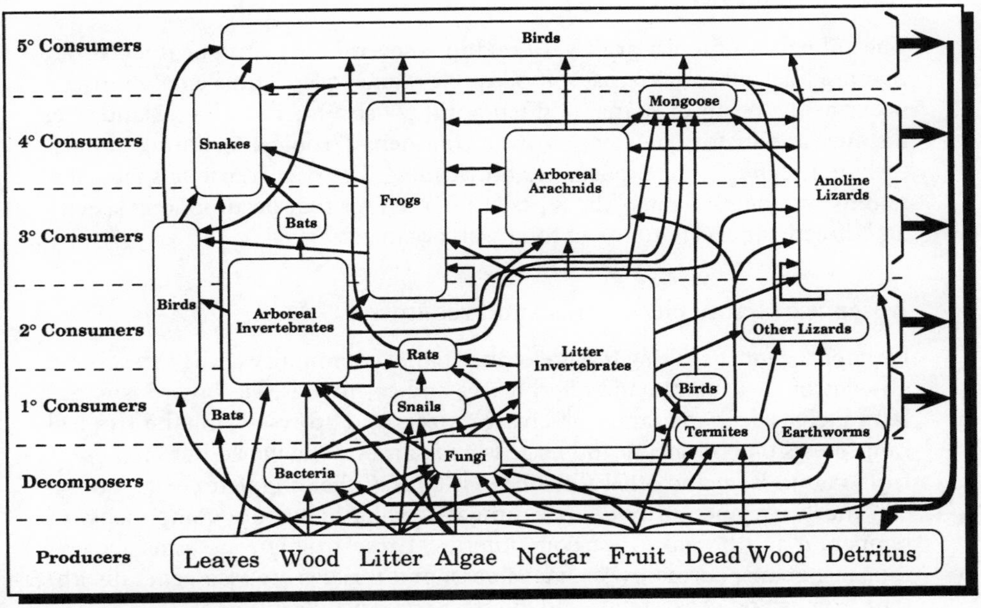

In the preceding twelve chapters the concept of the food web has been used as an organizing theme to help describe the relationships among organisms in a tropical forest community. Each chapter has focused on a particular assemblage of related taxa and the prey and predators of the species constituting those taxa. In this chapter, our focus shifts from the component organisms to a more macroscopic view of the food web. We begin by discussing the characteristics of the animal community from which we construct our food web. We define the emergent properties of the web and compare those emergent properties to other food webs. In this process, we point out where the results of this study may make contributions to the further development of ecological theory.

CHARACTERISTICS OF THE EL VERDE ANIMAL COMMUNITY

The animal community at El Verde exhibits several distinctive characteristics that are relevant to patterns of food web organization. These characteristics occur as the result of spatial and temporal constraints including island size, distance and history of isolation from continents, tropical location, and disturbance regime (e.g., exposure to hurricanes). The occurrence of such community attributes dictates the types of interactions that occur among species and directly affects patterns of food web organization.

Absence of Large Herbivores and Predators

Perhaps the most striking feature of the animal community of the forest is the absence of large mammalian herbivores and carnivores. Herbivorous ungulates such as deer and tapir, which are grazers and browsers, inhabit tropical rain forests on continents and large islands around the world but are absent from Puerto Rico and other Caribbean islands. Other ungulates (e.g., moose, elk, musk ox, and caribou) with similar feeding habits are components of temperate and boreal forest communities. Large frugivorous birds such as toucans, guans, currassows, chachalacas, and turkeys are also generally absent from island ecosystems, and Puerto Rico is no exception.

Continental and large insular forests in the tropics typically are inhabited by large predators including jaguar and cougar (Central and South America), leopard (Africa), and tiger (Southeast Asia); large predators such as grizzly bear, cougar, lynx, and wolf inhabit temperate and boreal forests. Large herbivores and their predators form a discrete set of predator and prey interactions that are missing from the El Verde food web.

The largest herbivores in El Verde forest are the fruit-eating bat *Artibeus jamaicensis* and birds such as the scaly-naped pigeon, ruddy–quail dove, and Puerto Rican parrot. Two lizard species (*Anolis cuvieri* and *A. evermanni*) consume some vegetable material, but are predominantly carnivorous. The largest predators in the El Verde forest are much smaller than top predators on continents and include introduced mammals (feral cats and Indian mongoose), birds (red-tailed hawk, Puerto Rican screech-owl, and Puerto Rican lizard cuckoo), and one reptile (Puerto Rican boa). The largest of these is the boa, which can achieve a length in excess of 2.5 m. The most abundant vertebrate predators in the forest are small lizards and frogs with individuals weighing only a few grams.

The absence of large predators and herbivores has several potential effects on food web structure. The proportions of top, intermediate, and basal predators are affected by the absence of predators and herbivores depending on the proportions of these groups in the missing components. The absence of top predators may release species at lower tropic levels from top-down control, resulting in increases in populations. Many of the patterns described below that involve lizards, frogs, and large invertebrates are a consequence of the high abundances of these groups.

Large mammalian predators and herbivores have low conversion efficiencies because they are endothermic. Most of the energy consumed by mammals is used to produce body heat, and only a small fraction is actually converted into biomass (Petrides et al. 1968; Short and Golley 1968; Pimm 1982). On the other hand, ectotherms such as reptiles, amphibians, and invertebrates have higher conversion efficiencies; a higher proportion of what they consume is converted into biomass, which in turn becomes available to the next trophic level (Moore et al. 1988). Food webs lacking large endotherms can potentially have longer food chains. Higher conversion efficiencies might also promote reciprocal predation (loops).

The absence of large herbivores can effect food chain length in another way. A large primary consumer requires a large predator. Most predator-prey relationships occur between prey and a large predator (Cohen et al. 1993). Very large primary consumers (e.g., elephants) are by definition top predators because no predator large enough to overcome them exists in the community. However, the same biomass of primary consumer packaged as smaller individuals can support more levels of predators without being constrained by size considerations. Large, endothermic herbivores and carnivores set clear constraints on the length of food chains by energy loss due to higher metabolic rates, large body size, and increased predation rates.

Low Faunal Richness Compared to Tropical Mainland

Lugo (1987) has shown that plant species richness is comparable between mainland and islands when similar habitats are compared on the same areal scale. This generalization does not apply to animals. Faunal species richness, including both vertebrates and invertebrates, is substantially lower in Puerto Rico than at comparable mainland locations.

Numerical comparisons are not possible for total invertebrates because of the extreme diversity in many tropical rain forests where most species remain undescribed (Erwin 1983b). Limited comparisons are possible for some of the better-known groups. Waide (1987) presented data showing that for nine mainland sites in the new and old world tropics, the number of butterfly species ranges from 125 to 550 compared to eighteen in Puerto Rico. He presented similar data comparing termites from four mainland sites with forty-three to eighty-five species compared to four in the forest at El Verde.

The only native mammals extant in Puerto Rico are bats, although one insectivore and six rodent species have become extinct in Puerto Rico since the Pleistocene (Willig, this volume). While little is known about the habitats of these extinct species, it is likely that at least some were rain forest inhabitants. The black (roof) rat and feral house cat, post-Columbian introductions, are conspicuous components of Puerto Rican forest communities. Another introduced species, the Indian mongoose, is not common within the forest. Monkeys, other primates, marsupials, and native rodents that are primary consumers and omnivores in continental rain forests are absent from Puerto Rico.

Bird, reptile, and amphibian faunas are also represented by relatively few species. Only forty bird, fourteen reptile, and thirteen amphibian species have been recorded from the forest at El Verde. Comparisons of these species numbers with four other Neotropical rain forests (Gentry 1990; table 14.1) show the low species richness at El Verde compared to mainland sites. Differences in the species richness of all vertebrate classes are substantial (seventy-seven at El Verde compared to 428 to 561 at the four mainland sites). The relatively low number of species at El Verde limits the number of possible feeding interactions compared to continental forests. However, the detailed food web derived from the El Verde community is one of the most diverse webs ever analyzed, and the number of possible interactions in the web is very great compared to most published webs. Investigation of the dynamics of the animal community in tabonuco forest is a first step towards understanding the trophic dynamics of species-rich tropical forest communities.

Table 14.1. Species richness of vertebrates in several rain forests

Locality	Mammals	Birds	Reptiles	Amphibians
El Verde, Puerto Rico	10	40	14	14
La Selva, Costa Rica	113	254	86	48
Barro Colorado, Panama	39	251	82	52
Central Amazonia, Brazil	51[a]	300	85[b]	42[c]
Rio Manu, Peru	95	332	54	80

Sources: Vertebrate data for Puerto Rico are from chapters in this volume. Mainland data are from Hartshorn and Poveda (1983) and from papers by the following authors in Gentry (1990): birds, Karr et al.; mammals, Wilson, Glanz, Janson and Emmons, Malcolm; amphibians and reptiles, Guyer, Rand and Myers, Rodriguez and Cadle, Zimmerman and Rodrigues.

Note: Figures represent numbers of species. Numbers are for upland forest and forest mosaic habitats and are approximate for some sites due to methods of data presentation.

[a] Excluding bats.

[b] Includes only snakes and lizards.

[c] Includes only frogs.

Superabundance of Frogs and Lizards

Shortly after dark on most nights throughout the year, one can hear numerous frogs calling from perches near ground level throughout the forest. On rainy nights when conditions are optimal for frog activity, the calls can be deafening. The abundance and ubiquitous distribution of coquis and prominence of their calls is a distinctive feature of the Puerto Rican landscape. The predominant species is *Eleutherodactylus coqui,* which like other members of the genus *Eleutherodactylus* in the West Indies deposits its eggs on land, and the young undergo direct development to the adult form within the egg. This life history feature allows the species to become distributed throughout the areal extent of the forest rather than be concentrated in the vicinity of standing water. As a result, populations of *E. coqui* can be dense. The population density estimate for *E. coqui* at El Verde is one of the highest densities known for any amphibian species (Stewart and Woolbright, this volume).

Anoline lizards are abundant and conspicuous components of the diurnal community at El Verde as they are on other islands throughout the Caribbean (Williams 1969, 1976; Moermond 1979a). Whereas anoles are conspicuous at ground level, they reach their highest densities in the forest canopy. Reagan (1992) estimated wet- and dry-season population densities of the canopy species, *Anolis stratulus,* at between 21,333 and 25,870 individuals ha^{-1}, the highest density reported for any lizard species.

As an explanation for the high density of some species on islands, Williamson (1981) describes density compensation, where reduction in species number within a group does not result in reduced cumulative density of that group. Although there are fewer species of frogs and lizards in the El Verde forest than in comparable mainland forests, the total number of individual lizards probably exceeds the total number of individuals for all lizard species in mainland forests. Thus it appears that density compensation only partially accounts for these high densities. Spatial and temporal factors contributing to the superabundance of both frogs and lizards are discussed in Stewart and Woolbright (this volume) and Reagan (this volume).

Discontinuities in the Community

The Aquatic-Terrestrial Boundary

The most dramatic discontinuity in the distribution of consumer species occurs at the boundary between aquatic and terrestrial habitats. The organisms occupying these two habitats form distinct communities, yet the food webs of the two communities share many common processes and resources. Stream consumers feed on detritus from terrestrial sources (Covich and McDowell, this volume) and on algae that are controlled in part by terrestrial sources of nutrients (C. Pringle pers. comm.). Trees scavenge nutrients from streams through floating root mats (K. Vogt pers. comm.). Microbial populations in riparian zones mediate the transfer of nitrogen between terrestrial and aquatic communities (W. McDowell pers. comm.). Stream organisms retain nutrients by consuming detritus before it can be washed out of the system.

Despite the tight interconnection of many ecosystem processes between aquatic and terrestrial systems, few consumers use resources of both habitats. The aquatic crab *Epilobocera* comes out onto land to feed on fallen fruits (Covich and McDowell, this volume), and the lizard *Anolis evermanni* feeds in part on stream insects (Reagan, this volume). The green heron (*Ardea herodias*) feeds on stream invertebrates. Two spiders (*Wendilgarda clara* and *Alcimosphenus borinquenae*) place their snares over small streams (Pfeiffer, chap. 7, this volume). Few other consumers have been observed to forage across the land-water barrier, indicating that this boundary separates two food webs that are faunistically distinct.

The Litter-Vegetation Boundary

Tabonuco forest presents two major discontinuous sources of consumable plant biomass to support the food web, live leaves, fruit, and nectar located from the understory to the emergent canopy and the litter-ground layer complex that provides both living and dead sources of food. The physical separation of the fauna into distinct strata within the forest has strong effects on the form of the food web at El Verde. Evidence from stomach analyses demonstrates that strong predator-prey links are most commonly found between species inhabiting the same stratum, and such links are less frequent between strata. For example, the vertical stratification of *Anolis* lizards and the ant species upon which they prey is shown by tight predator-prey couplings between *A. gundlachi* and *Pheidole moerens* at ground level and *A. stratulus* and *Myrmelachista ramulorum* in the canopy (Garrison and Willig, this volume). A similar trend was shown for *E. coqui* collected from the canopy. Canopy-foraging individuals consumed prey species that were never found in stomachs of frogs foraging at or near the ground (Stewart and Woolbright, this volume). The vertical distribution of other predators is also reflected in their diet. Puerto Rican todies commonly forage within 8 m of the ground, and their diet contains a high proportion of dipterans, which are the dominant taxa found in insect samples near the ground (Garrison and Willig, this volume).

Foraging strata reflect discontinuities in resources that allow the aggregation and specialization of prey and predators. In the case of most forest ecosystems, the major resource dichotomy is between canopy production and litter decomposition, and this difference has long been recognized through the description of separate predatory and saprophytic food chains in the ecological literature (e.g., Odum 1959). Pimm (1982) recognized the existence of compartments corresponding to major habitat divisions, but concluded that food webs were not compartmented within habitats. This conclusion corresponds to predictions based on dynamical constraints of models, but rests heavily on the definition of distinct habitats.

Frequent Disturbance

The implications of the disturbance regime in the LEF on the structure of the food web are difficult to interpret. Pimm (1982) concluded that food chain length in habitats that are regularly disturbed should be shorter than in those that are not. He reasoned that top predator populations in disturbed habitats had longer return times to equilibrial values, and thus their probability of being hit by another disturbance and then becoming extinct was higher. Pimm's simulations were based on classical Lotka-Volterra equations, where predator-prey relationships are restrictive because of the strength of the interaction. The prevalence of omnivory and donor-controlled predator-prey relationships in speciose ecosystems (Strong 1992) like El Verde allow the system to exhibit long food chains.

Three kinds of disturbance, hurricanes, landslides, and tree falls, create gaps in tabonuco forest. Of these, hurricanes cause the most widespread damage, with recurrence intervals for major storms of about sixty years (Scatena and Larsen 1991; Waide and Lugo 1992). Damage from hurricanes is not uniform over the landscape (Boose et al. 1994), and recurrent hurricanes thus create a mosaic of patches of varying size. The tabonuco forest ecosystem is rarely without disturbance for a sufficiently long period to approach equilibrium in species composition or forest structure. The animal community therefore must be adapted to frequent and large-scale disturbance and the resulting heterogeneous conditions. If predator-prey relationships were as tightly coupled as those predicted by Pimm (1982), then extinctions would be likely. A possible effect of the historical disturbance regime on the structure of the El Verde web is that this regime selects against tightly coupled predator-prey relationships and favors omnivores. The observed prevalence of omnivory and long chains indicates that classical Lotka-Volterra relationships (Pimm 1982) are not commonplace in the El Verde web.

The animal community at El Verde shows no consistent pattern of specialization with regard to small-scale disturbance. Response to gap size appears to be a function of the activity range (e.g., home range) of individuals within a species. Thus *E. coqui* (Stewart and Woolbright, this volume) and some species of snails and insects (Garrison and Willig, this volume) show a response to treefall gaps while species with larger home ranges (e.g., birds, anoles) show none but may respond to larger areas of disturbance such as landslides. From a food web perspective, perhaps the most important response is that of *E. coqui*. Immigration and rapid reproduction of coquis in gaps may make these areas focal points for understory predation and therefore sinks for prey populations. Otherwise, the occurrence of small gaps in the forest does not seem to foster much spatial heterogeneity in vertebrate populations. Because small gaps close over more rapidly than large gaps (Brokaw 1985), the effect on the food web of gap formation at the scale of a single tree is slight.

Temporal Heterogeneity

The El Verde web is a presented as a static, cumulative food web, as are most published webs (Cohen et al. 1990; Schoenly and Cohen 1991). Data from many years of observations taken at different times and places are combined into a single picture of food web structure. However, temporal changes clearly occur throughout the food web, and the nature of these changes influence the basic structure of the web. Predictable changes such as diurnal or seasonal cycles lead to resource discontinuities that encourage compartmentalization of the web.

Diurnal Dynamics

One of the most striking features of the web at El Verde is the complementary behavior of the two major groups of predators. Anoline lizards are the numerically dominant daytime predators whereas leptodactylid frogs are most active at night. This basic dichotomy in activity is reflected in the existence of distinct compartments (subwebs) within the web. Although these compartments overlap to some degree, certain taxa are clearly more likely to be preyed upon by diurnal predators and others by nocturnal predators. For example, the top avian predators in the web consume both lizards and frogs, but the diurnal species (red-tailed hawk, Puerto Rican lizard cuckoo) consume fewer frogs than the nocturnal Puerto Rican screech owl (Waide, this volume).

Day/night compartmentalization is less apparent at the bottom of the food chain. Basal species are equally available as prey during the day and night, and there is no barrier to the common consumption of these resources by both diurnal and nocturnal consumers. Most of the herbivory that occurs in the forest seems to be performed by nocturnal insects, but no quantitative assessment is available. Insect activity levels seem to be similar between day and night, but only 12% of insect species were captured during both day and night (Garrison and Willig, this volume). Little is known about the diel activity of detritivores feeding in the litter layer, but some species (e.g., snails) are more active at night (Garrison and Willig, this volume).

Seasonal Dynamics

During an annual cycle, the forest community experiences predictable fluctuations in both abiotic and biotic factors. Rainfall, temperature, humidity, and solar radiation change seasonally. Populations increase and decrease directly with these changes or indirectly through interactions with other populations. Intrinsic patterns of behavior, such as migration and reproduction, exert themselves and affect the interrelationships among species. For example, the seasonal presence of North American migrant birds increases the number of avian species by over 30%. Resident birds take increased quantities and different kinds of prey when feeding young compared to the rest of the year (Waide, this volume). These fluctuations act to form, break, strengthen, or weaken links among trophic species, which affects the structure of the food web.

Dynamics Over Successional Time

Changes in the species composition of the community over successional time are potentially important in food web dynamics at a site such as El Verde, where the occurrence of hurricanes periodically disrupts the physical structure of the forest (Walker et al. 1991). The cyclic nature of community dynamics in areas subject to hurricanes lends another dimension to the study of food webs, which has yet to be addressed adequately for lack of sufficient long-term data. At El Verde, surveys spanning nearly thirty years of successional time (1963–92) are available. Despite the occurrence of significant changes in vegetation composition over that period (Crow 1980), relatively little change has occurred in the composition of the vertebrate community at the site. Two frog species have become locally extinct, but otherwise the vertebrate fauna of the site remains nearly the same. Over the same period, the number of invertebrate species known from El Verde has continued to increase, but this reflects our accumulating knowledge of these organisms rather than ecological trends.

Population data exist for some species over this thirty-year period. Termites have shown a remarkable constancy in the number of colonies found through time (McMahan, this volume). Studies of anoline lizards suggest that these populations remain stable through time (Reagan, this volume). Most bird species have about the same density now as in earlier studies, with the exceptions discussed in Waide (this volume). Rat populations seem to be exceedingly variable on an annual basis, and this species may be subject to population cycles. However, despite the existence of population fluctuations, the basic structure of the web has not changed for the past thirty years. Food web parameters such as connectance may have fluctuated with the constituent populations, but in general it has been the strength of links that change rather than their existence. Preliminary data suggest that species turnover is low even in the very early stages of succession after a hurricane (Walker et al. 1991), although changes in vertical structure may alter feeding relationships.

ATTRIBUTES OF THE FOOD WEB

Having described the distinctive characteristics of the animal community, we now consider the properties of the food web and compare them with published generalizations. There have been several attempts to describe general patterns of food web organization (Gallopin 1972; Pimm 1979, 1980, 1982; Lawton and Warren 1988; Cohen et al. 1990; Pimm et al. 1991; Cohen et al. 1993). Several of these patterns have been criticized as being artifacts of the manner in which data were collected, analyzed, or presented (Paine 1983, 1988; May 1988; Lawton 1989).

We have defined the food web matrix (app. 14.A) to contain 156 "kinds of organisms" in the sense of Briand (1983), who used the term to represent a biological species, a stage in the life cycle or size-class within a single species, or a collection of functionally or taxonomically related species. The use of "kinds of organisms" instead of species places limitations on the strength of interpretations forthcoming from analyses of the matrix with respect to patterns of organization (e.g., lumping species may underrepresent the actual connectance of the web and omnivory; Havens 1992, 1993) but is unavoidable in the absence of additional information.

The 156 kinds of organisms in the food web represent 2,601 species of consumers, which is a minimum estimate of the number of heterotrophic species at El Verde (excluding most microorganisms). After nearly thirty years of collecting activity at this site, we believe that the number of vertebrate and macroinvertebrate species cataloged is close to the number existing. However, new records for some groups (e.g., insects and fungi) continue to accumulate.

Over 214 species of autotrophic plants occur in the El Verde area (R. F. Smith 1970a). For the purpose of our analysis, plant resources are classified into twenty structural (e.g., leaves, fruit, nectar, seeds, wood, dead wood, roots, litter, detritus, fungi) rather than taxonomic categories. Our justification for this practice is largely pragmatic; too little is known about the species composition of most herbivore diets to allow a more detailed treatment. However, recent studies of folivorous insects (Showalter unpublished) and litter arthropods (Diaz unpublished) at El Verde suggest that these groups forage on leaves of many species. If true, this contrasts greatly with the situation in mainland tropical forests, where tight consumer-host relationships are known to exist.

In addition to the community food web (app. 14.A), we describe more finely resolved webs that more closely resemble the complexity of real systems. The emerging consensus is that the taxonomic scale at which a web is resolved is also critical in resolving food web attributes (Cohen 1989; Martinez 1991; Havens 1992; Closs et al. 1993). Decreased taxonomic resolution results in decreased mean chain length, degree of omnivory, and species-to-trophic-level ratios (Hall and Raffaelli 1991; Martinez 1991). A review of fifty pelagic food webs found that the species-to-link ratio was not scale invariant and concluded that it is an artifact of taxonomic resolution (Havens 1992). Our purpose in aggregating at two levels of investigation is to determine if similar sensitivities exist in the El Verde web.

In large and diverse food webs (as the El Verde web is when compared to other published webs), the high numbers of species and interactions among species diffuse the effects of top-down forces and predispose the system to donor-controlled (i.e., bottom-up) interactions (Strong 1992). This situation, coupled with the large abundance of ectothermic frogs and lizards, establishes the conditions for the system to break with the conventional wisdom of food webs (i.e., the so-called regularities or laws of food web structure; Pimm 1982; Cohen et al. 1990; Pimm et al. 1991). The structure of the food web at El Verde is at variance with several of these laws (1–6 below).

1. *Feeding loops or cycles should be rare because they tend to destabilize the system.* Feeding loops are common in the El Verde web. When interactions that produce cycles are removed from the matrix in order to determine the distribution of food chain lengths, we estimate that 34.9% of the 19,804 chains observed (fig. 14.1) have at least one species that takes part in at least one loop. Some of our observed loops depend on reciprocal predation among different life history stages of the mutual predators (i.e., A eats juvenile B, B

eats juvenile A). These loops are considered more stable than loops which do not involve life history stages (Pimm and Rice 1987). Other observed loops are cannibalistic ("self-loop"; Reagan, this volume).

Figure 14.2 summarizes feeding relationships among four vertebrate and four invertebrate species involved in loops within the food web at El Verde. A variety of loops exists among these species, including one instance of A eats A (predation by male *E. coqui* frogs on eggs); eight instances of A eats B, B eats A; two loops where A eats B, B eats C, C eats A, and one loop where A eats B, B eats C, C eats D, and D eats A (fig. 14.2). Even longer loops are possible if some probable feeding relationships are considered in addition to those confirmed by observations. Twenty-three feeding interactions are shown of which eighteen have been observed. Of the others, two involved prey that could be identified only to family; and three were considered probable because both organisms are known to occur in the same habitat, and the prey are sufficiently small to permit consumption. Cross predation between *Eleutherodactylus* spp. and cannibalism by *E. coqui* on eggs has been observed. Eggs constitute a significant part of coqui diets, especially male diets (Stewart and Woolbright, this volume).

This is not the first time in which loops have been reported as common in a large terrestrial food web. Polis (1991a,b) indicates that this is a widespread phenomenon in the Coachella Valley food web, although he reports it occurring only among invertebrates. A striking aspect of cross predation at El Verde is the degree to which it occurs among distantly related taxa, par-

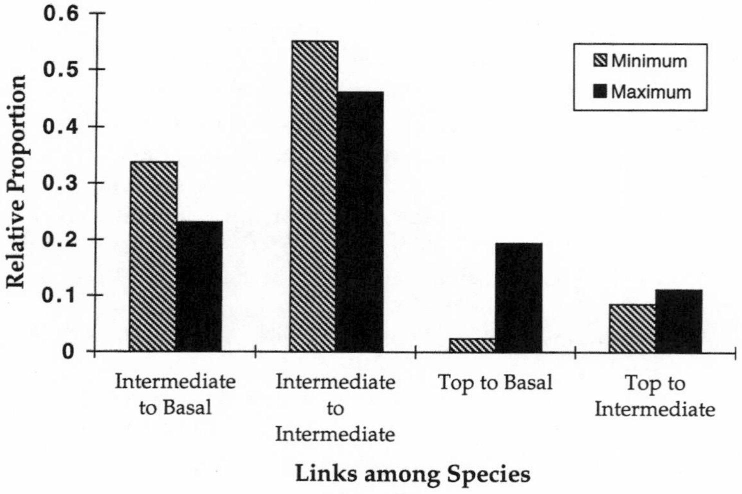

Figure 14.1. Effects of taxonomic resolution on relative proportion of links among trophic levels

ticularly between invertebrates and vertebrates. This phenomenon has been reported before. McCormick and Polis (1982) surveyed the literature and found over 200 references that reported arthropod predation on vertebrates. They defined predation as "the killing and consumption of many prey during a lifetime" and did not consider arthropods that consume vertebrates by other means, such as parasitism or scavenging. Their list included many combinations of predators and prey that have been documented as occurring within the food web at El Verde (fig. 14.2). The specific identification of lizard and frog prey and the detailed food habits known for lizard and frog species at El Verde confirms the existence of loops involving these groups, supporting McCormick and Polis (1982).

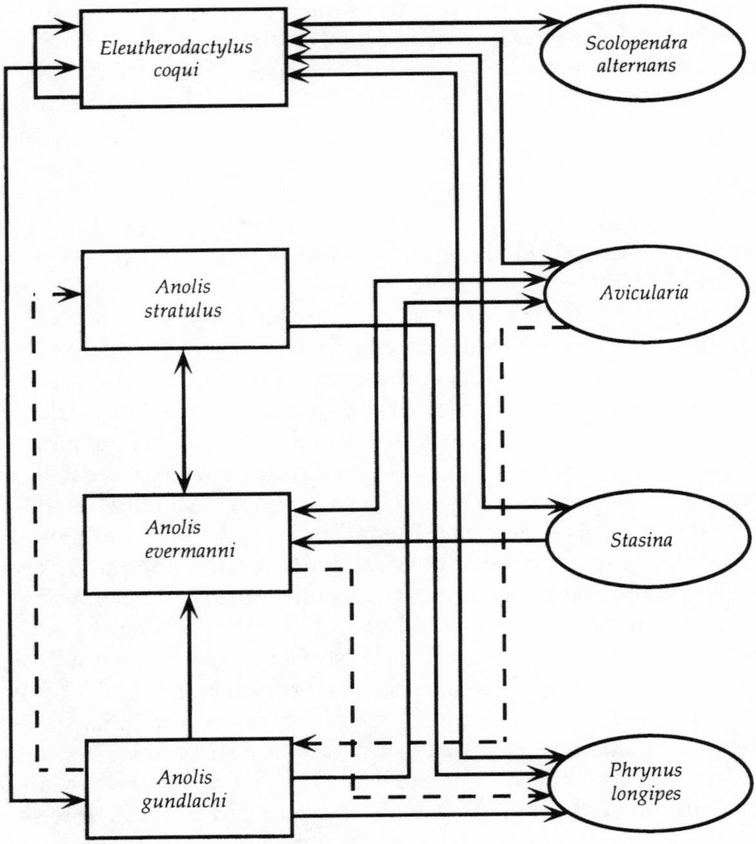

Figure 14.2. Diagram of loops among four invertebrate and four vertebrate species in the food web

Spiders and centipedes were previously reported as predators on vertebrates (McCormick and Polis 1982), but the order Amblypygida (tailless whip scorpions) was not. At El Verde amblypygids (*Phrynus longipes*) attain a body length of 33 cm, leg span of 450 mm, and body weight of more than 4 g (Reagan, unpublished), making them larger than the lizards and frogs on which they feed (fig. 14.3). Large amblypygids have been observed feeding on juvenile and adult anoles and on coquis. The prey were apparently caught as they moved into the range of the sit-and-wait amblypygids. One such encounter in which a large amblypygid (approximately 4 g) consumed an adult male *Anolis stratulus* (approximately 1.3 g) is shown in figure 14.3.

Formanowitz et al. (1981) investigated predation by crab spiders on juveniles of the frog *E. coqui*. Stewart and Woolbright (this volume) report crab spiders as prey of adult *E. coqui*. This particular type of cross predation is termed ontogenetic reversal by McCormick and Polis (1982), who report a spider-frog loop observed by Bhatnagar (1970).

McCormick and Polis (1982) pointed out three main attributes that facilitate invertebrate predation on vertebrates: (*a*) large relative size, (*b*) predatory adaptations such as venoms, and (*c*) social foraging. Of these three, large relative size is the most frequently employed by invertebrate predators. At El Verde, this is true for tailless whip scorpions, but spiders and centipedes employ both large size and venom to subdue vertebrate prey. Social foraging has not been observed as a tactic of the major invertebrate predators on vertebrates at El Verde.

The existence and pervasiveness of loops (also called cycles, reciprocal predation, and cross predation) as a general property of food webs has been debated at length. Many workers believe that while loops may exist, they are scarce and do not constitute a general property of most food webs (Pimm 1982; Pimm and Lawton 1978; Lawton and Warren 1988). Pimm (1982) states: "Biological constraints make it unlikely that two species can be simultaneously each other's predator and prey (a loop)." Support for this view comes from the general observation first stated by Elton (1927) and reiterated by subsequent authors that physical limits constrain both the upper and lower size of prey that a predator can consume. From this assumption it follows logically that one species would have difficulty in serving as both predator and prey for another species. This reasoning is intuitive, but fails to consider predation by the adult of one species on the eggs or juveniles of another species and vice versa. If adult and juvenile forms of the same species are considered as a single food web element, the existence of loops becomes biologically quite reasonable. Pimm and Rice (1987) found that multi-life-stage models (consuming different prey items at different life stages), or life history omnivory and loops, reduces stability (i.e., return times of population densities after disturbance are greater), but less so than did strict omnivory and loops.

Figure 14.3. *Phrynus* consuming an adult *Anolis stratulus*. (Photograph by D. Reagan)

In instances where species have distinct juvenile stages that differ in behavior and morphology from adults, a reduced possibility obtains for loops. Perhaps the best-known examples are for invertebrates such as insects where the larval stage (e.g., caterpillars and grubs) feed at different trophic levels and/or on different foods at the same trophic level in the same habitat as do the adults. In other examples, (e.g., mosquitoes) the larvae occur in freshwater aquatic habitats and the adults are terrestrial. Similar ontogenetic shifts between life stages exist for vertebrates such as amphibians where the larval forms are often aquatic herbivores (tadpoles) and the adults are carnivorous terrestrial predators.

An additional consideration in evaluating loops as a pattern involves terminology. Briand (1983) used his term "kind of organism" (defined above) as interchangeable with "species," adding to the existing confusion in defining exactly what is a species within a given food web. Food web diagrams use both approaches, some showing larval and adult forms as different food web elements and others grouping larvae and adults into a single element. For purposes of loop analysis, biological species (juveniles and adults) are often treated as discrete food web elements to retain the biologically meaningful differences between juvenile and adult forms. We have adopted this convention for insects that metamorphose, but have used a biological species definition for other animals.

Cross predation in food webs may occur more widely than previously documented because such relationships have been unsuspected (McCormick and Polis 1982) and because published food webs for large biological communities are generally incomplete and/or they emphasize vertebrates. Incidental observations of cross predation were considered novel by many investigators rather than a routine interaction. However, Martinez (1991) included detailed observations because he believed that ignoring these specialized feeding relationships would systematically distort the food web.

Pimm (1982) considered loops to be "unreasonable" in theoretical webs, yet McCormick and Polis (1982) documented such relationships between terrestrial invertebrates and vertebrates, and loops are a common feature of planktonic communities (Sprules 1972; Williams 1980). Sprules and Bowerman (1988) suggested that reciprocal predation and cannibalism may have a stabilizing effect on community dynamics and that the equilibrium assumptions of theoretical food webs rarely apply to many animal communities. Our results suggest that reciprocal predation may be widespread in tropical forest communities because the animals involved in loops at El Verde are widely distributed and are likely to have similar food habits throughout their ranges. However, unless feeding relationships are known for most species in the food web and the food web is evaluated at the community level, the existence of loops may be difficult to discern. We concur with McCormick and Polis

(1982); most food web studies have probably underestimated the complexity that is commonplace in most animal communities.

There are two theoretical benefits of significant reciprocal predation: the predators obtain food, and the predators decrease the future risk of predation for themselves and their progeny (McCormick and Polis 1982). Although there are few quantitative data on the contribution of vertebrates to the diets of invertebrates, the existence of loops is consistent with simple feedback mechanisms controlling populations.

Combining all of the feeding relationships that involve cross predation into a single diagram (fig. 14.2) provides a convincing argument that they are subsets of a food web composed of generalists (i.e., any of the organisms involved in cross-predation will consume any prey that it can reasonably capture and subdue). Generalized feeding has been confirmed for frogs and lizards where detailed gut analyses documented diverse prey and no apparent prey preference (see app. 8.A, this volume). This interpretation seems closer to biological reality and is consistent with the theoretical benefits of cross-predation.

Loops involving interactions between invertebrates and vertebrates may be important in the El Verde food web because of the comparatively small size and extreme abundance of frogs and lizards and the presence of several large invertebrate predators. The phenomenon may be observed commonly at El Verde because of the size similarity among the invertebrates and vertebrates, and because both large and small individuals of these species coexist in the forest.

2. *Community food webs have constant connectance.* Connectance represents the proportion of possible feeding relationships that actually occur in the community (Pimm 1982). It is also considered to be a measure of complexity in the community (May 1974). Several estimates of connectance have been proposed in the literature (Cohen et al. 1990; Martinez 1991, 1992). More recently, the most used and accepted is given by the formula

$$C = L/[S(S - 1)/2]$$

(Goldwasser and Roughgarden 1993a; Closs and Lake 1994), in which L is the number of links and S the number of species.

Across communities with increasing species richness, the number of links in the community food web increases with the following relationship,

$$L = 0.14S^2.$$

This relationship is also known as the constant connectance hypothesis (Martinez 1992). Another way of expressing this is that on the average, each

Table 14.2. Food web statistics from El Verde

	ECoWEB Mean	Minimum Resolution	Maximum Resolution
Tropho-species (*S*)	17	136	2,601
Number of links (*L*)	31	1,322	58,808
Links per species (*L/S*)	1.99	9.65	22.61
Min. chain length	2.22	3	19
Max. chain length	5.19 (15[a])	19	*
Mean chain length	2.71	8.56	*
Directed Connectance	0.273 (0.12[b])	0.144	0.0174
Connectance	0.136	0.072	0.0087
$S \times C$	4.64	19.58	45.26
Species Proportions			
Basal	0.190	0.272	0.670
Intermediate	0.525	0.662	0.327
Top	0.285	0.066	0.004
Link to Species Proportions			
Top to			
Basal	0.079 (0.079[a])	0.024	0.195
Intermediate	0.289 (0.358[a])	0.086	0.113
Intermediate to			
Basal	0.274 (0.274[a])	0.337	0.231
Intermediate	0.301 (0.289[a])	0.552	0.461

Sources: ECoWEB, Cohen et al. (1990); Goldwasser and Roughgarden (1993a).

[a] Predicted value by cascade model of Cohen et al. (1990).

[b] Predicted value from Martinez (1992).

* Unable to calculate because of large number of loops present in the web.

species in a community will interact with 14% of the other species in the community in a predator-prey fashion.

At its lowest resolution, the El Verde food web fits the constant connectance hypothesis (table 14.2, fig. 14.4). As resolution increases connectance drops quickly with the relationship

$$C = 7.79(S^{-0.87});\ [r^2 = 0.986].$$

As the number of trophic species increases, connectance drops in a power fashion. In fact, at its highest resolution the El Verde web connectance is almost an order of magnitude smaller than the expected value (table 14.2).

The constant connectance hypothesis predicts that the number of interactions that a species will have in a community is a fixed proportion independent of the number of species in the community (Martinez 1992). If species A is found on the mainland and on an island, the proportion of species with which it interacts will be roughly the same, around 14%. Thus, if the mainland community has 10,000 species and the island 100, it will interact with 1,400 on the mainland and fourteen on the island, give or take a few. This "feature" or "regularity" of food webs, like many others, is more a function of our lack of highly resolved data than of actual communities (Polis 1991a,b). Martinez (1991) found that sequential aggregation of a detailed lake food web resulted first in an increase in connectance until all trophically redundant species were aggregated, a gradual decline as further aggregation took place, and a final sharp increase in highly aggregated webs of less than thirty elements. This latter increase occurs in the size range of many of the food webs that have been used for previous analyses (e.g., Briand 1983) and was attributed by Martinez (1991) to a systematic overestimation of connectance in small webs.

Analysis of the detailed food web data matrix indicated that the entire community does not adhere to the constant connectance hypothesis (see above). Further analysis showed that the day and night sink food webs were structured in a fashion consistent with the constant connectance hypothesis and supporting the concept of day and night subwebs (see attribute 5, this chapter; Camilo, unpublished MS).

Martinez (1991) found that another measure of connectance, directed

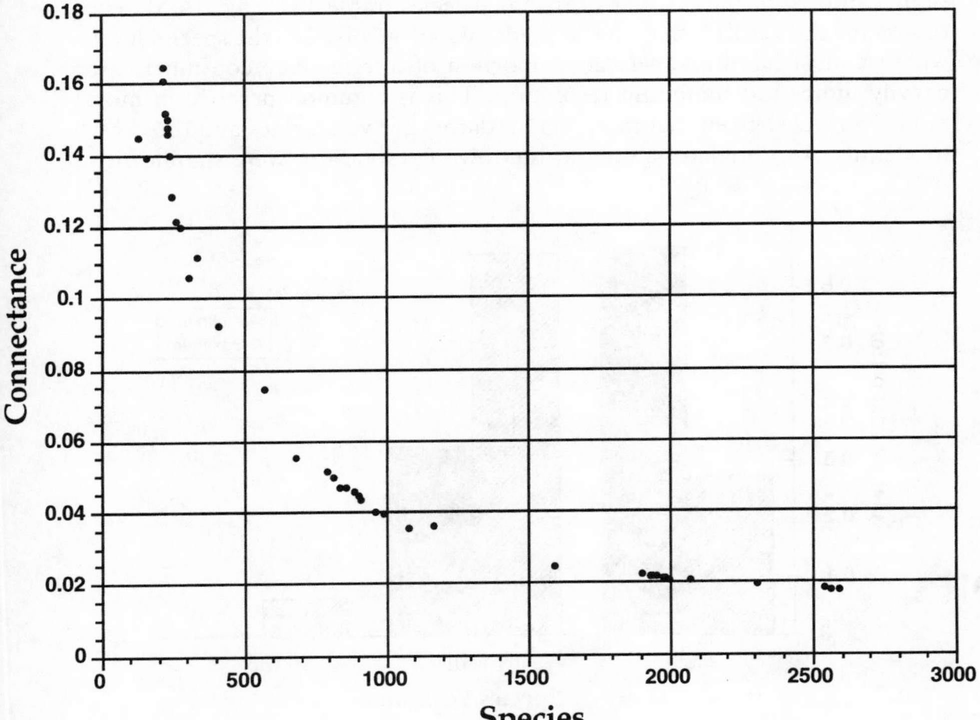

Figure 14.4. Connectance as a function of resolution in the El Verde food web

connectance (which he suggested is the most robust of the connectance measures), had a value of about 0.120 over a broad range of aggregation for his lake web. He also pointed out that the average directed connectances of 113 trophic-species webs (Briand and Cohen 1987) and eleven trophic-species webs of lakes (Briand 1985) were 0.127 ± 0.054 (mean ± 1 SD) and 0.120 ± 0.038, respectively. The directed connectance of the most resolved food web at El Verde is 0.0174, nearly an order of magnitude less than the directed connectance of the least aggregated of the Little Rock Lake webs (Martinez 1991).

3. *The ratios of basal to intermediate to top species and of links among top to intermediate to basal species are scale invariant.* That is, regardless of the number of species present in the community these proportions will deviate little from expected values (table 14.2). A corollary of this rule is that top predators are common in most, if not all, communities. In fact on average, 29% of all species in a given community should be top predators.

The results for the El Verde web demonstrate that both sets of ratios vary significantly with the number of trophic species (table 14.2, fig. 14.5). The reason for this result is that all top predators are resolved to the species level, whereas most basal and a large proportion of intermediate consumers are heavily lumped at minimum resolution. This is common practice in most published webs, given that most top predators are vertebrates and thus easy to identify. Top predators, even at the lowest resolution, were significantly

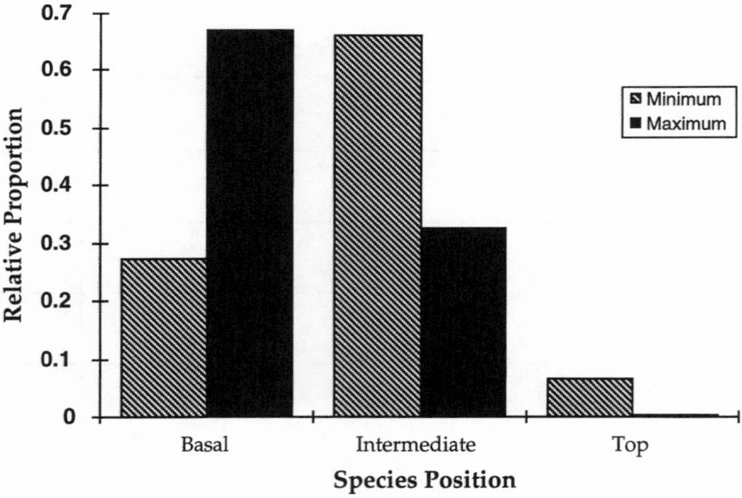

Figure 14.5. Effect of taxonomic resolution on the relative proportion of top, intermediate, and basal species

less common than expected in the El Verde web by at least an order of magnitude. Similar results were obtained by Polis (1991a,b) in the Coachella Valley food web, where top predators were "rare or nonexistent." On the other hand, Goldwasser and Roughgarden (1993a) did not find significant differences between a food web from the island of St. Martin and the mean from a collection of 213 webs (Cohen 1989b), either using lumped or segregated species. Nonetheless, the proportion of top predators in the St. Martin web was almost half (16.7%) of the expected 29%. Because ratios of top to intermediate to basal species vary greatly with taxonomic resolution, the concept of scale invariant ratios requires further investigation with more refined data sets.

We used the following definitions: basal species are prey, intermediate species are predators and prey, and top species have no predators (Lawton and Warren 1988). Nine (6.6%) of the 136 kinds of consumers in the El Verde web are top predations, ninety (66.2%) are intermediate predators, and thirty-seven (27.2%) are basal species. Calculation of the proportion of each of these groups is dependent on the taxonomic detail included in the web. In the El Verde web, vertebrates and some of the better-known invertebrates are represented as independent species. In contrast, most insects are grouped into orders, which are subdivided into adult and larval forms and herbivorous and carnivorous families. Because insects are intermediate species, this practice results in underestimating the proportion of intermediate species in the web. On the other hand, taxonomic resolution of basal species is poor, as it is in most webs, resulting in an underestimate of the number of basal species. The relative magnitude of these underestimates will determine the true proportions of top, intermediate, and basal species in the community web (*sensu* Martinez 1991).

We can estimate the true proportions by using the actual number of species determined for each of our food web elements. To do this, we must make the simplifying assumption that the food habits of all species within our elements are uniform. This allows us to calculate the proportion of top (0.04%), intermediate (32.7%), and basal (67.0%) species among the 2,601 species (both autotrophs and heterotrophs) in the community web directly (table 14.2), but the accuracy of our estimate is constrained by our simplifying assumption.

4. *Omnivores are rare.* Pimm and Lawton (1978) define an omnivore as a species that feeds on more than one trophic level. Although this definition differs from that found in most dictionaries and ecological texts, which define an omnivore as a species that feeds on both plant and animal material, it provides a basis for comparing real and randomly generated food webs.

In the El Verde web omnivory is pervasive, with some omnivores consuming at several trophic levels (app. 14.A). A typical case is that of anoles, which consume almost any prey item that they can fit in their mouths, but do not feed on those arthropods that are less than 2 mm in length (Dial 1992). Both *Anolis evermanni* and *A. stratulus* feed regularly on nectarivores, parasitoids, and predatory arthropods, as well as strictly phytophagous insects (Dial 1992; Reagan, this volume).

Many species of birds, frogs, and lizards feed, at least occasionally, at more than one trophic level. Approximately 25% of bird species at El Verde are thus omnivores (*sensu* Pimm and Lawton 1978; Waide, this volume). *Eleutherodactylus coqui* feeds predominantly on herbivorous arthropods, but its diet includes intermediate species such as arachnids, scorpions, and even lizards. The four species of *Anolis* for which food habits are known are all omnivorous. *Anolis cuvieri* feeds on basal (gastropod) and intermediate (anole) species and also feeds on plant material. Although the tendency toward herbivory by this large lizard is partially explicable on the basis of energetics, the smaller *A. evermanni* also consumes plant material. Intermediate species of invertebrates including spiders, tailless whip scorpions and centipedes that feed on vertebrates and on insect species are also omnivores by Pimm and Lawton's definition.

McCormick and Polis (1982) cite instances where invertebrate predation on vertebrates is known, but not generally recognized in constructing food webs. They also noted that some arthropod predators feed on several, often nonadjacent trophic levels. As noted earlier, this circumstance appears to be true for the food web at El Verde. In the food web at El Verde, omnivory for intermediate species is frequently linked with reciprocal predation, particularly between small vertebrates and large invertebrates.

The rationale for the "regularity" concerning omnivory is the observation that critical parameters of the Lotka-Volterra equation become unstable in communities with many omnivores (Pimm 1982; Pimm and Rice 1987; Camilo and Willig 1995). The conclusion from this observation is that food chains with omnivores are unstable. However, such results might be artifacts of the equations rather than attributes of the actual system. In contrast, it has been suggested that increased omnivory by top predators may decrease predatory pressures on prey populations, thereby increasing the stability of the system (Moore et al. 1988; Camilo and Willig in press). The prevalence of omnivory in insect-dominated webs has been proposed to result from multiple energy avenues and reduced connectance among avenues (Moore et al. 1988).

The proposed regularity in the food web literature that omnivores are rare is an artifact of lumping taxonomic species into trophic species in the compilation of food web data (Martínez 1991, 1993; Havens 1992). Finely resolved webs have much higher proportions of omnivorous species than do aggregated webs (Polis 1991a,b; Hall and Raffaelli 1991; Goldwasser and

Roughgarden 1993a,b). The lack of resolution in webs, especially within invertebrate groups that are extremely diverse taxonomically as well as functionally (e.g., nematodes, acari, and insects), has as a consequence that many predator-prey relationships are not taken into account in analyses (Polis 1991a,b). This in turn gives the impression that a predator is consuming only one "prey," when in fact it is consuming several at different trophic levels.

Yodzis (1984) analyzed the forty food webs published by Briand (1983) and concluded that the apparent rarity of omnivory may be a consequence of the conspicuous absence of species that feed both on plants and animals. He added that for biological reasons this is the most difficult form of omnivory and that omnivory is no more rare than one would expect on the basis of common sense.

5. *Food webs within the same habitat are not divided into compartments.* For any species, the definition of habitat depends in part on its size and activity (Elton 1966; Pimm 1982); however, the habitat of a group of interacting organisms, such as those in a community food web, generally corresponds to an ecosystem (e.g., forest, lake, old field). Defining these systems is somewhat arbitrary because of the variety of microhabitats embedded within the larger system (e.g., treefall gaps).

Pimm and Lawton (1980) investigated two extreme hypotheses regarding food web compartmentalization within habitats: the "reticulate" hypothesis, where species interactions would be homogeneous throughout the web, and the "compartmented" hypothesis, where only species within compartments would interact. These authors analyzed data from published food webs and found no evidence of food web compartmentalization. Their approach used the criterion that compartments existed if the interactions within the community web were grouped into subsystems and interacted little, if at all, with species outside of the compartment.

To investigate this pattern at El Verde, we defined the food web to comprise organisms inhabiting tabonuco forest. Small streams within the forest that could be considered as a subsystem of the forest were not included because of the relatively sharp delineation of their boundaries with the larger forest system. Data on the food habits of the major groups of organisms in the forest were analyzed to determine if separate compartments (subsystems) coexisted within the same forest type.

The two most abundant groups of secondary and tertiary consumers in the food web at El Verde are anoline lizards, which are diurnal, and leptodactylid frogs (coquis), which are nocturnal. The major anole predators are also active diurnally (Puerto Rican lizard-cuckoo, pearly-eyed thrasher, hawks), while the major coqui predators are nocturnal (tarantulas, crab spiders, amblypygids, Puerto Rican screech-owl). Thus, at the upper trophic levels, day-

Table 14.3. Number of taxa consumed by frogs and anoline lizards

Resolution	Frogs [a]	Anoles [b]	Total	Percentage shared
Class	13	12	14	79
Order	25	28	35	66
Family	63	89	115	32
Genus/species	75	104	158	13

[a] Four species, 246 individuals.

[b] Four species, 165 individuals.

night activity differences lead to different subwebs (compartments) within the same habitat type.

Anoles and coquis consume a variety of prey but are predominantly insectivorous. Data obtained from field observations and stomach content analysis of 246 frogs and 145 anoles at El Verde provided a relatively detailed inventory of food items, most of which were identified to the genus or species level (for details see Stewart and Woolbright, this volume, and Reagan, this volume). Where a prey item was identifiable only to a higher taxonomic level (e.g., insect order) and that order was consumed by one group only, a difference between groups also was counted at the family and genus/species level. These data were used to determine dietary overlap between *Anolis* and *Eleutherodactylus* at El Verde.

Results of prey analyses indicated that at the ordinal level of identification, the taxonomic level of identification commonly used in food habit studies, the overlap in anole and coqui diets was 66%. At the family level, the dietary overlap declined to 32%, and at the genus/species level, only 13% of the species consumed by anoles and coquis were shared (table 14.3). Nearly half (48%) of the prey species shared were ants. Only five (28%) of the eighteen families of Diptera and six (32%) of the nineteen families of Coleoptera were consumed by both groups. These results from finely resolved gut analyses provide convincing evidence for compartmentalization of the community food web into day and night subwebs.

Additional support for the existence of day and night compartments in the food web at El Verde comes from a day-night vertical transect study of invertebrates at El Verde (Garrison and Willig, this volume). The numbers of invertebrates collected during day and night were comparable, but notable differences existed in taxonomic composition. The results showed a decline in percent overlap moving from higher to lower taxonomic categories similar to that observed among anole and coqui prey. Overlap between day and night samples declined from 33% overlap at the ordinal level to 19% at the familial level and down to 12% at the species level. Differences between percent overlap at higher taxonomic levels are probably attributable to the fact that the sticky-trap study sampled mostly flying insects and completely missed species that do not fly, such as those that occurred in the litter. Within higher taxonomic groups, there was a strong tendency for species to be either diurnal or nocturnal (Blattodea—seven night, zero day; Orthoptera—eight night, zero day; Dipteran suborders Brachycera and Cyclorrapha [except Phoridae]—twenty-eight night, 114 day). Taxa (e.g., Blattodea and Orthoptera) that were found to be most active at night were also found as prey for coquis.

The ability to identify prey to the genus or species level was necessary to detect compartmentalization of the food web. These data are generally unavailable for community food webs. The results from El Verde provide convincing evidence of the need for detailed taxonomic information on feeding relationships for an appropriate analysis of food web patterns.

6. *Food chains are short, usually containing three to five links.* Because the laws of thermodynamics limit the efficiency of energy transfers, more energy exists at lower trophic levels than at higher ones. Consequently, the maximization of energy or nutrient intake in the diet is best accomplished by feeding at lower levels in the food chain. However, intense competition for prey at lower and intermediate trophic levels may influence predators to consume prey at higher trophic levels (Pimm 1982). Species feeding at lower trophic positions potentially experience interspecific competition, possibly constraining energetic gains. Feeding at higher trophic levels reduces the number of potential interspecific competitors with which one must contend. Actual trophic position may represent a compromise solution to maximize energy availability and minimize the intensity of interspecific competition (Camilo 1992).

The food chain length distribution was calculated using the algorithm presented in Cohen et al. (1990). The data matrix (app. 14.A) was divided into two matrices: in the first, all interactions above the diagonal that produced loops were eliminated; in the second matrix, the interactions below the diagonal that produced loops were removed. In both cases all elements in the diagonal (i.e., cannibalism) were set to zero. To estimate the proportion of chains with loops, we counted the number of chains that contained species involved in loops. The food chain length distribution of the highly resolved web was not calculated because it requires a degree of computing power not readily available.

The El Verde web has a mean food chain length of 8.6 links, and a maximum of nineteen (table 14.2, fig. 14.6). These results come from the "lumped" web matrix, which assumes that all organisms that constitute a trophic species share the same predators and prey (Cohen 1989a). As men-

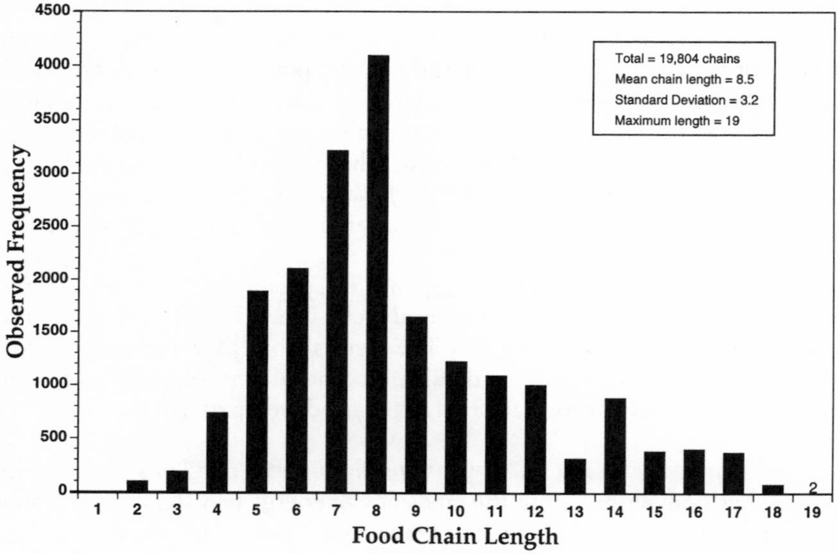

Figure 14.6. Frequency distribution of the lengths of food chains in the aggregated El Verde food web. The distribution shown does not include loops of any kind, or chains that contain loops. A rough estimate indicates that close to 35% of all chains have loops.

tioned above, this assumption is a significant simplification and represents a compromise made necessary by our lack of knowledge. As the degree of resolution increases, not only taxonomically but also ecologically, the observed distribution (fig. 14.6) will shift towards longer chain lengths.

CONCLUSIONS

The food web at El Verde exists in an environment of low variability in temperature, humidity, and day length. Some of the patterns that have emerged from this study can be traced to these characteristics of tropical ecosystems and hence should be present in other tropical webs. For example, the prominence of loops at El Verde may be due, in part, to the tropical breeding patterns that maintain various life stages of multiple species in the ecosystem throughout the annual cycle. The dominance of ectothermic organisms may be possible only because of the consistently high temperatures. Other conclusions may derive from factors specific to tabonuco forest in Puerto Rico, such as the pattern of disturbance and the relatively low species richness of the fauna. Species-rich tropical ecosystems subject to different disturbance regimes may exhibit very different food web patterns than those detected at El Verde.

In the community food web at El Verde, horizontal and vertical spatial heterogeneity and diurnal temporal variability produce aggregations of species interactions that define compartments of the web. These compartments are discrete when their spatial or temporal boundaries are sharp, as in the night and day compartments. In species-rich tropical communities, specialization driven by biological processes may result in similar discontinuities in food web structure. Under conditions of high diversity, compartmentalization of food webs may be encouraged by coevolution among predators and prey.

Many of the patterns observed in the El Verde web were only found because of the high taxonomic resolution of feeding relationships. The same resolution of data from a continental tropical forest could result in a list of species an order of magnitude greater than El Verde and a food web matrix with 100 times as many possible interactions. The extreme rarity of many species in tropical forests may limit the number of feasible interactions. Given the possible importance of these biotic constraints on food web structure, the ultimate challenge for students of food webs is the understanding of high-diversity tropical ecosystems. This book has attempted to provide an entry into these high-diversity communities by defining the features of a tropical forest food web, their environmental determinants, and their distinctiveness.

SUMMARY

In this chapter, we developed a macroscopic view of the food web derived from the community of plants and animals in tabonuco forest. We began by discussing the characteristics of the community, and then showed how these characteristics affect the attributes of the food web. Contrasts between the El Verde web and generalities developed from collections of other webs allowed us to identify forces influencing the structure of webs in general.

The absence of large herbivores and predators in Puerto Rico affects the proportions of top, intermediate, and basal predators in the food web. The absence of mammalian predators may release prey species from top-down control. Dominance of the community by ectotherms (particularly frogs and lizards) with high conversion efficiencies and small body sizes may promote longer food chains and loops within the web. Low faunal richness in tabonuco forest compared to mainland forests limits the number of possible interactions within the food web and may affect compartmentalization of the web. Physical discontinuities in the habitat promote similar discontinuities in the food web; stream and terrestrial organisms overlap little in diet and so do litter and canopy organisms. Frequent disturbance causes spatial heterogeneity within the forest, but the animal community shows little evidence of specialization for gaps. Stronger compartmentalization of the web is shown by day and night specialists. For example, dominant diurnal (lizards) and nocturnal (frogs) predators have very low overlap in diet when their prey is compared at the species level. Changes in the animal community between seasons has a potential effect on food web attributes, but relatively little change has been documented over thirty years of forest succession.

The food web for El Verde comprises 156 kinds of organisms (*sensu* Briand 1983), representing a minimum of 2,601 consumer species. This web is thus one of the most diverse to appear in the literature. Detailed and aggregated webs were analyzed to determine effects of taxonomic resolution on the calculation of food web parameters. The web's diversity, its domination by small ectotherms, and the absence of large predators in the community predispose the system to donor-controlled dynamics. Loops occur in 34.9% of the 19,804 chains derived from the web and involve both vertebrates and invertebrates. Connectance within the web drops as a power function as the number of species increases. At the highest taxonomic resolution, connectance is much lower than predicted from the constant connectance hypothesis (Martinez 1992). The directed connectance (Martinez 1991) of our detailed and aggregated webs are nearly identical (0.061 vs. 0.062). Top predators are an order of magnitude less common in the web (4%) than the concept of scale invariance dictates, possibly because of the taxonomic detail used in the web. Omnivory is common, as has been found in other highly resolved webs. Compartmentalization of the web into day/night and litter/canopy subwebs is apparent. Overlap in the diet of frogs and lizards is only 13%, despite foraging in the same habitat. The mean chain length of the aggregated web is 8.5 links, with a maximum chain length of nineteen; these values would be larger in the detailed web.

Detailed analysis of the El Verde food web has provided a comprehensive understanding of this insular tropical animal community and has produced new insights on general principles of food web organization. Food web studies at El Verde are continuing, building upon the existing long-term data base and addressing many of the questions raised by the investigations described in this book.

Appendix 14.A The Food Web Matrix

The food web matrix is based on records for all known feeding interactions between species at El Verde. Information was obtained from published studies of particular species or groups, from field observations of feeding individuals, and from examination of the digestive tracts of larger predators. Identification of prey to species level was made whenever possible, but much of this detail was not used in the analyses because of the uneven nature of our knowledge. Because prey could not be identified to species level in all cases, we have decided to group insect prey by order, deviating from this practice only when major behavioral differences exist within an order (e.g., Formicidae within the Hymenoptera, larvae and adults for some orders). For each element of the web, we indicate phylum, class, order, family (or number of families found at El Verde), genus, and species (or number of species found at El Verde). We also indicate whether each element is a top (T), intermediate (I), or basal (B) species. Known feeding relationships are presented in chapters 2 through 13 in full taxonomic detail.

Predators are shown at the top of the table and prey along the side. At the intersection between a predator column and a prey row, we put a "1" if the interaction has been observed in the field in the Luquillo Experimental Forest (LEF). A "9" indicates that the interaction has not been observed, but is likely based on interactions involving closely related species in the LEF, or observations or published accounts of the interaction outside the LEF. For example, the Puerto Rican screech owl preys heavily on the most common frog, *Eleutherodactylus coqui.* It is reasonable to assume that the screech owl also feeds on other, less common species of *Eleutherodactylus,* but such predation has not been observed. We therefore show these links as hypothetical in the food web. An "S" (for strong) indicates that the prey makes up more than 10% of the predator's diet.

EL VERDE FOOD WEB - FEBRUARY 1996

This table represents the community food web in tabonuco forest at the El Verde Field Station in Puerto Rico. All consumers whose diets are known from field data or published accounts are included in the web. Predators are shown at the top of the table and prey along the side. At the intersection between a predator column and a prey row, we put a "1" if the interaction has been observed in the field in the Luquillo Experimental Forest. A "9" indicates that the interaction has not been observed, but is likely based on 1) interactions involving closely related species in the LEF or 2) observations or published accounts of the interaction outside of the LEF. An "S" for strong indicates that the prey makes up more than 10% of the predator's diet.

For each element of the web, we indicate phylum, class, order, family, (or number of families found at El Verde), genus and species (or number of species found at El Verde). The column indicates whether the element is top (T), "intermediate" (I), or "basal" (B).

When possible, invertebrate orders are split into herbivorous and carnivorous elements. In some cases, insect families are further split when they contain behaviorally distinct sub-taxa or life stages or Diet for most insect families are inferred from published accounts. Elements are listed in descending order of trophic position except for parasites, which are placed together with the other insects. A complete list of invertebrate species is given in Appendix 6.

							B jamaicensis	B platypterus	E inornatus	F catus	A striatus	S vieilloti	O nudipes	A exiguum	A caeca	H auropunctatus
	CLASS	ORDER	FAMILY	SPECIES			1	2	3	4	5	6	7	8	9	10
C	AVES	FALCONIFORMES	ACCIPITRIDAE	Buteo jamaicensis	T	1	0									
C	AVES	FALCONIFORMES	ACCIPITRIDAE	Buteo platypterus	T	2		0								
C	REPTILIA	SQUAMATA	BOIDAE	Epicrates inornatus	T	3			0							
C	MAMMALIA	CARNIVORA	FELIDAE	Felis catus	T	4				0						
C	AVES	FALCONIFORMES	ACCIPITRIDAE	Accipiter striatus	T	5					0					
C	AVES	CUCULIFORMES	CUCULIDAE	Saurothera vieilloti	T	6						0				
C	AVES	STRIGIFORMES	STRIGIDAE	Otus nudipes	T	7							0			
C	REPTILIA	SQUAMATA	COLUBRIDAE	Arrhyton exiguum	T	8								0		
C	REPTILIA	SQUAMATA	AMPHISBAENIDAE	Amphisbaena caeca	T	9									0	
C	MAMMALIA	CARNIVORA	VIVERRIDAE	Herpestes auropunctatus	I	10	1									0
A	CHILOPODA	SCOLOPENDROMORPHA	SCOLOPENDRIDAE	Scolopendra alternans	I	11	S1	S1	1							S1
C	AMPHIBIA	ANURA	LEPTODACTYLIDAE	Eleutherodactylus coqui	I	12		S9	1			1	S1	9		9
C	AMPHIBIA	ANURA	LEPTODACTYLIDAE	E. richmondi	I	13		9	9			9	9	9		
C	AMPHIBIA	ANURA	LEPTODACTYLIDAE	E. portoricensis	I	14		9	9			9	9	9		
C	AMPHIBIA	ANURA	LEPTODACTYLIDAE	E. wightmanae	I	15		9	9			9	9	9		
C	AMPHIBIA	ANURA	LEPTODACTYLIDAE	E. eneidae	I	16		9	9			9	9	9		
C	AMPHIBIA	ANURA	LEPTODACTYLIDAE	E. hedricki	I	17		9	9			9	9	9		
C	AVES	PICIFORMES	PICIDAE	Melanerpes portoricensis	I	18	9	9								
C	AVES	CORACIIFORMES	TODIDAE	Todus mexicanus	I	19					9					
C	AVES	PASSERIFORMES	MUSCICAPIDAE	Mimocichla plumbea	I	20					1					
C	AVES	PASSERIFORMES	MIMIDAE	Margarops fuscatus	I	21	1									
C	REPTILIA	SQUAMATA	IGUANIDAE	Anolis cuvieri	I	22	1									9
C	REPTILIA	SQUAMATA	IGUANIDAE	A. evermanni	I	23	1	S9	9	9		S9	S9			9
C	REPTILIA	SQUAMATA	IGUANIDAE	A. stratulus	I	24	9	S9	9			S1	S9			9
C	REPTILIA	SQUAMATA	IGUANIDAE	A. gundlachi	I	25	1	S9	9	9		S1	S9			9
C	REPTILIA	SQUAMATA	COLUBRIDAE	Alsophis portoricensis	I	26	1						1			
C	AMPHIBIA	ANURA	LEPTODACTYLIDAE	Leptodactylus albilabris	I	27										9
A	ARACHNIDA	AMBLYPYGIDA	PHRYNIDAE	Phrynus longipes	I	28										
A	ARACHNIDA	ARANEAE	THERAPHOSIDAE	Avicularia laeta	I	29										
C	AVES	PASSERIFORMES	TYRANNIDAE	Myiarchus portoricensis	I	30					9					
C	AVES	PASSERIFORMES	VIREONIDAE	Vireo latimeri	I	31					9					
C	AVES	PASSERIFORMES	EMBERIZIDAE	Nesospingus speculiferus	I	32					9					
C	AVES	PASSERIFORMES	EMBERIZIDAE	Icterus dominicensis	I	33	9	9			9					
A	ARACHNIDA	ACARINA	17 FAMILIES	30 SPECIES	I	34									S1	
A	ARACHNIDA	ARANEAE	22 FAMILIES	56 SPECIES	I	35	1					S9			1	1
A	ARACHNIDA	ARANEAE	SPARASSIDAE	8 SPECIES	I	36										
A	ARACHNIDA	ARANEAE	MIMETIDAE	Mimetes portoricensis	I	37	9	9			9					
C	AVES	PASSERIFORMES	VIREONIDAE	Vireo altiloquus	I	38	9	9			1					
C	AVES	PASSERIFORMES	EMBERIZIDAE	Seiurus aurocapillus	I	39					9					
C	AVES	PASSERIFORMES	EMBERIZIDAE	S. motacilla	I	40					9					
C	REPTILIA	SQUAMATA	GEKKONIDAE	Sphaerodactylus klauberi	I	41								9		
C	REPTILIA	SQUAMATA	GEKKONIDAE	S. macrolepis	I	42								9		
C	REPTILIA	SQUAMATA	ANGUINIDAE	Diploglossus pleei	I	43										9
A	ARACHNIDA	SCORPIONIDA	BUTHIDAE	Tityus obtusus	I	44							S1			
A	INSECTA	HYMENOPTERA	FORMICIDAE	PONERINAE (8 SPECIES)	I	45										
A	CRUSTACEA	DECAPODA	POMAMONIDAE	Epilobocera situatifrons	I	46	1									
C	MAMMALIA	RODENTIA	MURIDAE	Rattus rattus	I	47	S1	1	1	9			1			1
C	AMPHIBIA	ANURA	BUFONIDAE	Bufo marinus	I	48	1									
A	CHILOPODA	GEOPHILOMORPHA	2 FAMILIES	2 species	I	49										
A	CHILOPODA	SCUTIGEROMORPHA	SCUTIGERIDAE	Antillora portoricensis	I	50										
A	INSECTA	HYMENOPTERA	VESPIDAE	2 SPECIES	I	51										
C	AVES	APODIFORMES	TROCHILIDAE	Chlorostilbon maugeus	I	52					9					
C	AVES	APODIFORMES	TROCHILIDAE	Anthracothorax viridis	I	53					9					

	Scolopendra	E coqui	E richmondi	E portoricensis	E wightmanii	E eneidae	E hedricki	Melanerpes	T mexicanus	M plumbea	M fuscatus	A cuvieri	A evermanni	A stratulus	A gundlachi	Alsophis	L albilabris	P longipes	A laeta	Myiarchus	V laimeri	N speculiferus	I dominicensis	Acarina 30 spp	Araneae 56 sp	Sparassidae 8 sp	Mimetes	V altiloquus	S aurocapllus	S motacilla	S klauberi	S macrolepis	D pleei	T obtusus
	11	12	13	14	15	16	17	18	19	20	21	22	23	24	25	26	27	28	29	30	31	32	33	34	35	36	37	38	39	40	41	42	43	44
1																																		
2																																		
3																																		
4																																		
5																																		
6																																		
7																																		
8																																		
9																																		
10																																		
11	0	1	S1	1	S1			1																										
12	1	1			9			1	1	S1	1				1	S1		.1	S1	1	9	9	9	9	1	1		9	1	9	1			1
13	9	9	0						9	9	9					9			9	9	9	9	9	9	9	9		9	9	9	9			9
14	9	9		0					9	9	9					9			9	9	9	9	9	9	9	9		9	9	9	9			9
15	9	9			0				9	9	9					9			9	9	9	9	9	9	9	9		9	9	9	9			9
16	9	9				0			9	9	9					9			9	9	9	9	9	9	9	9		9	9	9	9			9
17	9	9					0		9	9	9					9			9	9	9	9	9	9	9	9		9	9	9	9			9
18								0																										
19									0																									
20										0																								
21											0																							
22												0																						
23	9	9						9	9	S9	S1	9	0	1		9				9	9	9	9	1										
24	9	9						9	9	S9	9	9	1	0		1				9	9	9	9	9										
25	9	1						9	9	S9	S1	1	1		0	9	1	1	1	9	9	9	9	9										
26																0																		
27										1							0																	
28		1	1															0																
29		9	9	9	9	9	9						9		9				0															
30																				0														
31																					0													
32																						0												
33																							0											
34		1	1	1	1	1				1			1	1	1		1							0							S1	1		
35		1	S1	1	S1	1		1	S1	S1	1	1	1	1	1		1				S1	S1	S1		0		9	S1	1	1	S1	S1	S1	
36		1			1								1													0								
37																											0							
38																												0						
39																													0					
40																														0				
41	9									9						1		9							9						0			
42	9									9						1		9							9							0		
43	9																																0	
44		1						1			1												9							1				0
45		1		1	1							1	1	1	1																			
46		1									1																							
47																																		
48	1																																	
49		9	1	1		1				1	1		1									9			9						1		1	
50																					1													
51													1		1																			
52																																		
53																																		

	Ponerinae 8 spp	E situatifrons	R rattus	B marinus	Chilopoda 2 spp	Antillora	Vespidae 2 spp	C maugeus	A viridis	M varia	P americana	D tigrina	D caerulescens	D discolor	D angelae	S ruticilla	C flaveola	Loxigilla	T rostellatus	Pseudoscorpionida 2 sp	Opiliones 7 sp	Odonata 12 spp	G grisea	---------- larval Odonata	adult Neuroptera 9 spp	carnivorous adult Hemiptera	larval Coleoptera	adult Coleoptera	larval Diptera	--- adult Diptera 158 spp	------ larval Hymenoptera	adult Hymenoptera 109 spp	Myrmicinae 19 spp	Formicidae 24 spp	E fuscus
	45	46	47	48	49	50	51	52	53	54	55	56	57	58	59	60	61	62	63	64	65	66	67	68	69	70	71	72	73	74	75	76	77	78	79
1																																			
2																																			
3																																			
4																																			
5																																			
6																																			
7																																			
8																																			
9																																			
10																														9					
11																																			
12	1	1	1																											9			1		
13	9	9	9																											9					
14	9	9	9																											9					
15	9	9	9																											9					
16	9	9	9																											9					
17	9	9	9																											9					
18																														9					
19																														9					
20																														9					
21																													S1	9					
22																														9					
23																														9					
24																														9					
25																														9					
26																														9					
27																														9					
28																																			
29																																			
30																														9					
31																														9					
32																														9					
33																														9					
34																			1	1								9							
35					9			S1	S1	S1	S1		S1	S1		1	S1	1		9															
36																																			
37																																			
38																														9					
39																														9					
40																														9					
41																																			
42																																			
43																																			
44																																			
45	0																																		
46		0																																	
47			0																											9					
48		1		0																										9					
49					0																														
50						0																													
51							0																												
52								0																						9					
53									0																					9					

	L borealis	P parnelli	S zena	Schizomus	Orthoptera 23 spp	Diplopoda 18 spp	Thysanoptera 10 spp	Hirudinea	Philornis pici	Capillaria hepatica	Nematoda	Streblidae 7 spp	Labidocarpidae 5 spp	Spinturicidae 4 spp	Spelaeorhynchidae 3 spp	Macronyssidae 1 sp	Argasadae 2 spp	A jamaicensis	B cavernarum	E sezekorni	M redmani	S rufum	C squamosa	G montana	A vittata	E musica	--- Hemiptera 30 sp	-- larval Coleoptera	herbivorous adult Coleoptera 69 sp	larval Diptera	adult Diptera 314 sp
	80	81	82	83	84	85	86	87	88	89	90	91	92	93	94	95	96	97	98	99	100	101	102	103	104	105	106	107	108	109	110
1																															
2																															
3																															
4																															
5																															
6																															
7																															
8																															
9																															
10											9																				
11																															
12					1	1					9																				
13											9																				
14											9																				
15											9																				
16											9																				
17											9																				
18									9		9																				
19											9																				
20											9																				
21									1		9																				
22											9																				
23											9																				
24											9																				
25											9																				
26											9																				
27											9																				
28																															
29																															
30											9																				
31											9																				
32											9																				
33											9																				
34							9				1																				
35																															
36																															
37																															
38											9																				
39											9																				
40											9																				
41											9																				
42											9																				
43											9																				
44																															
45																															
46																															
47										1	9																				
48											9																				
49																															
50																															
51																															
52											9																				
53											9																				

	larval Hymenoptera	--- adult Hymenoptera 9 sp	Polydontes	C caracolla	Podocopa	Copepoda	P richmondi	Collembola 13 spp	Machilidae 1 sp	Blattellidae 17 sp	Blattidae 2 sp	Phasmatodea 4 spp	Isoptera 4 spp	Dermaptera 2 spp	Teratembiidae 1 sp	Psocoptera 13 spp	Homoptera 74 spp	larval Lepidoptera	adult Lepidoptera 234 spp	P hawayana	Oniscidae 4 spp	Ephemeroptera	Trichoptera 21 spp	Stylomatphora 38 spp	adult Coleoptera	larval Coleoptera	Fungi	Slime molds 9 spp	Bacteria
	111	112	113	114	115	116	117	118	119	120	121	122	123	124	125	126	127	128	129	130	131	132	133	134	135	136	137	138	139
1																													
2																													
3																													
4																													
5																													
6																													
7																													
8																													
9																													
10																													
11																													
12			1	1																									
13																													
14																													
15																													
16																													
17																													
18																													
19																													
20																													
21																													
22																													
23																													
24																													
25																													
26																													
27																													
28																													
29																													
30																													
31																													
32																													
33																													
34				1																									
35																													
36																													
37																													
38																													
39																													
40																													
41																													
42																													
43																													
44																													
45																													
46																													
47																													
48																													
49																													
50																													
51																													
52																													
53																													

	140	141	142	143	144	145	146	147	148	149	150	151	152	153	154	155	156	
1																		1
2																		2
3																		3
4																		4
5																		5
6																		6
7																		7
8																		8
9																		9
10																		10
11																		11
12																		12
13																		13
14																		14
15																		15
16																		16
17																		17
18																		18
19																		19
20																		20
21																		21
22																		22
23																		23
24																		24
25																		25
26																		26
27																		27
28																		28
29																		29
30																		30
31																		31
32																		32
33																		33
34																		34
35																		35
36																		36
37																		37
38																		38
39																		39
40																		40
41																		41
42																		42
43																		43
44																		44
45																		45
46																		46
47																		47
48																		48
49																		49
50																		50
51																		51
52																		52
53																		53

	CLASS		ORDER	FAMILY	SPECIES			1	2	3	4	5	6	7	8	9	10
C	AVES		PASSERIFORMES	EMBERIZIDAE	Mniotilta varia	I	54					9					
C	AVES		PASSERIFORMES	EMBERIZIDAE	Parula americana	I	55					9					
C	AVES		PASSERIFORMES	EMBERIZIDAE	Dendroica tigrina	I	56					9					
C	AVES		PASSERIFORMES	EMBERIZIDAE	D. caerulescens	I	57										
C	AVES		PASSERIFORMES	EMBERIZIDAE	D. discolor	I	58					9					
C	AVES		PASSERIFORMES	EMBERIZIDAE	D. angelae	I	59	9	9			9					
C	AVES		PASSERIFORMES	EMBERIZIDAE	Setophaga ruticilla	I	60					9		1			
C	AVES		PASSERIFORMES	EMBERIZIDAE	Coereba flaveola	I	61	1			9	1					
C	AVES		PASSERIFORMES	EMBERIZIDAE	Loxigilla portoricensis	I	62					9					
C	REPTILIA		SQUAMATA	TYPHLOPIDAE	Typhlops rostellatus	I	63										9
A	ARACHNIDA		PSEUDOSCORPIONIDA	2 FAMILIES	2 SPECIES	I	64										
A	ARACHNIDA		OPILIONES	3 FAMILIES	7 SPECIES	I	65									1	1
A	INSECTA		ODONATA	3 FAMILIES	12 SPECIES	I	66										
A	INSECTA		MANTODEA	MANTIDAE	Gonatista grisea	I	67										
A	INSECTA	c	ODONATA	COENAGRIONIDAE	Telebasis vulnerata	I	68										
A	INSECTA	a	NEUROPTERA	5 FAMILIES	9 SPECIES	I	69										
A	INSECTA	r	HEMIPTERA	8 FAMILIES	12 SPECIES	I	70						S1	S1			
A	INSECTA	n	COLEOPTERA	11 FAMILIES	LARVAE (32 SPECIES)	I	71									S1	
A	INSECTA	i	COLEOPTERA	15 FAMILIES	33 SPECIES	I	72						S1	S1		S1	1
A	INSECTA	v	DIPTERA	16 FAMILIES	LARVAE (72 SPECIES)	I	73										1
A	INSECTA	o	DIPTERA	12 FAMILIES	158 SPECIES	I	74							1		1	
A	INSECTA	r	HYMENOPTERA	19 FAMILIES	LARVAE (116 SPECIES)	I	75										1
A	INSECTA	e	HYMENOPTERA	19 FAMILIES	109 SPECIES	I	76										
A	INSECTA		HYMENOPTERA	FORMICIDAE	MYRMICINAE (19 SPECIES)	I	77										
A	INSECTA		HYMENOPTERA	FORMICIDAE	24 SPECIES	I	78									S1	
C	MAMMALIA		CHIROPTERA	VESPERTILIONIDAE	Eptesicus fuscus	I	79			9							
C	MAMMALIA		CHIROPTERA	VESPERTILIONIDAE	Lasiurus borealis	I	80			9							
C	MAMMALIA		CHIROPTERA	VESPERTILIONIDAE	Pteronotus parnelli	I	81			9							
C	AVES		PASSERIFORMES	EMBERIZIDAE	Spindalis zena	I	82	9	9			9					
A	ARACHNIDA		SCHIZOMIDA	SCHIZOMIDAE	Schizomus 2 spp.	I	83										
A	INSECTA		ORTHOPTERA	2 FAMILIES	23 SPECIES	I	84						1	S1			1
A	DIPLOPODA			10 FAMILIES	18 SPECIES	I	85									1	
A	INSECTA		THYSANOPTERA	2 FAMILIES	10 SPECIES	I	86									1	
E	HIRUDINEA			1 FAMILY	1 SPECIES	I	87										
A	INSECTA	p	DIPTERA	MUSCIDAE	Philornis sp.	I	88										
N	ADENOPHOREA	a	TRICHINELLIDA	TRICHURIDAE	Capillaria hepatica	I	89										
N	SECERNENTEA	r			Nematodes	I	90										
A	INSECTA	a	DIPTERA	STREBLIDAE	7 SPECIES	I	91										
A	INSECTA	s	DIPTERA	LABIDOCARPIDAE	5 SPECIES	I	92										
A	INSECTA	i	DIPTERA	SPINTURICIDAE	4 SPECIES	I	93										
A	INSECTA	t	DIPTERA	SPELAEORHYNCHIDAE	3 SPECIES	I	94										
A	ARACHNIDA	e	ACARINA	MACRONYSSIDAE	1 SPECIES	I	95										
A	ARACHNIDA	s	ACARINA	ARGASIDAE	2 SPECIES	I	96										
C	MAMMALIA		CHIROPTERA	PHYLOSTOMIDAE	Artibeus jamaicensis	B	97			9							
C	MAMMALIA		CHIROPTERA	PHYLOSTOMIDAE	Brachyphylla cavernarum	B	98			9							
C	MAMMALIA		CHIROPTERA	PHYLOSTOMIDAE	Erophylla sezekorni	B	99			9							
C	MAMMALIA		CHIROPTERA	PHYLOSTOMIDAE	Monophyllus redmani	B	100			9							
C	MAMMALIA		CHIROPTERA	PHYLOSTOMIDAE	Stenoderma rufum	B	101			9							
C	AVES		COLUMBIFORMES	COLUMBIDAE	Columba squamosa	B	102	1	9								
C	AVES		COLUMBIFORMES	COLUMBIDAE	Geotrygon montana	B	103				1						9
C	AVES		PSITTACIFORMES	PSITTACIDAE	Amazona vittata	B	104	9	9	1							
C	AVES		PASSERIFORMES	EMBERIZIDAE	Euphonia musica	B	105	9	9			9					
A	INSECTA	h	HEMIPTERA	5 FAMILIES	30 SPECIES	B	106						S1	S1			
A	INSECTA	e	COLEOPTERA	26 FAMILIES	LARVAE (86 SPECIES)	B	107									S1	
A	INSECTA	r	COLEOPTERA	26 FAMILIES	86 SPECIES	B	108						S1	S1		S1	1
A	INSECTA	b	DIPTERA	6 FAMILIES	LARVAE (429 SPECIES)	B	109										1
A	INSECTA	i	DIPTERA	25 FAMILIES	314 SPECIES	B	110							1		1	
A	INSECTA	v	HYMENOPTERA	6 FAMILIES	LARVAE (25 SPECIES)	B	111										1
A	INSECTA	o	HYMENOPTERA	4 FAMILIES	9 SPECIES	B	112										
M	GASTROPODA		STYLOMMATOPHORA	CAMAENIDAE	Polydontes sp.	B	113										
M	GASTROPODA		STYLOMMATOPHORA	CAMAENIDAE	Caracolus caracolla	B	114										
A	CRUSTACEA		PODOCOPA		1 SPECIES	B	115										
A	CRUSTACEA		COPEPODA		1 SPECIES	B	116										
A	CRUSTACEA		ISOPODA	ONISCIDAE	Philoscia richmondi	B	117										
A	ENTOGNATHA		COLLEMBOLA	4 FAMILIES	13 SPECIES	B	118									1	
A	INSECTA		MICROCORYPHIA	MACHILIDAE	1 SPECIES	B	119										
A	INSECTA		BLATTODEA	BLATTELLIDAE	17 SPECIES	B	120							1			1
A	INSECTA		BLATTODEA	BLATTIDAE	2 SPECIES	B	121										
A	INSECTA		PHASMATODEA	PHASMATIDAE	4 SPECIES	B	122										
A	INSECTA		ISOPTERA	2 FAMILIES	4 SPECIES	B	123									S1	
A	INSECTA		DERMAPTERA	2 FAMILIES	2 SPECIES	B	124						S9				
A	INSECTA		EMBIOPTERA	TERATEMBIIDAE	1 SPECIES	B	125										
A	INSECTA		PSOCOPTERA	9 FAMILIES	13 SPECIES	B	126										
A	INSECTA		HOMOPTERA	15 FAMILIES	74 SPECIES	B	127										
A	INSECTA		LEPIDOPTERA	22 FAMILIES	LARVAE (234 SPECIES)	B	128						S1	1			
A	INSECTA		LEPIDOPTERA	22 FAMILIES	234 SPECIES	B	129						1	1			1
E	OLIGOCHAETA		HAPLOTAXIDA	MEGASCOLECIDAE	P. hawayana	B	130									1	
A	CRUSTACEA		ISOPODA	ONISCIDAE	4 SPECIES	B	131										
A	INSECTA		EPHEMEROPTERA	LEPTOPHLEBIIDAE	1 SPECIES	B	132										1
A	INSECTA		TRICHOPTERA	10 FAMILIES	21 SPECIES	B	133										
M	GASTROPODA		STYLOMATOPHORA	17 FAMILIES	38 SPECIES	B	134									S1	

	11	12	13	14	15	16	17	18	19	20	21	22	23	24	25	26	27	28	29	30	31	32	33	34	35	36	37	38	39	40	41	42	43	44
54																																		
55																																		
56																																		
57																																		
58																																		
59																																		
60																																		
61																																		
62																																		
63			9																															
64		1	1	1	1	1							1		1		1																1	
65		1	1												1																1		1	
66								1	1						9					1										1				
67									1												1							1						
68													1																					
69		1		1					1		1		1	9						1														
70		1	1	1	1			1	S1		1		1	1	1		1			S1	1	1						1		1		1		
71		1	1	1	1	1	1	S1					1		1																S1		1	
72		1	S1	S1	1	S1	9	S1	S1	S1	1	1	S1	1	S1					S1	S1	S1	S1					S1	S1	1	S1	S1	S1	
73		1	1	1	1	1	1						1	1	1		1																1	
74		1	1	S1	1	S1			S1			1	S1	S1	S1		1			1	1		1		1			1		1	S1	1		
75		1											1	1	9		1																	
76		1	1	1	1	1			1	1	1	1	1	1	1		1			S1	1	1	1					1	1					
77		1	1	1	1								1	1	1																			
78		S1	S1	S1	S1	S1		S1	1	1	1	1	S1	S1	S1		S1			1		S1	1					1	S1		1	1	1	
79																																		
80																																		
81																																		
82																																		
83		9	9	9	9	9																			1	1								
84		S1	1	S1	1	1		1	S1	S1	1	1	1	1	S1		1	1		S1	S1	S1	S1					1	1		1	1		1
85		1	1	1	1				1		1				1		1														1		S1	
86		1	1	1	1								1		1																1	1		
87			1																															
88																																		
89																																		
90		1		1					1				1		1																			
91																																		
92																																		
93																																		
94																																		
95																																		
96																																		
97																																		
98																																		
99																																		
100																																		
101																																		
102																																		
103											1																							
104																																		
105																																		
106		1	1	1	1			1	S1		1		1	1	1		1			S1	1	1						1		1		1		
107		1	1	1	1	1	1	S1					1		1																S1		1	
108		1	S1	S1	1	S1	9	S1	S1	S1	1	1	S1	1	S1					S1	S1	S1	S1					S1	S1	1	S1	S1	S1	
109		1	1	1	1	1	1						1	1	1		1																1	
110		1	1	S1	1	S1			S1			1	S1	S1	S1		1			1	1		1		1			1		1	S1	1		
111		1											1	1	9		1																	
112		1	1	1	1	1			1	1	1	1	1	1	1		1			S1	1	1	1					1	1					
113																																		
114												1																						
115																																		
116																																		
117			1																															
118		1	1	1	1	1			1				1		1										1	1					S1	S1		
119		1											1																					
120		S1	1	S1	1			S1	1	9	1	9	1	1	1						1													
121		1													1																			
122		1		1					1		S1	1	1							1								1	1					
123		1	1	1	1	S1		1	9		1		1	9	1		1														1	1	S1	9
124		9						S1	S1	1	1		1	1	1		1				1		S1					1					S1	
125							1							1																				
126		1	1	1	1	1							1	1	1																1	1		
127		S1	1	1	S1	1			S1	1	1		S1	S1	1		1			S1	S1	1	S1		1			S1		1	1			
128		1	S1	1	S1	1		1	1	S1	1	S1	1	1	S1	1	1			S1	S1	S1	9					S1	1		1			
129		1	1	1	1				1		1		S1	S1	S1		1			1	S1	1	1								1		1	
130		1																																
131		1	1	1				1							1		1														S1	S1		
132		1							1				1		1																			
133		1		1									1		S1																1			
134		1	1	1	1	1		1		1		S1	1		1		1			1		1							S1	1	1		S1	

	45	46	47	48	49	50	51	52	53	54	55	56	57	58	59	60	61	62	63	64	65	66	67	68	69	70	71	72	73	74	75	76	77	78	79
54										0																				9					
55											0																			9					
56												0																		9					
57													0																	9					
58														0																9					
59															0															9					
60																0														9					
61																	0													9					
62																		0												9					
63																			0																
64																	9			0															
65																					0														
66																						0													
67																							0												
68																								0											
69																		1							0										
70											1		1									9				0			9		9	9			1
71																			S1							9	0								
72								S1	S1	S1	S1		S1	S1		1	S1	S1				9	9				9	0	9						1
73																					1								0		1	1			
74								S1	S1		S1		S1	1		S1						9	9							0					
75																													9		9	9			
76				1				S1	S1	1	1		1	1		S1																0			
77																																	0		
78									S1		1		1	1					S1															0	1
79																														9					0
80																														9					
81																														9					
82																														9					
83																																			
84										S1								1											9						
85																														1					
86									1								1																		
87																																			
88																																			
89																																			
90																																			
91																																			
92																																			
93																																			
94																																			
95																																			
96																																			
97																														9					
98																														9					
99																														9					
100																														9					
101																														9					
102																														9					
103																														9					
104																													1	9					
105																														9					
106											1		1									9							9		9	9			1
107																			S1							9									
108								S1	S1	S1	S1		S1	S1		1	S1	S1				9	9				9		9						1
109																					1										1	1			
110								S1	S1		S1		S1	1		S1						9	9												
111																													9		9	9			
112				1				S1	S1	1	1		1	1		S1																			
113																																			
114																																			
115																																			
116																																			
117																																			
118																				1	1					1		9					9		
119																																			
120																																			1
121																																			
122	1									1													S1							1		1			
123	1			1															S1									1	9			1			1
124										S1													1												
125																																			
126																																			
127								S1	S1	S1	S1		S1	S1		S1	S1	1				9	9						9				1		1
128										1	S1		1	S1			S1	1								9			S1		S1	9			
129										1	S1		1	1		S1						9	9												
130																																			
131																					1														
132																						9													
133																						9	S1												
134																	1	1	1																

	80	81	82	83	84	85	86	87	88	89	90	91	92	93	94	95	96	97	98	99	100	101	102	103	104	105	106	107	108	109	110
54											9																				
55											9																				
56											9																				
57											9																				
58											9																				
59											9																				
60											9																				
61											9																				
62											9																				
63																															
64																															
65																															
66																															
67																															
68																															
69																															
70																															
71																															
72		S1	1																												
73																															
74																															
75																															
76																								1							
77																															
78	1																														
79											9			1																	
80	0										9																				
81		0									9																				
82			0								9																				
83				0																											
84					0	1																									
85						0																									
86							0																								
87								0																							
88									0																						
89										0																					
90											0																				
91												0																			
92													0																		
93														0																	
94															0																
95																0															
96																	0														
97											9	1	1	1	1			0													
98											9	1	1			1			0												
99											9	1		1			1			0											
100											9	1		1	1						0										
101											9		1	1								0									
102									9		9												0								
103									9		9													0							
104									1		9														0						
105											9															0					
106																											0				
107																												0			
108		S1	1																										0		
109																														0	
110																															0
111																															
112																								1							
113																															
114																															
115																															
116																															
117																															
118																													1		
119																															
120																															
121																															
122																															
123																															
124				S1																											
125																															
126																															
127						1																									
128																															
129		1																													
130																															
131																															
132																															
133																															
134																															

	111	112	113	114	115	116	117	118	119	120	121	122	123	124	125	126	127	128	129	130	131	132	133	134	135	136	137	138	139
54																													
55																													
56																													
57																													
58																													
59																													
60																													
61																													
62																													
63																													
64																													
65																													
66																													
67																													
68																													
69																													
70																													
71																													
72																													
73																													
74																													
75																													
76																													
77																													
78																													
79																													
80																													
81																													
82																													
83																													
84																													
85																													
86																													
87																													
88																													
89																													
90																													
91																													
92																													
93																													
94																													
95																													
96																													
97																													
98																													
99																													
100																													
101																													
102																													
103																													
104																													
105																													
106																													
107																													
108																													
109																													
110																													
111	0																												
112		0																											
113			0																										
114				0																									
115					0																								
116						0																							
117							0																						
118								0																					
119									0																				
120										0																			
121											0																		
122												0																	
123													0																
124														0															
125															0														
126																0													
127																	0												
128																		0											
129																			0										
130																				0									
131																					0								
132																						0							
133																							0						
134																								0					

	140	141	142	143	144	145	146	147	148	149	150	151	152	153	154	155	156	
54																		54
55																		55
56																		56
57																		57
58																		58
59																		59
60																		60
61																		61
62																		62
63																		63
64																		64
65																		65
66																		66
67																		67
68																		68
69																		69
70																		70
71																		71
72																		72
73																		73
74																		74
75																		75
76																		76
77																		77
78																		78
79																		79
80																		80
81																		81
82																		82
83																		83
84																		84
85																		85
86																		86
87																		87
88																		88
89																		89
90																		90
91																		91
92																		92
93																		93
94																		94
95																		95
96																		96
97																		97
98																		98
99																		99
100																		100
101																		101
102																		102
103																		103
104																		104
105																		105
106																		106
107																		107
108																		108
109																		109
110																		110
111																		111
112																		112
113																		113
114																		114
115																		115
116																		116
117																		117
118																		118
119																		119
120																		120
121																		121
122																		122
123																		123
124																		124
125																		125
126																		126
127																		127
128																		128
129																		129
130																		130
131																		131
132																		132
133																		133
134																		134

	CLASS		ORDER	FAMILY	SPECIES			1	2	3	4	5	6	7	8	9	10
A	INSECTA		COLEOPTERA	11 FAMILIES	15 SPECIES (SCAVENGERS)	B	135						S1	S1		S1	1
A	INSECTA		COLEOPTERA	8 FAMILIES	VAE (SCAVENGERS; 12 SPEC	B	136										
	FUNGI		ALIVE		FUNGI		137										
	FUNGI		ALIVE		SLIME MOLDS		138										
	BACTERIA		ALIVE		BACTERIA		139										
	PLANTS		ALIVE		PLANTS		140										S1
	PLANTS		ALIVE		LIVE LEAVES		141										1
	PLANTS		ALIVE		LIVE WOOD		142										
	PLANTS		ALIVE		SAP		143										
	PLANTS		ALIVE		ROOTS		144										
	PLANTS		ALIVE		POLLEN		145										
	PLANTS		ALIVE		NECTAR		146										
	PLANTS		ALIVE		FRUIT		147										1
	PLANTS		ALIVE		SEEDS		148										
	PLANTS		ALIVE		FLOWERS		149										
	PLANTS		ALIVE		ALGAE		150										
	LICHENS		ALIVE		LICHENS		151										
	ORGANIC MATTER		DEAD		DEAD WOOD		152										
	ORGANIC MATTER		DEAD		DEAD LEAVES		153										
	ORGANIC MATTER		DEAD		SOM		154										
	ORGANIC MATTER		DEAD		DEAD ROOTS		155										
	ORGANIC MATTER		DEAD		DETRITUS		156										
								1	2	3	4	5	6	7	8	9	10

	11	12	13	14	15	16	17	18	19	20	21	22	23	24	25	26	27	28	29	30	31	32	33	34	35	36	37	38	39	40	41	42	43	44
135		1	S1	S1	1	S1	9	S1	S1	S1	1	1	S1	1	S1					S1	S1	S1	S1					S1	S1	1	S1	S1	S1	
136																																		
137											1													1										
138																																		
139																																		
140													1	1																				
141																																		
142								1																										
143																																		
144																																		
145																																		
146																																		
147								S1	1	S1	S1									S1	S1	S1						S1						
148		1			1			1	S1	1	1		1	1						1	1	S1	9						S1	1				
149		1																																
150																																		
151																																		
152																																		
153																																		
154																																		
155																																		
156																									1										
	11	12	13	14	15	16	17	18	19	20	21	22	23	24	25	26	27	28	29	30	31	32	33	34	35	36	37	38	39	40	41	42	43	44

	45	46	47	48	49	50	51	52	53	54	55	56	57	58	59	60	61	62	63	64	65	66	67	68	69	70	71	72	73	74	75	76	77	78	79
135				9				S1	S1	S1	S1		S1	S1		1	S1	S1				9	9				9		9						1
136				9																															
137																																	9		
138																																			
139																																			
140		9																			9														
141																					9														
142																										9									
143																																			
144																					9					9									
145																																			
146								S1	S1								S1																		
147			1										S1					S1			9					9							9	1	
148			1						1	1							1	1																1	
149																																			
150																																			
151																																			
152																																			
153																																		9	
154																																			
155																																			
156																																		9	
	45	46	47	48	49	50	51	52	53	54	55	56	57	58	59	60	61	62	63	64	65	66	67	68	69	70	71	72	73	74	75	76	77	78	79

	80	81	82	83	84	85	86	87	88	89	90	91	92	93	94	95	96	97	98	99	100	101	102	103	104	105	106	107	108	109	110
135		S1	1																												
136																															
137						S9	9				1																9		1	1	
138																															
139																															
140					1	9					9																				
141					1																			9			9	9	9	9	
142																												9	9		
143																														9	
144																											9	9	9	9	9
145																			1									9	9	9	
146																					1									9	9
147			S1															1	1	1		1	S1	S1	S1	S1		1	1	1	S1
148																								S1			9	9	9	9	9
149							9																					9	9	9	
150																															
151																															
152																															
153																															
154																															
155																															
156				9	9	9																								1	
	80	81	82	83	84	85	86	87	88	89	90	91	92	93	94	95	96	97	98	99	100	101	102	103	104	105	106	107	108	109	110

	111	112	113	114	115	116	117	118	119	120	121	122	123	124	125	126	127	128	129	130	131	132	133	134	135	136	137	138	139
135																									0				
136																										0			
137				1			S9	9									1			9	9						0		
138																												0	
139				1												1													0
140								9	9	9		1			9			1											
141				S1						9		S1		9	9			1									9		
142				S1									S1		9														
143	9	9								9							9		9										
144				1						9																	9		
145	9	9																											
146	9	9																	1										
147				1														9	1										
148	9	9		1						9								9	1										
149				1														9											
150				1																				9					
151				1												1													
152										9			1														9	9	
153				1						9																	9		
154																				9							9		
155																											9		
156					9	9	9	9	9	9	9					9				9									9
	111	112	113	114	115	116	117	118	119	120	121	122	123	124	125	126	127	128	129	130	131	132	133	134	135	136	137	138	139

	140	141	142	143	144	145	146	147	148	149	150	151	152	153	154	155	156	
135																		135
136																		136
137																		137
138																		138
139																		139
140	0																	140
141		0																141
142			0															142
143				0														143
144					0													144
145						0												145
146							0											146
147								0										147
148									0									148
149										0								149
150											0							150
151												0						151
152													0					152
153														0				153
154															0			154
155																0		155
156																	0	156
	140	141	142	143	144	145	146	147	148	149	150	151	152	153	154	155	156	

Appendix 14.B Mean Annual Population Densities and Biomass Estimates for Different Taxa or Functional Groups in the Tabonuco Forest, Luquillo Experimental Forest, Puerto Rico

Taxon/Functional group	Density (individuals m^{-2})	Biomass (g m^{-2})
Litter		588
Fungi in litter		1.5
Fungi in top 10 cm of soil		207
Bacteria in top 10 cm of soil		174–214
Mollusca		
Gastropoda		
Caracolus caracolla	3.1	
Nenia tridens	4.9	
Gaeotis nigrolineata	1.6	
Annelida		
Oligochaeta		4.2
Arthropoda		
Crustacea		
Isopoda	549	
Arachnida		
Schizomida	3.1	
Pseudoscorpionida	262	0.015
Acarina		
Mesostigmata	1390	
Orbatida	2000	
Amblypygida	1.8	
Opiliones		
Stygnomma spp.	146	0.09
Araneae	356	
Modisimus montanus	142	
Theotima sp.	74	
Masteria petrunkevitchi	36	
Pseudosparianthus jayuyae	21	
Corythalia gloriae (litter)	20	
Heteroonops spinimanus	14	
Ochyrocera sp.	9.9	
Phrurolithus sp.	9.4	
Corinna jayuyae	8.7	
Lygromma sp.	4.9	
Oligoctenus ottleyi	2.9	
Caponina spp.	2.4	
Clubiona portoricensis (litter)	1.7	
Ischnocolus culebrae	1.5	
Stasina portoricensis (litter)	1.1	
Psalistops corozali	1.1	
Dysderina sp.	0.8	
Nops sp.	0.7	
Triaeris stenopsis	0.6	
Emanthis sp.	0.5	
Oonops sp.	0.4	
Hahnia ernesti	0.2	
Leucauge regnyi	1.8	
Miagrammopes animotus	0.8	

Taxon/Functional group	Density (individuals m^{-2})	Biomass (g m^{-2})
Modisimus signatus	0.5	
Theridiosoma nechdomae	0.3	
Wendilgarda clara	0.08	
Alcimosphenus borinquenae	0.02	
Edricus crassicauda	0.01	
Micromerys dalei	8×10^{-3}	
Emanthis spp.	6×10^{-3}	
Corythalia gloriae (understory)	5×10^{-3}	
Stasina portoricensis (understory)	3×10^{-3}	
Argyrodes spp.	2×10^{-3}	
Clubiona portoricensis (understory)	1×10^{-3}	
Micrathena militaris	1×10^{-3}	
Mimetus portoricensis	1×10^{-3}	
Aysha tenuis	6×10^{-4}	
Capichameta hamata	6×10^{-4}	
Epicaudus mutchleri	6×10^{-4}	
Eriophora edax	6×10^{-4}	
Gasteracantha cancriformis	6×10^{-4}	
Nephila clavipes	6×10^{-4}	
Unidentified	4.4×10^{-3}	
Chilopoda	5.6	
Pauropoda	5.6	
Diplopoda	300	
Docodesmus maldonadoi	143	
Prostemmiulus heatwoli	102	
Spirobolellus richmondi	20	
Glomeridesmus marmoreus	17	
Agenodesmus reticulatus	6.7	
Lophoturus niveus	5.9	
Liomus obscurus	2.1	
Liomus ramosus	0.6	
Styraxodesmus juliogarciai	0.6	
Siphonophora portoricensis	0.4	
Ricodesmus stejneri	0.1	
Ellipura		
Collembola	1292	
Insecta		
Entotrophi	24	
Microcoryphia	1.5	
Dictyoptera and Orthoptera	115	
Phasmatodea (in litter)	7.9	
Lamponius portoricensis	0.4–1.0	
Agamemnon iphimedeia	0.15	
Orthoptera		
Gryllidae	6.8	
Isoptera		
Nasutitermes costalis	450	
Parvitermes discolor	374	
Psocoptera	15	
Thysanoptera	34	
Homoptera-Hemiptera	521	
Neuroptera	1.6	
Lepidoptera	96	

Taxon/Functional group	Density (individuals m^{-2})	Biomass ($g\ m^{-2}$)
Coleoptera	236	
Diptera	618	
Hymenoptera		
wasps (various families)	49	
Formicidae	1209	
Solenopsis spp.	428	
Pheidole moerens	221	
Brachymyrmex heeri	142	
Paratrechina near *vividula*	122	
Ochetomyrmex auropunctata	88	
Strumigenys gundlachi	50	
Strumigenys rogeri	42	
Odontomachus brunneus	38	
Iridomyrmex melleus	24	
Cyphomyrmex minutus	19	
Anochetus mayri	16	
Hypoponera sp.	7.3	
Paratrechina microps	3.7	
Paratrechina steinheili	2.9	
Pheidole sp.	1.2	
Pachycondyla stigma	1.1	
Monomorium floricola	0.2	
Vertebrata		
Amphibia		
Bufo marinus	0.01	
Eleutherodactylus coqui	2.07	0.33
E. portoricensis	0.08	0.03
E. wightmanae	0.04	0.04
E. hedricki	0.02	0.007
E. richmondi	0.01	0.003
Reptilia		
Sphaerodactylus klauberi	1	
Anolis gundlachi	0.2	10.8 (wet weight)
A. evermanni	0.15	6.2 (wet weight)
A. stratulus	2.15	38.7 (wet weight)
Aves	3.3×10^{-3}	0.017
Todus mexicanus	7.7×10^{-4}	
Coereba flaveola	7.2×10^{-4}	
Vireo altiloquus	6.0×10^{-4}	
Geotrygon montana	4.5×10^{-4}	
Margarops fuscatus	1.8×10^{-4}	
Chlorostilbon maugaeus	7.3×10^{-5}	
Spindalis zena	5.5×10^{-5}	
Melanerpes portoricensis	4.0×10^{-5}	
Loxigilla portoricensis	4.0×10^{-5}	
Buteo jamaicensis	2.0×10^{-5}	
Saurothera vieilloti	2.0×10^{-5}	
Myiarchus antillarum	2.0×10^{-5}	
Turdus plumbeus	1.5×10^{-5}	
Nesospingus speculiferus	1.9×10^{-5}	

Note: Biomass is calculated using dry weight unless otherwise indicated.

Glossary

abscission. The dropping of leaves, flowers, fruits, or other plant parts.

acarids. Referring to the order of arthropods which include mites and ticks.

achlorophyllous. Referring to plants lacking chlorophyll for obtaining energy from sunlight through photosynthesis.

actinomycetes. Prokaryotic microorganisms related to bacteria but resembling fungal mycelia in form.

agaric. Basidiomycete fungi generally referred to as mushrooms.

alates. Winged adults of the termite reproductive caste which swarm from the nest at the beginning of the nuptial flight.

allocation pattern. The way in which an organism distributes resources among structural parts or functional processes.

ambush predator. A species that waits quietly for prey to approach (a sit-and-wait hunting style) rather than actively seeking it.

ampodeids. Referring to one family of primitive, wingless arthropods in the order Diplura. Diplurans are small (1–7 mm), long, white, soil-dwelling arthropods which have antennae and two posterior filaments.

andesite. A class of volcanic rocks formed by extrusion onto the earth's surface.

Annelida. Phylum of invertebrates which include many soil-dwelling worms. The most conspicuous examples are earthworms.

anthropogenic. Resulting from or influenced by man's activities.

antiserial. Serum antigen action.

anuran. Name given to members of the order Anura of the class Amphibia; refers to frogs and toads.

apterous. Without wings

arboreal. Inhabiting trees

aril. A spongy or fleshy outgrowth on a seed.

ascomycete. A group of true fungi (Ascomycotina) that produce their sexual spores in a sack, called an ascus.

assimilation efficiency. The percentage of energy taken in that is absorbed into the bloodstream; similar to digestive efficiency.

auchenorrhynchous. Pertaining to the Aurchenorrhyncha, a suborder of the insect order Homoptera which includes cicadas, leafhoppers, treehoppers, and plant-hoppers. Insects in this suborder are active, jumping species with small bristlelike antennae.

aufwuchs. A community of aquatic organisms, and associated detritus, adhering to and forming a surface coating on submerged stones, plants, and other objects; periphyton.

autoecological. The ecology of individual species.

autotroph. Organism capable of synthesizing carbohydrates from simpler carbon sources like carbon dioxide, and which do not require other organisms as food.

basidiomycetous. Fungi belonging to the Basidiomycotina that produce their sexual spores on a clublike structure, called a basidium.

biodiversity. The diversity of life, including genetic diversity, species richness, and species diversity (species richness and evenness).

biomass. The total mass of organisms of populations or other taxonomic units within a given area at a given time.

biome. A major regional ecological community characterized by distinctive life forms.

Brachycera. A suborder of flies (Diptera) which includes all advanced flies. All adult Brachycera have fewer than six antennal segments, and the larvae have mouth parts which work in a vertical plane. One large group of the suborder comprises larvae which lack a head capsule (maggots). Examples of Brachycera include Asilidae (robber flies) and Muscidae (house flies).

bromeliad. A member of the Bromeliaceae or pineapple family; frequent epiphytes on tree limbs or trunks; elongate straplike leaves.

caecilians. A wormlike burrowing amphibian of the order Gymnophiona.

camaenid. Pertaining to Camaenidae, a large family of primarily plant- or algae-feeding snails.

capromyid. Species (e.g., hutias, agouties, cavies, Indian coneys) in the Caribbean rodent family Capromyidae.

carbon mineralization. The conversion of organic carbon into an inorganic form.

carnivore. A meat-eating organism.

catadromous. Living in fresh water but breeding in salt water.

catchment. The drainage basin above any given point in a stream; in American usage synonymous with watershed.

chela (pl. chelae). The tip of an arthropod appendage bearing a claw. The "pincers" of a scorpion are examples of chelae.

chelipod. Derived from chelae, an appendage on crustaceans.

clamp connection. A specialized structure allowing passage of nuclei around septations and only found in some but not all basidiomycetes.

cline. A gradual change in some feature of a population along an environmental gradient.

collembolan. Referring to an order of mostly small, terrestrial or litter-inhabiting arthropods, called springtails, which have a jumping apparatus (furcula) located on the underside of the fourth abdominal segment. Springtails are often abundant components of the litter.

comminutory. Breakdown into smaller particles by mechanical action.

compartmentalization. The segregation of a large food web into subsets or blocks of species, with higher connectance within a block than among blocks.

competitive release. In situations where the number of potential competitors is small, as on islands, particular species may enjoy elevated local densities compared to situations in which they compete with a larger suite of species, as on the mainland.

connectance. The actual, divided by the possible number of interspecific interactions.

crepuscular. Becoming active at twilight or during the early morning hours.

cryptic. Hidden or not readily visible by virtue of secretive habits.

D-Vac. A machine used for sampling small animals. It usually consists of a motor to which is attached a large flexible funnel. This machine is worn on the back of an individual. This funnel is maneuvered by the individual over leaves, ground, etc. and is used to suck up small mobile organisms.

dealate. Reproductive individuals of social insects that have shed their wings.

deme. A small or local population.

depauperate. Impoverished.

detritivore. Organism that feeds on decomposing organic matter.

detritus. Fragmented organic matter derived from the decay of plant and animal remains.

diaspidids. Referring to a large family of the insect order Homoptera. Diaspididae consist of the armored scales so-called because the sessile females secrete a thin, strong, cover which serves to protect the animal. Armored scales are totally phytophagous and occur on leaves and plant stems.

digestive efficiency. The percentage of the energy intake that is not subsequently excreted by the organism; similar to assimilation efficiency.

diplopods. Millipedes.

dipsocorid hemipterans. Referring to a small family (Dipsicoridae) of the insect order Hemiptera (true bugs). Dipsocorids are minute litter dwelling insects with piercing-sucking mouthparts. They are thought to be predacious.

disinfest. To remove contaminating organisms from a surface.

disperser. An organism that spreads propagules of another species from the point of origin.

echimyids. Species (e.g., spiny rats and casiraguas) in the tropical rodent family Echimyidae.

eclosion. Emergence of arthropods to the adult stage from an immature stage.

ecological efficiency. The percentage of the biomass produced at one trophic level that is incorporated into biomass produced by the next highest trophic level.

ecophysiological. Relating to the physical adaptations of organisms to habitat or environment.

ectomycorrhizal fungi. Higher fungi belonging to the basidiomycetes and ascomycetes that form a symbiotic relationship with plant roots. The fungus covers the root tip with mycelia and surrounds the outer cortex cells of its host.

ectoparasite. A parasite which lives externally on its host.

edentate. Name given to members of the order Edentata of the class Mammalia; refers to sloths, armadillos, and ant eaters.

endemic. Native and restricted to a particular area.

endocarp. The inner part of the fruit wall.

endoparasites. Animals which derive benefit (usually in the form of food) at the expense of the host. Endoparasites occur inside the host.

endophyte. In the broad sense, any microorganism growing within a live plant, but more usually restricted to those that do not cause disease and may be beneficial to the host plant.

entomobryids. Referring to a large family of the arthropod order Collembola (springtails). They are characterized by their elongate body form, well-developed prothorax, and possession of scales covering the body.

epigeic. Occurring on the surface of the soil or litter.

epiphyllic. Growing on the surfaces of leaves.
epiphytic. An organism that is attached to another with benefit to the former but not the latter.
equilibrium. State at which the system does not tend to undergo any further change of its own accord.
euryphagic. Referring to animals that eat a broad range of foods.
eurytopic. Occupying a wide range of habitats.
evapotranspiration. The sum of transpiration from plants and evaporation from soil and bodies of water.
exocarp. The outer covering of a fruit.
exoskeleton. The outer covering (skeleton) of insects, spiders, mites, and all other arthropods.
exotic. An organism that has been introduced into an area.
extramatrical. Used here in reference to hyphae that extend into soil from a mycorrhiza and that function in nutrient uptake.
facultative parasites. Microorganisms that have both saprophytic and parasitic capabilities.
folivore. A leaf-eating organism.
foraging range. The area an animal traverses while foraging; that portion of the home range within which an animal forages.
formicids. Referring to the family of the insect order Hymenoptera which includes the ants.
fossorial. Specialized for burrowing to the extent that some adaptations such as vestigial eyes render the animal poorly adapted for surface existence. The most extremely fossorial species make persistent burrows. Less fossorial species force their way through loose soil, and semifossorial species burrow in leaf-litter and humus but also move about on the surface.
frass. Solid (dry) excrement.
frugivore. A fruit-eating organism.
fulgoroid. Pertaining to a large assemblage of planthoppers belonging to the superfamily Fulgoroidea. They include all the small jumping forms except cicadas (Cicadidae), spittlebugs (Cercopidae), leafhoppers (Cicadellidae), and treehoppers (Membracidae).
fungivore. Organism that consumes fungi.
gamma source. A source of gamma irradiation, similar to X-rays.
generalist. A species that has a broad range of tolerance for various conditions.
geophagous. Feeding on soil matter (usually including decayed plant and animal remains).
gleaning animalivore. A feeding guild of flying animals whose members primarily consume vertebrates and invertebrates that are situated on the surface of living vegetation.
granivore. A seed-eating organism.
greenfall. Loss of green foliage, usually produced by pathogens or herbivores; insect-induced defoliation.
gryllids. Referring to the family of the insect order Orthoptera, which includes the crickets and their allies.

headwater stream. Stream at its origin, upstream of any incoming tributaries.

hectare. A metric unit of areal measurement equal to 2.47 acres.

heliothermic. Achieving operational body temperature by basking or shuttling from sun to shade.

hemimetabolous. Pertaining to insects which undergo an incomplete metamorphosis including only an egg, larval, and adult stage. In hemimetabolous forms, the larva (also called nymph or naiad, if aquatic) is similar in morphology an ecology to its adult. Mayflies, dragonflies, bugs, cicadas, grasshoppers, and crickets are hemimetabolous insects.

hemipterans. Referring to the insect order Hemiptera, or true bugs. All true bugs have piercing-sucking mouth parts. Many are plant feeders, others feed on other insects, and a few are ectoparasites.

herbivore. A plant-eating organism.

herpetofauna. Referring to reptiles and amphibians.

heptaxodontids. Species (e.g., quemi and giant hutias) in the Caribbean rodent family, Heptaxodontidae; all taxa are now extinct.

heterotroph. Organism that requires carbon in an organic form; organism that feeds from other organisms, their parts or products.

holometabolous. Pertaining to insects which undergo a complete metamorphosis, including egg, larva, pupa, and adult stage. In most holometabolous insects, the larval stage is morphologically and ecologically different from the adult. Flies, moths and butterflies, wasps, and beetles are holometabolous insects.

home range. The area an animal traverses while conducting its usual daily activities such as foraging, raising young, resting, or procuring mates.

homothalism. A mating system in which the organisms are self-fertile and therefore have only one sex.

humicolous. Referring to microorganisms that decompose humic materials, which are partially decayed organic materials.

hyperparasite. Organism that parasitizes a parasite.

hyphae. Linear, tubelike organs of which most fungi are formed, as opposed to cells.

hyphomycetous. Fungi of unknown affinities because they lack a sexual stage.

ideoroncids. Referring to the family Ideoroncidae of the arachnid order Pseudoscorpiones (pseudoscorpions). These small, litter-dwelling arthropods resemble scorpions but lack the stinger. They are all predacious.

insectary. An insect-rearing facility.

insectivore. An organism that consumes insects.

instar. The stage of an arthropod between molts. For example, an insect which has five larval stages can be termed to have five larval instars.

isopods. Referring to the order Isopoda of the class Crustacea. Isopods are common, multisegmented, soil and litter-dwelling arthropods which are often called pill bugs. Isopods are detritivores but they may also feed on plants. Most isopods are marine dwelling.

isotomids. Referring to a large family of the arthropod order Collembola (springtails). They are characterized by their elongate body form, well-developed prothorax, and lack of scales covering the body.

keystone mutualist. A species that is critical to the function or structure of an ecosystem as a consequence of its mutualistic association with other taxa. In particular, species which pollinate flowers or disperse seeds may affect the spatial organization and genetic structure of other dominant or keystone species.

leaf area index. A measure of the area of photosynthetic surface expanded over a given area of ground; normally given in m^2 per m^2.

leptodactylid. A frog belonging to the family Leptodactylidae.

light transmission. Method used in microscopy by which light is passed through a condenser and then through the observed object; microscope for transmitted light.

lunar phobia. A behavioral strategy that reduces activity during times of full moon, ostensibly as a mechanism to avoid predation.

macrofungi. Fungi with fruiting bodies that are readily visible without magnification.

macroarthropod. A relatively large arthropod which is readily visible to the unaided eye. See microarthropod.

macrosaprophagic. Feeding on comparatively large pieces of decaying organic material.

menthid. Referring to the family Menthidae of the arachnid order Pseudoscorpiones (pseudoscorpions). See Ideoroncids.

mesostigmatid mites. Referring to a suborder of mites which include many free-living and parasitic species. They are characterized by having a pair of spiracular (breathing) openings on the side of the body near the third or fourth pair of legs.

metabolic expenditure. The amount of energy used in metabolism of an organism.

microfungi. Fungi with fruiting bodies that are generally not visible without magnification.

microepiphyte. A very small plant which grows in trees.

microarthropod. Relatively small arthropod which includes specimens readily visible only under some degree of magnification. See macroarthropod.

microbial catabolism. The breakdown of complex molecules into simpler ones by the metabolic action of microorganisms.

microbitrophic. Organisms that feed on microorganisms.

microbivores. Organisms that feed on microscopic organisms.

microclimate. Referring to the climate immediately near the vicinity of a target organism.

microflora. Relatively small plants which include specimens readily visible only under some degree of magnification.

microhabitat. A small specialized habitat; microenvironment.

microlepidoptera. A commonly used informal term referring to small moths, such as the families Tineidae, Tortricidae, Gelechiidae, and others the knowledge of whose taxonomy is poor compared to that of other families of Lepidoptera.

molossid insectivore. A guild of bats, all in the Molossidae, that primarily forage for insects that fly well above the canopy. These bats have relatively long and narrow wings that provide aerodynamic capabilities for fast and high flight at the cost of reduced maneuverability.

monophagous. Feeding on only one species of prey.

morpho-species. A taxonomic classification of organisms based on morphologically distinguishable characteristics, but not necessarily identified to known species by scientific names.

mycelium. The vegetative body of a fungus which is composed of hyphae.

mycetophilid. Referring to a common family of primitive flies, the Mycetophilidae. Also called fungus gnats because the immature stages are often found in moldy habitats.
mycorrhizal. Forming a usually mutually beneficial symbiotic relationship between plant roots and fungi.
mymercophage. A predator on ants and/or termites.
nasutes. Soldier termites of some species in the Nasutitermitinae, possessing a snout-like organ (nasus) used to eject toxic or sticky substances at intruders.
necromass. The total mass of dead organic matter per unit area.
necrophagic. Feeding on carrion.
necropsy. Anatomical examination of a dead organism.
nectarivore. An organism that feeds on nectar.
Neotropical. The biogeographical region that comprises Central and South America (from Aguas Calientes in Mexico to Tierra del Fuego in Argentina) and the West Indies.
niche. The ecological role of a species in an ecosystem.
nocturnal subweb. Those species of the food web and their prey that forage at night.
nonmycorrhizal. Plants or fungi that do not enter into symbiotic mycorrhizal relationships.
nutrient flux. The transport of nutrients from one location to another.
oligophagous. Feeding on only a few species of prey.
omnivore. An organism that consumes plant and animal material.
opilionids. Referring to the arachnid order Opiliones. Also called harvestmen, or daddy-long-legs.
orb-weaver. Common name referring to a large family of spiders (Araneidae) which construct relatively large circular or spiraling webs used to ensnare flying or crawling insects and other arthropods.
orthopterans. Referring to the insect order Orthoptera, which includes crickets, katydids, and grasshoppers.
pantropical. Found worldwide throughout the tropics.
parasite. An organism which derives benefit (usually food) to the detriment of the host organism. Parasites usually harm but do not kill their host, and each parasite may use more than one host during its lifetime.
parasitism. Obtaining nutrition directly from a living host.
parasitoid. An organism which derives benefit (food) by killing the host. Parasitoids usually undergo their life cycle on or in the host, thereby killing it. Parasitoids are widely used as biocontrol agents.
particulate. Contained in particles.
pathogenic. Causing disease.
pathogens. Organisms or substances that cause disease.
podurid. Referring to a large family (Poduridae) of the arthropod order Collembola (springtails). They are characterized by their elongate form and reduced prothorax.
pedipalps. The second pair of appendages on the arachnids (scorpions, etc.) sometimes used for crushing prey.
pelletization. The production of fecal pellets by arthropods in order to minimize the loss of water via excretion.
pericarp. Fruit wall derived from the ovary.

periphyton. Attached algae.
peritrophic. A special type of mycorrhizal or other symbiotic relationship in which the microorganism surrounds but does not penetrate the roots of its host plant.
phasmatid. Referring to the insect order Phasmatodea, or walking sticks.
phragmotic. Referring to the truncated head of some termite soldiers, particularly in the genus *Crytotermes;* used as living stoppers to close nest openings.
physical comminution. See comminutory.
physogastric. Referring to an exaggerated swelling of a queen termite's abdomen due to the growth of fat bodies and/or ovaries.
pierid. Referring to a family of butterflies in the insect order Lepidoptera. Pierids are small to large butterflies which are predominantly white or yellow.
piscivore. An organism that feeds on fish.
polyethism. Division of labor among members of an insect society. It may be based on caste, age, sex, or size.
polyphagous. Feeding on several different species of prey.
ponerines. Referring to the subfamily Ponerinae of the insect family Formicidae (ants).
preoral digestion. Breakdown and digestion of tissue by saliva injected into captured prey. This type of digestion is practiced by spiders and other arachnids. After breakdown of these tissues into a liquid form they are ingested into the oral cavity.
predator. An animal which captures, traps, or ensnares other animals for food.
primary producer. An organism that synthesizes complex organic substances from inorganic substrates.
proctodeal feeding. The solicitation of fluid hind-gut contents from one colony member by another.
production efficiency. The proportion of assimilated energy that is transformed into new biomass.
production. That part of assimilated energy converted to biomass per unit time per unit area (net production).
prostemmiulid. Referring to the family Prostemmiulidae of the millipedes.
prostigmatids. Referring to a suborder of mites which include many free-living and parasitic species. They are characterized by having a pair of spiracular (breathing) openings between the bases of the chelicerae or on the shoulders of the body in front of the first pair of legs. Some of the most economically important mites such as the spider mites (Tetranychidae) and eriophyid mites (Eriophyoidea) belong to this group.
proteolytic. Capable of degrading proteins.
pselaphid. Referring to the family Pselaphidae, or short-winged mold beetles. They comprise a few small litter-dwelling species, some of which are predaceous.
sanguinivore. A carnivorous organism that feeds on blood, such as the vampire bat.
saprovore. An organism which feeds on rotting or decaying material.
sapsucking. Process of extracting liquid nutrients from plants by means of piercing, sucking mouthparts. Aphids, stink bugs, and mealybugs are sapsucking insects.
scansorial. Climbing or adapted for climbing.
scatophagic. Feeding on fecal material.
sclerotized. Composed of a hardened material within the exoskeleton of arthropods.
scolytids. Referring to the large family of beetles, the Scolytidae, within the insect

order Coleoptera. Scolytids are often called bark beetles, and several are destructive pests of forest trees. Others are seed feeders.

secondary production. The assimilation of organic matter by a primary consumer.

senescent. Declining and approaching death.

septate. Referring to nonprimitive fungi that have cross-walls in their hyphae.

sit-and-wait. A foraging behavior in which the predator remains quiet until the prey approaches; ambush style of predation.

sminthurids. Referring to a large family of the arthropod order Collembola (springtails). Sminthurids differ from other families by their oval or globular-shaped body.

snout-vent length. A measurement of body length of an organism measured from the tip of the snout to the vent or end of the backbone.

sociotomy. Colony multiplication through fission; a portion of the colony departs with one or more reproductive forms, leaving comparable units behind in the parental nest.

specialist. A species with highly specific requirements for many of its needs; e.g., a predator that eats only a subset of its available prey.

speciose. Comprised of comparatively many species.

spectral light quality. Refers to the wavelengths of light available for photosynthesis at different points within the vegetation.

spot maps. A technique for estimating bird populations by counting singing males in a fixed area.

sporophore. Any structure that bears spores.

standing stock. The total biomass of a group of organisms present at a given time, expressed per unit of area or volume.

stenotopic. Occupying a narrow range of habitats.

sternorrhynchous. Pertaining to the suborder Sternorrhyncha of the order Homoptera; includes aphids, armored scales, mealybugs, and psyllids. Insects in this suborder are usually less active (or may be sessile) forms with longer, filamentous antennae.

stomodeal feeding. Mouth-to-mouth exchange of nutrients among members of an insect society.

stylettiform chelicerae. A modification of the pincerlike chelicerae into a piercing stylet. Many mites have stylettiform chelicerae.

subdominants. Species that are dominant to an inferior or partial degree with respect to other species in an assemblage.

symbiont. A participant in the relationship between two interacting organisms or populations.

taxon (pl. taxa). Any named category of plants or animals. Examples of taxa are species, genus, and family.

tergal plates. The portions of an insect's dorsal body wall bounded by sutures.

termitophile. An invertebrate that has become adapted to living in the nests of termites in such close association that it cannot survive independently.

translocation. The transport of material within a plant.

trophallaxis. The exchange of alimentary fluids among colony members and guests.

trophic. Pertaining to nourishment.

-trophic. Suffix indicating the mode of nutrition of an organism, e.g., autotrophic, heterotrophic.

Tullgren extraction. A passive means of collecting soil or litter animals using a device which gradually heats the soil thereby forcing the organisms away from the heat source. A receptacle usually, with a preservative such as alcohol, is fastened at the other end. The organisms fleeing the drying conditions fall through a collecting funnel and are preserved in the receptacle.

vesicular-arbuscular endomycorrhizal. Refers to a usually mutually beneficial symbiosis in which the fungi are primitive water molds lacking septate hyphae. These fungi form treelike structures (arbuscles) for nutrient exchange within the host plant's root cortex cells and vesicles which are used by the fungus for storage.

xylophagous. Wood-eating.

Contributors

Gerardo R. Camilo
Department of Biology
St. Louis University
3507 Laclede Ave.
St. Louis, MD 63103-2010

Alan P. Covich
Department of Fisheries and Wildlife Biology
135 Wagar Hall
Colorado State University
Fort Collins, Colorado 80523

Ava Gaa Kessler
Plaza 29, MN-22, Urb. Claro
Bo. Hato Tejas
Bayamon, Puerto Rico 00619

Michael R. Gannon
Department of Biology
The Pennsylvania State University
3000 Ivyside Park
Altoona, PA 16601

Rosser W. Garrison
Terrestrial Ecology Division
University of Puerto Rico
GPO Box 363682
San Juan, Puerto Rico 00936

William T. Lawrence, Jr.
Department of Geography
LeFrak Hall, Room 1113
University of Maryland
College Park, Maryland 20742

D. Jean Lodge
Center for Forest Mycology Research
Forest Products Lab
USDA-Forest Service
P.O. Box B
Palmer, Puerto Rico 00721

William H. McDowell
Department of Natural Resources
215 James Hall
University of New Hampshire
Durham, New Hampshire 03824

Elizabeth A. McMahan
Department of Biology
University of North Carolina
Chapel Hill, North Carolina 27599-3280

William J. Pfeiffer
Terrestrial Ecology Division
Center for Energy and Environmental Research
University of Puerto Rico
San Juan, Puerto Rico 00936

Douglas P. Reagan
Woodward-Clyde Consultants
Stanford Place III, Suite 1000
4582 S. Ulster Street
Denver, Colorado 80237
and
Terrestrial Ecology Division
University of Puerto Rico
GPO Box 363682
San Juan, Puerto Rico 00936

Margaret M. Stewart
Department of Biological Sciences
State University of New York at Albany
Albany, New York 12222

Richard Thomas
Biology Department
University of Puerto Rico
P.O. Box 23360
University of Puerto Rico-Rio Piedras
San Juan, Puerto Rico 00931-3360

Robert B. Waide
Terrestrial Ecology Division
University of Puerto Rico
GPO Box 363682
San Juan, Puerto Rico 00936

Michael R. Willig
Ecology Program
Department of Biological Sciences and the Museum
Texas Tech University
Lubbock, Texas 79409-3131

Lawrence L. Woolbright
Biology Department
Siena College
Loudonville, New York 12211

Bibliography

Abe, T. 1979. Studies on the distribution and ecological role of termites in a lowland rain forest of West Malaysia. 2. Food and feeding habits of termites in Pasoh Forest Reserve. *Japanese Journal of Ecology* 29:121–35.

Abe, T. 1980. Studies on the distribution and ecological role of termites in a lowland rain forest of West Malaysia. 4. The role of termites in the process of wood decomposition in Pasoh Forest Reserve. *Revue d'Ecologie et de Biologie du Sol* 17:23–40.

Abe, T. 1982. Ecological role of termites in a tropical rain forest. In *The biology of social insects: Proceedings of the Ninth Congress of the International Union for the Study of Social Insects,* Boulder, Colorado, August 1982, ed. M. D. Breed, C. D. Michener, and H. E. Evans, 71–75. Boulder, Colo.: Westview Press.

Abe, T., and T. Matsumoto. 1979. Studies on the distribution and ecological role of termites in a lowland rain forest of West Malaysia: 3. Distribution and abundance of termites in Pasoh Forest Reserve. *Japanese Journal of Ecology* 29:337–51.

Adams, J. 1984. The habitat and feeding ecology of woodland harvestmen (Opiliones) in England. *Oikos* 42:361–70.

Addison, J., and D. Parkinson. 1978. Influence of collembolan feeding activities on soil metabolism at a high arctic site. *Oikos* 30:529–38.

Aguayo, C. G. 1961. Aspecto general de la fauna malacológica puertorriqueña. *Caribbean Journal of Science* 1:89–105.

Aide, T. M. 1993. Patterns of lead development and herbivory in a tropical understory community. *Ecology* 74:455–66.

Albert, A. M. 1979. Chilopoda as part of the predatory macroarthropod fauna in forests: Abundance, biomass, and metabolism. In *Myriapod biology,* ed. M. Camatini, 215–31. New York: Academic Press.

Albert, A. M. 1983. Energy budgets for populations of long-lived arthropod predators (Chilopoda: Lithobiidae) in an old beech forest. *Oecologia* 56:292–305.

Albert, R. 1976. Struktur und dynamik der spinnen populationen in Buchenwäldern des Solling. *Verhandlungen für die gesammte Ökologie* (Göttingen) 1976:83–91.

Alexander, C. P. 1932. The craneflies of Puerto Rico (Diptera). *Journal of the Department of Agriculture of Porto Rico* 16:349–87.

Alexopoulos, C. J. 1970. Rain forest myxomycetes. In *A tropical rain forest: A Study of irradiation and ecology at El Verde, Puerto Rico,* ed. H. T. Odum and R. F. Pigeon, F21–F23. Oak Ridge, Tenn.: U.S. Atomic Energy Commission.

Allan, J. D., and A. S. Flecker. 1993. Biodiversity conservation in running waters. *BioScience* 43:32–43.

Allan, J. D., L. W. Barnthouse, R. A. Prestbye, and D. R. Strong. 1973. On foliage arthropod communities of Puerto Rican second growth vegetation. *Ecology* 54:628–32.

Allee, W. C., and M. Torvik. 1927. Factors affecting animal distributions in a small stream of the Panama rainforest in the dry season. *Journal of Ecology* 15:66–71.

Alvarez, J. 1991. Effects of treefall gaps on a tropical land snail community. M.S. thesis, Texas Tech University, Lubbock.

Alvarez, J. A., and M. R. Willig. 1993. Effects of treefall gaps on the density of land snails in the Luquillo Experimental Forest of Puerto Rico. *Biotropica* 25:100–110.

American Ornithologists' Union. 1983. *Check-list of North American birds.* 6th ed. Washington, D.C.: American Ornithologists' Union.

Anderson, D. J. 1982. The home range: A new nonparametric estimation technique. *Ecology* 63:103–12.

Anderson, J. F. 1970. Metabolic rates of spiders. *Comparative Biochemistry and Physiology* 33:51–72.

Anderson, J. F. 1974. Responses in starvation in the spiders *Lycosa lenta* Hentz and *Filistata hibernalis* (Hentz). *Ecology* 55:576–85.

Anderson, J. F., and K. N. Prestwich. 1982. Respiratory gas exchange in spiders. *Physiological Zoology* 55:72–90.

Anderson, J. M. 1973. The breakdown and decomposition of sweet chestnut (*Castanea sativa* Mill.) and beech (*Fagus sylvatica* L.) leaf litter in two deciduous woodland soils: II. Changes in carbon, hydrogen, nitrogen, and polyphenol content. *Oecologia* (Berlin) 12:275–88.

Anderson, J. M. 1975. Succession, diversity and trophic relationships of some soil animals in decomposing leaf litter. *Journal of Animal Ecology* 44:475–95.

Anderson, J. M., and D. E. Bignell. 1980. Bacteria in the food, gut contents and faeces of the litter-feeding millipede *Glomeris marginata* (Villers). *Soil Biology and Biochemistry* 12:251–4.

Anderson, J. M., and I. N. Healey. 1972. Seasonal and interspecific variation in the major components of the gut contents of some woodland Collembola. *Journal of Animal Ecology* 41:359–68.

Anderson, J. M., and M. J. Swift. 1983. Decomposition in tropical forests. In *Tropical rain forest: Ecology and management,* ed. S. L. Sutton, T. C. Whitmore, and A. C. Chadwick, 287–309. Boston: Blackwell Scientific Publications.

Anderson, J. P. E., and K. H. Domasch. 1975. Measurement of bacterial and fungal contribution to respiration of selected agricultural and forest soils. *Canadian Journal of Microbiology* 21:314.

Anderson, P. K. 1960. Ecology and evolution in island populations of salamanders in the San Francisco Bay region. *Ecological Monographs* 30:359–85.

Anderson, R. M. 1982. Host parasite population biology. In *Parasites—Their world and ours,* ed. D. F. Mettrick, 303–12. Amsterdam: Elsevier Biomedical Press.

Anderson, R. V., D. C. Coleman, C. V. Cole, and E. T. Elliot. 1981. Effect of the nematodes *Acrobeloides* and *Mesodiplogaster iheritieri* on substrate utilization and nitrogen and phosphorus mineralization in soil. *Ecology* 62:549–55.

Andrews, R. M. 1976. Growth rate in island and mainland anoline lizards. *Copeia* 1976:477–82.

Andrews, R. M. 1979. Evolution of life histories: A comparison of *Anolis* lizards from matched island and mainland habitats. *Breviora* 454:1–51.

Andrews, R. M. 1991. Population stability of a tropical lizard. *Ecology* 72:1204–17.

Anthony, H. E. 1918. Indigenous land mammals of Puerto Rico, living and extinct. *Memoirs of the American Museum of Natural History,* New Series 2:331–435.

Anthony, H. E. 1925–26. Mammals of Puerto Rico, living and extinct. In *Scientific survey of Porto Rico and the Virgin Islands.* Vol. 9, Pts. 1–2, 1–241. New York: New York Academy of Sciences.

Archer, A. F. 1965. Nuevos argiopidos de las Antillas. *Caribbean Journal of Science* 5:129–33.

Arendt, W. J. 1983. The effects of dipteran ectoparasitism on the growth and development of nestlings of the Pearly-eyed Thrasher (*Margarops fuscatus*) in the Luquillo Mountains, Puerto Rico. M.A. thesis, University of Missouri, Columbia.

Arendt, W. J. 1985. *Philornis* ectoparasitism of Pearly-eyed Thrashers. II. Effects on adults and reproduction. *Auk* 102:281–92.

Arnett, R. H. 1973. *The beetles of the United States.* Ann Arbor, Mich.: American Entomological Institute.

Arnett, R. H. 1985. *American insects: A handbook of the insects of America north of Mexico.* New York: Van Nostrand Reinhold.

Arsuffi, T. L., and K. Suberkropp. 1984. Leaf processing capabilities of aquatic hyphomycetes: Interspecific differences and influence on shredder feeding preferences. *Oikos* 42:144–54.

Asbury, C. E., and D. J. Lodge. 1987. Rhizomorph fungi retard downhill export of fallen leaves in a tropical montane rain forest. *Bulletin of the Ecological Society of America* 68:257.

Ashe, J. S. 1984. Generic revision of the subtribe Gyrophaenina (Coleoptera: Staphylinidae: Aleocharinae) with review of the described subgenera and major features of evolution. *Quaestiones Entomologicae* 20:129–349.

Ashford, A. E., and W. G. Allaway. 1982. A sheathing mycorrhiza *Pisonia grandis* R. Br. (Nyctaginaceae) with development of transfer cells rather than a Hartig net. *The New Phytologist* 90:511–9.

Ashford, A. A., and W. G. Allaway. 1985. Transfer cells and Hartig net in the root epidermis of the sheathing mycorrhiza of *Pisonia grandis* R. Br. from Seychelles. *New Phytologist* 100:595–612.

Askins, R. A., and D. N. Ewert. 1991. Impact of Hurricane Hugo on bird populations on St. John, U.S. Virgin Islands. *Biotropica* 23:481–7.

Atkey, P. T., and D. A. Wood. 1983. An electron microscope study of wheat straw composted as a substrate for the cultivation of the edible mushroom (*Agaricus bisporus*). *Journal of Applied Bacteriology* 55:293–304.

Atkinson, W. D. 1985. Coexistence of Australian rainforest Diptera breeding in fallen fruit. *Journal of Animal Ecology* 54:507–18.

Auclair, J. L. 1958. Honeydew excretion in the pea aphid, *Acyrthosiphum pisum* (Harr.) (Homoptera: Aphididae). *Journal of Insect Physiology* 2:330–7.

Auclair, J. L. 1959. Feeding and excretion of the pea aphid, *Acyrthosiphum pisum* (Harr.), reared on different varieties of peas. *Entomologia Experimentalist et Applicata* 2:279–86.

Auerbach, M. J. 1984. Stability, probability, and the topology of food webs. in *Ecological communities: Conceptual issues and evidence,* ed. D. R. Strong, D. Simberloff, L. G. Abele, and A. B. Thistle, 413–36. Princeton, N.J.: Princeton University Press.

Augspurger, C. K. 1983. Seed dispersal of the tropical tree, *Platypodium elegans,* and the escape of its seedlings from fungal pathogens. *Journal of Ecology* 71:759–71.

Ausmus, B. S. 1977. Regulation of wood decomposition rates by arthropod and annelid populations. In *Soil organisms as components of ecosystems,* ed. V. Lohm and T. Persson, 180–92. *Ecological Bulletins 25.* Stockholm: Swedish Natural Science Research Council.

Ausmus, B. S., and M. Witkamp. 1973. *Litter and soil microbial dynamics in a deciduous forest stand.* EDFB-IBP-73-10. Oak Ridge, Tenn.: Oak Ridge National Laboratory.

Bååth, E., and B. E. Söderström. 1979. The significance of hyphal diameter in calculation of fungal biovolume. *Oikos* 33:11–14.

Bååth, E., U. Lohm, B. Lundgren, T. Rosswall, B. Söderström, B. Sohlenius, and A. Wiren. 1978. The effect of nitrogen and carbon supply on the development of soil organism populations and pine seedlings: A microcosm experiment. *Oikos* 31:153–63.

Baker, R. J., and H. H. Genoways. 1978. Zoogeography of Antillean bats. In *Zoogeography in the Caribbean,* 53–97. Academy of Natural Sciences of Philadelphia Special Publication 13. Philadelphia: Academy of Natural Sciences of Philadelphia.

Baker, R. J., P. V. August, and A. A. Steuter. 1980. Erophylla sezekorni. *Mammalian Species* 115:1–5.

Banks, N. 1896. New North American spiders and mites. *Transactions of the American Entomological Society* 23:54–91.

Banks, N. 1914. New West Indian spiders. *Bulletin of the American Museum of Natural History* 33:637–42.

Bannister, B. A. 1970. Ecological life cycle of *Euterpe globosa* Gaertn. In *A tropical rain forest: A Study of irradiation and ecology at El Verde, Puerto Rico,* ed. H. T. Odum and R. F. Pigeon, B299–B314. Oak Ridge, Tenn.: U.S. Atomic Energy Commission.

Barbault, R. 1967. Recherches écologiques dans la savane de Lamto (Côte-d'Ivoire): Le cycle annuel de la biomasse des amphibiens et des lézards. *Revue d'Ecologie: La Terre et la Vie* 21:297–318.

Barbault, R. 1970. Recherches écologiques dans la savane de Lamto (Côte-d'Ivoire): Les traits quantitatifs du peuplement des ophidiens. *Revue d'Ecologie: La Terre et la Vie* 24:94–107.

Barbault, R. 1976. Structure et dynamique d'un peuplement d'amphibiens en savane protegee de feu (Lamto, Côte-d'Ivoire). *Revue d'Ecologie: La Terre et la Vie* 30:246–63.

Barber, G. 1939. Insects of Puerto Rico, and the Virgin Islands: Hemiptera-Heteroptera except the Miridae and Coreidae. In *Scientific survey of Porto Rico and the Virgin Islands.* Vol. 14, Pt. 3, 263–441. New York: New York Academy of Sciences.

Barcant, M. 1970. *Butterflies of Trinidad and Tobago.* London: Collins.

Bardate, R. J., R. T. Prentiki, and T. Fenchel. 1974. Phosphorus cycle of model ecosystems: Significance for decomposer food chains and effect of bacterial grazers. *Oikos* 25:239–51.

Barlocher, F., and B. Kendrick. 1973. Fungi and food preferences of *Gammarus pseudolimnaeus. Archiv für Hydrobiologie* 72:501–16.

Barlocher, F., and B. Kendrick. 1975a. Assimilation efficiency of *Gammarus pseudolimnaeus* (Amphipoda) feeding on fungal mycelium of autumn-shed leaves. *Oikos* 26:55–59.

Barlocher, F., and B. Kendrick. 1975b. Leaf-conditioning by microorganisms. *Oecologia* (Berlin) 20:359–62.

Barnes, R. D. 1953. The ecological distribution of spiders in non-forest maritime communities at Beaufort, North Carolina. *Ecological Monographs* 23:315–37.

Barnish, G. 1984. The freshwater shrimps of Saint Lucia, West Indies (Decapoda, Natantia). *Crustaceana* 47:314–20.

Baskin, J., and E. E. Williams. 1966. The lesser Antillean *Ameiva* (Sauria: Teiidae), re-evaluation, zoogeography and the effects of predation. *Studies on the Fauna of Curacao and other Caribbean Islands* 23:144–76.

Basnet, K. 1992. Effect of topography on the pattern of trees in tabonuco (*Dacryodes excelsa*) dominated rain forest of Puerto Rico. *Biotropica* 24:31–42.

Baumann, R. W. 1982. Plecoptera. In *Aquatic biota of Mexico, Central America and the West Indies,* ed. S. H. Hurlbert and A. Villalobos-Figueroa, 278–9. San Diego, Calif.: S. H. Hurlbert, Dept. of Biology, San Diego State University.

Bazzaz, F. A. 1983. Dynamics of wet tropical forests and their species strategies. In *Physiological ecology of plants of the wet tropics,* ed. E. Medina, H. A. Mooney, and C. Vásquez-Yánes, 233–43. The Hague: Dr. W. Junk Publishers.

Bazzaz, F. A., and S. T. A. Pickett. 1980. Physiological ecology of tropical succession: A comparative review. *Annual Review of Ecology and Systematics* 11:287–310.

Beard, J. S. 1944. Climax vegetation in tropical America. *Ecology* 25:127–58.

Beard, J. S. 1949. Natural vegetation of the Windward and Leeward Islands. *Oxford Forestry Memoirs* 21. Oxford: Clarendon Press.

Beck, L. 1968. Sobre a biologia de alguns Aracnideos na floresta tropical de Reserva Ducke (I. N. P. A., Manaus/Brasil). *Amazoniana* 1:247–50.

Beck, L. 1971. Bodenzoologische gliederung und charakterisierung des amazonischen regenwaldes. *Amazoniana* 3:69–132.

Bedford, G. O. 1978. Biology and ecology of the Phasmatodea. *Annual Review of Entomology* 23:125–49.

Beebe, W. 1916. Fauna of four square feet of jungle debris. *Zoologica* (New York) 2:107–19.

Beebe, W. 1925. Studies of a tropical jungle: One quarter of a square mile of jungle at Kartabo, British Guiana. *Zoologica* (New York) 6:1–193.

Beinroth, F. H. 1971. The general pattern of the soils of Puerto Rico. In *Transactions of the Fifth Caribbean Geological Conference,* 225–30. Port-of-Spain, Trinidad: Queens College Press.

Bell, C. R. 1970. Seed distribution and germination experiment. In *A tropical rain forest: A study of irradiation and ecology at El Verde, Puerto Rico,* ed. Odum, H. T. and R. F. Pigeon, D177–D182. Oak Ridge, Tenn.: U.S. Atomic Energy Commission.

Bell, H. L. 1982. A bird community of New Guinean lowland rainforest. 3. Vertical distribution of the avifauna. *Emu* 82:143–62.

Belsky, A. J. 1986. Does herbivory benefit plants? A review of the evidence. *American Naturalist* 127:870–92.

Benedict, F. F. 1976. Herbivory rate and leaf properties in four forests of Puerto Rico and Florida. M.S. thesis, University of Florida, Gainesville.

Benemann, J. R. 1973. Nitrogen fixation in termites. *Science* 181:164–5.

Benke, A. C. 1979. A modification of the Hynes method for estimating secondary production with particular significance for multivoltine populations. *Limnology and Oceanography* 24:168–71.

Benke, A. C. 1984. Secondary production of aquatic insects. In *The ecology of aquatic insects,* ed. V. H. Resh and D. M. Rosenberg, 91–164. New York: Praeger Publishers.

Bentley, B. L., and E. J. Carpenter. 1980. Effects of desiccation and rehydration on nitrogen fixation by epiphylls in a tropical rainforest. *Microbial Ecology* 6:109–14.

Berthet, P. 1967. The metabolic activity of oribatid mites (Acarina) in different forest floors. In *Secondary productivity of terrestrial ecosystems,* ed. K. Petrusewicz, 709–25. Warszawa: Panstowe Wydawn, Naukowe.

Best, T. L., and K. G. Castro. 1981. Synopsis of Puerto Rican mammals. *Studies in Natural Sciences* 2:1–12.

Beuchat, C. A., F. H. Pough, and M. M. Stewart. 1984. Response to simultaneous dehydration and thermal stress in three species of Puerto Rican frogs. *Journal of Comparative Physiology* B 154:579–85.

Bevege, D. I., G. D. Bowen, and M. F. Skinner. 1975. Comparative carbohydrate physiology of ecto- and endo-mycorrhizas. In *Endomycorrhizas,* ed. F. E. Sanders, B. Mosse, and P. B. Tinker, 149–74. London: Academic Press.

Bhajan, W. R., M. Canals, J. A. Colon, A. M. Block, and R. G. Clements. 1978. A limnological survey of the Rio Espiritu Santo Watershed, Puerto Rico (1976–1977). Unpublished report, Center for Energy and Environment Research, Rio Piedras, P.R.

Bhatnagar, R. K. 1970. The spider *Lycosa carmichaeli* Graveley as a predator of small frogs. *Journal of the Bombay Natural History Society* 67:589.

Biaggi, V., Jr. 1955. The Puerto Rican Honeycreeper (Reinita) *Coereba flaveola portoricensis* (Bryant). Special Publication. Río Piedras: University of Puerto Rico. Agricultural Experiment Station.

Bierregaard, R. O., Jr. 1990. Species composition and trophic organization of the understory bird community in a Central Amazonian terra firme forest. In *Four Neotropical rainforests,* ed. A. Gentry, 217–36. New Haven, Conn.: Yale University Press.

Binns, E. S. 1981. Fungus gnats (Diptera: Mycetophilidae/Sciaridae) and the role of mycophagy in soil: A review. *Revue d'Écologie et de Biologie du Sol* 18:77–90.

Bishop, J. E. 1973. *Limnology of a small Malayan river, Sungai Gombak.* The Hague: W. Junk.

Blackwell, M., T. G. Laman, and R. L. Gilberston. 1982. Spore dispersal of *Fuligo septica* (Myxomycetes) by Lathridiid beetles. *Mycotaxon* 14:58–60.

Blair, J. M., and D. A. Crossley, Jr. 1988. Litter decomposition, nitrogen dynamics and litter microarthropods in a southern appalachian hardwood forest 8 years following clearcutting. *Journal of Applied Ecology* 25:683–98.

Blake, D. H. 1941. New species of *Chaetocnema*, and other chrysomelids (Coleoptera) from the West Indies. *Proceedings of the Entomological Society of Washington* 43:171–80.

Blake, D. H. 1943. New species of the genus *Hadropoda* Suffrian from the West Indies. *Bulletin of the Museum of Comparative Zoology* 92:413–41.

Blake, D. H. 1948. Six new species of West Indian Chrysomelidae. *Proceedings of the Entomological Society of Washington* 50:121–7.

Blake, D. H. 1950. A new genus of flea-beetles from the West Indies. *Psyche* 57:10–25.

Blake, D. H. 1951. A revision of the beetles of the genus *Chalcosicya* Blake (Chrysomelidae) from the West Indies. *Bulletin of the Museum of Comparative Zoology* 106:287–312.

Blake, D. H. 1952. Six new species of *Megistops* with keys to the known species (Coleoptera). *Psyche* 59:1–12.

Blake, D. H. 1953. The chrysomelid beetles of the genus *Strabala* Chevrolat. *Proceedings of the U.S. National Museum* 103:121–34.

Blake, D. H. 1964. Notes on new and old species of Alticinae (Coleoptera) from the West Indies. *Proceedings of the U.S. National Museum* 115:8–29.

Blake, D. H. 1970. A review of the beetles of the genus *Metachroma* Chevrolat (Coleoptera: Chrysomelidae). *Smithsonian Contributions to Zoology* 57:1–109.

Blake, J. G., F. G. Stiles, and B. A. Loiselle. 1990. Birds of La Selva Biological Station: Habitat use, trophic composition, and migrants. In *Four Neotropical rainforests,* ed. A. Gentry, 161–82. New Haven, Conn.: Yale University Press.

Bloomfield, J., K. A. Vogt, and D. J. Vogt. 1993. Decay rate and substrate quality of fine roots and foliage of two tropical trees species in the Luquillo Experimental Forest, Puerto Rico. *Plant Soil* 150:233–45.

Blower, J. G. 1970. The millipedes of a Cheshire wood. *Journal of Zoology* (London) 160:455–96.

Blower, J. G. 1979. The millipede faunas of two British limestone woods. In *Myriapod biology,* ed. M. Camatini, 203–14. New York: Academic Press.

Blower, J. G., and P. D. Gabbutt. 1964. Studies on the millipedes of a Devon oak wood. *Proceedings of the Zoological Society of London* 143:143–76.

Bohart, R., and L. Strange. 1965. A revision of the genus *Zethus* Fabricius in the Western Hemisphere (Hymenoptera: Eumenidae). *University of California Publications in Entomology 40.* Berkeley: University of California Press.

Bohnsack, K. K. 1954. A study of the forest floor arthropods of an oak hickory woods in southern Michigan. Ph.D. Diss., University of Michigan, Ann Arbor.

Bonaccorso, F. J. 1975. Foraging and reproductive ecology in a community of bats in Panama. Ph.D. diss. University of Florida, Gainesville.

Boon, P. J. 1988. Notes on the distribution and biology of *Smicridea* (Tricheptera: Hydropsychidae) in Jamaica. *Archiv für Hydrobiologie* 111:423–33.

Boon, P. J., B. P. Jupp, and D. G. Lee. 1986. The benthic ecology of rivers in the Blue Mountains (Jamaica) prior to construction of a water regulation scheme. *Archiv für Hydrobiologie* Supplementband 74:315–55.

Boose, E. R., D. R. Foster, and M. Fluet. 1994. Hurricane impacts to tropical and temperate forest landscapes. *Ecological Monographs* 64:369–400.

Borgmeier, T. 1969. New or little-known phorid flies, mainly of the Neotropical region (Diptera, Phoridae). *Studia Entomologica* 2:33–132.

Bornebusch, C. H. 1930. The fauna of the forest floor. *Forstlige Forsøgsväsen i Danmark* 11:1–224.

Borror, D. J., and D. M. DeLong. 1971. *An introduction to the study of insects.* New York: Holt, Rinehart, and Winston.

Borror, D. J., C. A. Triplehorn, and N. F. Johnson. 1981. *An introduction to the study of insects.* 6th ed. Philadelphia: Saunders.

Bourgeron, P. S. 1983. In *Tropical rain forest ecosystems: Structure and function,* ed. F. B. Golley, 29–47. *Ecosystems of the world 14A.* Amsterdam: Elsevier Scientific Publishing Co.

Bourliére, F. 1983. Animal species diversity in tropical forests. In *Tropical rain forest ecosystems: Structure and function,* ed. F. B. Golley, 77–92. *Ecosystems of the world* 14A. Amsterdam: Elsevier Scientific Publishing Co.

Bouysse, P. 1984. The Lesser Antilles island arc: Structure and geodynamic evolution. Initial Reports, Deep Sea Drilling Project, 78A:83–103. United States Government Printing Office.

Bouysse, P., P. Andreieff, and D. Westercamp. 1980. Evolution of the Lesser Antilles island arc: New data from the submarine geology. *Transactions of the 9th Caribbean Geological Conference* (Santo Domingo), 75–88.

Bouysse, P., P. Andreieff, M. Richard, J. C. Baubron, A. Mascle, R. C. Maury, and D. Westercamp. 1985. Aves Swell and northern Lesser Antilles Ridge: Rock-dredging results from ARCANTE 3 cruise. In *Caribbean geodynamics,* ed. A. Mascle, 65–76. Paris: Editions Technip.

Bowen, G. D. 1980. Mycorrhizal roles in tropical plants and ecosystems. In *Mycorrhizal roles in tropical plants and ecosystems,* ed. P. Mikola, 165–90. Oxford: Clarendon Press.

Bowen, S. H. 1983. Detritivory in neotropical fish communities. *Environmental Biology of Fishes* 9:137–44.

Bowles, J. B., P. D. Heideman, and K. R. Erickson. 1990. Observations on six species of free-tailed bats (Molossidae) from Yucatan, Mexico. *Southwestern Naturalist* 35:151–7.

Branch, V. 1976. Habits and food of *Anolis equestris* in Florida. *Copeia* 1976:187–9.

Bray, J. R. 1964. Primary consumption in three forest canopies. *Ecology* 45:165–7.

Breznak, J. A., W. J. Brill, J. W. Mertins, and H. C. Coppel. 1973. Nitrogen fixation in termites. *Nature* 244:577–80.

Brian, M. V. 1977. *Ants.* London: Collins.

Brian, M. V., ed. 1978. Production ecology of ants and termites. *International Biological Programme 13.* Cambridge: Cambridge University Press.

Briand, F. 1983. Environmental control of food web structure. *Ecology* 64:253–63.

Briand, F. 1985. Structural singularities of freshwater food webs. *Internationale Vereinigung für Theoretische und Angewandte Limnologie, Verhandlungen* 22: 3356–64.

Briand, F. 1990. Environmental control of food web structure. In *Community food webs: Data and theory,* ed. J. E. Cohen, F. Briand, and C. M. Newman, 49–54. Berlin: Springer-Verlag.

Briand, F., and J. E. Cohen. 1984. Community food webs have a scale invariant structure. *Nature* 307:264–7.

Briand, F., and J. E. Cohen. 1987. Environmental correlates of food chain length. *Science* 238:956–60.

Briand, F., and J. E. Cohen. 1990. Environmental correlates of food chain length. In *Community food webs: Data and theory,* ed. J. E. Cohen, F. Briand, and C. M. Newman, 55–62. Berlin: Springer-Verlag.

Briggs, J. C. 1984. Freshwater fishes and biogeography of Central America and the Antilles. *Systematic Zoology* 33:428–35.

Briggs, J. C. 1987. *Biogeography and plate tectonics.* New York: Elsevier.

Bright, D. 1985. Studies on West Indian Scolytidae (Coleoptera). *Entomologische Arbeiten aus dem Museum George Frey Tutzing Bei Muenchen* 34:169–87.

Brignoli, P. M. 1983. *A catalogue of the Araneas described between 1940 and 1981.* Manchester: Manchester University Press.

Brinson, M. M. 1976. Organic matter losses from four watersheds in the humid tropics. *Limnology and Oceanography* 21:572–82.

Briscoe, C. B. 1966. Weather in the Luquillo Mountains of Puerto Rico. U.S. Forest Service. Institute for Tropical Forestry. Research Paper ITF-3. Rio Piedras, P.R.: U.S. Forest Service. Institute for Tropical Forestry.

Broadhead, E. 1983. The assessment of faunal diversity and guild size in tropical forests with particular reference to the Psocoptera. In *Tropical rain forest: Ecology and management,* ed. S. L. Sutton, T. C. Whitmore, and A. C. Chadwick, 107–19. Oxford: Blackwell Scientific Publications.

Brockie, R. E., and A. Moeed. 1986. Animal biomass in a New Zealand forest compared with other parts of the world. *Oecologia* 70:24–34.

Brokaw, N. V. L. 1985. Treefalls, regrowth, and community structure in tropical forests. In *The ecology of natural disturbance and patch dynamics,* ed. S. T. A. Pickett and P. S. White, 53–69. Orlando, Fla.: Academic Press.

Brokaw, N. V. L., and J. S. Grear. 1991. Forest structure before and after Hurricane Hugo at three elevations in the Luquillo Mountains, Puerto Rico. *Biotropica* 23:386–92.

Brown, F. M., and B. Heineman. 1972. *Jamaica and its butterflies.* London: E. W. Classey Ltd.

Brown, L. H., and D. Amadon. 1968. *Eagles, hawks, and falcons of the world.* London: Country Life Books.

Brown, S., A. E. Lugo, S. Silander, and L. Liegel. 1983. Research history and opportunities in the Luquillo Experimental Forest. U.S. Forest Service. Southern Forest Experiment Station. General Technical Report SO-44. New Orleans, La.: U.S. Forest Service. Southern Forest Experiment Station.

Brown, S., and A. E. Lugo. 1984. Biomass of tropical forests: A new estimate based on forest volumes. *Science* 223:1288–93.

Brown, S., and A. E. Lugo. 1986. Role of tropical forests in the global carbon cycle. An annual report to Oak Ridge National Laboratory, subcontracts 19B-07762C and 19X-43326C. University of Illinois, Dept. Forestry, Urbana, Ill.

Brown, S., A. E. Lugo, and B. Liegel, eds. 1980. The role of tropical forests in the world carbon cycle: A symposium held at the Institute of Tropical Forestry in Rio Piedras, Puerto Rico on March 19, 1980. *Carbon Dioxide Effects Research and Assessment Program* 7. Washington, D.C.: U.S. Dept. of Energy. Assistant Secretary for Environment. Office of Health and Environmental Research.

Brown, W. L., Jr. 1954. The ant genus *Strumigenys* Fred. Smith in the Ethiopian and Malagasy regions. *Bulletin of the Museum of Comparative Zoology* 112:1–34.

Brown, W. L., Jr. 1959. The Neotropical species of the ant genus *Strumigenys* Fr. Smith: Group of *gundlachi* (Roger). *Psyche* 66:37–52.

Brown, W. L., Jr. 1976. Contributions toward a reclassification of the Formicidae. Part VI. Ponerinae, Tribe Ponerini, Subtribe Odontomachiti. Section A. Introduction, subtribal characters. Genus *Odontomachus*. *Studia Entomologica* 19:67–171.

Brown, W. L., Jr. 1978. Contributions toward a reclassification of the Formicidae. Part VII. Ponerinae, Tribe Ponerini, Subtribe Odontomachiti. Section B. Genus *Anochetus* and bibliography. *Studia Entomologica* 20:549–638.

Brues, C. T. 1932. Phoridae associated with ants and termites in Trinidad. *Psyche* 39:134–8.

Bruning, E. F. 1983. Vegetation structure and growth. In *Tropical rain forest ecosystems: Structure and function,* ed. F. B. Golley, 49–75. *Ecosystems of the world* 14A. Amsterdam: Elsevier Scientific Publishing Co.

Bryant, E. B. 1940. Cuban spiders in the Museum of Comparative Zoology. *Bulletin of the Museum of Comparative Zoology* (Harvard) 86:249–532.

Bryant, E. B. 1942. Additions to the spider fauna of Puerto Rico. *Journal of the Department of Agriculture, Puerto Rico* 26:1–19.

Bryant, E. B. 1943. The salticid spiders of Hispaniola. *Bulletin of the Museum of Comparative Zoology* (Harvard) 92:443–522.

Bryant, E. B. 1945. The Argiopidae of Hispaniola. *Bulletin of the Museum of Comparative Zoology* (Harvard) 95:357–422.

Bryant, E. B. 1947a. Spiders of Mona Island. *Psyche* 54:86–99.

Bryant, E. B. 1947b. Notes on spiders from Puerto Rico. *Psyche* 54:183–93.

Bryant, E. B. 1948. The spiders of Hispaniola. *Bulletin of the Museum of Comparative Zoology* (Harvard) 100:329–447.

Bucher, T. L., M. J. Ryan, and G. A. Bartholomew. 1982. Oxygen consumption during resting, calling and nest-building in the frog, *Physalaemus pustulosus*. *Physiological Zoology* 55:10–22.

Bultman, T. L., and G. W. Uetz. 1982. Abundance and community structure of forest floor spiders following litter manipulation. *Oecologia* (Berlin) 55:34–51.

Bultman, T. L., G. W. Uetz, and A. R. Brady. 1982. A comparison the cursorial spider communities along a successional gradient. *Journal of Arachnology* 10:23–33.

Burgess, G. H., and R. Franz. 1989. Zoogeography of the Antillean freshwater fish fauna. In *Biogeography of the West Indies: Past, present, and future,* ed. C. A. Woods, 263–305. Gainesville, Fla.: Sandhill Crane Press.

Burke, K., P. J. Fox, and A. Sengor. 1978. Buoyant ocean floor and the evolution of the Caribbean. *Journal of Geophysical Research* 83:3949–54.

Burton, T. M. 1976. An analysis of the feeding ecology of the salamanders (Amphibia, Urodela) of the Hubbard Brook Experimental Forest, New Hampshire. *Journal of Herpetology* 10:187–204.

Burton, T. M., and G. E. Likens. 1975a. Energy flow and nutrient cycling in salamander populations in the Hubbard Brook Experimental Forest, New Hampshire. *Ecology* 56:1068–80.

Burton, T. M., and G. E. Likens. 1975b. Salamander populations and biomass in the Hubbard Brook Experimental Forest, New Hampshire. *Copeia* 1975:541–6.

Buskirk, R. E., and W. H. Buskirk. 1976. Changes in arthropod abundance in a highland Costa Rican forest. *American Midland Naturalist* 95:288–99.

Butcher, J. W., R. Snider, and R. J. Snider. 1971. Bioecology of edaphic Collembola and Acarina. *Annual Review of Entomology* 16:249–88.

Buxton, R. D. 1981. Termites and the turnover of dead wood in an arid tropical environment. *Oecologia* 51:379–84.

Caldwell, J. S. 1942. New Psyllidae from Puerto Rico with notes on others (Homoptera). *Journal of Agriculture of the University of Puerto Rico* 26:28–33.

Caldwell, J. S., and L. F. Martorell. 1950a. A review of the Auchenorrhynchous Homoptera of Puerto Rico. Part I. Cicadellidae. *Journal of Agriculture of the University of Puerto Rico* 34:1–132.

Caldwell, J. S., and L. F. Martorell. 1950b. A review of the Auchenorrhynchous Homoptera of Puerto Rico, Part II. The Fulgoroidea except Kinnaridae. *Journal of Agriculture of the University of Puerto Rico* 34:133–269.

Caldwell, J. S., and L. F. Martorell. 1951a. A brief review of the Psyllidae of Puerto Rico (Homoptera). *Annals of the Entomological Society of America* 44:603–13.

Caldwell, J. S., and L. F. Martorell. 1951b. New leafhoppers from Puerto Rico (Cicadellidae: Homoptera). *Journal of Agriculture of the University of Puerto Rico* 35:88–89.

Cameron, G. N., and T. W. LaPoint. 1978. Effects of tannins on the decomposition of Chinese tallow leaves by terrestrial and aquatic invertebrates. *Oecologia* (Berlin) 32:349–66.

Camilo, G. R. 1992. Food web analysis of microbenthic riffle communities. Ph.D. diss. Texas Tech University, Lubbock.

Camilo, G. R., and M. R. Willig. 1994. Dynamics of a food chain model from an arthropod-dominated lotic community. *Ecological Modelling* 79:121–9.

Campbell, H. W. 1973. Ecological observations on *Anolis lionotus* and *Anolis poecilopus* (Reptilia, Sauria) in Panama. *American Museum Novitates* 2516:1–29.

Canals, M. 1979. Some ecological aspects of the biology of *Macrobrachium crenulatum* (Holthius, 1950) Palemonidae, Decapoda in the Puerto Rico including notes on its taxonomy. *Science-Ciencia* 6:130–32.

Carlberg, U. 1986. Phasmatodea: A biological review. *Zoologischer Anzeiger* 216: 1–18.

Carpenter, C. C., and J. C. Gillingham. 1984. Giant centipede (*Scolopendra alternans*) attacks marine toad (*Bufo marinus*). *Caribbean Journal of Science* 20:71–72.

Carpenter, C. R. 1958. Territoriality: A review of concepts and problems. In *Behavior and evolution,* ed. A. Roe and G. G. Simpson, 224–50. New Haven, Conn.: Yale University Press.

Carpenter, S. R., J. F. Kitchell, and J. R. Hodgson. 1985. Cascading trophic interactions and lake productivity. *Bioscience* 35:634–9.

Carroll, G. C. 1980. Forest canopies: Complex and independent subsystems. In *Forest: Fresh perspectives from ecosystem analysis,* ed. R. H. Waring, 87–108. Corvallis: Oregon State University Press.

Cary, J. F. 1992. Ecology, homerange, and foraging of *Caracolus caracolla* in the Luquillo Experimental Forest, Puerto Rico. M.S. thesis, Texas Tech University, Lubbock.

Case, J. E., and T. L. Holcombe. 1980. Geologic-tectonic map of the Caribbean region. Miscellaneous Investigations Series Map I-1100. Reston, Va.: U.S. Geological Survey.

Case, T. J. 1975. Species numbers, density compensation, and the colonizing ability of lizards on islands in the Gulf of California. *Ecology* 56:3–18.

Chace, F. A., Jr., and H. H. Hobbs, Jr. 1969. The freshwater and terrestrial decapod crustaceans of the West Indies with special reference to Dominica. *U.S. National Museum Bulletin* 292:1–258.

Chalumeau, F. 1978. Contribution à l'etude des Scarabaeoidea des Antilles. II. Remarques et observations, descriptions de nouveaux taxa. *Bulletin de la Societe Entomologique de Mulhouse* Oct.-Dec.:41–55.

Chalumeau, F. 1982. Contribution a l'etude des Scarabaeoidea des Antilles (III). *Nouvelle Revue d'Entomologique* 12:321–45.

Chalumeau, F. 1985. *Phyllophaga* Harris 1826 (Melolonthinae): désignation de types et peuplement des Iles Sous-le-Vent (Antilles) (Coleoptera, Scarabaeidae). *Nouvelle Revue d'Entomologique,* n.s. 2:21–34.

Chalumeau, F., and L. Gruner. 1976. Scarabaeoides des Antilles, Pt. 2: Melolonthinae et Rutelinae. *Annales de la Societe Entomologique de France,* n.s. 12:83–112.

Chandler, A. C. 1918. The western newt or water-dog (*Notophthalmus torosus*) a natural enemy of mosquitoes. *Oregon Agricultural College Experiment Station Bulletin* 152:1–24.

Chapin, E. A. 1940. A revision of the West Indian beetles of the scarabaeid subfamily Aphodiinae. *Proceedings of the U.S. National Museum* 89:1–41.

Charles-Dominique, P. 1986. Inter-relations between frugivorous vertebrates and pioneer plants: *Cecropia,* birds and bats in French Guyana. In *Frugivores and seed dispersal,* ed. A. Estrada, T. H. Fleming, C. Vásquez-Yánes, and R. Dirzo, 119–35. The Hague: W. Junk.

Chiarello, N., J. C. Hickman, and H. A. Mooney. 1982. Endomycorrhizal role for interspecific transfer of phosphorus in a community of annual plants. *Science* 217:941–3.

Chickering, A. M. 1945. Hypotypes of *Accola spinosa* (Petrunkevitch) (Dipluridae) from Panama. *Transactions of the Academy of Arts and Sciences of Connecticut* 36:159–67.

Chickering, A. M. 1964. Two new species of the genus *Accola. Psyche* 71:174–80.

Chickering, A. M. 1968a. The genus *Miagrammops* (Araneae, Uloboridae) in Panama and the West Indies. *Breviora* 289:1–28.

Chickering, A. M. 1968b. The genus *Dysderina* in Central America and the West Indies. *Breviora* 296:1–37.

Chickering, A. M. 1968c. The genus *Ischothyreus* (Araneae, Oonopidae) in Central America and the West Indies. *Psyche* 75:135–56.

Chickering, A. M. 1968d. The genus *Triaeris* Simon in Central America and the West Indies. *Psyche* 75:351–9.

Chickering, A. M. 1969a. The genus *Stenoonops* (Araneae, Oonopidae) in Panama and the West Indies. *Breviora* 339:1–35.

Chickering, A. M. 1969b. The family Oonopidae (Araneae) in Florida. *Psyche* 76: 144–62.

Chickering, A. M. 1970. The genus *Oonops* in Panama and the West Indies. Part 1. *Psyche* 77:487–512.

Chickering, A. M. 1972a. The genus *Oonops* in Panama and the West Indies. Part 2. *Psyche* 78:203–14.

Chickering, A. M. 1972b. The genus *Oonops* in Panama and the West Indies. *Psyche* 78:104–15.

Chickering, A. M. 1972c. The spider genus *Trachelus* (Araneae, Clubionidae) in the West Indies. *Psyche* 79:215–30.

Choate, J. R., and E. C. Birney. 1968. Sub-recent Insectivora and Chiroptera from Puerto Rico, with description of a new bat of the genus *Stenoderma*. *Journal of Mammalogy* 49:400–412.

Christiansen, K. 1964. Bionomics of Collembola. *Annual Review of Entomology* 9:147–78.

Christophe, T., and P. Blandin. 1977. The spider community in the litter of a coppiced chestnut woodland (Forêt de Montmorency, Val d'Oise, France). *Bulletin of the British Arachnological Society* 4:132–40.

Cintron, G. 1970. Niche separation of tree frogs in the Luquillo Forest. In *A tropical rain forest: A study of irradiation and ecology at El Verde, Puerto Rico,* ed. H. T. Odum and R. Pigeon, E51–E53. Oak Ridge, Tenn.: U.S. Atomic Energy Commission.

Clarke, R. D., and P. R. Grant. 1968. An experimental study of the role of spiders as predators in a forest litter community. Part 1. *Ecology* 49:1152–4.

Clay, K. 1988. Fungal endophytes of grasses: A defensive mutualism between plants and fungi. *Ecology* 69:10–16.

Closs, G. 1991. Multiple definitions of food web statistics: An unnecessary problem for food web research. *Australian Journal of Ecology* 16:413–15.

Closs, G. P., and P. S. Lake. 1994. Spatial and temporal variation in the structure of an intermittent-stream food web. *Ecological Monographs* 64:1–21.

Closs, G., G. A. Watterson, P. J. Donnelly. 1993. Constant predator-prey ratios: An arithmetical artifact? *Ecology* 74:238–43.

Cloudsley-Thompson, J. L. 1968. *Spiders, scorpions, centipedes and mites.* New York: Pergamon.

Coddington, J. A. 1986. The genera of the spider family Theridiosomatidae. *Smithsonian Contributions to Zoology* 422:1–96. Washington, D.C.: Smithsonian Institution Press.

Coddington, J., and C. E. Valerlo. 1980. Observations on the web and behavior of *Wendilgarda* spiders (Araneae: Theriodiosomatidae). *Psyche* 87:93–105.

Cohen, J. E. 1989a. *Ecologist's co-operative web bank version 1.0. Machine-readable data base of food webs.* New York: Rockefeller University.

Cohen, J. E. 1989b. Food webs and community structure. In *Perspectives in ecologi-*

cal theory, ed. J. Roughgarden, R. M. May, and S. A. Levin, 181–202. Princeton, N.J.: Princeton University Press.

Cohen, J. E. 1990. Food webs and community structure. In *Community food webs: Data and theory,* ed. J. E. Cohen, F. Briand, and C. M. Newman, 1–14. Berlin: Springer-Verlag.

Cohen, J. E., and F. Briand. 1984. Trophic links of community food webs. *Proceedings of the National Academy of Sciences of the United States of America* 81:4105–9.

Cohen, J. E., and C. M. Newman. 1985. A stochastic theory of community food webs. I. Models and aggregated data. *Proceedings of the Royal Society of London* B 224:421–48.

Cohen, J. E., and Z. J. Palka. 1990. A stochastic theory of community food webs. V. Intervality and triangulation in the trophic niche overlap graph. *American Naturalist* 135:435–63.

Cohen, J. E., C. M. Newman, and F. Briand. 1985. A stochastic theory of community food webs. II. Individual webs. *Proceedings of the Royal Society of London* B 224:449–61.

Cohen, J. E., F. Briand, and C. M. Newman. 1986. A stochastic theory of community food webs. III. Predicted and observed length of food chains. *Proceedings of the Royal Society of London* B 228:317–53.

Cohen, J. E., F. Briand, and C. M. Newman. 1990. *Community food webs: Data and theory.* Heidelberg, Germany: Springer-Verlag.

Cohen, J. E., R. Beaver, S. Cousins, D. DeAngelis, L. Goldwasser, K. Heong, R. Holt, A. Kohn, J. Lawton, N. Martinez, R. O'Malley, L. Page, B. Patten, S. Pimm, G. Polis, M. Rejmánek, T. Schoener, K. Schoenly, W. G. Sprules, J. Teal, R. Ulanowicz, P. Warren, H. Wilbur, and P. Yodzis. 1993. Improving food webs. *Ecology* 74:252–8.

Colebourn, P. H. 1974. The influence of habitat structure on the distribution of *Araneus diadematus* Clerck. *Journal of Animal Ecology* 43:401–9.

Coleman, D. C., C. P. P. Reid, and C. V. Cole. 1983. Biological strategies of nutrient cycling in soil systems. In *Advances in ecological research.* Vol. 13, ed. A. Macfadyen and E. D. Ford, 1–53. New York: Academic Press.

Coley, P. D. 1980. Effects of leaf age and plant life history patterns on herbivory. *Nature* 284:545–6.

Coley, P. D. 1982. Rates of herbivory on different tropical trees. In *The ecology of a tropical forest,* ed. E. G. Leigh, Jr., A. S. Rand, and D. M. Windsor, 123–32. Washington, D.C.: Smithsonian Institution Press.

Coley, P. D. 1983. Herbivory and defensive characteristics of tree species in a lowland tropical forest. *Ecological Monographs* 53:209–33.

Coley, P. D. 1987a. Between-species differences in leaf defenses of tropical trees. In *Proceedings of Symposium on Plant-Herbivore Interactions,* Snowbird, 7–9 August 1985, comp. D. Provenza, T. Flinders, and E. Durant McArthur, 30–35. U.S. Forest Service. Intermountain Research Station. General Technical Report INT-222. Ogden, Utah: U.S. Forest Service, Intermountain Research Station.

Coley, P. D. 1987b. Patrones en las defensas de las plantas: ¿Porqué los herbíveros prefieren ciertas especies? *Revista de Biologia Tropical* 35:151–64.

Coley, P. D., J. P. Bryant, and F. S. Chapin. 1985. Resource availability and plant antiherbivore defense. *Science* 230:895–9.

Collins, N. M. 1979. Observations on the foraging activity of *Hospitalitermes umbrinus* (Haviland), (Isoptera: Termitidae) in the Gunong Mulu National Park, Sarawak. *Ecological Entomology* 4:231–8.

Collins, N. M. 1980. The distribution of soil macrofauna on the west ridge of Gunung (Mount) Mulu, Sarawak. *Oecologia* (Berlin) 44:263–75.

Collins, N. M. 1981a. Populations age structure and survivorship of colonies of *Macroterms bellicosus* (Isoptera, Macrotermitinae). *Journal of Animal Ecology* 50:293–311.

Collins, N. M. 1981b. The role of termites in the decomposition of wood and leaf litter in the Southern Guinea Savanna of Nigeria. *Oecologia* 51:389–99.

Collins, N. M. 1983. Termite populations and their role in litter removal in Malaysian rain forests. In *Tropical rain forest: Ecology and management,* ed. S. L. Sutton, T. C. Whitmore, and A. C. Chadwick, 311–25. Boston: Blackwell Scientific Publications.

CSIRO. 1991. *The insects of Australia: A textbook for students and research workers.* Carlton South, Australia: Melbourne University Press.

Comstock, W. P. 1944. Insects of Porto Rico and the Virgin Islands: Lepidoptera (Suborder) Rhopalocera (Superfamily) Papilionoidea (True Butterflies) (Superfamily) Hesperioidea (Skippers). In *Scientific survey of Porto Rico and the Virgin Islands.* Vol. 12, Pt. 4, 421–622. New York: New York Academy of Sciences.

Connell, J. H. 1983. On the prevalence and relative importance of interspecific competition: Evidence from field experiments. *American Naturalist* 122:661–96.

Cooke, F. P., J. P. Brown, and S. Mole. 1984. Herbivory, foliar enzyme inhibitors, nitrogen and leaf structure of young and mature leaves in a tropical forest. *Biotropica* 16:257–63.

Corbet, P. S. 1980. Biology of Odonata. *Annual Review of Entomology* 25:189–217.

Corbet, P. S. 1981. Seasonal incidence of Anisoptera in light traps in Trinidad, West Indies. *Odonatologia* 10:179–87.

Corujo Flores, I. N. 1980. A study of fish populations in the Espiritu Santo River estuary. M.S. thesis, University of Puerto Rico, Rio Piedras.

Cotgreave, P. 1993. The relationship between body size and population abundance in animals. *Trends in Ecology and Evolution* 8:244–8.

Covich, A. P. 1988a. Atyid shrimp in the headwaters of the Luquillo Mountains, Puerto Rico: Filter feeding in natural and artificial streams. *Internationale Vereiningung für Theoretische und Angewandte Limnologie, Verhandlungen* 23: 2108–13.

Covich, A. P. 1988b. Geographical and historical comparisons of neotropical streams: Biotic diversity and detrital processing in highly variable habitats. *Journal of the North American Benthological Society* 7:361–86.

Covich, A. P., and T. A. Crowl. 1990. Effects of hurricane storm flow on transport of woody debris in a rain forest stream (Luquillo Experimental Forest, Puerto Rico). In *Tropical hydrology and Caribbean water resources: Proceedings of the International Symposium on Tropical Hydrology and Fourth Caribbean Islands Water Resources Congress.* San Juan, Puerto Rico, July 22–27, 1990, ed. J. Hari

Krishna, V. Quiñones-Aponte, F. Gómez-Gómez, and G. L. Morris, 197–205. Bethesda, Md.: American Water Research Association.

Covich, A. P., and J. H. Thorp. 1991. Crustacea: Introduction and Peracarida. In *Ecology and classification of North American freshwater invertebrates,* ed. J. H. Thorp and A. P. Covich, 669–93. New York: Academic Press.

Covich, A. P., T. A. Crowl, S. L. Johnson, D. Varza, and D. L. Certain. 1991. Post-Hurricane Hugo increases in atyid shrimp abundances in a Puerto Rican montane stream. *Biotropica* 23:448–54.

Covich, A. P., T. A. Crowl, S. L. Johnson, M. Pyron. In review. Distribution and abundance of a shrimp assemblage along a stream corridor: Spatial and temporal responses to disturbance (Luquillo Experimental Forest, Puerto Rico). *Biotropica.*

Cowley, G. T. 1970a. Effect of radiation on the microfungal populations of six litter species in the Luquillo Experimental Forest. In *A tropical rainforest: A study of irradiation and ecology at El Verde, Puerto Rico,* ed. H. T. Odum and R. F. Pigeon, F25–F28. Oak Ridge, Tenn.: U.S. Atomic Energy Commission.

Cowley, G. T. 1970b. Fleshy fungi in relation to irradiation and cutting in the Luquillo Experimental Forest. In *A tropical rain forest: A study of irradiation and ecology at El Verde, Puerto Rico,* ed. H. T. Odum and R. F. Pigeon, F9–F13. Oak Ridge, Tenn.: U.S. Atomic Energy Commission.

Cowley, G. T. 1970c. Vertical study of microfungal populations on leaves of *Dacryodes excelsa* and *Manilkara bidentata.* In *A tropical rain forest: A study of irradiation and ecology at El Verde, Puerto Rico,* ed. H. T. Odum and R. F. Pigeon, F-41–F-42. U.S. Atomic Energy Commission. Available from NTIS, U.S. Dept. of Commerce, Springfield, VA 22161, USA.

Coyle, F. A. 1981. Effects of clear cutting on the spider community of a southern Appalachian forest. *Journal of Arachnology* 9:285–98.

Craig, C. L. 1987. The significance of spider size to the diversification of spider-web architectures and spider reproductive modes. *American Naturalist* 129:47–68.

Critchley, B. R., A. G. Cook, U. Critchley, T. J. Perfect, A. Russell-Smith, and R. Yeadon. 1979. Effects of bush cleaning and soil cultivation on the invertebrate fauna of a forest soil in the humid tropics. *Pedobiologia* 19:425–38.

Critchlow, R. E., Jr., and S. C. Stearns. 1982. The structure of food webs. *American Naturalist* 120:478–99.

Cromack, K., Jr., P. Sollins, R. L. Todd, D. A. Crossley, Jr., W. M. Fender, R. Fogel, and A. W. Todd. 1977. Soil microorganism-arthropod interactions: Fungi as major calcium and sodium sources. In *The role of arthropods in forest ecosystems,* ed. W. J. Mattson, 78–84. New York: Springer-Verlag.

Cromack, K., Jr., P. Sollins, W. C. Graustein, K. Speidel, A. W. Todd, G. Spycher, C. Y. Li, and R. L. Todd. 1979. Calcium oxalate accumulation and soil weathering in mats of the hypogeous fungus *Hysterangium crassum. Soil Biology and Biochemistry* 11:463–8.

Cromroy, H. L. 1958. A preliminary survey of the plant mites of Puerto Rico. *Journal of Agriculture of the University of Puerto Rico* 42:39–144.

Crossley, D. A., Jr. 1977. The roles of terrestrial saprophagous arthropods in forest soils: Current status of concepts. In *The role of arthropods in forest ecosystems,* ed. W. J. Mattson, 49–56. New York: Springer-Verlag.

Crossley, D. A., Jr., and K. K. Bohnsack. 1960. Long-term ecological study in the Oak Ridge area: III. The oribatid mite fauna in pine litter. *Ecology* 41:628–38.

Crossley, D. A., Jr., J. T. Callahan, C. S. Gist, J. R. Maudsley, and J. B. Waide. 1976. Compartmentalization of arthropod communities in forest canopies at Coweeta. *Journal of the Georgia Entomological Society* 11:44–49.

Crow, T. R. 1980. A rain forest chronicle: A 30-year record of change in structure and composition at El Verde, Puerto Rico. *Biotropica* 12:42–55.

Crow, T. R., and D. F. Grigal. 1979. A numerical analysis of arborescent communities in the rain forest of the Luquillo Mountains, Puerto Rico. *Vegetatio* 40:135–46.

Crow, T. R., and P. Weaver. 1977. Tree growth in a moist tropical forest of Puerto Rico. U.S. Forest Service. Institute of Tropical Forestry. Research Paper ITF-22. Rio Piedras, P.R.: U.S. Forest Service. Institute for Tropical Forestry.

Crowell, K. L. 1962. Reduce interspecific competition among the birds of Bermuda. *Ecology* 43:75–88.

Crowl, T. A. and A. P. Covich. 1994. Responses of a freshwater shrimp to chemical and tactile stimuli from a large decapod predator. *Journal of the North American Benthological Society* 13:291–8.

Cruz, A. 1987. Avian community organization in a mahogany plantation on a neotropical island. *Caribbean Journal of Science* 23:286–96.

Cruz, A. 1988. Avian resource use in a Caribbean pine plantation. *Journal of Wildlife Management* 52:274–9.

Cruz, A., and C. A. Delannoy. 1983. Status, breeding biology, and conservation needs of the Puerto Rican Sharp-shinned Hawks, *Accipiter striatus venator.* Report on work elements 4B, D, and E. U.S. Fish and Wildlife Service Contract no. 14-16-0004-82-047.

Cruz, A., and C. A. Delannoy. 1984a. Ecology of the elfin woods warbler (*Dendroica angelae*) I. Distribution, habitat usage, and population densities. *Caribbean Journal of Science* 20:89–96.

Cruz A., and C. A. Delannoy. 1984b. Ecology of the elfin woods warbler (*Dendroica angelae*) II. Feeding ecology of the elfin woods warbler and associated insectivorous birds in Puerto Rico. *Caribbean Journal of Science* 20:153–62.

Cruz, G. A. 1987. Reproductive biology and feeding habits of cuyamel, *Joturus pichardi* and tepemechin, *Agonostomus monticula* (Pisces; Mugilidae) from Rio Platano, Mosquita, Honduras. *Bulletin of Marine Science* 40:63–72.

Cuevas, E., S. Brown, and A. E. Lugo. 1991. Above- and belowground organic matter storage and production in a tropical pine plantation and a paired broadleaf secondary forest. *Plant and Soil* 135:257–68.

Culver, D. C. 1974. Species packing in Caribbean and north temperate ant communities. *Ecology* 55:974–88.

Cummins, K. W. 1988. The study of stream ecosystems: A functional view. In *Concepts of ecosystem ecology: A comparative view,* ed. L. R. Pomeroy and J. J. Alberts, 247–62. New York: Springer Verlag.

Cummins, K. W., and M. J. Klug. 1979. Feeding ecology of stream invertebrates. *Annual Review of Ecology and Systematics* 10:147–72.

Curran, C. H. 1928. Insects of Porto Rico, and the Virgin Islands. Diptera or two-winged flies. In *Scientific survey of Porto Rico and the Virgin Islands.* Vol. 11, Pt. 1, 1–118. New York: New York Academy of Sciences.

Curran, C. H. 1931. First supplement to the Diptera of Puerto Rico, and the Virgin Islands. *American Museum Novitates* 456:1–23.

Dalrymple, G. H. 1980. Comments on the density and diet of a giant anole *Anolis equestris*. *Journal of Herpetology* 14:412–5.

Darnell, R. M. 1956. Analysis of a population of the tropical freshwater shrimp *Atya scabra* (Leach). *American Midland Naturalist* 55:131–8.

Darwin, C. 1859. *The origin of species*. London: John Murray.

Davis, W. T. 1928. The cicadas of Puerto Rico with a description of a new genus and species. *Journal of the New York Entomological Society* 36:29–34.

DeAngelis, D. L. 1975. Stability and connectance in food web models. *Ecology* 56:238–43.

DeAngelis, D. L., S. M. Bartell, and A. L. Brenkert. 1989. Effects of nutrient recycling and food chain length on resilience. *Nature* 134:778–805.

DeAngelis, D. L. 1992. Dynamics of nutrient cycling and food webs. *Population and Community Biology Series* 9. London: Chapman and Hall.

Delannoy, C. A. 1992. Status surveys of the Puerto Rican Sharp-shinned Hawk (*Accipiter striatus venator*) and Puerto Rican Broad-winged hawk (*Buteo platypterus brunnescens*). Final Status Report to the U.S. Fish and Wildlife Service.

Deligne, J., A. Quennedy, and M. S. Blum. 1981. The enemies and defense mechanisms of termites. In *Social insects*. Vol. II, ed. H. R. Hermann, 1–76. New York: Academic Press.

Dengo, G., and J. E. Case, eds. 1990. *The Caribbean region*. Vol. H, *The Decade of North American Geology*. Boulder, Co.: Geological Society of America.

Devoe, N. N. 1990. Differential seedling and regeneration rates in openings and under closed canopy in subtropical wet forests. Ph.D. Diss. Yale University, New Haven, Conn.

Dexter, R. R. 1932. The food habits of the imported toad *Bufo marinus* in the sugar cane sections of Puerto Rico. *Bulletin International Society of Sugar Cane Technology* 74:1–6.

Deyrup, M., J. Trager, and N. Carlin. 1985. The genus *Odontomachus* in the southeastern United States (Hymenoptera: Formicidae). *Entomological News* 96:188–95.

Dial, R. 1992. A food web for a tropical rain forest: The canopy view from *Anolis*. Ph.D. diss., Stanford University, Stanford, Calif.

Diamond, J. M. 1970. Ecological consequences of island colonization by southwest Pacific birds. II. The effect of species diversity on total population density. *Proceedings of the National Academy of Sciences of the United States of America* 67:1715–21.

Diaz, H., and G. Rodriguez. 1977. The branchial chamber of some terrestrial and semiterrestrial crabs. *Biological Bulletin* (Woods Hole) 153:485–504.

Dietrich, W. E., D. M. Windsor, and T. Dunne. 1982. Geology, climate, and hydrology of Barro Colorado Island. In *The ecology of a tropical forest,* ed. E. G. Leigh, Jr., A. S. Rand, and D. M. Windsor, 21–46. Washington, D.C.: Smithsonian Institution Press.

Dietrick, E. J. 1961. An improved back pack motor fan for suction sampling of insect populations. *Journal of Economic Entomology* 54:394–5.

Dirzo, R. 1984. Insect-plant interactions: Some ecophysiological consequences. In

The physiological ecology of plants of the wet tropics, ed. E. Medina, H. A. Mooney, and C. Vásquez-Yánes, 209–24. The Hague: W. Junk.

Dirzo, R. 1987. Estudios sobre interacciones planta-herbívoro en "Los Tuxtlas," Veracruz. *Revista de Biologia Tropical* 35 (suppl. 1): 119–31.

Dirzo, R., and C. Domínguez. 1986. Seed shadows, seed predation and the advantages of dispersal. In *Frugivores and seed dispersal,* ed. A. Estrada, T. H. Fleming, C. Vásquez-Yánes, and R. Dirzo, 237–49. Tasks for Vegetative Science 15. The Hague: W. Junk.

Dixon, A. F. G., P. Kindlmann, J. Leps, and J. Holman. 1987. Why there are so few species of aphids, especially in the tropics. *American Naturalist* 129: 580–92.

Donnelly, M. A. 1991. Feeding patterns of the strawberry poison frog, *Dendrobates pumilio* (Anura: Dendrobatidae). *Copeia* 1991: 723–30.

Dowding, P. 1985. The influence of water on fungal community development. *Bulletin of the British Mycological Society* 19 (Suppl. 1): 2–3.

Doyle, T. W. 1980. Modelling of succession in the tabonuco forest. M.S. thesis, University of Tennessee, Knoxville.

Doyle, T. W. 1981. The role of disturbance in the gap dynamics of a montane rain forest: An application of a tropical forest successional model. In *Forest succession: Concepts and application,* ed. D. C. West, H. H. Shugart, and D. B. Botkin, 56–73. New York: Springer-Verlag.

Drake, C., and J. Maldonado. 1954. Puerto Rican water-striders (Hemiptera). *Proceedings of the Biological Society of Washington* 67: 219–22.

Drewry, G. E. 1969a. Animal diversity. In *The rain forest project annual report,* ed. C. F. Jordan and G. E. Drewry, 107–12. PRNC-129. Río Piedras, P.R.: Puerto Rico Nuclear Center.

Drewry, G. E. 1969b. Appendix A. Key to Dolichopodidae of the rain forest at El Verde. In *The rain forest project annual report,* ed. C. F. Jordan and G. E. Drewry, 113–17. PRNC-129. Río Piedras, P.R.: Puerto Rico Nuclear Center.

Drewry, G. E. 1969c. Appendix B. Key to Muscidae (*sensus latus*) of the rain forest at El Verde (Anthomyiidae and Muscidae). In *The rain forest project annual report,* ed. C. F. Jordan and G. E. Drewry, 118–21. PRNC-129. Rio Piedras, P.R.: Puerto Rico Nuclear Center.

Drewry, G. E. 1969d. Species diversity. In *The rain forest project annual report,* ed. C. F. Jordan, and G. E. Drewry, 91–121. PRNC-129. Rio Piedras, P.R.: Puerto Rico Nuclear Center.

Drewry, G. 1970a. Factors affecting activity of rain forest frog populations as measured by electrical recording of sound pressure levels. In *A tropical rain forest: A study of irradiation and ecology at El Verde, Puerto Rico,* ed. H. T. Odum and R. F. Pigeon, E55–E68. Oak Ridge, Tenn.: U.S. Atomic Energy Commission.

Drewry, G. E. 1970b. A list of insects from El Verde, Puerto Rico. In *A tropical rain forest: A study of irradiation and ecology at El Verde, Puerto Rico,* ed. H. T. Odum and R. F. Pigeon, E129–E150. Oak Ridge, Tenn.: U.S. Atomic Energy Commission.

Drewry, G. E. 1970c. The role of amphibians in the ecology of Puerto Rican rain forest. In *The rain forest project annual report,* ed. R. G. Clements, G. E. Drewry, and R. J. Lavigne, 16–54. PRNC-147. Rio Piedras, P.R.: Puerto Rico Nuclear Center.

Drewry, G. E., and A. S. Rand. 1983. Characteristics of an acoustic community: Puerto Rican frogs of the genus *Eleutherodactylus*. *Copeia* 1983:941–53.

Drift, J. van der. 1951. Analysis of the animal community in a beech forest floor. *Tijdschrift voor Entomologie* 94:1–168.

Drift, J. van der. 1963. A comparative study of the soil fauna in forests and cultivated land on sandy soils in Suriname. In *Studies on the fauna of Suriname and other Guyanas.* Vol. 6, No. 19, ed. D. C. Geijskes and P. Wagenaar Hummelinck, 1–42. The Hauge: Martinus Nijhoff.

Drift, J. van der, and E. Jansen. 1977. Grazing of springtails on hyphal mats and its influence on fungal growth and respiration. In *Soil organisms as components of ecosystems,* ed. U. Lohm and T. Persson, 203–9. *Ecological Bulletins* (Stockholm) 25. Stockholm: Swedish Natural Science Research Council.

Drift, J. van der, and M. Witkamp. 1960. The significance of the breakdown of oak litter by *Eniocyla fusilla* (Burm.). *Archives Néerlandaises de Zoologie* 13:486–92.

Dudgeon, D. 1987. Niche specificities of four fish species (Homalopteridae, Cobitidae, and Gobiidae) in a Hong Kong forest stream. *Archiv für Hydrobiologie* 108:349–64.

Duellman, W. E., and L. Trueb. 1986. *Biology of amphibians.* New York: McGraw-Hill.

Duncan, R. A., and R. Hargraves. 1984. Plate tectonic evolution of the Caribbean region in the mantle reference frame. *Geological Society of America Memoir* 162:81–93.

Duncan, R. A., and M. A. Richards. 1991. Hotspots, mantle plumes, flood basalts, and true polar wander. *Reviews of Geophysics* 29:31–50.

Dziabaszewski, A. 1976. Arachnoidea (Aranei, Opiliones, Pseudoscorpionidea) on crowns of trees: An ecological faunistic study. *Seria Zoologia* (Poznán) 4:1–218.

Eberhard, W. G. 1980. Spider and fly play cat and mouse. *Natural History* 89:57–60.

Edgar, W. D. 1971. Aspects of the ecological energetics of the wolf spider *Pardosa* (*Lycosa*) *lugubris* (Walckenaer). *Oecologia* 7:136–54

Edmisten, J. 1970a. Preliminary studies of the nitrogen budget of a tropical rainforest. In *A tropical rain forest: A study of irradiation and ecology at El Verde, Puerto Rico,* ed. H. T. Odum, and R. F. Pigeon, H211–H215. Oak Ridge, Tenn.: U.S. Atomic Energy Commission.

Edmisten, J. 1970b. Soil studies in the El Verde rain forest. In *A tropical rain forest: A study of irradiation and ecology at El Verde, Puerto Rico,* ed. H. T. Odum and R. F. Pigeon. H79–H87. Oak Ridge, Tenn.: U.S. Atomic Energy Commission.

Edmisten, J. 1970c. Studies of *Phytolacca icosandra.* In *A tropical rain forest: A study of irradiation and ecology at El Verde, Puerto Rico,* ed. H. T. Odum and R. F. Pigeon, D183–D188. Oak Ridge, Tenn.: U.S. Atomic Energy Commission.

Edmisten, J. 1970d. Survey of mycorrhiza and nodules in the El Verde forest. In *A tropical rain forest: A study of irradiation and ecology at El Verde, Puerto Rico,* ed. H. T. Odum and R. F. Pigeon, F15–F20. Oak Ridge, Tenn.: U.S. Atomic Energy Commission.

Edmunds, G. F., Jr. 1982. Ephemeroptera. In *Aquatic biota of Mexico, Central America and the West Indies,* ed. S. H. Hurlbert and A. Villalobos-Figueroa, 242–8. San Diego, Calif.: S. H. Hurlbert, Dept. of Biology, San Diego State University.

Edwards, C. A., and G. W. Heath. 1975. Studies in leaf litter breakdown. II. The influence of leaf age. *Pedobiologia* 15:348–54.

Edwards, C. A., D. E. Reichle, and D. A. Crossley, Jr. 1970. The role of soil invertebrates in turnover of organic matter and nutrients. In *Analysis of temperate forest ecosystems,* ed. D. E. Reichle, 147–72. New York: Springer-Verlag.

Ehrlich, P. R., and L. E. Gilbert. 1973. Population structure and dynamics of the tropical butter-fly *Heliconius ethilla. Biotropica* 5:69–82.

Eisner, T., I. Kriston, and D. J. Aneshansley. 1976. Defensive behavior of a termite (*Nasutitermes exitiosus*). *Behavioral Ecology and Sociobiology* 1:83–125.

Elliott, J. M. 1977. Some methods for the statistical analysis of samples of benthic invertebrates. *Freshwater Biological Association Scientific Publication No. 25.* Kendal, U.K.: Titus Wilson and Son, Ltd.

Elton, C. 1927. *Animal Ecology.* New York: MacMillan.

Elton, C. S. 1966. *The pattern of animal communities.* London: Chapman and Hall.

Elton, C. S. 1973. The structure of invertebrate populations inside Neotropical rain forest. *Journal of Animal Ecology* 42:55–104.

Elton, C. S. 1975. Conservation and the low population density of invertebrates inside Neotropical rain forest. *Biological Conservation* 7:3–15.

Emerson, A. E. 1929. Ecological relationships between termites and termitophiles in British Guiana. In *Proceedings of the 10th International Congress of Zoology.* Budapest, 1927, Pt. 2, 1008–9. Budapest: Imprimerie Stephaneum.

Emerson, S. B. 1985. Skull shape in frogs—correlations with diet. *Herpetologica* 41:177–88.

Emlen, J. T. 1971. Population densities of birds derived from transect counts. *Auk* 88:323–42.

Emlen, J. T. 1977. Land bird communities of Grand Bahama Island: The structure and dynamics of an avifauna. *Ornithological Monographs 24.* Washington, D.C.: American Ornithologists' Union.

Emlen, J. T. 1978. Density anomalies and regulatory mechanisms in land bird populations on the Florida peninsula. *American Naturalist* 112:265–86.

Emmel, T. C., and G. T. Austin. 1990. The tropical rain forest butterfly fauna of Rondonia, Brazil: Species diversity and conservation. *Tropical Lepidoptera* 1:1–12.

Emsley, M. G. 1969. The Schizopteridae (Hemiptera: Heteroptera) with the description of new species from Trinidad. *Memoirs of the American Entomological Society* 25:1–154.

Enders, F. 1974. Vertical stratification in orb-web spiders (Araneidae, Araneae) and a consideration of other methods of coexistence. *Ecology* 55:317–28.

Englemann, M. D. 1961. The role of soil arthropods in the energetics of an old field community. *Ecological Monographs* 31:221–38.

Enrique de Jesus, J. 1987. Taxonomy, distribution and habitat utilization by land snails in the Caribbean National Forest. Unpublished manuscript.

Equihua-Martínez, A., and T. H. Atkinson. 1987. Catálogo de Platypodidae (Coleoptera) de Norte y Centroamérica. *Folia Entomologica Mexicana* 72:5–31.

Erdman, D. S. 1961. Notes on the biology of the gobiid fish *Sicydium plumieri* in Puerto Rico. *Bulletin of Marine Science of the Gulf and Caribbean* 11:448–56.

Erdman, D. S. 1972. *Inland game fishes of Puerto Rico.* San Juan, P.R.: Commonwealth of Puerto Rico. Department of Agriculture.

Erdman, D. S. 1984. Exotic fishes in Puerto Rico. In *Distribution, biology, and management of exotic fishes,* ed. W. R. Courtney, Jr. and J. R. Stouffer, 162–76. Baltimore: Johns Hopkins University Press.

Erdman, D. S. 1986. The green stream goby, *Sicydium plumieri,* in Puerto Rico. *Tropical Fish Hobbyist* 34 (Feb.): 70–74.

Erwin, T. L. 1982. Tropical forests: Their richness in Coleoptera and other arthropod species. *Coleopterist's Bulletin* 36: 74–75.

Erwin, T. L. 1983a. Beetles and other insects of tropical forest canopies at Manaus, Brazil, sampled by insect fogging. In *Tropical rain forest: Ecology and management,* ed. S. L. Sutton, T. C. Whitmore, and A. C. Chadwick, 59–75. Palo Alto, Calif.: Blackwell Scientific Publications.

Erwin, T. L. 1983b. Tropical forest canopies: The last biotic frontier. *Bulletin of the Entomological Society of America* 29: 14–19.

Erwin, T. L., and J. C. Scott. 1980. Seasonal and size patterns, trophic structure, and richness of Coleoptera in the tropical arboreal ecosystem: The fauna of the tree *Leubea seemannii* Triana and Planch in the Canal Zone of Panama. *Coleopterist's Bulletin* 34: 305–22.

Estrada Pinto, A. 1970. Phenological studies of trees at El Verde. In *A tropical rain forest: A study of irradiation and ecology at El Verde, Puerto Rico,* ed. H. T. Odum and R. F. Pigeon, D237–D269. Oak Ridge, Tenn.: U.S. Atomic Energy Commission.

Etheridge, R. 1960. The relationships of the anoles (Sauria, Iguanidae): An interpretation based on skeletal morphology. Ph.D. Diss., University of Michigan, Ann Arbor.

Etheridge, R. 1965. The abdominal skeleton of lizards in the family Iguanidae. *Herpetologica* 21: 161–8.

Evans, G. C. 1972. *The Quantitative Analysis of Plant Growth.* Berkeley: University of California Press.

Ewel, J. J., and J. L. Whitmore. 1973. The ecological life zones of Puerto Rico and the U.S. Virgin Islands. U.S. Forest Service. Institute of Tropical Forestry. Research Paper ITF-18. Rio Piedras, P.R.: U.S. Forest Service. Institute of Tropical Forestry.

Ewel, J., F. Benedict, C. Berish, and B. Brown. 1982. Leaf area, light transmission, roots and leaf damage in nine tropical plant communities. *Agro-Ecosystems* 7: 305–26.

Exline, H., and H. W. Levi. 1962. American spiders of the genus *Argyrodes* (Araneae, Theridiidae). *Bulletin of the Museum of Comparative Zoology (Harvard)* 127: 75–204.

Faaborg, J. 1979. Qualitative patterns of avian extinction on neotropical land-bridge islands: Lessons for conservation. *Journal of Applied Ecology* 16: 88–107.

Faaborg, J. 1982. Avian population fluctuations during drought conditions in Puerto Rico. *Wilson Bulletin* 94: 20–30.

Faaborg, J., W. J. Arendt, and M. S. Kaiser. 1984. Rainfall correlates of bird population fluctuations in a Puerto Rican dry forest: A nine-year study. *Wilson Bulletin* 96: 575–93.

Faeth, S. H. 1984. Density compensation in vertebrates and invertebrates: A review and an experiment. In *Ecological communities: Conceptual issues and the evi-*

dence, ed. D. R. Strong, Jr., D. Simberloff, L. G. Abele, and A. B. Thistle, 491–509. Princeton, N.J.: Princeton University Press.

Fager, E. W. 1968. The community of invertebrates in decaying oak wood. *Journal of Animal Ecology* 37:121–42.

Fain, A., G. Anastos, J. Camin, and D. Johnston. 1967. Notes on the genus *Spelaeorhynchus:* Description of *S. praecursor* Neuman and of two new species. *Acarologia* 9:535–56.

Farlow, J. O. 1976. A consideration of the trophic dynamics of a late Cretaceous large-dinosaur community (Oldman formation). *Ecology* 57:841–57.

Fauth, J. E., B. I. Crother, and J. B. Slowinski. 1989. Elevational patterns of species richness, evenness, and abundance of the Costa Rican leaf-litter herpetofauna. *Biotropica* 21:178–85.

Felgenhauer, B. E., and L. G. Abele. 1983. Ultrastructure and functional morphology of feeding and associated appendages in the tropical freshwater shrimp *Atya innocous* with notes on its ecology. *Journal of Crustacean Biology* 3:336–63.

Felgenhauer, B. E., and L. G. Abele. 1985. Feeding structures of two atyid shrimps with comments on caridean phylogeny. *Journal of Crustacean Biology* 5:397–419.

Ferguson, D. C. 1971. Bombycoidea. Saturniidae (Part). In *The moths of America north of Mexico.* Fasc. 20:2A, ed. Dominick, R. B., T. Dominick, D. C. Ferguson, J. G. Franclemont, R. W. Hodges, and E. G. Munroe, 1–153. London: E. W. Classey, Ltd.

Fermor, T. R., and D. A. Wood. 1981. Degradation of bacteria by *Agaricus bisporus* and other fungi. *Journal of General Microbiology* 126:377–87.

Fernandez, D. S., and N. Fetcher. 1991. Changes in light availability following Hurricane Hugo in a subtropical montane forest in Puerto Rico. *Biotropica* 23:393–9.

Ferrar, P. 1982. Termites of South African savanna. IV. Subterranean populations, mass determinations and biomass estimates. *Oecologia* 52:147–51.

Ferrington, L. D., K. M. Buzby, and E. C. Masteller. 1993. Composition and temporal abundance of Chironomidae emergence from a tropical rainforest stream at El Verde, Puerto Rico. *Journal of the Kansas Entomological Society* 66:167–80.

Fetcher, N., B. L. Haines, R. A. Cordero, D. J. Lodge, L. R. Walker, D. S. Fernandez, and W. T. Lawrence. Responses of tropical plants to mineral nutrients on a landslide in Puerto Rico. *Oecologia* (in press).

Fink, L. K., Jr. 1972. Bathymetric and geologic studies of the Guadeloupe region, Lesser Antilles arc. *Marine Geology* 12:267–88.

Fitch, H. S. 1975. Sympatry and interrelationships in Costa Rican anoles. *Occasional Papers, University of Kansas Museum of Natural History 40.*

Fitch, H. S. 1982. Resources of a snake community in prairie woodland habitat of northeastern Kansas. In *Herpetological communities: A symposium of the Society for the Study of Amphibians and Reptiles and the Herpetologists League,* August 1977, ed. N. J. Scott, Jr., 83–97. U.S. Fish and Wildlife Service. Wildlife research report 13. Washington, D.C.: U.S. Fish and Wildlife Service.

Fittkau, E. J., and H. Klinge. 1973. On biomass and trophic structure of the central Amazonian rain forest ecosystem. *Biotropica* 5:2–14.

Fitzpatrick, L. C. 1973a. Energy allocation in the Allegheny salamander, *Desmognathus ochrophaeus. Ecological Monographs* 43:43–58.

Fitzpatrick, L. C. 1973b. Influence of seasonal temperatures on the energy budget and metabolic rates of the northern two-lined salamander, *Eurycea bislineata. Comparative Biochemistry and Physiology* 45A:807–18.

Fleming, T. H. 1973. Numbers of mammal species in North and Central American forest communities. *Ecology* 54:555–63.

Fleming, T. H. 1982. Foraging strategies of plant-visiting bats. In *Ecology of bats,* ed. T. H. Kunz, 287–325. New York: Plenum Press.

Fleming, T. H. 1985. Coexistence of five sympatric *Piper* (Piperaceae) species in a tropical dry forest. *Ecology* 66:688–700.

Fleming, T. H., and E. R. Heithaus. 1981. Frugivorous bats, seed shadows, and the structure of tropical forests. *Biotropica* 13 (Suppl.):45–53.

Fleming, T. H., E. T. Hooper, and D. E. Wilson. 1972. Three Central American bat communities: Structure, reproductive cycles, and movement patterns. *Ecology* 53:555–69.

Flint, O. S. 1992. New species of caddisflies from Puerto Rico (Trichoptera). *Proceedings of the Entomological Society of Washington* 94:379–89.

Flint, O. S., Jr. 1964. The caddisflies of Puerto Rico. Puerto Rico Agricultural Experiment Station. Technical Paper 40. Rio Piedras: University of Puerto Rico. Agricultural Experiment Station.

Flint, O. S., Jr. 1978. Probable origins of the West Indian Trichoptera and Odonata faunas. In *Proceedings of the second international symposium on Trichoptera,* ed. M. I. Crichton, 215–23. The Hague: W. Junk.

Flint, O. S., Jr., and E. C. Masteller. 1993. Emergence, composition, and phenology of Trichoptera emergence from a tropical rainforest stream at El Verde, Puerto Rico. *Journal of the Kansas Entomological Society* 66:140–50.

Flores, S., and D. W. Schemske. 1984. Dioecy and monoecy in the flora of Puerto Rico and the Virgin Islands: Ecological correlates. *Biotropica* 16:132–9.

Forbes, W. T. 1930. Insects of Porto Rico, and the Virgin Islands. Heterocera or moths (excepting Noctuidae, Geometridae and Pyralidae). *Scientific survey of Porto Rico and the Virgin Islands.* Vol. 12, Pt. 1, 1–171. New York: New York Academy of Sciences.

Ford, J. I., and R. A. Kinzie III. 1982. Life crawls upstream. *Natural History* 91: 60–67.

Formanowicz, D. R., Jr., M. M. Stewart, K. Townsend, F. H. Pough, and P. F. Brussard. 1981. Predation by giant crab spiders on the Puerto Rican frog *Eleutherodactylus coqui. Herpetologica* 37:125–9.

Foster, J. W. 1949. *Chemical activities of fungi.* New York: Academic Press.

Fox, I. 1946. A review of the species of biting midges or *Culicoides* from the Caribbean region. *Annals of the Entomological Society of America* 39:248–58.

Fox, R. L. 1982. Some highly weathered soils of Puerto Rico. 3. Chemical properties. *Geoderma* 27:139–77.

Frangi, J. L. and A. E. Lugo. 1985. Ecosystem dynamics of a subtropical flood plain forest. *Ecological Monographs* 55:351–69.

Frankland, J. C. 1975. Estimation of live fungal biomass. *Soil Biology and Biochemistry* 7:339–40.

Frankland, J. C. 1982. Biomass and nutrient cycling by decomposer Basidiomycetes. In *Decomposer Basidiomycetes,* ed. J. C. Frankland, J. N. Hedger, and M. J. Swift,

241–61. British Mycological Society Symposium 4. Cambridge: Cambridge University Press.

Frankland, J. C., D. K. Lindley, and M. J. Swift. 1978. A comparison of two methods for the estimation of mycelial biomass in leaf litter. *Soil Biology and Biochemistry* 10:323–33.

Franks, N. R., and W. H. Bossert. 1983. The influence of swarm raiding army ants on the patchiness and diversity of a tropical leaf litter ant community. In *Tropical rain forest: Ecology and management,* ed. S. L. Sutton, T. C. Whitmore, and A. C. Chadwick, 151–63. Boston: Blackwell Scientific Publications.

Fraser, D. F. 1976. Coexistence of salamanders in the genus *Plethodon:* A variation of the Santa Rosalia theme. *Ecology* 57:238–51.

Fredrickson, A. G., and G. Stephanopoulos. 1981. Microbial competition. *Science* 213:972–4.

Fretwell, S. D. 1977. The regulation of plant communities by the food chains exploiting them. *Perspectives in Biological Medicine* 20:169–85.

Fretwell, S. D. 1987. Food chain dynamics: The central theory of ecology? *Oikos* 50:291–301.

Frith, D. W. 1975. A preliminary study of insect abundance on West Island, Aldabra Atoll, Indian Ocean. *Transactions of the Royal Entomological Society of London* 127:209–29.

Fryer, G. 1977. Studies on the functional morphology and ecology of the atyid prawns of Dominica. *Philosophical transactions of the Royal Society London,* B 277:57–129.

Funke, W. 1973. Rolle der Tiere in Wald-Ökosystemen des Solling. In *Ökosystemforschung,* ed. H. Ellenberg, 143–64. Berlin: Springer-Verlag.

Gaa-Ojeda, A. 1983. Aspects of the life history of two species of *Sphaerodactylus* (Gekkonidae) in Puerto Rico. M.S. thesis, University of Puerto Rico, Rio Piedras.

Gabbutt, P. D. 1956. The spiders of an oak wood in south-east Devon. *Entomologist's Monthly Magazine* 92:351–8.

Gabbutt, P. D. 1967. Quantitative sampling of the pseudoscorpion *Chthonius ischnocheles* from beech litter. *Journal of Zoology* (London) 151:469–78.

Gadgil, R. L., and G. D. Gadgil. 1971. Mycorrhizae and litter decomposition. *Nature* 233:133.

Gallopin, G. C. 1972. Structural properties of food webs. In *Systems analysis and simulation in ecology,* ed. B. C. Patten, 241–82. New York: Academic Press.

Gannon, M. R. 1991. Foraging ecology, reproductive biology, and systematics of the Red Fig-eating bat (*Stenoderma rufum*) in the Tabonuco Rainforest of Puerto Rico. Ph.D. diss., Texas Tech University, Lubbock.

Gannon, M. R., and M. R. Willig. 1992. Bat reproduction in the Luquillo Experimental Forest of Puerto Rico. *Southwestern Naturalist* 37:414–19.

Gannon, M. R., and M. R. Willig. 1994a. Effects of Hurricane Hugo on bats of the Luquillo Experimental Forest of Puerto Rico. *Biotropica* 26:320–31.

Gannon, M. R., and M. R. Willig. 1994b. Records of bat ectoparasites from the Luquillo Experimental Forest of Puerto Rico. *Caribbean Journal of Science* 30: 281–3.

Gannon, M. R., and M. R. Willig. Ecology of ectoparasites from tropical bats. *Environmental Entomology* (in press).

Gannon, M. R., and M. R. Willig. Home range and foraging behavior of the red fig-eating bat (*Stenoderma rufum*). *Journal of Mammalogy* (in review).

Gans, C. 1969. Amphisbaenians—reptiles specialized for a burrowing existence. *Endeavour* 28:146–51.

García Montiel, D. 1991. The effect of human activity on the structure and composition of a tropical forest in Puerto Rico. M.S. thesis, University of Puerto Rico, Rio Piedras.

Garcia-Diaz, J. 1938. An ecological survey of the freshwater insects of Puerto Rico. *Journal of Agriculture of the University of Puerto Rico* 22:43–97.

Garrison, R. W. 1986. *Diceratobasis melanogaster* spec. nov., a new damselfly from the Dominican Republic (Zygoptera: Coenagrionidae), with taxonomic and distributional notes on the Odonata of Hispaniola and Puerto Rico. *Odonatologica* 15:61–76.

Garrison, R. W. 1989. A synopsis of the genus *Hetaerina* with descriptions of four new species (Odonata: Calopterygidae). *Transactions of the American Entomological Society* 116:175–259.

Gay, F. J., T. Greaves, F. G. Holdaway, and A. H. Wetherly. 1955. Standard laboratory colonies of termites for evaluating the resistance of timber, timber preservatives, and other materials to termite attack. *Australia CSIRO Bulletin* 277:1–60.

Genoways, H. H., and R. J. Baker. 1972. Stenoderma rufum. *Mammalian Species* 18:1–4.

Genoways, H. H., and R. J. Baker. 1975. A new species of *Eptesicus* from Guadeloupe, Lesser Antilles (Chiroptera: Vespertilionidae). *Occasional Papers. The Museum, Texas Tech University* 34:1–7.

Gentry, A. H. 1982. Patterns of neotropical plant species diversity. *Evolutionary Biology* 15:1–84.

Gentry, A. H. 1986. Endemism in tropical versus temperate plant communities. In *Conservation Biology: The science of scarcity and diversity,* ed. M. E. Soule, 153–81. Sunderland, Mass.: Sinauer Associates.

Gentry, A. H. 1990. Floristic similarities and differences between southern Central America and upper and central Amazonia. In *Four Neotropical rainforests,* ed. A. H. Gentry, 141–57. New Haven, Conn.: Yale University Press.

Gentry, A. H., ed. 1990. *Four Neotropical rainforests.* New Haven, Conn.: Yale University Press.

Geoffroy, J. J. 1981. Étude d'un écosystéme forestier mixte. V. Traits généraux du peuplement de Diplopodes édaphiques. *Revue d'Écologie et de Biologie du Sol* 18:357–72.

Gerdemann, J. W. 1968. Vesicular-arbuscular mycorrhiza and plant growth. *Annual Review of Phytopathology* 6:397–418.

Gertsch, W. J., and S. E. Riechert. 1976. The spatial and temporal partitioning of a desert spider community with descriptions of new species. *American Museum Novitates* 2604:1–25.

Gibbons, J. R. H. 1985. The biogeography and evolution of Pacific Island reptiles and amphibians. In *Biology of Australasian frogs and reptiles,* ed. G. Grigg, R. Shine, and H. Ehmann, 125–42. Chipping Norton, N.S.W.: Royal Zoological Society of New South Wales.

Gilbert, G. S., S. P. Hubbell, and R. B. Foster. 1994. Density and distance-to-adult effects of a canker disease of trees in a moist tropical forest. *Oecologia* 98:100–08.

Gilbert, L. E. 1980. Food web organization and the conservation of neotropical diversity. In *Conservation biology: An evolutionary-ecological perspective,* ed. M. E. Soulé and B. A. Wilcox, 11–33. Sutherland, Mass.: Sinauer Press.

Gilbert, L. E. 1984. The biology of butterfly communities. In *The biology of butterflies,* ed. R. I. Vane-Wright and P. R. Ackery, 41–54. London: Academic Press.

Gimpel, W. F., Jr., D. R. Miller, and J. A. Davidson. 1974. A systematic revision of the wax scales, genus *Ceroplastes* in the United States (Homoptera: Coccoidea: Coccoidae). Maryland Agricultural Experiment Station. Miscellaneous Publication 841. College Park, Md.: University of Maryland.

Giovanetti, M., and B. Mosse. 1980. An evaluation of techniques for measuring vesicular-arbuscular mycorrhizal infection in roots. *New Phytologist* 84: 489–500.

Gist, C. S., and D. A. Crossley, Jr. 1975a. The litter arthropod community in a southern Appalachian hardwood forest: Numbers, biomass and mineral element content. *American Midland Naturalist* 93:107–22.

Gist, C. S., and D. A. Crossley, Jr. 1975b. A model of mineral-element cycling for an invertebrate food-web in a southeastern hardwood forest litter community. In *Mineral cycling in southeastern ecosystems: Proceedings of a symposium held at Augusta, Georgia,* May 1–3, 1974, ed. F. G. Howell, J. B. Gentry, and M. H. Smith, 84–106. ERDA Symposium Series. Oak Ridge, Tenn.: U.S. Research and Development Administration. Technical Information Center. Office of Public Affairs.

Glanz, W. E. 1990. Neotropical mammal densities: How unusual is the community on Barro Colorado, Panama? In *Four Neotropical rainforests,* ed. A. H. Gentry, 287–313. New Haven, Conn.: Yale University Press.

Glaser, J. W. 1979. The role of predation in shaping and maintaining the structure of communities. *American Naturalist* 113:631–41.

Goffinet, G. 1975. Écologie édaphique des milieux naturels de Haut-Shaba (Zaïre). I. Caractéristiques, écotopiques et synécologie comparée des zoocénoses intercaliques. *Revue d'Écologie et de Biologie du Sol* 12:691–722.

Goffinet, G. 1976. Écologie édaphique des milieux naturels de Haut-Shaba (Zaïre). II. Phenologie et fluctuations démographiques au niveau des groupes zoologiques dominants et de quelques populations d'Arthropodes. *Ecological Bulletins* (Stockholm) 7:335–52.

Goldwasser, L. and J. Roughgarden. 1993a. Construction and analysis of a large Caribbean food web. *Ecology* 74:1216–33.

Goldwasser, L., and J. Roughgarden. 1993b. Sampling effects and the estimation of food web properties. Unpublished manuscript.

Golley, F. B. 1968. Secondary productivity in terrestrial communities. *American Zoologist* 8:53–59.

Golley, F. B. 1983a. The abundance of energy and chemical elements. In *Tropical rain forest ecosystems: Structure and function,* ed. F. B. Golley, 101–15. *Ecosystems of the world* 14A. Amsterdam: Elsevier Scientific Publishing Co.

Golley, F. B. 1983b. Decomposition. In *Tropical rain forest ecosystems: Structure and function,* ed. F. Golley, 157–66. *Ecosystems of the world* 14A. Amsterdam: Elsevier Scientific Publishing Co.

Golley, F. B., J. T. McGinnis, and R. G. Clements. 1968. A final report to Batelle Memorial Institute, Columbus Laboratories, on Terrestrial Ecology, in the investigation of the bioenvironmental feasibility of a proposed sea-level canal. Athens: University of Georgia, Institute of Ecology.

Gonser, R. A., and L. L. Woolbright. 1995. Homing behavior of the Puerto Rican frog, *Eleutherodactylus coqui. Journal of Herpetology* 29:481–4.

Gorman, G. C. 1977. Comments on ontogenetic color change in *Anolis cuvieri* (Reptilia, Lacertilia, Iguanidae). *Journal of Herpetology* 11:221.

Gorman, G. C., and L. Atkins. 1969. The zoogeography of Lesser Antillean *Anolis* lizards: An analysis based on chromosomes and lactic dehydrogenase. *Bulletin of the Museum of Comparative Zoology* 138:53–80.

Gorman, G. C., and R. Harwood. 1977. Notes on population density, agility, and activity patterns of the Puerto Rican grass lizard, *Anolis pulchellus* (Reptilia, Lacertilia, Iguanidae). *Journal of Herpetology* 11:363–8.

Gorman, G. C., and S. Hillman. 1977. Physiological basis for climatic niche partitioning in two species of Puerto Rican *Anolis* (Reptilia, Lacertilia, Iguanidae). *Journal of Herpetology* 11:337–40.

Gosz, J. R., G. E. Likens, and F. H. Bormann. 1972. Nutrient content of litter fall on the Hubbard Brook experimental forest, New Hampshire. *Ecology* 53:769–84.

Grant, C. 1931. A new frog from Puerto Rico. *Copeia* 1931:55–56.

Grant, P. R. 1966a. The density of land birds on the Tres Marías Islands in Mexico. I. Numbers and biomass. *Canadian Journal of Zoology* 44:391–400.

Grant, P. R. 1966b. The density of land birds on the Tres Marías Islands in Mexico. II. Distribution of abundances in the community. *Canadian Journal of Zoology* 44:1023–30.

Grant, P. R. 1968. Bill size, body size, and the ecological adaptations of bird species to competitive situations on islands. *Systematic Zoology* 17:319–33.

Greenstone, M. H. 1984. Determinants of spider web species diversity: Vegetation structural diversity vs. prey availability. *Oecologia* (Berlin) 62:299–304.

Greenstone, M. H., and A. F. Bennett. 1980. Foraging strategy and metabolic rate in spiders. *Ecology* 61:1255–9.

Griffiths, T. A., and D. Klingener. 1988. On the distribution of Greater Antillean bats. *Biotropica* 20:240–51.

Guariguata, M. R. 1990. Landslide disturbance and forest regeneration in the upper Luquillo Mountains of Puerto Rico. *Journal of Ecology* 78:814–32.

Gulmon, S. L., and H. A. Mooney. 1986. Costs of defense and their effects on plant productivity. In *On the economy of plant form and function: Proceedings of the Sixth Annual Maria Moors Cabot Symposium on Primary Productivity and Adaptive Patterns of Energy Capture in Plants,* ed. T. J. Givnish, 681–98. New York: Cambridge University Press.

Guyer, C. 1990. The herpetofauna of La Selva, Costa Rica. In *Four Neotropical rainforests,* ed. A. H. Gentry, 371–85. New Haven, Conn.: Yale University Press.

Hacskaylo, E., and J. A. Vozzo. 1967. Inoculation of *Pinus caribea* with pure cultures of mycorrhizal fungi in Puerto Rico. In *Proceedings of the 14th Congress of the*

International Union of Forestry Research Organizations, Munich 1967, Section 24, 139–48. Freiburg: Deutscher Verband Forstlicher Forschungsanstalten.

Hagelstein, R. 1932. Revision of the Myxomycetes. In *Scientific survey of Porto Rico and the Virgin Islands.* Vol. 8, Pt. 2, 241–8. New York: New York Academy of Sciences.

Hagstrum, D. W. 1970. Ecological energetics of the spider *Tarentula kochi* (Araneae: Lycosidae). *Annals of the Entomological Society of America* 63:1297–1304.

Hairston, N. G., F. E. Smith, and L. B. Slobodkin. 1960. Community structure, population control, and competition. *American Naturalist* 94:421–4.

Hall, S. J., and D. Raffaelli. 1991. Food-web patterns: Lessons from a species-rich web. *Journal of Animal Ecology* 60:823–42.

Handley, C. O., D. E. Wilson, and A. L. Gardner. 1991. Demography and natural history of the common fruit bat, *Artibeus jamaicensis,* on Barro Colorado Island, Panama. *Smithsonian Contributions to Zoology* 511:1–173.

Hanlon, R. D. G. 1981a. Some factors influencing microbial growth on soil animal faeces. I. Bacterial and fungal growth on particulate oak leaf litter. *Pedobiologia* 21:257–63.

Hanlon, R. D. G. 1981b. Some factors influencing microbial growth on soil animal faeces. II. Bacterial and fungal growth on soil animal faeces. *Pedobiologia* 21:264–70.

Hanlon, R. D. G., and J. M. Anderson. 1979. The effects of Collembola grazing on microbial activity in decomposing leaf litter. *Oecologia* (Berlin) 38:93–9.

Hanlon, R. D. G., and J. M. Anderson. 1980. The influence of macroarthropod feeding activities on microflora in decomposing oak leaves. *Soil Biology and Biochemistry* 12:255–61.

Harcombe, P. A. 1977. The influence of fertilization on some aspects of succession in humid tropical forest. *Ecology* 58:1375–83.

Harley, J. L., and S. E. Smith. 1983. *Mycorrhizal symbiosis.* London: Academic Press.

Harrelson, M. A. 1969. Nitrogen fixation in the epiphyllae. Ph.D. diss., University of Georgia, Athens.

Harrison, A. D., and J. J. Rankin. 1976. Hydrobiological studies of Eastern Lesser Antillean Islands. II. St. Vincent freshwater fauna, its distribution, tropical river zonation and biogeography. *Archiv für Hydrobiologie* Supplementband 50:275–311.

Harrison, J. L. 1962. The distribution of feeding habits among animals in a tropical rain forest. *Journal of Animal Ecology* 31:53–63.

Hart, C. W., Jr. 1961. The freshwater shrimps (Atyidae and Palaemonidae) of Jamaica, W. I., with a discussion of their relation to the ancient geography of the western Caribbean area. *Proceedings of the Academy of Natural Sciences of Philadelphia* 13:61–80.

Hart, C. W., Jr. 1964. A contribution to the limnology of Jamaica and Puerto Rico. *Caribbean Journal of Science* 4:331–4.

Hartshorn, G. S. 1980. Neotropical forest dynamics. *Biotropica* 12 (Suppl.):23–31.

Hartshorn, G. S., and L. J. Poveda. 1983. Checklist of trees. In *Costa Rican natural history,* ed. D. H. Janzen, 158–83. Chicago: University of Chicago Press.

Harvey, A. E., M. F. Jurgensen, and M. J. Larsen. 1978. Seasonal distribution of ectomycorrhizae in a mature Douglas-fir/larch forest in Western Montana. *Forest Science* 24:203–8.

Haselwandter, K., O. Bobleter, and D. J. Read. 1990. Degradation of ^{14}C-labelled lignin and dehydro-polymer of coniferyl alcohol by ericoid and ectomycorrhizal fungi. *Archives of Microbiology* 153:352–4.

Hassall, M., and S. P. Rushton. 1984. Feeding behaviour of terrestrial isopods in relation to plant defences and microbial activity. In *The biology of terrestrial isopods,* ed. S. L. Sutton and D. Holdich, 487–505. *Proceedings of the Symposium of the Zoological Society of London, No. 53.* New York: Academic Press.

Hassall, M., and S. P. Rushton. 1985. The adaptive significance of coprophagous behaviour in the terrestrial isopod *Porcellio scaber. Pedobiologia* 28:169–75.

Hatley, C. L., and J. A. MacMahon. 1980. Spider community organization: Seasonal variation and role of vegetation architecture. *Environmental Entomology* 9: 632–9.

Hauge, E. 1977. The spider fauna of two forest habitats in northern Norway. *Astarte* 10:93–102.

Havens, C. 1993. Effect of scale on food web structure: Response to Martinez. *Science* 260:243.

Havens, K. 1992. Scale and structure in natural food webs. *Science* 257:1107–9.

Haverty, M. I. 1974. The significance of the subterranean termite, *Heterotermes aureus* (Snyder), as a detritivore in a desert grassland ecosystem. Ph.D. Diss., University of Arizona, Tucson.

Haverty, M. I. 1977. The proportion of soldiers in termite colonies: A list and bibliography. *Sociobiology* 2:199–216.

Hayes, M. P. 1983. Predation on the adults and prehatching stages of glass frogs (Centrolenidae). *Biotropica* 15:74–76.

Healey, I. N., and A. Russell-Smith. 1970. The extraction of fly larvae from woodland soils. *Soil Biology and Biochemistry* 2:119–29.

Healey, I. N., and A. Russell-Smith. 1971. Abundance and feeding preferences of fly larvae in two woodland soils. In *Organismes du sol et production primaire,* 177–92. *Proceedings of the 4th Colloquium of Soil Zoology.* Paris: INRA.

Heatwole, H. 1962. Environmental factors influencing local distribution and activity of the salamander, *Plethodon cinereus. Ecology* 43:460–72.

Heatwole, H. 1982. A review of structuring in herpetological assemblages. In *Herpetological communities: A symposium of the Society for the Study of Amphibians and Reptiles and the Herpetologists League,* August 1977, ed. N. J. Scott, Jr., 1–19. Wildlife research report 13. Washington, D.C.: U.S. Fish and Wildlife Service.

Heatwole, H., and A. Heatwole. 1978. Ecology of the Puerto Rican camaenid treesnails. *Malacologia* 17:241–315.

Heatwole, H., and F. MacKenzie. 1967. Herpetogeography of Puerto Rico. IV. Paleogeography, faunal similarity, and endemism. *Evolution* 21:429–38.

Heatwole, H., F. Torres, S. Blasini de Austin, and A. Heatwole. 1969. Studies on anuran water balance. I. Dynamics of evaporative water loss by the coquí, *Eleutherodactylus portoricensis. Comparative Biochemistry and Physiology* 28:245–69.

Hedger, J. 1990. Fungi in the tropical forest canopy. *Mycologist* 4:200–02.

Hedger, J. N. 1985. Tropical agarics: Resource relations and fruiting periodicity. In

Developmental biology of higher fungi, ed. D. Moore, L. A. Casselton, D. A. Wood, and J. C. Frankland, 41–86. Cambridge: Cambridge University Press.

Hedges, S. B., and R. Thomas. 1991. Cryptic species of snakes (Typhlopidae: *Typhlops*) from the Puerto Rico bank detected by protein electrophoresis. *Herpetologica* 46:448–59.

Heeley, W. 1941. Observation on the life histories of some terrestrial isopods. *Proceedings of the Zoological Society of London* B 111:79–149.

Heithaus, E. R. 1982. Coevolution between bats and plants. In *Ecology of bats,* ed. T. H. Kunz, 327–67. New York: Plenum Press.

Henderson, R. W., and B. Crother. 1989. Biogeographic patterns of predation in West Indian colubrid snakes. In *Biogeography of the West Indies: Past, present and future,* ed. C. A. Woods, 479–518. Gainesville, Fla.: Sandhill Crane Press.

Henderson, R. W., T. A. Noeske-Hallin, J. A. Ottenwalder, and A. Schwartz. 1987. On the diet of *Epicrates striatus* on Hispaniola, with notes on *E. fordi* and *E. gracilis. Amphibia-Reptilia* 8:251–8.

Herd, R. M. 1983. *Pteronotus parnellii. Mammalian Species* 209:1–5.

Hering, T. F. 1982. Decomposing activity of Basidiomycetes in forest litter. In *Decomposer Basidiomycetes: Their biology and ecology,* ed. J. C. Frankland, J. N. Hedger, and M. J. Swift, 213–25. Cambridge: Cambridge University Press.

Hernández-Prieto, E. 1986. Pollination and nectar-robbing of *Tabebuia rigida* Bignoniaceae in an elfin forest in the Luquillo Mountains of Puerto Rico. M.S. thesis, University of Puerto Rico, Rio Piedras.

Hertz, P. E. 1977. Responses to dehydration in *Anolis* lizards sampled along altitudinal transects. *Copeia* 1980:440–6.

Hespenheide, H. A. 1975. Prey characteristics and predator niche width. In *Ecology and evolution of communities,* ed. M. Cody and J. Diamond, 158–80. Cambridge, Mass.: Harvard University Press.

Heyer, W. R., and K. A. Berven. 1973. Species diversities of herpetofaunal samples from similar microhabitats at two tropical sites. *Ecology* 54:642–5.

Hildrew, A. G., C. R. Townsend, and A. Hasham. 1985. The predatory Chironomidae of an iron-rich stream: Feeding ecology and food web structure. *Ecological Entomology* 10:403–13.

Hlavac, T. F. 1969. A review of the species of *Scarites* (*Antilliscaris*) (Coleoptera: Carabidae) with notes on their morphology and evolution. *Psyche* 76:1–17.

Hobbs, H. H., Jr., and C. W. Hart, Jr. 1982. The shrimp genus *Atya* (Decapoda: Atyidae). *Smithsonian Contributions in Zoology* 364:1–143.

Hodgkinson, K. C., and H. G. Baas Becking. 1977. Effect of defoliation on root growth of some arid perennial plants. *Australian Journal of Agricultural Research* 29:31–42.

Hoenicke, R. 1983. The effects of leaf-cutter ants on populations of *Astyanax fasciatus* (Characidae) in three tropical lowland wet forest streams. *Biotropica* 15:237–9.

Holdridge, L. 1967. *Life zone ecology.* San Jose, Costa Rica: Tropical Science Center.

Holdridge, L. R., W. C. Grenke, W. H. Hathaway, T. Liang, and J. Tosi. 1971. *Forest environments in tropical life zones: A pilot study.* New York: Pergamon Press.

Holland, E. A., and D. C. Coleman. 1987. Litter placement effects on microbial and organic matter dynamics in an agroecosystem. *Ecology* 68:425–33.

Hölldobler, B., and E. O. Wilson. 1990. *The ants.* Cambridge, Mass.: Belknap Press.

Holler, J. R. 1966. Microfungi of soil, root, and litter of a Puerto Rican lower montane rain forest. M.S. thesis, University of South Carolina, Columbia.

Holler, J. R., and G. T. Cowley. 1970. Response of soil, root, and litter microfungal populations to radiation. In *A tropical rain forest: A study of irradiation and ecology at El Verde, Puerto Rico,* ed. H. T. Odum and R. F. Pigeon, F35–F39. Oak Ridge, Tenn.: U.S. Atomic Energy Commission.

Holmes, J. C. 1983. Impact of infectious disease agents on population growth and geographical distribution of animals. In *Population biology of infectious diseases,* ed. R. M. Anderson and R. M. May, 37–51. Berlin: Springer-Verlag.

Holmes, R. T., and F. W. Sturges. 1973. Annual energy expenditure by the avifauna of a northern hardwoods ecosystem. *Oikos* 24:24–29.

Holmes, R. T., and F. W. Sturges. 1975. Bird community dynamics and energetics in a northern hardwoods ecosystem. *Journal of Animal Ecology* 45:175–200.

Holt, J. A. 1985. Acari and Collembola in the litter and soil of three north Queensland rainforests. *Australian Journal of Ecology* 10:57–65.

Homan, J. A., and J. K. Jones, Jr. 1975. Monophyllus redmani. *Mammalian Species* 57:1–3.

Hovore, F. T. 1989a. The Cerambycidae (Coleoptera) of La Selva Biological Reserve. Unpublished manuscript.

Hovore, F. T. 1989b. The Cerambycidae (Coleoptera) of Monteverde Cloud Forest and surrounding environs: A species inventory. Unpublished manuscript.

Howard-Williams, C., and W. J. Junk. 1977. The chemical composition of Central Amazonian macrophytes with special reference to their role in the ecosystem. *Archiv für Hydrobiologie* 79:446–64.

Howe, H. F. 1981. Dispersal of a neotropical nutmeg (*Virola sebifera*) by birds. *Auk* 98:88–98.

Howe, H. F., and D. DeSteven. 1979. Fruit production, migrant bird visitation, and seed dispersal of *Guarea glabra* in Panama. *Oecologia* 39:185–96.

Howe, H. F., and G. F. Estabrook. 1977. On intraspecific competition for avian dispersers in tropical trees. *American Naturalist* 111:817–32.

Howe, H. F., and R. Primack. 1975. Differential seed dispersal by birds of the tree *Casearia nitida* (Flacourtiaceae). *Biotropica* 7:278–83.

Howe, H. F., and G. A. Vande Kerckhove. 1979. Fecundity and seed dispersal of a tropical tree. *Ecology* 60:180–9.

Howe, H. F., and G. A. Vande Kerckhove. 1981. Removal of wild nutmeg (*Virola surinamensis*) crops by birds. *Ecology* 62:1093–1106.

Huey, R. B., and E. R. Pianka. 1981. Ecological consequences of foraging mode. *Ecology* 62:991–9.

Huey, R. B., and T. P. Webster. 1976. Thermal biology of *Anolis* lizards in a complex fauna: The *cristatellus* group on Puerto Rico. *Ecology* 57:985–94.

Huhta, V. 1965. Ecology of spiders in the soil and litter of Finnish forests. *Annales Zoologici Fennici* 2:260–308.

Huhta, V. 1971. Succession in the spider communities of the forest floor after clearcutting and prescribed burning. *Annales Zoologici Fennici* 8:483–542.

Huhta, V., and A. Koskenniemi. 1975. Numbers, biomass and community respiration of soil invertebrates in spruce forests at two latitudes in Finland. *Annales Zoologici Fennici* 12:164–82.

Humphreys, W. F. 1977. Variables influencing laboratory energy budgets of *Geolycosa godeffroyi* (Araneae). *Oikos* 28:225–33.

Humphreys, W. F. 1978. Ecological energetics of *Geolycosa godeffroyi* (Araneae: Lycosidae) with an appraisal of production efficiency in ectothermic animals. *Journal of Animal Ecology* 47:627–52.

Hunt, H. W., D. C. Coleman, E. R. Ingham, R. E. Ingham, E. T. Elliott, J. C. Moore, S. L. Rose, C. P. P. Reid, and C. R. Morley. 1987. The detrital food web in a shortgrass prairie. *Biology and Fertility of Soils* 3:57–68.

Hunt, H. W., E. T. Elliot, and D. E. Walter. 1989. Inferring tropic levels from pulse-dynamics in detrital food webs. *Plant and Soil* 115:4247–59.

Hunte, W. 1975. *Atya lanipes* Holthuis in Jamaica, including taxonomic notes and a description of the first larval stage (Decapoda, Atyidae). *Crustaceana* 28: 66–72.

Hunte, W. 1978. The distribution of fresh water shrimps (Atyidae and Palaemonidae) in Jamaica. *Zoological Journal of the Linnean Society of London* 64:135–50.

Hunte, W. 1979. The rediscovery of the freshwater shrimp *Macrobrachium crenulatum* in Jamaica. *Studies of the Fauna of Curacao and other Caribbean Islands* 183:69–74.

Hunter, M. D., and P. Price. 1992. Playing chutes and ladders: heterogeneity and the relative roles of bottom-up and top-down forces in natural communities. *Ecology* 73:724–32.

Hurlbert, S. H., and A. Villalobos-Figueroa, eds. 1982. Aquatic biota of Mexico, Central America, and the West Indies. San Diego State University, San Diego, California.

Hutchinson, G. E. 1959. Homage to Santa Rosalia or why are there so many kinds of animals? *American Naturalist* 93:145–59.

Hutto, R. L. 1980. Wider habitat distribution of migratory land birds in western Mexico, with special reference to small foliage-gleaning insectivores. In *Migrant birds in the Neotropics: Ecology, behavior, distribution, and conservation,* ed. A. Keast and E. S. Morton, 181–203. Washington, D.C.: Smithsonian Institution Press.

Hutton, R. S., and R. A. Rasmussen. 1970. Microbiological and chemical observations in a tropical forest. In *A tropical rain forest: A study of irradiation and ecology at El Verde, Puerto Rico,* ed. H. T. Odum and R. F. Pigeon, F43–F56. Oak Ridge, Tenn.: U.S. Atomic Energy Commission.

Hynes, H. B. N. 1971. Zonation of the invertebrate fauna in a West Indian stream. *Hydrobiologia* 38:1–18.

Illies, J. 1969. Biogeography and ecology of Neotropical freshwater insects, especially those from running waters. In *Biogeography and ecology in South America,* ed. E. J. Fittkau, 685–708. The Hague: W. Junk.

Inchaustegui, S. J., A. Schwartz, and R. W. Henderson. 1985. Hispaniolan giant *Diploglossus* (Sauria: Anguidae): Description of a new species and notes on the ecology of *Diploglossus warreni. Amphibia-Reptilia* 6:195–202.

Inger, R. F. 1980a. Densities of floor-dwelling frogs and lizards in lowland forests of Southeast Asia and Central America. *American Naturalist* 115:761–70.

Inger, R. F. 1980b. Relative abundances of frogs and lizards in forests of Southeast Asia. *Biotropica* 12:14–22.

Iniguez Davalos, L. I. 1993. Patrones ecologicos en la communidad de murciélagos de La Sierra de Manantlan. In *Avances en el estudio de los mamíferos de México,* ed. R. A. Medellin and G. Ceballos, 355–70. Publicaciones Especiales, vol. 1, Associacion Mexicana de Mastozoologia, Distrito Federal, México.

International Bird Census Committee. 1970. An international standard for a mapping method in bird census work recommended by the International Bird Census Committee. *Audubon Field Notes* 24:722–6.

Iverson, T. M. 1973. Decomposition of autumn-shed leaves in a spring brook and its significance for the fauna. *Archiv für Hydrobiologie* 72:305–12.

Jaeger, R. G. 1978. Plant climbing by salamanders: Periodic availability of plant-dwelling prey. *Copeia* 1978:686–91.

Jaeger, R. G. 1979. Seasonal spatial distributions of the terrestrial salamander *Plethodon cinereus. Herpetologica* 35:90–93.

Jaeger, R. G. 1980. Fluctuations in prey availability and food limitation for a terrestrial salamander. *Oecologia* 44:335–41.

Janos, D. P. 1980a. Mycorrhizae influence tropical succession. *Biotropica* 12 (Suppl.): 56–64.

Janos, D. P. 1980b. Vesicular-arbuscular mycorrhizae affect lowland tropical rain forest plant growth. *Ecology* 61:151–62.

Janos, D. P. 1983. Tropical mycorrhizas, nutrient cycles and plant growth. In *Tropical rain forest: Ecology and management,* ed. S. L. Sutton, T. C. Whitmore, and A. C. Chadwick, 327–45. Special Publication of the British Ecological Society, No. 2. Oxford: Blackwell Scientific Publications.

Janse, J. M. 1896. Les endophytes radicaux de quelques plantes Javanaises. *Annales du Jardin Botanique de Buitenzorg* 14:53–212.

Janson, C. H., and L. H. Emmons. 1990. Ecological structure of the nonflying mammal community at Cocha Cashu Biological Station, Manu National Park, Peru. In *Four Neotropical rainforests,* ed. A. H. Gentry, 314–38. New Haven, Conn.: Yale University Press.

Janzen, D. H. 1972. Association of a rainforest palm and seed-eating beetles in Puerto Rico. *Ecology* 53:258–61.

Janzen, D. H. 1973a. Sweep samples of tropical foliage insects: Description of study sites, with data on species abundances and size distributions. *Ecology* 54:659–87.

Janzen, D. H. 1973b. Sweep samples of tropical foliage insects: Effects of seasons, vegetation types, elevation, time of day, and insularity. *Ecology* 54:687–708.

Janzen, D. H. 1981. Patterns of herbivory in a tropical deciduous forest. *Biotropica* 13:271–82.

Janzen, D. H. 1983a. Food webs: Who eats what, why, how, and with what effects in a tropical forest. In *Tropical rain forest ecosystems: Structure and function,* ed. F. B. Golley, 167–82. *Ecosystems of the world* 14A. Amsterdam: Elsevier Scientific Publishing Co.

Janzen, D. H. 1983b. Insects. In *Costa Rican natural history,* ed. D. H. Janzen, 619–45. Chicago: University of Chicago Press.

Janzen, D. H. 1988. Ecological characterization of a Costa Rican dry forest caterpillar fauna. *Biotropica* 20:120–35.

Janzen, D. H., and T. W. Schoener. 1968. Differences in insect abundance and diversity between wetter and drier sites during a tropical dry season. *Ecology* 49: 96–110.

Jennings, D. H. 1982. The movement of *Serpula lacrimans* from substrate to substrate over nutritionally inert surfaces. In *Decomposer Basidiomycetes: Their biology and ecology,* ed. J. C. Frankland, J. N. Hedger, and M. J. Swift, 91–108. Cambridge: Cambridge University Press.

Jensen, V. 1974. Decomposition of angiosperm tree leaf litter. In *Biology of plant litter decomposition.* Vol. 1, ed. C. H. Dickinson and G. J. E. Pugh, 68–104. New York: Academic Press.

Jocqué, R. 1973. The spider fauna of adjacent woodland areas with different humus types. *Biologisch Jaarbock* 41:153–78.

Johnson, K. A., and E. G. Whitford. 1975. Foraging ecology and relative importance of subterranean termites in Chihuahuan Desert ecosystems. *Environmental Entomology* 4:66–70.

Johnson, S. L., A. P. Covich, T. A. Crowl, A. Estrada-Pinto, J. Bithorn. Do seasonality and disturbance influence reproduction in freshwater atyid shrimp in headwater streams, Puerto Rico? *Verhandlungen Internationale Vereinigung fur theoretische und angewandt Limnologie* (in press).

Johnston, A. 1949. Vesicular-arbuscular mycorrhiza in Sea Island cotton and other tropical plants. *Tropical Agriculture, Trinidad* 26:118–21.

Jolly, G. M. 1965. Explicit estimates from capture-recapture data with both death and immigration-stochastic model. *Biometrika* 52:225–47.

Jones, J. K., Jr., H. H. Genoways, and R. J. Baker. 1971. Morphological variation in *Stenoderma rufum. Journal of Mammalogy* 52:244–7.

Jones, K. L. 1982. Prey patterns and trophic niche overlap in four species of Caribbean frogs. In *Herpetological communities: A symposium of the Society for the Study of Amphibians and Reptiles and the Herpetologists League, August 1977,* ed. N. J. Scott, Jr., 49–55. U.S. Fish and Wildlife Service. Wildlife research Report 13. Washington, D.C.: U.S. Fish and Wildlife Service.

Jones, P. T. C., and J. E. Mollison. 1948. A technique for the quantitative estimation of soil micro-organisms. *Journal of General Microbiology* 2:54–69.

Jones, R. J. 1979. Expansion of the nest of *Nasutitermes costalis. Insectes Sociaux* 26:322–42.

Jordan, C. F. 1983. Productivity of tropical rain forest ecosystems and the implications for their use as future wood and energy sources. In *Tropical rain forest ecosystems: Structure and function,* ed. F. B. Golley, 117–36. *Ecosystems of the world* 14A. Amsterdam: Elsevier Scientific Publishing Co.

Jordan, C. F. 1985. *Nutrient cycling in tropical forest ecosystems.* Chichester, England: John Wiley & Sons.

Jordan, C. F., and G. Escalante. 1980. Root productivity in an Amazonian rain forest. *Ecology* 61:14–18.

Jordan, C. F., R. L. Todd, and G. Escalante. 1979. Nitrogen conservation in a tropical rain forest. *Oecologia* 39:123–8.

Josens, G. 1973. Observations sur les bilans energetiques dans deux populations de termites A Lamto (Côte d'Ivoire). *Annales de la Société Royale Zoologique de Belgique* 103:169–76.

Karr, J. 1971. Structure of avian communities in selected Panama and Illinois habitats. *Ecological Monographs* 41:207–33.

Karr, J. R. 1975. Production, energy pathways, and community diversity in forest birds. In *Tropical ecological systems: Trends in terrestrial and aquatic research,* ed. F. B. Golley and E. Medina, 161–78. New York: Springer-Verlag.

Karr, J. R. 1976. On the relative abundances of migrants from the north temperate zone in tropical habitats. *Wilson Bulletin* 88:433–58.

Karr, J. R. 1982a. Avian extinction on Barro Colorado Island, Panama: A reassessment. *American Naturalist* 119:220–39.

Karr, J. R. 1982b. Population variability and extinction in the avifauna of a tropical land bridge island. *Ecology* 63:1975–8.

Karr, J. R. 1990. The avifauna of Barro Colorado Island the Pipeline Road, Panama. In *Four Neotropical rainforests,* ed. A. Gentry, 183–98. New Haven, Conn.: Yale University Press.

Karr, J. R., and F. C. James. 1975. Eco-morphological configurations and convergent evolution in species and communities. In *Ecology and evolution of communities,* ed. M. L. Cody and J. M. Diamond, 258–91. Cambridge, Mass.: Harvard University Press.

Karr, J. R., S. Robinson, J. G. Blake, and R. O. Bierregaard, Jr. 1990. Birds of four Neotropical forests. In *Four Neotropical rainforests,* ed. A. Gentry, 237–69. New Haven, Conn.: Yale University Press.

Kaushik, N. K., and H. B. N. Hynes. 1971. The fate of dead leaves that fall into streams. *Archiv für Hydrobiologie* 68:465–515.

Kiester, A. R., G. C. Gorman, and D. C. Arroyo. 1975. Habitat selection behavior of three species of *Anolis* lizards. *Ecology* 56:220–5.

Kepler, A. K. 1977. Comparative study of todies (Todidae) with emphasis on the Puerto Rican tody, *Todus mexicanus. Nuttall Ornithological Club Publication No. 16.* Cambridge, Mass.: Nuttall Ornithological Club.

Kepler, C. B., and A. K. Kepler. 1970. Preliminary comparison of bird species diversity and density in Luquillo and Guanica forests. In *A tropical rain forest: A study of irradiation and ecology at El Verde, Puerto Rico,* ed. H. T. Odum and R. F. Pigeon, E183–E191. Oak Ridge, Tenn.: U.S. Atomic Energy Commission.

Kepler, C. B., and K. C. Parkes. 1972. A new species of warbler (Parulidae) from Puerto Rico. *Auk* 89:1–18.

Kevan, D. K. McE. 1962. *Soil animals.* New York: Philosophical Library.

Kirkland, A. H. 1904. Usefulness of the American toad. U.S. Department of Agriculture Farmers' Bulletin No. 196. Washington, D.C.: U.S. Government Printing Office.

Kistner, D. H. 1969. The biology of termitophiles. In *Biology of termites,* Vol. I., ed. K. Krishna and F. Weesner, 525–57. New York: Academic Press.

Kistner, D. H. 1979. Social and evolutionary significance of social insect symbionts. In *Social insects.* Vol. I, ed. H. R. Hermann, 339–413. New York: Academic Press.

Kistner, D. H. 1982. The social insects' bestiary. In *Social insects.* Vol. III, ed. H. R. Hermann, 1–224. New York: Academic Press.

Kitazawa, Y. 1967. Community metabolism of soil invertebrates in forest ecosystems in Japan. In *Secondary productivity of terrestrial ecosystems,* ed. K. Petrusewicz, 649–61. Warszawa: Panstwowe Wydawn, Naukowe.

Kitchell, J. F., R. V. O'Neill, D. Webb, G. W. Gallepp, S. M. Bartell, J. F. Koonce, and B. S. Ausmus. 1979. Consumer regulation of nutrient cycling. *BioScience* 29: 28–34.

Klots, E. B. 1932. Insects of Porto Rico and the Virgin Islands. Odonata or dragon flies. *Scientific survey of Porto Rico and the Virgin Islands,* Vol. 14, Pt. 1, 1–107. New York: New York Academy of Sciences.

Kodric-Brown, A., J. H. Brown, G. S. Byers, and D. F. Gori. 1984. Organization of a tropical island community of hummingbirds and flowers. *Ecology* 65:1358–68.

Kormanik, P. P., and A. C. McGraw. 1982. Quantification of vesicular-arbuscular mycorrhizae in plant roots. In *Methods and principles of mycorrhizal research,* ed. N. C. Schenck, 37–54. St. Paul, Minn.: American Phytopathology Society.

Kreisel, H. 1971. Ektotrophe Mykorrhiza bei *Coccoloba uvifera* in Kuba. *Biologische Rundschau* 9:97–98.

Krishna, K. 1969. Introduction. In *Biology of termites.* Vol. I, ed. K. Krishna and F. Weesner, 1–17. New York: Academic Press.

Krishna, K. 1970. Taxonomy, phylogeny, and distribution of termites. In *Biology of termites,* vol. II, ed. K. Krishna and F. Weesner, 127–152. New York: Academic Press.

Kritsky, G. R. 1979. A revision of the genus *Enicocephalus* (Hemiptera: Enicocephalidae). *Journal of Agriculture of the University of Puerto Rico* 63:91–99.

Kuno, E., and N. Hoyko, 1970. Comparative analysis of the population dynamics of rice leafhoppers, *Nephotettix cincticeps* Uhler and *Nilaparvata lugens* Stål, with special reference to natural regulation of their numbers. *Researches on Population Ecology* 12:154–84.

La Caro, F. 1982. Leaf litter disappearance in a tropical montane rain forest. Ph.D. diss., University of California, Davis.

La Caro, F., and R. L. Rudd. 1985. Leaf litter disappearance rates in Puerto Rican montane rain forest. *Biotropica* 17:269–76.

Lack, D. 1976. Island biology illustrated by the land birds of Jamaica. *Studies in Ecology 3.* Oxford: Blackwell Scientific Publications.

Laessøe, T., and D. J. Lodge. 1994. Three host-specific *Xylaria* species. *Mycologia* 86:436–46.

La Fage, J. P., and W. L. Nutting. 1978. Nutrient dynamics of termites. In *Production ecology of ants and termites,* ed. M. V. Brian, 165–232. *International Biological Programme 13.* London: Cambridge University Press.

Lahmann, E. J., and C. M. Zúñiga. 1981. Use of spider threads as resting places by tropical insects. *Journal of Arachnology* 9:339–41.

Lam, P. K. S., and D. Dudgeon. 1985. Breakdown of *Ficus fistulosa* (Moraceae) leaves in Hong Kong, with special reference to dynamics of elements and the effects of invertebrate consumers. *Journal of Tropical Ecology* 1:249–64.

Lampert, W. 1981. Inhibitory and toxic effects of blue-green algae on *Daphnia.* Internationale Revue der Gesamten. *Hydrobiologie* 66:285–98.

Langenheim, J. H. 1984. The roles of plant secondary chemicals in wet tropical ecosystems. In *The physiological ecology of plants of the wet tropics,* ed. E. Medina, H. A. Mooney, and C. Vásquez-Yánes, 189–208. The Hague: W. Junk.

Larsen, M. C., and A. J. Torres-Sánchez. 1990. Rainfall-soil moisture relations in landslide-prone areas of a tropical rain forest, Puerto Rico. In *Tropical hydrology and Caribbean water resources,* ed. J. H. Krishna, V. Quiñones-Aponte, F. Gómez-

Gómez and G. L. Morris, 121–30. *Proceedings of the International Symposium on Tropical Hydrology and Fourth Caribbean Islands Water Resources Congress.*

Lasebikan, B. A. 1974. Preliminary communication on microarthropods from a tropical rain forest in Nigeria. *Pedobiologia* 14:402–11.

Lasiewski, R. C., and W. R. Dawson. 1967. A re-examination of the relation between standard metabolic rate and body weight in birds. *Condor* 69:13–23.

La Val, R. K., and H. S. Fitch. 1977. Structure, movement, and reproduction in three Costa Rican bat communities. *Occasional Papers, Museum of Natural History, University of Kansas* 69:1–28.

Lavelle, P. 1983a. *Agastrodrilus* Omodeo and Vaillaud, a genus of carnivorous earthworms from the Ivory Coast. In *Earthworm ecology: From Darwin to vermiculture,* ed. J. E. Satchell, 425–9. New York: Chapman and Hall.

Lavelle, P. 1983b. The structure of earthworm communities. In *Earthworm ecology: From Darwin to vermiculture,* ed. J. E. Satchell, 449–66. New York: Chapman and Hall.

Lavelle, P. 1984. The soil system in the humid tropics. *Biology International* 9:2–17.

Lavelle, P., and Kohlmann, B. 1984. Étude quantitative de la macrofaune du sol dans une forêt tropicale humide du Mexique (Bonampak, Chiapas). *Pedobiologia* 27:377–93.

Lavigne, R. J. 1970a. Appendix I. Key to Formicidae (workers) of Luquillo Experimental Forest. In *The rain forest project annual report,* ed. R. G. Clements, G. E. Drewry, and R. J. Lavigne, 78–82. PRNC-147. Rio Piedras, P.R.: Puerto Rico Nuclear Center.

Lavigne, R. J. 1970b. The ecology of ants of Luquillo forest in the vicinity of El Verde field station: Preliminary work. In *The rain forest project annual report,* ed. R. G. Clements, G. E. Drewry, and R. J. Lavigne, 74–82. PRNC-147. Rio Piedras, P.R.: Puerto Rico Nuclear Center.

Lavigne, R. J. 1970c. The role of insects in the food web of the tropical wet forest: Preliminary report. In *The rain forest project annual report,* ed. R. G. Clements, G. E. Drewry, and R. J. Lavigne, 83–85. PRNC-147. Rio Piedras, P.R.: Puerto Rico Nuclear Center.

Lavigne, R. J. 1977. Notes on the ants of Luquillo Forest, Puerto Rico (Hymenoptera: Formicidae). *Proceedings of the Entomological Society of Washington* 79: 216–37.

Lavigne, R. J., and G. E. Drewry. 1970. Feeding behavior of the frogs and lizards in the tropical wet forest: Preliminary report. In *The rain forest project annual report,* ed. R. G. Clements, G. E. Drewry, and R. J. Lavigne, 64–73. PRNC-147. Rio Piedras, P.R.: Puerto Rico Nuclear Center.

Lawson, D. L., M. J. Klug, and R. W. Merritt. 1984. The influence of the physical, chemical, and microbiological characteristics of decomposing leaves on the growth of the detritivore *Tipula abdominalis* (Diptera: Tipulidae). *Canadian Journal of Zoology* 62:2339–43.

Lawton, J. H. 1989. Food webs. In *Ecological concepts,* ed. J. M. Cherrett, 43–78. Oxford: Blackwell Scientific Publications.

Lawton, J. H., and S. P. Rallison. 1979. Stability and diversity in grassland communities. *Nature* 279:351.

Lawton, J. H., and P. H. Warren. 1988. Static and dynamic explanations for patterns in food webs. *Trends in Ecology and Evolution* 3:242–5.

Layton, B. W. 1986. Reproductive chronology and habitat use by black rats (*Rattus rattus*) in Puerto Rican parrot (*Amazona vittata*) nesting habitat. M.S. thesis, Louisiana State University, Baton Rouge.

Leal, M., and R. Thomas. 1992. *Eleutherodactylus coqui* (Puerto Rican coqui) prey. *Herpetological Review* 23:79–80.

Lee, K. E. 1983. Earthworms of tropical regions: Some aspects of their ecology and relationships with soils. In *Earthworm ecology: From Darwin to vermiculture,* ed. J. E. Satchell, 179–93. New York: Chapman and Hall.

Legler, J. M., and J. L. Sullivan. 1979. The application of stomach-flushing to lizards and anurans. *Herpetologica* 35:107–10.

Lehtinen, P. T. 1967. Classification of the cribellate spiders and some allied families, with notes on the evolution of the suborder Araneomorpha. *Annales Zoologici Fennici* 4:199–468.

Leibold, M. A. 1989. Resource edibility and the effects of predators and productivity on the outcome of trophic interactions. *American Naturalist* 134:922–49.

Leigh, E. G., Jr. 1975. Structure and climate in tropical rain forest. *Annual Review of Ecology and Systematics* 6:67–85.

Leigh, E. G., Jr., and N. Smythe. 1978. Leaf production, leaf consumption and regulation of folivory on Barro Colorado Island. In *The ecology of arboreal folivores,* ed. G. G. Montgomery, 33–50. Washington, D.C.: Smithsonian Institution Press.

Lenz, J. 1992. Microbial loop, microbial food web and classical food chain: Their significance in pelagic marine ecosystems. *Arch. Hydrobiol. Beih. Ergebn. Limnol.* 37:265–78.

Lepage, M. 1974. Les termites d'une savane sahelienne (Ferlo Septentrional, Senegal): Peuplement, populations, consommation, role dans l'ecosysteme. Ph.D. diss., University of Dijon.

Lepage, M. 1984. Distribution, density and evolution of *Macrotermes bellicosus* nests (Isoptera: Macrotermitinae) in the northeast of Ivory Coast. *Journal of Animal Ecology* 53:107–17.

Levi, H. W. 1953. Observations on two species of pseudoscorpions. *Canadian Entomologist* 85:55–62.

Levi, H. W. 1955a. The spider genera *Chrysso* and *Tidarren* in America (Araneae, Theridiidae). *Journal of the New York Entomological Society* 43:59–81.

Levi, H. W. 1955b. The spider genera *Oronota* and *Stemmops* in North America and the West Indies. *Annals of the Entomological Society of America* 48:333–42.

Levi, H. W. 1957. The spider genera *Crustulina* and *Steatoda* in North America, Central America, and the West Indies (Araneae, Theridiidae). *Bulletin of the Museum of Comparative Zoology* (Harvard) 117:366–424.

Levi, H. W. 1959. The spider genus *Colesoma* (Araneae, Theridiidae). *Breviora* 110:1–8.

Levi, H. W. 1962. The genera of the spider family Theridiidae. *Bulletin of the Museum of Comparative Zoology* (Harvard) 127:1–73.

Levi, H. W. 1963a. American spiders of the genera *Audifia, Euryopis* and *Dipoena* (Araneae, Theridiidae). *Bulletin of the Museum of Comparative Zoology* (Harvard) 129:121–85.

Levi, H. W. 1963b. American spiders of the genera *Achaearanea* and the new genus *Echinotheridion* (Araneae, Theridiidae). *Bulletin of the Museum of Comparative Zoology* (Harvard) 129:187–240.

Levi, H. W. 1963c. American spiders of the genus *Theridion* (Araneae, Theridiidae). *Bulletin of the Museum of Comparative Zoology* (Harvard) 129:481–592.

Levi, H. W. 1971. The orb-weaver genus *Neoscona* in North America (Araneae, Araneidae). *Bulletin of the Museum of Comparative Zoology* (Harvard) 141: 465–500.

Levi, H. W. 1977. The American orb-weaver genera *Cyclosa, Metazygia* and *Eustala* north of Mexico (Araneae, Araneidae). *Bulletin of the Museum of Comparative Zoology* (Harvard) 148:61–127.

Levi, H. W. 1978. The American orb-weaver genera *Colphepeira, Micrathena* and *Gasteracantha* north of Mexico (Araneae, Araneidae). *Bulletin of the Museum of Comparative Zoology* (Harvard) 148:417–42.

Levi, H. W. 1980. The American orb-weaver genus *Mecynogea,* the subfamily Metinae and the genera *Pachygnatha, Glenognatha* and *Azila* of the subfamily Tetragnathinae north of Mexico (Araneae, Araneidae). *Bulletin of the Museum of Comparative Zoology* (Harvard) 149:1–74.

Levi, H. W. 1981. The American orb-weaver genera *Dolichognatha* and *Tetragnatha* north of Mexico. *Bulletin of the Museum of Comparative Zoology* (Harvard) 149:271–318.

Levi, H. W. 1986a. The Neotropical orb-weaver genera *Chrysometa* and *Homalometa* (Araneae, Tetragnathidae). *Bulletin of the Museum of Comparative Zoology* (Harvard) 151:91–215.

Levi, H. W. 1986b. The orb-weaver genus *Wixia* (Araneae, Araneidae). *Psyche* 93: 35–46.

Levi, H. W., and D. E. Randolph. 1975. A key and checklist of American spiders of the family Theridiidae north of Mexico. *Journal of Arachnology* 3:31–51.

Levi, M. P., and E. B. Cowling. 1969. Role of nitrogen in wood deterioration. VII. Physiological adaptation of wood-destroying and other fungi to substrates deficient in nitrogen. *Phytopathology* 59:460–8.

Lévieux, J. 1983. The soil fauna of tropical savannas. IV. The ants. In *Tropical savannas,* ed. F. Bourliére, 525–40. New York: Elsevier Scientific Publishing Co.

Levings, S. C. 1983. Seasonal, annual, and among-site variation in the ground ant community of a deciduous tropical forest: Some causes of patchy species distributions. *Ecological Monographs* 53:435–55.

Levings, S. C., and D. M. Windsor. 1982. Seasonal and annual variation in litter arthropod populations. In *The ecology of a tropical forest,* ed. E. G. Leigh, Jr., A. S. Rand, and D. M. Windsor, 355–87. Washington, D.C.: Smithsonian Institution Press.

Levings, S. C., and D. M. Windsor. 1985. Litter arthropod populations in a tropical deciduous forest: Relationships between years and arthropod groups. *Journal of Animal Ecology* 54:61–69.

Levins, R., M. L. Pressick, and H. Heatwole. 1973. Coexistence patterns in insular ants. *American Scientist* 61:463–72.

Lewis, A. R. 1986. Body size and growth in two populations of the Puerto Rican ground lizard (Teiidae). *Journal of Herpetology* 20:190–95.

Lewis, E. R., and P. M. Narins. 1985. Do frogs communicate with seismic signals? *Science* 227:187–9.

Lewis, J., G. Draper, C. Bourdon, C. Bowin, P. Mattson, F. Maurrasse, F. Nagle, and

G. Pardo. 1990. Geology and tectonic evolution of the northern Caribbean margin. In *The Caribbean region,* ed. G. Dengo and J. E. Case, 77–140. *The decade of North American geology.* Vol. H. Boulder, Co.: Geological Society of America.

Licht, P. 1974. Response of *Anolis* lizards to food supplementation in nature. *Copeia* 1974:215–21.

Liebermann-Jaffe, S. 1981. Ecology of web-building spiders at Corcovado National Park, Costa Rica A preliminary study. *Studies on Neotropical Fauna and Environment* 16:99–106.

Liebherr, J. K., ed. 1988. *Zoogeography of Caribbean insects.* Ithaca, N.Y.: Comstock Publishing Assoc.

Light, S. F. 1933. Termites of western Mexico. *University of California Publications in Entomology* 6. Pt. 5, 79–164. Berkeley: University of California Press.

Lindeman, R. L. 1941. Seasonal food cycle dynamics in a senescent lake. *American Midland Naturalist* 26:636–73.

Lindeman, R. L. 1942. The trophic-dynamic aspect of ecology. *Ecology* 23:399–418.

Linsley, E. G. 1959. Ecology of Cerambycidae. *Annual Review of Entomology* 4: 99–138.

Linsley, E. G. 1961. The Cerambycidae of North America. Part I. Introduction. *University of California Publications in Entomology 18.* Berkeley: University of California Press.

Liogier, A. 1965. Nomenclature changes and additions to Britton and Wilson's "Flora of Porto Rico and the Virgin Islands." *Rhodora* 67:315–61.

Lister, B. C. 1976. The nature of niche expansion in West Indian *Anolis* lizards. I. Ecological consequences of reduced competition. *Evolution* 30:659–76.

Lister, B. C. 1981. Seasonal niche relationships of rain forest anoles. *Ecology* 62: 1548–60.

Little, E. L., Jr., and F. H. Wadsworth. 1964. *Common Trees of Puerto Rico and the Virgin Islands.* U.S. Department of Agriculture. Agriculture Handbook No. 249. Washington, D.C.: U.S. Government Printing Office.

Little, E. L., Jr., and R. O. Woodbury. 1976. Trees of the Caribbean National Forest, Puerto Rico. U.S. Forest Service. Institute of Tropical Forestry. Research Paper ITF-20. Río Piedras, P.R.: U.S. Forest Service. Institute of Tropical Forestry.

Little, E. L., R. O. Woodbury, and F. H. Wadsworth. 1974. Trees of Puerto Rico and the Virgin Islands. Agricultural Handbook no. 449. U.S. Department of Agriculture, Forest Service, Washington, D.C.

Lloyd, J. E. 1965. Aggressive mimicry in *Photinus:* Firefly femmes fatales. *Science* 149:653–4.

Lodge, D. J. 1987a. Nutrient concentrations, percentage moisture and density of field-collected fungal mycelia. *Soil Biology and Biochemistry* 19:727–33.

Lodge, D. J. 1987b. Resurvey of mycorrhizal associations in the El Verde rain forest in Puerto Rico. In *Mycorrhizae in the next decade: Practical applications research priorities,* ed. D. M. Sylvia, L. L. Hung, and J. H. Graham, 127. *Proceedings of the 7th North American Conference on Mycorrhizae,* May 3–8, 1987, Gainesville: University of Florida. Institute of Food and Agricultural Sciences.

Lodge, D. J. 1988. Three new species of *Mycena* (Basidiomycota: Tricholomataceae) from Puerto Rico. *Transactions of the British Mycological Society* 91:109–16.

Lodge, D. J. 1993. Nutrient cycling by fungi in wet tropical forests. In *Aspects of*

Tropical Mycology, ed. S. Isaac, J. C. Frankland, R. Watling, and A. J. S. Whalley. *British Mycological Society Symposium Series* 19:37–58. Cambridge: Cambridge University Press.

Lodge, D. J., and C. E. Asbury. 1988. Basidiomycetes reduce export of organic matter from forest slopes. *Mycologia* 80:888–90.

Lodge, D. J., and E. R. Ingham. 1991. A comparison of agar film techniques for estimating fungal biovolumes in litter and soil. *Agriculture, Ecosystems and Environment* 34:131–44.

Lodge, D. J., and D. N. Pegler. 1990. The Hygrophoraceae of the Luquillo Mountains of Puerto Rico. *Mycological Research* 94:443–56.

Lodge, D. J., F. N. Scatena, C. E. Asbury, and M. J. Sánchez. 1991. Fine litterfall and related nutrient inputs resulting from Hurricane Hugo in subtropical wet and lower montane rain forests of Puerto Rico. *Biotropica* 23:336–42.

Lodge, D. J., W. H. McDowell, and C. P. McSwiney. 1994. The importance of nutrient pulses in tropical forests. *Trends in Ecology and Evolution* 9:384–7.

Loiselle, B. A., and W. G. Hoppes. 1983. Nest predation in insular and mainland lowland rainforest in Panama. *Condor* 85:93–95.

Loman, J. 1979. Food, feeding rates and prey-size selection in juvenile and adult frogs, *Rana arvalis* Nilss. and *R. temporaria* L. *Ekologia Polska* 27:581–601.

Losos, J. B., M. R. Gannon, W. J. Pfeiffer, and R. B. Waide. 1990. Notes on the ecology and behavior of *Anolis cuvieri* (Lacertilia: Iguanidae) in Puerto Rico. *Caribbean Journal of Science* 26:65–66.

Lowman, M. D. 1984. An assessment of techniques for measuring herbivory: Is rainforest defoliation more intense than we thought? *Biotropica* 16:264–8.

Lowman, M. D. 1985a. Insect herbivory in Australian rain forests: Is it higher than in the Neotropics? *Proceedings of the Ecological Society of Australia* 14:109–19.

Lowman, M. D. 1985b. Temporal and spatial variability in insect grazing of the canopies of five Australian rainforest tree species. *Australian Journal of Ecology* 10:7–24.

Lubin, Y. D. 1978. Seasonal abundance and diversity of web-building spiders in relation to habitat structure on Barro Colorado Island, Panamá. *Journal of Arachnology* 6:31–51.

Lubin, Y. D., and O. P. Young. 1977. Food resources of anteaters (Edentata: Myremecophagidae). 1. A year's census of arboreal nests of ants and termites on Barro Colorado Island, Panama Canal Zone. *Biotropica* 9:26–36.

Lubin, Y. D., W. G. Eberhard, and G. G. Montgomery. 1978. Webs of *Miagrammopes* (Araneae: Uloboridae) in the Neotropics. *Psyche* 85:1–23.

Lugo, A. E. 1974. Evaluation of research and applications in the Neotropics. In *Fragile ecosystems,* ed. E. G. Farnsworth and F. B. Golley, 67–111. New York: Springer-Verlag.

Lugo, A. E. 1986. Water and the ecosystems of the Luquillo Experimental Forest. U.S. Forest Service. Southern Forest Experiment Station. General Technical Report SO-63. New Orleans, LA.

Lugo, A. E. 1987. Are island ecosystems different from continental ecosystems? *Acta Cientifica* 1:48–54.

Lugo, A. E. 1988. Estimating reductions in the diversity of tropical forest species. In *Biodiversity,* ed. E. O. Wilson and F. M. Peter, 58–70. Washington, D.C.: National Academy Press.

Lugo, A. E., and S. Brown. 1981a. Ecological monitoring in the Luquillo Forest Reserve. *Ambio* 10:102–7.
Lugo, A. E., and S. Brown. 1981b. Tropical lands: Popular misconceptions. *Mazingira* 5:10–19.
Lugo, A. E., and S. Brown. 1981c. Tropical lands: popular misconceptions. *Mazingira* 5:11–19.
Lugo, A. E., and J. L. Frangi. 1993. Fruit fall in the Luquillo Experimental Forest. *Biotropica* 25:73–84.
Lugo, A. E., J. A. Gonzalez-Liboy, B. Cintrón, and K. Dugger. 1978. Structure, productivity, and transpiration of a subtropical dry forest in Puerto Rico. *Biotropica* 10:278–91.
Luxton, M. 1982a. General ecological influence of the soil fauna on decomposition and nutrient circulation. *Oikos* 39:355–7.
Luxton, M. 1982b. Quantitative utilization of energy by the soil fauna. *Oikos* 39: 342–54.
Luxton, M. 1982c. Substrate utilization by the soil fauna. *Oikos* 39:340–1.
Lynch, J. F. 1991. Effects of Hurricane Gilbert on birds in a dry tropical forest in the Yucatan Peninsula. *Biotropica* 23:488–96.
Lyons, J., and D. W. Schneider. 1990. Factors influencing fish distribution and community structure in a small coastal river in southwestern Costa Rica. *Hydrobiologia* 203:1–14.
MacArthur, R. H. 1955. Fluctuations of animal populations, and a measure of community stability. *Ecology* 36:533–6.
MacArthur, R. H. 1972. *Geographical ecology: Patterns in the distribution of species.* New York: Harper and Row.
MacArthur, R. H., and E. O. Wilson. 1967. *The theory of island biogeography.* Princeton, N.J.: Princeton University Press.
MacArthur, R. H., J. M. Diamond, and J. R. Karr. 1972. Density compensation in island faunas. *Ecology* 53:330–42.
MacArthur, R., H. Recher, and M. Cody. 1966. On the relation between habitat selection and species diversity. *American Naturalist* 100:319–32.
MacArthur, R. H., J. MacArthur, D. MacArthur, and A. MacArthur. 1973. The effect of island area on population densities. *Ecology* 54:657–8.
Macfadyen, A. 1963. The contribution of the microfauna to total soil metabolism. In *Soil organisms: Proceedings of the colloquium on soil fauna, soil microflora, and their relationships,* ed. J. Doeksen and J. van der Drift, 3–17. Amsterdam: North-Holland Publishing Co.
Malcolm, J. R. 1990. Estimation of mammalian densities in continuous forest north of Manaus. In *Four Neotropical rainforests,* ed. A. H. Gentry, 339–57. New Haven, Conn.: Yale University Press.
Maldonado, A. A., and E. Ortiz. 1966. Electrophoretic patterns of serum proteins of some West Indian *Anolis* (Sauria, Iguanidae). *Copeia* 1966:179–82.
Maldonado-Capriles, J. 1969. The Miridae of Puerto Rico. Puerto Rico Agricultural Experiment Station. Technical Paper 45. Rio Piedras, P.R.: University of Puerto Rico. Agricultural Experiment Station.
Mann, P., K. Burke, and T. Matumoto. 1984. Neotectonics of Hispaniola: Plate motion, sedimentation, and seismicity at a restraining bend. *Earth and Planetary Science Letters* 70:311–24.

Mares, M. A., and D. Williams. 1977. Experimental support for food particle size resource allocations in heteromyid rodents. *Ecology* 58:1186–90.

Mares, M. A., M. R. Willig, K. E. Streilein, and T. E. Lacher, Jr. 1981. The mammals of northeastern Brazil: A preliminary assessment. *Annals of the Carnegie Museum* 50:81–137.

Mari Mutt, J. A. 1976. The genera of Collembola (Insecta) in Puerto Rico. *Journal of Agriculture of the University of Puerto Rico* 60:112–28.

Mari Mutt, J. A. 1987. Puerto Rican species of Paronellidae (Insecta: Collembola). *Caribbean Journal of Science* 23:400–416.

Marquis, R. J. 1984. Leaf herbivores decrease fitness of a tropical plant. *Science* 226:537–9.

Martínez, N. D. 1991. Artifacts or attributes? Effects of resolution on the Little Rock Lake food web. *Ecological Monographs* 61:367–92.

Martínez, N. D. 1992. Constant connectance in community food webs. *American Naturalist* 139:1208–18.

Martínez, N. D. 1993. Effects of scale on food web structure. *Science* 260:242–3.

Martorell, L. F. 1941. Some notes on forest entomology, Part IV. *Caribbean Forester* 2:80–82.

Martorell, L. F. 1945. A survey of the forest insects of Puerto Rico. Part I–II. *Journal of Agriculture of the University of Puerto Rico* 29:70–608.

Martorell, L. F. 1955. *Acanthochila spinacosta* Van Duzee, a new tingitid for Puerto Rico. *Journal of Agriculture of the University of Puerto Rico* 39:47–48.

Martorell, L. F. 1963. *Leptopharsa constricta* (Champion) a new lacewing bug (Tingidae: Hemiptera) for Puerto Rico. *Journal of Agriculture of the University of Puerto Rico* 47:56.

Martorell, L. F. 1975. *Annotated food plant catalog of the insects of Puerto Rico.* Mayaqüez: University of Puerto Rico. Agricultural Experiment Station.

Mascle, A., and P. Letouzey. 1990. Geological map of the Caribbean. Institut Francais de Pétrole. Paris: Editions Technip.

Mason, W. H., and E. P. Odum. 1969. The effect of coprophagy on retention and bioelimination of radionuclides by detritus-feeding animals. In *Proceedings of the Second National Symposium on Radioecology,* ed. D. J. Nelson and F. C. Evans, 721–74. Washington, D.C.: U.S. Atomic Energy Commission.

Masteller, E. C., and K. M. Buzby. 1993. Composition and temporal abundance of aquatic insect emergence from a tropical rainforest stream at El Verde, Puerto Rico. *Journal of the Kansas Entomological Society* 66:133–9.

Matsumoto, T. 1976. The role of termites in an equatorial rain forest ecosystem of West Malaysia. I. Population density, biomass, carbon, nitrogen, and calorific content and respiration rate. *Oecologia* 22:153–78.

Matthews, E. G. 1965. The taxonomy, geographical distribution and feeding habits of the Canthonines of Puerto Rico. *Transactions of the American Entomological Society* 91:431–65.

Matthews, E. G. 1966. A taxonomic and zoogeographic survey of the Scarabaeinae of the Antilles (Coleoptera: Scarabaeidae). *Memoirs of the American Entomological Society* 21:1–134.

Mattson, W. J. 1980. Herbivory in relation to plant nitrogen content. *Annual Review of Ecology and Systematics* 11:119–61.

Mattson, W. J., and N. D. Addy. 1975. Phytophagous insects as regulators of forest primary production. *Science* 190:515–22.

Mattson, W. J., and R. A. Haack. 1987. The role of drought in outbreaks of plant-eating insects. *BioScience* 37:110–18.

Maurin, M., and J. Levieux. 1984. Données sur le peuplement en Diplopodes d'une forêt tropicale humide après un arasement rècent: Comparaison avec une plantation industrielle. *Revue d'Écologie et de Biologie du Sol* 21:21–35.

May, R. M. 1972. Will a large complex system be stable? *Nature* 238:413–4.

May, R. M. 1973. Stability and complexity in model ecosystems. *Monographs in Population Biology,* 6. Princeton, N.J.: Princeton University Press.

May, R. M. 1974. *Stability and complexity in model ecosystems.* Princeton, N.J.: Princeton University Press.

May, R. M. 1983a. Parasite infections as regulators of animal populations. *American Scientist* 71:36–45.

May, R. M. 1983b. The structure of food webs. *Nature* 301:566–8.

May, R. M. 1988. How many species are there on earth? *Science* 241:1441–9.

McBrayer, J. F. 1973. Exploitation of deciduous leaf litter by *Apheloria montana* (Diplopoda: Eurydesmidae). *Pedobiologia* 13:90–98.

McBrayer, J. F., and D. E. Reichle. 1971. Trophic structure and feeding rates of forest soil invertebrate populations. *Oikos* 22:381–8.

McCormick, S., and G. A. Polis. 1982. Arthropods that prey on vertebrates. *Biological Reviews of the Cambridge Philosophical Society* 57:29–58.

McDowell, W. H., and C. E. Asbury. 1994. Export of carbon, nitrogen, and major ions from three tropical montane watersheds. *Limnology and Oceanography* 39:111–25.

McDowell, W. H., and A. Estrada-Pinto. 1988. Rainfall at the El Verde Field Station, 1964–1986. CEER-T-228. Rio Piedras, P.R.: Center for Energy and Environment Research.

McDowell, W. H., C. Gines-Sanchez, C. E. Asbury, and C. R. Perez. 1990. Influence of sea salt aerosols and long range transport on precipitation chemistry at El Verde, Puerto Rico. *Atmospheric Environment* 24A:2813–21.

McElravy, E., V. H. Resh, H. Wolda, and O. S. Flint, Jr. 1981. Diversity of Trichoptera in a "non-seasonal" tropical environment. In *Proceedings of the Third International Symposium on Trichoptera,* ed. G. Moretti, 149–56. The Hague: W. Junk.

McElravy, E., H. Wolda, and V. H. Resh. 1982. Seasonality and annual variability of caddisfly adults (Trichoptera) in a "non-seasonal" tropical environment. *Archiv für Hydrobiologie* 94:302–17.

McGill, W. B., E. A. Paul, J. A. Shields, and W. E. Lowe. 1973. Turnover of microbial populations and their metabolites in soil. In *Modern methods in the study of microbial ecology,* ed. T. Rosswall, 293–301. Bulletins from the Ecological Research Committee, NFR no. 17. Stockholm: Statens Naturvetenskapliga Forskningsråd.

McIlveen, W. D., and H. Cole, Jr. 1976. Spore dispersal of Endogonaceae by worms, ants, wasps, and birds. *Canadian Journal of Botany* 54:1486–9.

McKaye, W. 1980. Comments on breeding biology of *Gobiomorus dormitor. Copeia* 3:542–4.

McKey, D. 1975. The ecology of coevolved seed dispersal systems. In *Coevolution of animals and plants,* ed. L. E. Gilbert and P. H. Raven, 159–91. Austin: University of Texas Press.

McLaughlin, J. F., and J. Roughgarden. 1989. Avian predation on *Anolis* lizards in the northeastern Caribbean: An inter-island contrast. *Ecology* 70:617–28.

McMahan, E. A. 1970a. Polyethism in workers of *Nasutitermes costalis* (Hölmgren). *Insectes Sociaux* 17:113–20.

McMahan, E. A. 1970b. Radiation and the termites at El Verde. In *A tropical rain forest: A study of irradiation and ecology at El Verde, Puerto Rico,* ed. H. T. Odum and R. F. Pigeon, E105–E122. Oak Ridge, Tenn.: U.S. Atomic Energy Commission.

McMahan, E. A. 1970c. The termites at El Verde: 1969 Survey. In *The rain forest project annual report,* ed. R. G. Clements, G. E. Drewry, and R. J. Lavigne, 111–17. PRNC-147. Rio Piedras: Puerto Rico Nuclear Center.

McMahan, E. A. 1974. Non-aggressive behavior of large soldiers of *Nasutitermes exitiosus* (Hill). (Isoptera: Termitidae). *Insectes Sociaux* 21:95–106.

McMahan, E. A., and C. M. Blanton. 1993. Effects of Hurricane Hugo on a population of the termite *Nasutitermes costalis* in the Luquillo Experimental Forest, Puerto Rico. *Caribbean Journal of Science* 29:202–8.

McMahan, E. A., and N. E. Sollins. 1970. Diversity of microarthropods after irradiation. In *A tropical rain forest: A study of irradiation and ecology at El Verde, Puerto Rico,* ed. H. T. Odum and R. F. Pigeon, E151–E158. Oak Ridge, Tenn.: U.S. Atomic Energy Commission.

McMillan, J. H. 1975. Interspecific and seasonal analyses of the gut contents of three Collembola. *Revue d'Écologie et de Biologie du Sol* 12:449–57.

McNaughton, S. J. 1993. Biodiversity and function of grazing ecosystems. In *Biodiversity and ecosystem function,* ed. E.-D. Schulze and H. A. Mooney, 361–83. New York: Springer-Verlag.

McQueen, D. J., M. R. S. Johannes, J. R. Post, T. J. Stewart, and D. R. S. Lean. 1989. Bottom-up and top-down impacts on freshwater pelagic community structure. *Ecological Monographs* 1989.

Medellin, R. A. 1993. Estructura y diversidad de una comunidad de murciélagos en el tropico humido mexicano. In *Avances en el estudio de los mamíferos de México,* ed. R. A. Medellin and G. Ceballos, 333–54. Publicaciones Especiales, vol. 1, Associacion Mexicana de Mastozoologia, Distrito Federal, México.

Medina, E., and H. Klinge. 1983. Productivity of tropical forests and tropical woodlands. In *Encyclopedia of plant physiology.* Vol. 12D, *Physiological plant ecology IV,* ed. O. L. Lange, P. S. Nobel, C. B. Osmond, and H. Ziegler, 281–303. Berlin: Springer-Verlag.

Medina, E., E. Cuevas, and P. L. Weaver. 1981. Composición foliar y transpiración de especies leñosas de Pico del Este, Sierra de Luquillo, Puerto Rico. *Acta Cientifica Venezolana* 32:159–65.

Medina-Gaud, S. 1961. The Thysanoptera of Puerto Rico. Puerto Rico Agricutural Experiment Station. Technical Paper 32. Rio Piedras: University of Puerto Rico. Agricultural Experimental Station.

Medina-Gaud, S. 1963. A new species of *Heterothrips* (Thysanoptera: Heterothripidae) from Puerto Rico. *Journal of Agriculture of the University of Puerto Rico* 45:164–68.

Menge, B. A., and J. P. Sutherland. 1976. Species diversity gradients: Synthesis of the roles of predation, competition, and temporal heterogeneity. *American Naturalist* 110:351–69.

Mercado, N. 1970. Leaf growth, leaf survival, leaf holes, color of cambium, and terminal bud condition. In *A tropical rain forest: A study of irradiation and ecology at El Verde, Puerto Rico,* ed. H. T. Odum and R. F. Pigeon, D271–D286. Oak Ridge, Tenn.: U.S. Atomic Energy Commission.

Meserve, P. L. 1977. Three-dimensional home ranges of cricetid rodents. *Journal of Mammalogy* 58:549–58.

Michaud, E. J., and J. R. Dixon. 1989. Prey items of 20 species of the Neotropical snake genus *Liophis. Herpetological Review* 20:39–41.

Miller, R. R. 1966. Geographical distribution of Central American freshwater fishes. *Copeia* 1966:773–802.

Miller, R. R. 1982. Pisces. In *Aquatic biota of Mexico, Central America, and the West Indies,* ed. S. H. Hurlbert and A. Villalobos-Figueroa, 486–501. San Diego, Calif.: S. H. Hurlbert, Dept. of Biology, San Diego State University.

Miller, R. R., and M. L. Smith. 1986. Origin and geography of the fishes of Central Mexico. In *The zoogeography of North American freshwater fishes,* ed. C. H. Hocutt and E. O. Wiley, 487–517. New York: John Wiley & Sons.

Milstead, W. W. 1957. Some aspects of competition in natural populations of whiptail lizards (genus *Cnemidophorus*). *Texas Journal of Science* 9:410–47.

Milstead, W. W., A. S. Rand, and M. M. Stewart. 1974. Polymorphism in cricket frogs: An hypothesis. *Evolution* 28:489–91.

Misra, R. 1968. Energy transfer along terrestrial food chains. *Tropical Ecology* 9: 105–18.

Mitchell, M. J. 1977. Population dynamics of oribatid mites (Acari, Cryptostigmata) in an aspen woodland soil. *Pedobiologia* 17:305–19.

Mittebach, G. C. 1988. Competition among refuging sunfishes and effects of fish density on littoral zone invertebrates. *Ecology* 69:614–23.

Mittler, T. E. 1958. The excretion of honeydew by *Tuberolachnus salignus* (Gmelin) (Homoptera: Aphididae). *Proceedings of the Royal Entomological Society of London* Series A 33:49–55.

Mittler, T. E. 1970. Uptake rates of plant sap and synthetic diet by the aphid *Myzus persicae. Annals of the Entomological Society of America* 63:1701–5.

Miyashita, K. 1969. Effects of the locomotor activity, temperature and hunger on the respiratory rate of *Lycosa* T-*insignata* Boes. et Str. (Araneae: Lycosidae). *Applied Entomology and Zoology* 4:105–13.

Moermond, T. C. 1973. Patterns of habitat utilization in *Anolis* lizards. Ph.D. Diss., Harvard University, Cambridge, Mass.

Moermond, T. C. 1979a. Habitat constraints on the behavior, morphology, and community structure of *Anolis* lizards. *Ecology* 60:152–64.

Moermond, T. C. 1979b. The influence of habitat structure on *Anolis* foraging behavior. *Behavior* 70:147–67.

Moll, A. G. 1978. Abundance studies on the *Anolis* lizards and insect populations in altitudinally different tropical forest habitats. CEER-11. Rio Piedras, P.R.: Center for Energy and Environment Research.

Molofsky, J., C. A. S. Hall, and N. Myers. 1986. A comparison of tropical forest surveys. TR032, DOE/NBB-0078, Washington, D.C.: U.S. Dept. of Energy.

Mooney, H. A., O. Björkman, A. E. Hall, E. Medina, and P. B. Tomlinson. 1980. The study of the physiological ecology of tropical plants: Current status and needs. *BioScience* 30:22–26.

Moore, A. M., and L. Burns. 1970. Appendix C: Preliminary observations on the earthworm populations of the forest soils of El Verde. In *A tropical rain forest: A study of irradiation and ecology at El Verde, Puerto Rico,* ed. H. T. Odum and R. F. Pigeon, I283–I284. Oak Ridge, Tenn.: U.S. Atomic Energy Commission.

Moore, B. P. 1969. Biochemical studies in termites. In *Biology of termites.* Vol. I, ed. K. Krishna and F. Weesner, 407–32. New York: Academic Press.

Moore, J. C., and P. C. de Ruiter. 1991. Temporal and spatial heterogeneity of trophic interactions within below-ground food webs. *Agriculture, Ecosystems and Environment* 34:371–97.

Moore, J. C., D. C. Walter, and W. H. Hunt. 1988. Arthropod regulation of micro- and mesobiota in below-ground detrital food webs. *Annual Review Entomology* 33:419–39.

Moore, J. C., P. C. de Ruiter, H. W. Hunt. 1993. Influence of productivity on the stability of real and model ecosystems. *Science* 261:906–8.

Morgan, G. S., and C. A. Woods. 1986. Extinction and the zoogeography of West Indian land mammals. *Biological Journal of the Linnean Society* 28:167–203.

Moriarty, C. 1978. Eels: A natural and unnatural history. New York: Universe Books.

Morrison, D. W. 1978a. Foraging ecology and energetics of the frugivorous bat *Artibeus jamaicensis. Ecology* 59:716–23.

Morrison, D. W. 1978b. Influence of habitat on the foraging distances of the fruit bat, *Artibeus jamaicensis. Journal of Mammalogy* 59:622–4.

Morrison, D. W. 1978c. Lunar phobia in a Neotropical fruit bat, *Artibeus jamaicensis* (Chiroptera: Phyllostomidae). *Animal Behaviour* 26:852–5.

Morrison, D. W. 1978d. On the optimal searching strategy of refuging predators. *American Naturalist* 112:925–34.

Morrison, D. W. 1979. Apparent male defense of tree hollows in the fruit bat, *Artibeus jamaicensis. Journal of Mammalogy* 60:11–15.

Morrison, D. W. 1980. Efficiency of food utilization by fruit bats. *Oecologia* 45: 270–3.

Morrow, P. A. 1984. Assessing the effects of herbivory. In *The physiological ecology of plants of the wet tropics,* ed. E. Medina, H. A. Mooney, and C. Vásquez-Yánes, 225–31. The Hague: W. Junk.

Moulder, B. C., and D. E. Reichle. 1972. Significance of spider predation in the energy dynamics of forest-floor arthropod communities. *Ecological Monographs* 42:473–98.

Moyle, P. B., and F. R. Senanayake. 1984. Resource partitioning among the fishes of rainforest streams in Sri Lanka. *Journal of Zoology* 202:195–224.

Mueller-Dumbois, D., and F. G. Howarth. 1981. Niche and life-form integration in island communities. In *Island ecosystems: Biological organization in selected Hawaiian communities,* ed. D. Mueller-Dumbois, K. W. Bridges, and H. L. Carson, 337–64. Stroudsburg, Pa.: Hutchinson Ross Publishing Co.

Mulkern, G. B. 1967. Food selection of grasshoppers. *Annual Review of Entomology* 12:59–79.

Muma, M. H. 1967. *Arthropods of Florida and neighboring land areas: Scorpions, whip scorpions and wind scorpions of Florida.* Gainesville: Florida Department of Agriculture.

Muniz-Melendez, E. 1978. Demographic analysis of the life history of *Inga vera* subspecies *vera.* M.S. thesis, University of Tennessee, Knoxville.

Myers, N. 1983. Conversion rates in tropical moist forests. In *Tropical rain forest ecosystems: Structure and function,* ed. F. B. Golley, 289–300. Amsterdam: Elsevier Scientific Publishing Co.

Nakahara, S., and C. E. Miller. 1981. A list of the Coccoidea species (Homoptera) of Puerto Rico. *Proceedings of the Entomological Society of Washington* 83:28–39.

Narins, P. M., and S. L. Smith. 1986. Clinal variation in anuran advertisement calls: Basis for acoustic isolation? *Behavioral Ecology and Sociobiology* 19:135–41.

Nellis, D. W. 1971. Additions of the natural history of *Brachyphylla* (Chiroptera). *Caribbean Journal of Science* 11:91.

Nellis, D. W., and C. P. Ehle. 1977. Observations on the behavior of *Brachyphylla cavernarum* (Chiroptera) in the Virgin Islands. *Mammalia* 41:403–9.

Nelson, C. D. 1964. The production and translocation of photosynthate C^{14} in conifers. In *Formation of wood in forest trees,* ed. M. H. Zimmermann, 243–57. New York: Maria Moors Cabot Foundation.

Neuhauser, E. F., and R. Hartenstein. 1978. Phenolic content and palatability of leaves and wood to soil isopods and diplopods. *Pedobiologia* 18:99–109.

Nevling, L. I., Jr. 1971. The ecology of an elfin forest in Puerto Rico. 16. The flowering cycle and an interpretation of its seasonality. *Journal of the Arnold Arboretum* 52:586–613.

Newell, K. 1984. Interaction between two decomposer Basidiomycetes and a collembolan under Sitka spruce: Grazing and its potential effects on fungal distribution and litter decomposition. *Soil Biology and Biochemistry* 16:235–9.

Newell, S. Y., and A. Statzell-Tallman. 1982. Factors for conversion of fungal biovolume values to biomass, carbon and nitrogen: Variation with mycelial ages, growth conditions, and starins of fungi from a salt marsh. *Oikos* 39:261–8.

Newman, C. M., and J. E. Cohen. 1986. A stochastic theory of community food webs. IV. Theory of food chain lengths in large webs. *Proceedings of the Royal Society of London* B 228:354–77.

Newman, E. I. 1988. Mycorrhizal links between plants: Functioning and ecological significance. *Advances in Ecological Research* 18:243–70.

Newman, R. M., W. C. Kerfoot, and Z. Hansom, III. 1990. Watercress and amphipods: Potential chemical defense in a spring stream macrophyte. *Journal of Chemical Ecology* 16:245–59.

Nicholson, P. B., K. L. Bocock, and O. W. Neal. 1966. Studies on the decomposition of the faecal pellets of a millipede [*Glomeris marginata* (Villers)]. *Journal of Ecology* 54:755–66.

Nielsen, B. O. 1978. Above ground food resources and herbivory in a beech forest ecosystem. *Oikos* 31:273–9.

Norton, R. A. 1985. Aspects of the biology and systematics of soil arachnids, particularly saprophagous and mycophagous mites. *Quaestiones Entomologicae* 21: 523–31.

Novo, J., A. R. Estrada, and G. Alayon. 1985. *Eleutherodactylus* (Anura: Leptodactylidae) depredado por un araneido. *Miscelanea Zoologica Academia Ciencias de Cuba* 28:1–2.

Nowak, R. S., and M. M. Caldwell. 1984. A test of compensatory photosynthesis in the field: Implications for herbivory tolerance. *Oecologia* 61:311–18.

O'Brien, L. B., and S. W. Wilson. 1985. Planthopper systematics and external morphology. In *The leafhoppers and planthoppers,* ed. L. R. Nault and J. G. Rodriguez, 61–102. New York: John Wiley & Sons.

O'Neill, R. V. 1989. Perspectives in hierarchy and scale. In *Perspectives in ecological theory,* ed. J. Roughgarden, R. M. May, and S. A. Levin, 140–56. Princeton, N.J.: Princeton University Press.

O'Neill, R. V., D. L. DeAngelis, J. B. Waide, and T. F. H. Allen. 1986. A hierarchical concept of ecosystems. *Monographs in Population Ecology, 23.* Princeton, N.J.: Princeton University Press.

Odum, E. P. 1959. *Fundamentals of ecology.* 2d ed. Philadelphia: Saunders.

Odum, E. P. 1971. *Fundamentals of ecology.* 3d ed. Philadelphia: Saunders.

Odum, H. T. 1956. Efficiencies, size of organisms, and community structure. *Ecology* 37:592–7.

Odum, H. T. 1957. Trophic structure and productivity of Silver Springs, Florida. *Ecological Monographs* 27:55–112.

Odum, H. T. 1970a. The El Verde study area and the rain forest systems of Puerto Rico. In *A tropical rain forest: A study of irradiation and ecology at El Verde, Puerto Rico,* ed. H. T. Odum and R. F. Pigeon, B3–B32. Oak Ridge, Tenn.: U.S. Atomic Energy Commission.

Odum, H. T. 1970b. Rain forest structure and mineral-cycling homeostasis. In *A tropical rain forest: A study of irradiation and ecology at El Verde, Puerto Rico,* ed. H. T. Odum and R. F. Pigeon, H3–H52. Oak Ridge, Tenn.: U.S. Atomic Energy Commission.

Odum, H. T. 1970c. Summary: An emerging view of the ecological system at El Verde. In *A tropical rain forest: A study of irradiation and ecology at El Verde, Puerto Rico,* ed. H. T. Odum and R. F. Pigeon, I191–I281. Oak Ridge, Tenn.: U.S. Atomic Energy Commission.

Odum, H. T. 1983. *Systems ecology: An introduction.* New York: John Wiley & Sons.

Odum, H. T., and C. F. Jordan. 1970. Metabolism and evapotranspiration of the lower forest in a giant plastic cylinder. In *A tropical rainforest: A study of irradiation and ecology at El Verde, Puerto Rico,* ed. H. T. Odum and R. F. Pigeon, I165–I189. Oak Ridge, Tenn.: U.S. Atomic Energy Commission.

Odum, H. T., and J. Ruiz-Reyes. 1970. Holes in leaves and the grazing control mechanism. In *A tropical rain forest: A study of irradiation and ecology at El Verde, Puerto Rico,* ed. H. T. Odum and R. F. Pigeon, I69–I80. Oak Ridge, Tenn.: U.S. Atomic Energy Commission.

Odum, H. T., and R. F. Pigeon, eds. 1970. *A tropical rain forest: A study of irradiation and ecology at El Verde, Puerto Rico.* Oak Ridge, Tenn.: U.S. Atomic Energy Commission.

Odum, H. T., B. J. Copeland, and R. Z. Brown. 1963. Direct and optical assay of leaf mass of the lower montane rain forest of Puerto Rico. *Proceedings of the National Academy of Sciences of the United States of America* 49:429–34.

Odum, H. T., W. Abbott, R. K. Selander, F. B. Golley, and R. F. Wilson. 1970a.

Estimates of chlorophyll and biomass of the tabonuco forest of Puerto Rico. In *A tropical rain forest: A study of irradiation and ecology at El Verde, Puerto Rico,* ed. H. T. Odum and R. F. Pigeon, I3–I19. Oak Ridge, Tenn.: U.S. Atomic Energy Commission.

Odum, H. T., G. Drewry, and J. R. Kline. 1970b. Climate at El Verde, 1963–1966. In *A tropical rain forest: A study of irradiation and ecology at El Verde, Puerto Rico,* ed. H. T. Odum and R. F. Pigeon, B347–B418. Oak Ridge, Tenn.: U.S. Atomic Energy Commission.

Odum, H. T., G. Drewry, and E. A. McMahan. 1970c. Introduction to Section E. In *A tropical rain forest: A study of irradiation and ecology at El Verde, Puerto Rico,* ed. H. T. Odum and R. F. Pigeon, E3–E15. Oak Ridge, Tenn.: U.S. Atomic Energy Commission.

Odum, H. T., A. Lugo, G. Cintrón, and C. F. Jordan. 1970d. Metabolism and evapotranspiration of some rain forest plants and soil. In *A tropical rain forest: A study of irradiation and ecology at El Verde, Puerto Rico,* ed. H. T. Odum and R. F. Pigeon, I103–I164. Oak Ridge, Tenn.: U.S. Atomic Energy Commission.

Ogle, C. J. 1970. Pollen analysis of selected sphagnum-bog sites in Puerto Rico. In *A tropical rain forest: A study of irradiation and ecology at El Verde, Puerto Rico,* ed. H. T. Odum and R. F. Pigeon, B135–B145. Oak Ridge, Tenn.: U.S. Atomic Energy Commission.

Ohiagu, C. E. 1979. Nest and soil populations of *Trinervitermes* spp. with particular reference to *T. geminatus* (Wasmann), (Isoptera) in Southern Guinea savanna near Mokwa, Nigeria. *Oecologia* 40:167–78.

Oksanen, L., S. D. Fretwell, J. Arruda, and P. Niemela. 1981. Exploitation ecosystems in gradients of primary productivity. *American Naturalist* 118:240–61.

Olson, J. S., J. A. Watts, and L. J. Allison. 1983. *Carbon in live vegetation of major world ecosystems.* TR004, DOE/NBB-0037, Washington, D.C.: U.S. Dept. of Energy.

Opell, B. D. 1979. Revision of the genera and tropical American species of the spider family Uloboridae. *Bulletin of the Museum of Comparative Zoology* (Harvard) 148:443–549.

Opell, B. D. 1981. New Central and South American Uloboridae (Arachnida, Araneae). *Bulletin of the American Museum of Natural History* 170:219–28.

Opell, B. D. 1984. Phylogenetic review of the genus *Miagrammopes* (*sensu lato*) (Araneae, Uloboridae). *Journal of Arachnology* 12:229–40.

Orians, G. H. 1969. The number of bird species in some tropical forests. *Ecology* 50:793–801.

Ortiz Carrasquillo, W. 1980. Resumen historico de la introducción de los peces de agua dulce en los lagos artificiales de Puerto Rico desde 1915 hasta 1975. *Science-Ciencia* 7:95–107.

Osborn, H. 1935. Insects of Porto Rico and the Virgin Islands. Homoptera (excepting the Stenorhynchi). *Scientific survey of Porto Rico and the Virgin Islands.* Vol. 14, Pt. 2, 111–260. New York: New York Academy of Sciences.

Otte, D. 1981. *The North American grasshoppers. Acrididae: Gomphocerinae and Acridinae.* Vol. 1. Cambridge, Mass.: Harvard University Press.

Ovaska, K. 1991. Diet of the frog *Eleutherodactylus johnstonei* (Leptodactylidae) in Barbados, West Indies. *Journal of Herpetology* 25:486–88.

Overton, W. S. 1971. Estimating the numbers of animals in wildlife populations. In *Wildlife Management Techniques,* ed. R. H. Giles, Jr., 403–56. The Wildlife Society, Washington, D.C.

Ovington, J. D., and J. S. Olson. 1970. Biomass and chemical content of El Verde lower montane rain forest plants. In *A tropical rain forest: A study of irradiation and ecology at El Verde, Puerto Rico,* ed. H. T. Odum and R. F. Pigeon, H53–H77. Oak Ridge, Tenn.: U.S. Atomic Energy Commission.

Owen, D. F. 1983. The abundance and biomass of forest animals. In *Tropical rain forest ecosystems: Structure and function,* ed. F. B. Golley, 93–100. *Ecosystems of the world* 14A. New York: Elsevier Scientific Publishing Co.

Pacala, S., and J. Roughgarden. 1982. Resource partitioning and interspecific competition in two two-species insular *Anolis* communities. *Science* 217:444–6.

Pacala, S., and J. Roughgarden. 1984. Control of arthropod abundance by *Anolis* lizards on St. Eustatius (Neth. Antilles). *Oecologia* (Berlin) 64:160–2.

Padgett, D. 1975. The contribution of aquatic hyphomycetes in the decomposition of submerged leaf litter. Ph.D. diss., Ohio State University, Columbus.

Padgett, D. E. 1976. Leaf decomposition by fungi in a tropical rainforest stream. *Biotropica* 8:166–78.

Paine, R. T. 1966. Food web complexity and species diversity. *American Naturalist* 100:65–75.

Paine, R. T. 1969. The *Pisaster-Tegula* interaction: Prey patches, predator food preference, and intertidal community structure. *Ecology* 50:950–61.

Paine, R. T. 1980. Food webs: Linkage, interaction strength, and community structure. *Ecology* 61:950–61.

Paine, R. T. 1983. Intertidal food webs: Does connectance describe their essence? In *Current trends in food web theory: Report on a food web workshop,* ed. D. L. DeAngelis, W. M. Post, and G. Sugihara, 11–16. ORNL-5983. Oak Ridge, Tenn.: Oak Ridge National Laboratory.

Paine, R. T. 1988. Food webs: Road maps of interactions or grist for theoretical development. *Ecology* 69:1648–54.

Pardieck, K., and R. B. Waide. 1992. Mesh size as a factor in avian community studies using mist nets. *Journal of Field Ornithology* 63:250–5.

Park, O. 1942. A study in Neotropical Pselaphidae. *Northwestern University Studies in the Biological Sciences and Medicine 1*. Evanston, Ill.: Northwestern University.

Park, O. 1964. Observations upon the behavior of the myrmecophilous pselaphid beetles. *Pedobiologia* 4:129–37.

Parrotta, J. A., and D. J. Lodge. 1991. Fine root dynamics in subtropical wet forest following hurricane disturbance. *Biotropica* 23:343–7.

Parton, W. J., R. L. Sanford, and J. W. B. Stewart. 1989. Modeling soil organic matter dynamics in tropical soils. In *Dynamics of soil organic matter in tropical ecosystems,* ed. D. C. Coleman, J. M. Oades, and G. Uehama, 753–71. Honolulu: University of Hawaii Press.

Pasteels, J. M. 1965. Polyethisme chez les ouvriers de *Nasutitermes lujae* (Termitidae Isopteres). *Biologia Gabonica* 1:191–205.

Paulson, D. R. 1982. Odonata. In *Aquatic biota of Mexico, Central America and the West Indies,* ed. S. H. Hurlbert and A. Villalobos-Figueroa, 249–77. San Diego, Calif.: S. H. Hurlbert, Dept. of Biology, San Diego State University.

Payne, A. I. 1986. *The ecology of tropical lakes and rivers.* New York: John Wiley & Sons.

Payne, W. J. 1970. Energy yields and growth of heterotrophs. *Annual review of Microbiology* 24:17–52.

Peck, S. B. 1970. The Catopinae (Coleoptera: Leiodidae) of Puerto Rico. *Psyche* 77: 237–42.

Peck, S. B. 1972. Leiodidae and Catopinae (Coleoptera: Leiodidae) from Jamaica and Puerto Rico. *Psyche* 79:49–57.

Pegler, D. N. 1983. Agaric flora of the Lesser Antilles. *Kew Bulletin* Additional Series 9. London: Her Majesty's Stationery Office.

Pegler, D. N., and J. P. Fiard. 1979. Taxonomy and ecology of *Lactarius* (Agaricales) in the Lesser Antilles (West Indies). *Kew Bulletin* 33:601–28.

Penny, N. D., and J. R. Arias. 1982. *Insects of an Amazon forest.* New York: Columbia University Press.

Perel, T. S., L. O. Karpachevsky, and E. V. Yegorova. 1971. The role of Tipulidae (Diptera) larvae in decomposition of forest litterfall. *Pedobiologia* 11:66–70.

Perfecto, I., and G. R. Camilo. Effect of hurricanes on ant communities: The case of Nicaragua and Puerto Rico. *Biotropica* (in press).

Perry, J. J. 1970. The survival of actinomycetes in a radiation field. In *A tropical rainforest: A study of irradiation and ecology at El Verde, Puerto Rico,* ed. H. T. Odum and R. F. Pigeon, F67–F68. Oak Ridge, Tenn.: U.S. Atomic Energy Commission.

Persson, T., E. Bååth, M. Clarholm, H. Lundkvist, B. E. Söderström, and B. Sohlenius. 1980. Trophic structure, biomass dynamics and carbon metabolism of soil organisms in a Scots pine forest. In *Structure and function of northern coniferous forests: An ecosystem study,* ed. T. Persson, 419–59. *Ecological Bulletins* (Stockholm) 32. Stockholm: Swedish Natural Science Research Council.

Pescador, M. L., E. C. Masteller, and K. M. Buzby. 1993. Composition and temporal abundance of Ephemeroptera emergence from a tropical rainforest stream at El Verde, Puerto Rico. *Journal of the Kansas Entomological Society* 66:151–9.

Peters, W. L. 1971. A revision of the Leptophlebiidae of the West Indies (Ephemeroptera). *Smithsonian Contributions to Zoology* 62:1–48.

Petersen, H. 1982a. Structure and size of soil animal populations. *Oikos* 39:306–29.

Petersen, H. 1982b. The total soil fauna biomass and its composition. *Oikos* 39: 330–9.

Petersen, H., and M. Luxton. 1982. A comparative analysis of soil fauna populations and their role in decomposition processes. *Oikos* 33:287–388.

Petersen, R. H. 1992. Further notes on mating systems in *Melanotus. Mycotaxon* 45:331–41.

Petrides, G. A., F. B. Golley, and I. L. Brisbin. 1968. Energy flow and secondary productivity. In *A practical guide to the study of the productivity of large herbivores,* ed. F. B. Golley and H. K. Buechner, 9–17. International Biological Programme Handbook no. 7. Oxford: Blackwell Scientific Publications.

Petrunkevitch, A. 1929. The spiders of Porto Rico. Part I. *Transactions of the Connecticut Academy of Arts and Sciences* 30:1–158.

Petrunkevitch, A. 1930a. The spiders of Porto Rico. Part II. *Transactions of the Connecticut Academy of Arts and Sciences* 30:159–355.

Petrunkevitch, A. 1930b. The spiders of Porto Rico. Part III. *Transactions of the Connecticut Academy of Arts and Sciences* 31:1–191.

Pfeiffer, W. J. 1988. Dynamics and energetics of the arthropod grazing food web associated with *Spartina alterniflora* Loisel. in a Georgia intertidal grassland. Ph.D. diss., University of Georgia, Athens.

Philibosian, R. 1975. Territorial behavior and population regulation in the lizards, *Anolis acutus* and *A. cristatellus*. *Copeia* 1975:428–44.

Phillipson, J. 1966. *Ecological energetics*. New York: St. Martin's Press.

Pianka, E. R. 1966. Latitudinal gradients in species diversity: A review of concepts. *American Naturalist* 100:33–46.

Pianka, E. R. 1973. The structure of lizard communities. *Annual Review of Ecology and Systematics* 4:53–74.

Picchi, V. 1977. A systematic review of the genus *Aneurus* of North and Middle America and the West Indies (Hemiptera: Aradidae). *Quaestiones Entomologicae* 13:255–308.

Pickett, S. T. A., and P. S. White. 1985. *The ecology of natural disturbance and patch dynamics*. San Diego: Academic Press.

Pimentel, D. 1955. Biology of the Indian mongoose in Puerto Rico. *Journal of Mammalogy* 36:62–68.

Pimm, S. L. 1979. The structure of food webs. *Theoretical Population Biology* 16: 144–58.

Pimm, S. L. 1980. Properties of food webs. *Ecology* 61:219–25.

Pimm, S. L. 1982. *Food webs*. New York: Chapman and Hall.

Pimm, S. L. 1983. The causes of foodweb structure: Dynamics, energy flow, and natural history. In *Current trends in food web theory: Report on a food web workshop*, ed. D. L. DeAngelis, W. M. Post, and G. Sugihara, 45–49. ORNL-5983. Oak Ridge, Tenn.: Oak Ridge National Laboratory.

Pimm, S. L. 1988. Energy flow and trophic structure. In *Concepts of ecosystem ecology: A comparative view*, ed. L. R. Pomeroy and J. J. Alberts, 263–78. New York: Springer-Verlag.

Pimm, S. L. 1991. *The balance of nature*. Chicago: University of Chicago Press.

Pimm, S. L., and R. L. Kitching. 1987. The determinants of food chain lengths. *Oikos* 50:302–7.

Pimm, S. L., and J. H. Lawton. 1977. Number of trophic levels in ecological communities. *Nature* 268:329–31.

Pimm, S. L., and J. H. Lawton. 1978. On feeding on more than one trophic level. *Nature* 275:542–4.

Pimm, S. L., and J. H. Lawton. 1980. Are food webs divided into compartments? *Journal of Animal Ecology* 49:879–98.

Pimm, S. L., and J. C. Rice. 1987. The dynamics of multispecies, multilife-stage models of aquatic food webs. *Theoretical Population Biology* 32:303–25.

Pimm, S. L., J. H. Lawton, and J. E. Cohen. 1991. Food web patterns and their consequences. *Nature* 350:669–74.

Pindell, J. L., and S. F. Barrett. 1990. Geological evolution of the Caribbean region: A plate-tectonic perspective. In *The Caribbean region*, ed. G. Dengo and J. E. Case, 405–32. *The decade of North American geology*. Vol. H. Boulder, Colo.: Geological Society of America.

Pindell, J. L., and J. F. Dewey. 1982. Permo-Triassic reconstruction of western Pangea and the evolution of the Gulf of Mexico/Caribbean region. *Tectonics* 1:179–211.

Pinet, B., D. Lajat, P. Le Quellec, and P. Bouysee. 1985. Structure of Aves Ridge and Grenada Basin from multichannel seismic data. In *Caribbean geodynamics,* ed. A. Mascle, 53–64. Paris: Editions Technip.

Pipkin, S. B. 1965. The influence of adult and larval food habits on population size of Neotropical ground-feeding *Drosophila. American Midland Naturalist* 74:1–27.

Platnick, N. I. 1974. The spider family Anyphaenidae in America north of Mexico. *Bulletin of the Museum of Comparative Zoology* (Harvard) 146:205–66.

Plowman, K. P. 1979. Litter and soil fauna of two Australian subtropical forests. *Journal of Ecology* 4:87–104.

Plowman, K. P. 1981. Distribution of Cryptostigmata and Mesostigmata (Acari) within the litter and soil layers of two subtropical forests. *Australian Journal of Ecology* 6:365–74.

Polenec, A. 1964. Ökologische Untersuchungen der Arachidenfauna in Anemome-Fagetum. *Bioloski Vêstnik* 12:133–46.

Polenec, A. 1974. Ökologisch-faunistische untersuchungen der spinnenfauna in Querceto-Carpinetum in Slovenske Gorice. *Bioloski Vêstnik* 22:235–40.

Polis, G. A. 1991a. Complex trophic interactions in deserts: An empirical critique of food-web theory. *American Naturalist* 138:123–155.

Polis, G. A. 1991b. Food webs in desert communities: Complexity via diversity and omnivory. In *Ecology of desert communities,* ed. G. A. Polis, 383–438. Tucson: University of Arizona Press.

Porter, K. G. 1977. The plant-animal interface in freshwater ecosystems. *American Scientist* 65:159–70.

Pough, F. H. 1980. The advantages of ectothermy for tetrapods. *American Naturalist* 115:92–112.

Pough, F. H., T. L. Taigen, M. M. Stewart, and P. F. Brussard. 1983. Behavioral modification of evaporative water loss by a Puerto Rican frog. *Ecology* 64:244–52.

Pough, F. H., E. M. Smith, D. H. Rhodes, and A. Collazo. 1987. The abundance of salamanders in forest stands with different histories of disturbance. *Forest Ecology and Management* 20:1–9.

Powell, J. A., D. W. Belitsky, and G. B. Rathbun. 1981. Status of the West Indian manatee (*Trichechus manatus*) in Puerto Rico. *Journal of Mammalogy* 62:642–6.

Power, M. E. 1984. Depth distributions of armored catfish: predator-induced resource avoidance. *Ecology* 65:523–8.

Power, M. E. 1990. Effects of fish on river food webs. *Science* 250:811–4.

Power, M. E. 1992. Top-down and bottom-up forces in food webs: Do plants have primacy? *Ecology* 73:733–46.

Pregill, G. K., and S. L. Olson. 1981. Zoogeography of West Indian vertebrates in relation to Pleistocene climatic cycles. *Annual Review of Ecology and Systematics* 12:75–98.

Prestwich, G. D., B. L. Bentley, and E. J. Carpenter. 1980. Nitrogen sources for Neotropical nasute termites: Fixation and selective foraging. *Oecologia* 46:397–401.

Price, P. W., C. E. Bouton, P. Gross, B. A. McPherson, J. N. Thompson, and A. E. Weis. 1980. Interactions among three trophic levels: Influence of plants on interactions between insect herbivores and natural enemies. *Annual Review of Ecology and Systematics* 11:41–65.

Pringle, C. M., G. A. Blake, A. P. Covich, K. M. Buzby, and A. Finley. 1993. Effects of omnivorous shrimp in a montane tropical stream: Sediment removal, disturbance of sessile invertebrates and enhancement of understory algal biomass. *Oecologia* 93:1–11.

Pritchard, G. 1983. Biology of Tipulidae. *Annual Review of Entomology* 28:1–22.

Proctor, J. 1984. Tropical forest litterfall II: The data set. In *A tropical rain forest: The Leeds Symposium,* ed. A. C. Chadwick and S. L. Sutton, 83–113. Leeds, U.K.: Leeds Philosophical and Literary Society.

Pyke, G. H., H. R. Pulliam, and E. L. Charnov. 1977. Optimal foraging: A selective review of theory and tests. *Quarterly Review of Biology* 52:137–54.

Pyron, M., and A. P. Covich. Ecology and migration of snails in rivers of eastern Puerto Rico. *Journal of the North American Benthological Society* (in review).

Quintero, D., Jr. 1981. The amblypygid genus *Phrynus* in the Americas (Amblypygi, Phrynidae). *Journal of Arachnology* 9:117–66.

Raffaele, H. A. 1983. A guide to the birds of Puerto Rico and the Virgin Islands. San Juan, P.R.: Fondo Educativo Interamericano.

Raffaele, H. A., M. J. Velez, R. Cotte, J. J. Whelan, E. R. Keil, and W. Cupiano. 1973. Rare and endangered animals of Puerto Rico: A committee report. Hyattsville, Md.: U.S. Soil Conservation Service.

Ramos, J. A. 1957. A review of the Auchenorrhynchous Homoptera of Puerto Rico. *Journal of Agriculture of the University of Puerto Rico* 41:38–117.

Ramos, J. A. 1982. Checklist of the butterflies of Puerto Rico (Lepidoptera, Rhopalocera, West Indies). *Caribbean Journal of Science* 17:59–68.

Ramos, J. A. 1988. Zoogeography of the Auchenorrhynchous Homoptera of the greater Antilles (Hemiptera). In *Zoogeography of Caribbean insects,* ed. J. K. Liebherr, 61–70. Ithaca, N.Y.: Comstock Publishing Associates.

Rand, A. S. 1964. Ecological distribution in anoline lizards of Puerto Rico. *Ecology* 45:745–52.

Rand, A. S. 1967. Ecology and social organization in the Iguanid lizard *Anolis lineatopus. Proceedings of the U.S. National Museum* 122:1–79.

Rand, A. S., and R. Andrews. 1975. Adult color dimorphism and juvenile pattern in *Anolis cuvieri. Journal of Herpetology* 9:257–60.

Rand, A. S., and C. W. Myers. 1990. The herpetofauna of Barro Colorado Island, Panama: An ecological summary. In *Four Neotropical rainforests,* ed. A. H. Gentry, 386–409. New Haven, Conn.: Yale University Press.

Rand, A. S., and W. M. Rand. 1982. Variation in rainfall on Barro Colorado Island. In *The ecology of a tropical forest,* ed. E. G. Leigh, Jr., A. S. Rand, and D. M. Windsor, 47–59. Washington, D.C.: Smithsonian Institution Press.

Ratcliffe, B. C. 1976. A revision of the genus *Strategus* (Coleoptera: Scarabaeidae). *Bulletin of the University of Nebraska State Museum* 10:93–204.

Rayner, A. D. M., and N. K. Todd. 1979. Population and community structure and dynamics of fungi in decaying wood. *Advances in Botanical Research* 7:333–420.

Read, D. J., J. R. Leake, and A. R. Langdale. 1989. The nitrogen nutrition of mycorrhizal fungi and their host plants. In *Nitrogen, phosphorus and sulphur utilization by fungi,* ed. L. Boddy, R. Marchant, and D. J. Read. Cambridge: Cambridge University Press.

Reagan, D. P. 1984. Ecology of the Puerto Rican boa (*Epicrates inornatus*) in the Luquillo Mountains of Puerto Rico. *Caribbean Journal of Science* 20:119–27.

Reagan, D. P. 1986. Foraging behavior of *Anolis stratulus* in a Puerto Rican rain forest. *Biotropica* 18:157–60.

Reagan, D. P. 1991. The response of *Anolis* lizards to hurricane-induced habitat changes in a Puerto Rican rain forest. *Biotropica* 23:468–74.

Reagan, D. P. 1992. Congeneric species distribution and abundance in a three-dimensional habitat: The rain forest anoles of Puerto Rico. *Copeia* 1992:392–403.

Reagan, D. P. 1995. Lizard ecology in the canopy of an island rain forest. In *Forest Canopies,* ed. M. D. Lowman and N. M. Nadkarni. San Diego: Academic Press.

Reagan, D. P., and C. P. Zucca. 1982. Inventory of the Puerto Rican boa (*Epicrates inornatus*) in the Caribbean National Forest. Final Report to the Forest Service, U.S. Department of Agriculture. CEER-T-136. Rio Piedras, P.R.: Center for Energy and Environment Research.

Reagan, D. P., R. W. Garrison, J. E. Martínez, R. B. Waide, and C. B. Zucca. 1982. Tropical rain forest cycling and transport program: Phase I report. CEER-T-137. Rio Piedras, P.R.: Center for Energy and Environment Research.

Recher, H. F. 1970. Population density and seasonal changes of the avifauna in a tropical forest before and after gamma irradiation. In *A tropical rain forest: A study of irradiation and ecology at El Verde, Puerto Rico,* ed. H. T. Odum and R. F. Pigeon, E69–E93. Oak Ridge, Tenn.: U.S. Atomic Energy Commission.

Recher, H. F., and J. T. Recher. 1966. A contribution to the knowledge of the avifauna of the Sierra de Luquillo, Puerto Rico. *Caribbean Journal of Science* 6: 151–62.

Redhead, J. F. 1968. Mycorrhizal associations in some Nigerian forest trees. *Transactions of the British Mycological Society* 51:377–87.

Redhead, J. F. 1980. Mycorrhiza in natural tropical forests. In *Tropical mycorrhiza research,* ed. P. Mikola, 127–42. Oxford: Clarendon Press.

Rees, C. J. C. 1983. Microclimate and flying Hemiptera fauna of a primary lowland rain forest in Sulawesi. In *Tropical rain forest: Ecology and management,* ed. S. L. Sutton, T. C. Whitmore, and A. C. Chadwick, 121–26. Oxford: Blackwell Scientific Publications.

Rehn, J. A. G., and M. Hebard. 1927. The Orthoptera of the West Indies: No. 1. Blattidae. *Bulletin of the American Museum of Natural History* 54:1–320.

Reichle, D. E. 1977. The role of soil invertebrates in nutrient cycling. In *Soil organisms as components of ecosystems,* ed. U. Lohm and T. Persson, 145–56. *Ecological Bulletins 25.* Stockholm: Swedish Natural Science Research Council.

Reichle, D. E., and D. A. Crossley, Jr. 1967. Investigations on heterotrophic productivity in forest insect communities. In *Secondary productivity of terrestrial ecosystems,* ed. K. Petrusewicz, 563–87. Warszawa: Panstowe Wydawn, Naukowe.

Reichle, D. E., M. H. Shanks, and D. A. Crossley, Jr. 1969. Calcium, potassium and sodium content of forest floor arthropods. *Annals of the Entomological Society of America* 62:57–62.

Reichle, D. E., R. A. Goldstein, R. I. Van Hook, and G. J. Dodson. 1973. Analysis of insect consumption in a forest canopy. *Ecology* 54:1076–84.

Reid, C. P. P. 1979. Mycorrhizae and water stress. In *Root physiology and symbiosis,* ed. A. Riedacker and J. Gagnaire-Michard, 392–408. *Proceedings of IUFRO Symposium,* Compte Rendus, Nancy, France, Sept. 11–15, 1978.

Renne, P. R., J. M. Mattinson, C. W. Hatten, M. Somin, T. C. Onstott, G. Millan, and E. Linares. 1989. $^{40}Ar/^{39}$ and U-Pb evidence for late proterozoic (Grenville-age) continental crust in northcentral Cuba and regional tectonic implications. *Precambrian Research* 42:325–41.

Reyes, V. G., and J. M. Tiedje. 1976. Ecology of the gut microbiota of *Tracheoniscus rathkei* (Crustacea, Isopoda). *Pedobiologia* 16:67–74.

Rice, B., and M. Westoby. 1983. Plant species richness at the 0.1 ha scale in Australia vegetation compared to other continents. *Vegetatio* 52:129–40.

Richards, P. W. 1952. *The tropical rain forest: An ecological study.* Cambridge: Cambridge University Press.

Ricklefs, R. E. 1973. *Ecology.* 2d ed. New York: Chiron Press.

Riechert, S. E. 1974. Thoughts on the ecological significance of spiders. *BioScience* 24:352–6.

Riechert, S. E. 1978. Energy-based territoriality in populations of the desert spider *Agelenopsis aperta* (Gertsch). *Arachnology,* ed. P. Merrett, 211–22. *Symposium of the Zoological Society of London Number 42.* New York: Academic Press.

Riechert, S. E. 1981. The consequences of being territorial: Spiders, a case study. *American Naturalist* 117:871–92.

Riechert, S. E., and R. G. Gillespie. 1986. Habitat choice and utilization in web-building spiders. In *Spiders: Webs, behavior, and evolution,* ed. W. A. Shear, 23–48. Stanford, Calif.: Stanford University Press.

Riechert, S. E., and T. Lockley. 1984. Spiders as biological control agents. *Annual Review of Entomology* 29:299–320.

Riley, N. D. 1975. *A field guide to the butterflies of the West Indies.* London: Collins.

Risley, L. S., and D. A. Crossley, Jr. 1988. Herbivore-caused greenfall in the southern Appalachians. *Ecology* 69:1118–27.

Rivero, J. A. 1963. *Eleutherodactylus hedricki,* a new species of frog from Puerto Rico. *Breviora* 185:81–85.

Rivero, J. A. 1978. *Los anfibios y reptiles de Puerto Rico.* Rio Piedras: Universidad de Puerto Rico. Editorial Universitario.

Rivero, J. A., and D. Segui-Crespo. 1992. *Anfibios y reptiles en neustro folklore.* Mayagüez: J. A. Rivero.

Rivero, J. A., J. Maldonado, and H. Mayorga. 1963. On the habits and food of *Eleutherodactylus karlschmidti* Grant. *Caribbean Journal of Science* 3:25–27.

Robinson, M. H., and B. Robinson. 1974. A census of web-building spiders in a coffee plantation at Wau, New Guinea and an assessment of their insecticidal effect. *Tropical Ecology* 15:95–107.

Robinson, M. H., and B. Robinson. 1976. A tipulid associated with spider webs in Papua, New Guinea. *Entomologist's Monthly Magazine* 112:1–3.

Robinson, M. H., Y. D. Lubin, and B. Robinson. 1974. Phenology, natural history and species diversity of web-building spiders on three transects at Wau, New Guinea. *Pacific Insects* 16:117–63.

Robinson, S. K., and J. Terborgh. 1990. Bird communities of the Cocha Cashu Biological Station in Amazonian Peru. In *Four Neotropical rainforests,* ed. A. Gentry, 199–216. New Haven, Conn.: Yale University Press.

Rodríguez, G. A., and D. P. Reagan. 1984. Bat predation by the Puerto Rican boa, *Epicrates inornatus. Copeia* 1984:219–20.

Rodríguez, L. B., and J. E. Cadle. 1990. A preliminary overview of the herpetofauna of Cocha Cashu, Manu National Park, Perú. In *Four Neotropical rainforests,* ed. A. H. Gentry, 410–25. New Haven, Conn.: Yale University Press.

Rodríguez-Duran, A. 1984. Community structure of a bat colony at Cueva Cucaracha. M.S. thesis, University of Puerto Rico, Mayaguez, Puerto Rico.

Rodríguez-Duran, A., and T. H. Kunz. 1992. Pteronotus quadridens. *Mammalian Species* 395:1–4.

Rodríguez-Robles, J. A., and M. Leal. 1993. Effects of prey type on the feeding behavior of *Alsophis portoricensis* (Serpentes: Colubridae). *Herpetologica* 27: 163–8.

Rodríguez-Robles, J. A., and R. Thomas. 1992. Venom function in the Puerto Rican racer, *Alsophis portoricensis* (Serpentes: Colubridae). *Copeia* 1992:62–68.

Rodríguez-Vidal, J. A. 1959. Puerto Rican Parrot (*Amazona vittata vittata*) study. *Monographs of the Department of Agriculture and Commerce, No. 1.* San Juan: Commonwealth of Puerto Rico.

Roewer, C. F. 1951. Neue Namen einiger Aranee-Arten. *Abheilung naturwissenschafte Ver Bremen* 32:437–56.

Rogers, J. D., T. Laessøe, and D. J. Lodge. 1991. *Camillea:* New combinations and a new species. *Mycologia* 83:224–7.

Rohlf, F. J., and R. R. Sokal. 1969. *Statistical tables.* San Francisco: W. H. Freeman and Co.

Rolle, F. J. 1963. Life history of the red-legged thrush (*Mimocichla plumbea ardodiacea*) in Puerto Rico. *Studies on the Fauna of Curacao and other Caribbean Islands* 14:1–40.

Romell, L. G. 1935. An example of myriapods as mull formers. *Ecology* 16: 67–71.

Romero, J. L., and R. F. Ruppel. 1973. A new species of *Silba* (Diptera: Lonchaeidae) from Puerto Rico. *Journal of Agriculture of the University of Puerto Rico* 57: 165–8.

Roughgarden, J. 1993. Anolis *lizards of the Caribbean: Ecology, evolution, and plate tectonics.* Oxford: Oxford University Press.

Roughgarden, J., and S. Pacala. 1989. Taxon cycle among *Anolis* lizard populations: Review and evidence. In *Speciation and its consequences,* ed. D. Otte and J. Endler, 403–32. Sunderland, Mass.: Sinauer Associates.

Roughgarden, J., D. Heckel, and E. R. Fuentes. 1983. Coevolutionary theory and the biogeography of community structure of *Anolis.* In *Lizard ecology: Studies of a model organism,* ed. R. B. Huey, E. R. Pianka, and T. W. Schoener, 371–410. Cambridge, Mass.: Harvard University Press.

Roughgarden, J., J. Rummel, and S. Pacala. 1983. Experimental evidence of strong present-day competition between the *Anolis* populations of the Anguilla Bank: A preliminary report. In *Advances in herpetology and evolutionary biology: Essays in honor of Ernest Williams,* ed. A. Rhodin and K. Miyata, 499–506. Cambridge, Mass.: Museum of Comparative Zoology.

Rowell, C. H. F., M. Rowell-Rahier, H. E. Braker, G. Cooper-Driver, and L. D. Gomez P. 1983. The palatability of ferns and the ecology of two tropical forest grasshoppers. *Biotropica* 15:207–16.

Rudnick, A. 1960. A revision of the mites of the family Spinturnicidae (Acarina).

University of California Publications in Entomology 17:157–283. Berkeley: University of California Press.

Ruibal, R., and R. Philibosian. 1970. Eurythermy and niche expansion in lizards. *Copeia* 1970:645–53.

Rummel, J. D., and J. Roughgarden. 1985. Effects of reduced perch-height separation on competition between two *Anolis* lizards. *Ecology* 66:430–44.

Rushton, S. P., and M. Hassall. 1983. Food and feeding rates of the terrestrial isopod *Armadillidium vulgare* (Latreille). *Oecologia* (Berlin) 57:415–9.

Russell, A., and N. McWhirter. 1987. *Guinness book of world records.* New York: Bantam Books.

Russell-Smith, A. 1981. Seasonal activity and diversity of ground-living spiders in two African savanna habitats. *Bulletin of the British Arachnological Society* 5: 145–54.

Russell-Smith, A., and P. Swann. 1972. The activity of spiders in coppiced chestnut woodland in southern England. *Bulletin of the British Arachnological Society* 2: 99–103.

Rypstra, A. L. 1983. The importance of food and space in limiting web-spider densities: A test using field enclosures. *Oecologia* (Berlin) 59:312–6.

Rypstra, A. L. 1984. A relative measure of predation on web-spiders in temperate and tropical forests. *Oikos* 43:129–32.

Rypstra, A. L. 1986. Web spiders in temperate and tropical forests: Relative abundance and environmental correlates. *American Midland Naturalist* 115:42–51.

Safir, G. R., J. S. Boyer, and J. W. Gerdeman. 1972. Nutrient status and mycorrhizal enhancement of water transport in soybean. *Plant Physiology* 43:700–703.

Salick, J., R. Herrera, and C. F. Jordan. 1983. Termitaria: Nutrient patchiness in nutrient-deficient rain forests. *Biotropica* 15:1–7.

Sandlin, E. A., and M. R. Willig. 1993. Effects of age, sex, prior experience and intraspecific food variation on diet composition of a tropical folivore (Phasmatodea: Phasmatidae). *Environmental Entomology* 22:625–33.

Sandlin-Smith, E. A. 1989. Foraging ecology of a neotropical folivore, *Lamponius portoricensis* Rehn (Phasmatodea: Phasmatidae). M.S. thesis, Texas Tech University, Lubbock.

Sanford, R. L., Jr., W. J. Parton, D. S. Ojima, and D. J. Lodge. 1991. Hurricane effects on soil organic matter dynamics and forest production in the Luquillo Experimental Forest, Puerto Rico: Results of simulation modeling. *Biotropica* 23:364–72.

Santana, C. E., and S. A. Temple. 1988. Breeding biology and diet of red-tailed hawks in Puerto Rico. *Biotropica* 20:151–60.

Sastre-De Jesus, I. 1979. Ecological life cycle of *Buchenavia capitata* (Vahl.) Eichl., a late secondary successional species in the rain forest of Puerto Rico. M.S. thesis, University of Tennessee, Knoxville.

Satchell, J. E. 1974. Litter-interface of animate/inanimate matter. In *Biology of plant litter decomposition,* ed. C. H. Dickinson and G. J. F. Pugh, xiii–xliv. New York: Academic Press.

Saunders, J. L. 1965. The *Xyleborus-Ceratocystis* complex of Cacao. *Cacao* 10: 7–13.

Scatena, F. N. 1989. An introduction to the physiography and history of the Bisley Experimental Watersheds in the Luquillo Mountains of Puerto Rico. U.S. Forest

Service. Southern Forest Experiment Station. General Technical Report SO-72. New Orleans, LA.

Scatena, F. N., and M. C. Larsen. 1991. Physical aspects of hurricane damage in Puerto Rico. *Biotropica* 23:317–23.

Scatena, F. N., W. Silver, T. Siccama, A. Johnson, and M. J. Sanchez. 1993. Biomass and nutrient content of the Bisley Experimental Watersheds, Luquillo Experimental Forest, Puerto Rico, before and after Hurricane Hugo, 1989. *Biotropica* 25: 15–27.

Schaefer, D. A., and W. G. Whitford. 1981. Nutrient cycling by the subterranean termite *Gnathamitermes tubiformans* in a Chihuahuan desert ecosystem. *Oecologia* 48:277–83.

Schal, C., and W. J. Bell. 1986. Vertical community structure and resource utilization in Neotropical forest cockroaches. *Ecological Entomology* 11:411–23.

Schaus, W. 1940a. Insects of Porto Rico and the Virgin Islands. Moths of the families Geometridae and Pyralidae. *Scientific survey of Porto Rico and the Virgin Islands.* Vol. 12, Pt. 3, 291–417. New York: New York Academy of Sciences.

Schaus, W. 1940b. Insects of Porto Rico and the Virgin Islands. Moths of the family Noctuidae. *Scientific survey of Porto Rico and the Virgin Islands.* Vol. 12, Pt. 2, 177–290. New York: New York Academy of Sciences.

Schlesinger, W. H. 1977. Carbon balance in terrestrial detritus. *Annual Review of Ecology and Systematics* 8:51–81.

Schmidt, K. P. 1928. Amphibians and land reptiles of Porto Rico, with a list of those reported from the Virgin Islands. *Scientific survey of Porto Rico and the Virgin Islands.* Vol. 10, Pt. 1, 1–160. New York: New York Academy of Sciences.

Schneider, D. W., and T. M. Frost. 1986. Massive upstream migrations by a tropical freshwater neritid snail. *Hydrobiologia* 137:153–7.

Schoener, T. W. 1968. The *Anolis* lizards of Bimini: Resource partitioning in a complex fauna. *Ecology* 49:704–26.

Schoener, T. W. 1969a. Models of optimal size for solitary predators. *American Naturalist* 103:277–313.

Schoener, T. W. 1969b. Optimal size and specialization in constant and fluctuating environments. *Brookhaven Symposium in Biology* 22:103–14.

Schoener, T. W. 1969c. Size patterns in West Indian *Anolis* lizards. I. Size and species diversity. *Systematic Zoology* 18:386–401.

Schoener, T. W. 1971a. An empirically based estimate of home range. *Theoretical Population Biology* 20:281–325.

Schoener, T. W. 1971b. Theory of feeding strategies. *Annual Review of Ecology and Systematics* 2:369–404.

Schoener, T. W. 1974. Resource partitioning in ecological communities. *Science* 185: 27–31.

Schoener, T. W. 1975. Are lizard population sizes unusually constant through time? *American Naturalist* 126:633–41.

Schoener, T. W. 1981. An empirically based estimate of home range. *Theoretical Population Biology.* 20:281–325.

Schoener, T. W. 1985. Some comments on Connell's and my reviews of field experiments on interspecific competition. *American Naturalist* 125:730–40.

Schoener, T. W. 1989. Food webs from the small to the large. *Ecology* 70:1559–89.

Schoener, T. W., and G. C. Gorman. 1968. Some niche differences among three species of Lesser Antillean anoles. *Ecology* 49:819–30.

Schoener, T. W., and A. Schoener. 1971. Structural habitats of West Indian *Anolis* lizards II. Puerto Rico uplands. *Breviora* 375:1–39.

Schoener, T. W., and D. Spiller. 1987. Effect of lizards on spider populations: Manipulative reconstruction of a natural experiment. *Science* 236:949–52.

Schoener, T. W., and C. A. Toft. 1983. Spider populations: Extraordinarily high densities on islands without top predators. *Science* 219:1353–5.

Schoenly, K., and J. E. Cohen. 1991. Temporal variation in food web structure: 16 empirical cases. *Ecological Monographs* 61:267–98.

Schowalter, T. D. 1994. Invertebrate community structure and herbivory in a tropical rain forest canopy in Puerto Rico following Hurricane Hugo. *Biotropica* 26: 312–9.

Schowalter, T. D., J. W. Webb, and D. A. Crossley, Jr. 1981. Community structure and nutrient content of canopy arthropods in clearcut and uncut forest ecosystems. *Ecology* 62:1010–19.

Schubart, H. O.-R., and L. Beck. 1968. Zur Coleopterenfauna amazonisches Boden. *Amazoniana* 1:311–22.

Schwalm, P. A., P. H. Starrett, and R. W. McDiarmid. 1977. Infrared reflectance in leaf-sitting neotropical frogs. *Science* 196:1225–7.

Schwartz, A. 1958. Another new large *Eleutherodactylus* (Amphibia: Leptodactylidae) from western Cuba. *Proceedings of the Biological Society of Washington* 71: 37–42.

Schwartz, A. 1978. Some aspects of the herpetogeography of the West Indies. In *Zoogeography in the Caribbean,* ed. F. B. Gill, 31–51. Academy of Natural Sciences of Philadelphia Special Publication 13. Philadelphia: Academy of Natural Sciences of Philadelphia.

Schwartz, A., and R. W. Henderson. 1991. *Amphibians and reptiles of the West Indies: Descriptions, distributions and natural history.* Gainesville: University of Florida Press.

Schwartz, A., and R. Thomas. 1975. *A check-list of West Indian amphibians and reptiles.* Carnegie Museum of Natural History Special Publication 1. Pittsburgh: Carnegie Museum of Natural History.

Scogin, R. 1982. Dietary observations on the red, fig-eating bat (*Stenoderma rufum*) in Puerto Rico. *ALISO* 10:259–61.

Scott, N. J., Jr. 1976. The abundance and diversity of the herpetofaunas of tropical forest litter. *Biotropica* 8:41–58.

Scott, N. J., Jr. 1982. The herpetofauna of forest litter plots from Cameroon, Africa. In *Herpetological communities: A Symposium of the Society for the Study of Amphibians and Reptiles and the Herpetologists League,* August 1977, ed. N. J. Scott, Jr., 145–50. U.S. Fish and Wildlife Service Wildlife Research Report 13. Washington, D.C.: U.S. Fish and Wildlife Service.

Seastedt, T. R. 1984. The role of microarthropods in decomposition and mineralization processes. *Annual Review of Entomology* 29:25–46.

Seastedt, T. R., and D. A. Crossley, Jr. 1984. The influence of arthropods on ecosystems. *BioScience* 34:157–61.

Seaver, F. J., and C. E. Chardon. 1926. Mycology. *Scientific survey of Porto Rico and the Virgin Islands.* Vol. 8, Pt. 1, 1–208. New York: New York Academy of Sciences.

Seiders, V. M. 1971. Geologic map of the El Yunque quadrangle, Puerto Rico. U.S. Geological Survey Miscellaneous Geological Investigations Map I-658. Washington, D.C.: U.S. Department of the Interior.

Sexton, O. J., J. Bauman, and E. Ortleb. 1972. Seasonal food habits of *Anolis limnifrons. Ecology* 53:182–6.

Shear, W. A. 1978. Taxonomic notes on the armoured spiders of the families Tetrablemmidae and Pacullidae. *American Museum of Natural History Novitates* 2650:1–46.

Shelly, T. E. 1984. Prey selection by the Neotropical spider *Micrathena schreibersi* with notes on web-site tenacity. *Proceedings of the Entomological Society of Washington* 86:493–502.

Shelly, T. E. 1988. Relative abundance of day-flying insects in treefall gaps vs. shaded understory in a Neotropical forest. *Biotropica* 20:114–9.

Shields, J. A., E. A. Paul, W. E. Lowe, and D. Parkinson. 1973. Turnover of microbial tissue in soil under field conditions. *Soil Biology and Biochemistry* 5:753–64.

Shiroya, T., G. R. Lister, V. Slankis, G. Krotov, and C. D. Nelson. 1962. Translocation of products of photosynthesis to roots of pine seedlings. *Canadian Journal of Botany* 40:1125–36.

Short, H. L., and F. B. Golley. 1968. Metabolism. In *A practical guide to the study of the productivity of large herbivores,* ed. F. B. Golley and H. K. Buechner, 95–105. International Biological Programme Handbook no. 7. Oxford: Blackwell Scientific Publications.

Shump, K. A., and A. U. Shump. 1982. *Lasiurus borealis. Mammalian Species* 183:1–6.

Sieverding, E. 1984. Influence of soil water regimes on VA mycorrhiza. III. Comparison of the mycorrhizal fungi and their influence on transpiration. *Zeitschrift für Acker-und Pflanzenbau* 153:52–61.

Sieving, K. E. 1992. Nest predation and differential insular extinction among selected forest birds of central Panama. *Ecology* 73:2310–28.

Silva-Taboada, G. 1976. Historia y actualizacion taxonomica de algunas especies antillanas de murcielagos de los generos *Pteronotus, Brachyphylla, Lasiurus,* y *Antrozous* (Mammalia: Chiroptera). *Poeyana* 153:1–24.

Silva-Taboada, G. 1979. *Los murcielagos de Cuba.* La Habana: Academia de Ciencias de Cuba.

Silva-Taboada, G., and R. H. Pine. 1969. Morphological and behavioral evidence for the relationship between the bat genus *Brachyphylla* and the Phyllonycterinae. *Biotropica* 1:10–19.

Silver, W. L., and K. A. Vogt. 1993. Fine root dynamics following single and multiple disturbances in a subtropical wet forest ecosystem. *Journal of Ecology* 81: 729–38.

Simberloff, D. S. 1970. Taxonomic diversity of island biotas. *Evolution* 24:23–47.

Singer, R., and I. Araujo. 1979. Litter decomposition and ectomycorrhiza in Amazonian forests. *Acta Amazonica* 41:549–51.

Singer, R., and D. J. Lodge. 1988. New tropical species in the Paxillaceae. *Mycologia Helvetica* 3:207–13.

Singer, R., and J. L. Morello. 1960. Ectotrophic forest tree mycorrhizae and forest communities. *Ecology* 41:549–51.

Skaife, S. H. 1961. *Dwellers in darkness.* New York: Doubleday.

Slater, J. A. 1983. The *Ozophora* of Panama, with descriptions of thirteen new species (Hemiptera, Lygaeidae). *American Museum Novitates* 2765:1–29.

Slater, J. A., and R. M. Baranowski. 1978. *How to know the true bugs (Hemiptera-Heteroptera).* Dubuque, Iowa: C. Brown Co. Publishers.

Sleep, N. H. 1990. Hotspots and mantle plumes: Some phenomenology. *Journal of Geophysical Research* 95B:6715–36.

Smith, C. F. 1960. A new species of Aphidae: Homoptera from Puerto Rico. *Journal of Agriculture of the University of Puerto Rico* 44:157–62.

Smith, C. F. 1970. Notes on the genus *Picturaphis* and related genera with a new species of *Picturaphis* from Puerto Rico (Aphididae: Homoptera). *Journal of Agriculture of the University of Puerto Rico* 54:683–8.

Smith, C. F., L. F. Martorell, and M. E. Perez-Escolar. 1963. Aphididae of Puerto Rico. Puerto Rico. Agricultural Experiment Station. Technical Paper 37. Rio Piedras: University of Puerto Rico. Agricultural Experiment Station.

Smith, C. F., L. F. Martorell, M. E. Perez-Escolar, and S. Medina-Gaud. 1971. Additions and corrections to the Aphididae of Puerto Rico. *Journal of Agriculture of the University of Puerto Rico* 55:192–258.

Smith, D. R., and R. J. Lavigne. 1973. Two new species of ants of the genera *Tapinoma* Foerster and *Paratrechina* Motschoulsky from Puerto Rico (Hymenoptera: Formicidae). *Proceedings of the Entomological Society of Washington* 75:181–7.

Smith, J. D. 1972. Systematics of the chiropteran family Mormoopidae. University of Kansas, Museum of Natural History, Miscellaneous Publication 56. Lawrence: University of Kansas.

Smith, M. R. 1936. The ants of Puerto Rico. *Journal of Agriculture of the University of Puerto Rico* 20:819–75.

Smith, R. F. 1970a. List of common plant species at El Verde. In *A tropical rain forest: A study of irradiation and ecology at El Verde, Puerto Rico,* ed. H. T. Odum and R. F. Pigeon, B59–B61. Oak Ridge, Tenn.: U.S. Atomic Energy Commission.

Smith, R. F. 1970b. The vegetation structure of a Puerto Rican rain forest before and after short-term gamma irradiation. In *A tropical rain forest: A study of irradiation and ecology at El Verde, Puerto Rico,* ed. H. T. Odum and R. F. Pigeon, D103–D140. Oak Ridge, Tenn.: U.S. Atomic Energy Commission.

Snow, B. K., and D. W. Snow. 1971. The feeding ecology of tanagers and honeycreepers in Trinidad. *Auk* 88:291–322.

Snow, D. W. 1965. A possible selective factor in the evolution of fruiting seasons in tropical forest. *Oikos* 15:274–81.

Snyder, F. M. 1957. Puerto Rican *Neodexiopsis* (Diptera: Muscidae: Coenosiinae). *Journal of Agriculture of the University of Puerto Rico* 41:207–29.

Snyder, N. F. R., and J. W. Wiley. 1976. Sexual size dimorphism in hawks and owls of North America. Ornithological Monographs 20. Washington, D.C.: The American Ornithologists' Union.

Snyder, N. F. R., J. W. Wiley, and C. B. Kepler. 1987. The parrots of Luquillo: Natural history and conservation of the Puerto Rican parrot. Los Angeles: Western Foundation of Vertebrate Zoology.

Snyder, T. E. 1925. Descriptions of new species and hitherto unknown castes of termites from America and Hawaii. *Proceedings of the U.S. Natural History Museum* 64:1–40.

Sokal, R. R., and F. J. Rohlf. 1969. *Biometry.* San Francisco: W. H. Freeman Co.

Soma, K., and T. Saitô. 1983. Ecological studies of soil organisms with references to the decomposition of pine needles. II. Litter feeding and breakdown by the woodlouse, *Porcellio scaber. Plant and Soil* 75:139–51.

Sorensen, L. H. 1983. The influence of stress treatments on the microbial biomass and the rate of decomposition of humified matter in soils containing different amounts of clay. *Plant and Soil* 75:107–20.

Speed, R. C., L. C. Gerhard, and E. H. McKee. 1979. Ages of deposition, deformation, and intrusion of Cretaceous rocks, eastern St. Croix, Virgin Islands. *Geological Society of America Bulletin I* 90:629–32.

Sprules, W. G. 1972. Effects of size-selective predation and food competition on high altitude zooplankton communities. *Ecology* 53:375–86.

Sprules, W. G., and J. E. Bowerman. 1988. Omnivory and food chain length in zooplankton food webs. *Ecology* 69:418–26.

Stamps, J. A. 1977. The relationship between resource competition, risk, and aggression in a tropical territorial lizard. *Ecology* 58:349–58.

Stanton, N. L. 1979. Patterns of species diversity in temperate and tropical litter mites. *Ecology* 60:295–304.

Stark, N. 1972. Nutrient cycling pathways and litter fungi. *BioScience* 22: 355–60.

Stark, N., and M. Spratt. 1977. Root biomass and nutrient storage in rain forest oxisols near San Carlos de Rio Negro. *Tropical Ecology* 18:1–9.

Steigen, A. L. 1975. Energetics in a population of *Pardosa palustris* (L.) (Araneae: Lycosidae) on Hardanervidda. In *Fennscandian tundra ecosystems: analysis and systems ecology* ed. F. Welgolaski. Berlin: Springer-Verlag.

Stevenson, J. A. 1975. Fungi of Puerto Rico and the American Virgin Islands. *Contributions of Reed Herbarium 23*. Baltimore, Md.: Reed Herbarium.

Stewart, M. M. 1974. Parallel pattern polymorphism in the genus *Phrynobatrachus* (Amphibia, Ranidae). *Copeia* 1974:823–32.

Stewart, M. M. 1979. The role of introduced species in a Jamaican frog community. In *IV Symposium Internacional de Ecologia Tropical,* ed. H. Wolda, 113–46. *Proceedings of the 4th International Symposium on Tropical Ecology,* March 7–11, 1977, Panama City.

Stewart, M. M. 1985. Arboreal habitat use and parachuting in a subtropical forest frog. *Journal of Herpetology* 19:391–401.

Stewart, M. M. 1995. Climate driven population fluctuations in rain forest frogs. *Journal of Herpetology* 29:437–46.

Stewart, M. M., and G. E. Martin. 1980. Coconut husk piles—a unique habitat for Jamaican terrestrial frogs. *Biotropica* 12:107–16.

Stewart, M. M., and F. H. Pough. 1983. Population density of tropical forest frogs: Relation to retreat sites. *Science* 221:570–2.

Stewart, M. M., and A. S. Rand. 1992. Diel variation in the use of aggressive calls by the frog *Eleutherodactylus coqui*. *Herpetologica* 48:49–56.

Stiles, F. G. 1975. Ecology, flowering phenology, and hummingbird pollination of some Costa Rica *Heliconia* species. *Ecology* 56:285–301.

Stiles, F. G. 1978a. Ecological and evolutionary implication of bird pollination. *American Zoologist* 18:715–29.

Stiles, F. G. 1978b. Temporal organization of flowering among the hummingbird food plants of a tropical wet forest. *Biotropica* 10:194–210.

St. John, T. V. 1980a. A survey of mycorrhizal infection in an Amazonian rain forest. *Acta Amazonica* 10:527–33.

St. John, T. V. 1980b. Uma lista de especies de plantas tropicais brasileiras naturalmente infectadus com micorrhiza vesicular-arbuscular. *Acta Amazonica* 10: 229–34.

St. John, T. V., and C. Uhl. 1983. Mycorrhizae at San Carlo de Rio Negro, Venezuela. *Acta Cientifica Venezolana* 34:233–7.

Stock, J. H. 1986. Caribbean biogeography and a biological calendar for geological events. In *Crustacean biogeography*, ed. R. H. Gore and K. L. Heck, Rotterdam: A. A. Balkema.

Stout, J. 1980. Leaf decomposition rates in Costa Rican lowland tropical rainforest streams. *Biotropica* 12:264–72.

Strickland, A. H. 1944. The arthropod fauna of some tropical soils with notes on the techniques applicable to entomological soil surveys. *Tropical Agriculture* 21: 107–14.

Strong, D. R. 1983. *Chelobasis bicolor* (Abejon de Platanillo, rolled-leaf hispine). In *Costa Rican natural history*, ed. D. H. Janzen, 708–11. Chicago: University of Chicago Press.

Strong, D. R. 1992. Are trophic cascades all wet? Differentiation and donor-control in speciose ecosystems. *Ecology* 73:747–54.

Stuart, M. K., and M. H. Greenstone. 1990. Beyond ELISA: a rapid, sensitive, specific immunodot assay for identification of predator stomach contents. *Annals of the Entomological Society of America* 83:1101–7.

Surface, H. A. 1913. First report on the economic features of the amphibians of Pennsylvania. *Zoological Bulletin of the Division of Zoology of the Pennsylvania Department of Agriculture* 3:66–152.

Sutton, S. 1980. *Woodlice*. New York: Pergamon Press.

Sutton, S. L. 1983. The spatial distribution of flying insects in tropical rain forests. In *Tropical rain forest: Ecology and management*, ed. S. L. Sutton, T. C. Whitmore, and A. C. Chadwick, 77–91. Oxford: Blackwell Scientific Publications.

Sutton, S. L., and P. J. Hudson. 1980. The vertical distribution of small flying insects in the lowland rain forest of Zaïre. *Zoological Journal of the Linnean Society* 68:111–23.

Swanepoel, P., and H. H. Genoways. 1983. Brachyphylla cavernarum. *Mammalian Species* 205:1–6.

Sweet, M. H. 1960. The seed bugs: A contribution to the feeding habitats of the Lygaeidae (Hemiptera: Heteroptera). *Annals of the Entomological Society of America* 53:317–21.

Swift, M. J. 1977. The ecology of wood decomposition. *Science Progress, Oxford* 64:175–99.

Swift, M. J., O. W. Heal, and J. M. Anderson. 1979. *Decomposition in terrestrial ecosystems*. Berkeley: University of California Press.

Szelistowski, W. A. 1985. Unpalatability of the poison arrow frog *Dendrobates pumilio* to the ctenid spider *Cupiennius coccineus*. *Biotropica* 17:345–6.

Taigen, T. L., and F. H. Pough. 1983. Prey preference, foraging behavior, and metabolic characteristics of frogs. *American Naturalist* 122:509–20.

Taigen, T. L., and K. D. Wells. 1985. Energetics of vocalization by an anuran amphibian (*Hyla versicolor*). *Journal of Comparative Physiology* 155:163–70.

Taigen, T. L., F. H. Pough, and M. M. Stewart. 1984. Water balance of terrestrial anuran (*Eleutherodactylus coqui*) eggs: Importance of parental care. *Ecology* 65:248–55.

Tamsitt, J. R., and I. Fox. 1970a. Mites of the family Listrophoridae in Puerto Rico. *Canadian Journal of Zoology* 48:398–9.

Tamsitt, J. R., and I. Fox. 1970b. Records of bat ectoparasites from the Caribbean region (Siphonaptera, Acarina, and Diptera). *Canadian Journal of Zoology* 48: 1093–7.

Tamsitt, J. R., and D. Valdivieso. 1970. Observations on bats and their ectoparasites. In *A tropical rain forest: A study of irradiation and ecology at El Verde, Puerto Rico,* ed. H. T. Odum and R. F. Pigeon, E123–E128. Oak Ridge, Tenn.: U.S. Atomic Energy Commission.

Tanaka, K., and Y. Ito. 1982. Decrease in respiratory rate in a wolf spider, *Pardosa astrigera* (L. Koch), under starvation. *Researches on Population Biology* 24: 360–74.

Tanaka, L. K., and S. K. Tanaka. 1982. Rainfall and seasonal changes in arthropod abundance on a tropical ocean island. *Biotropica* 14:114–23.

Taylor, C. M. 1994. *Annotated checklist of the flowering plants of the El Verde Field Station, Puerto Rico.* Report to the Terrestrial Ecology Division, University of Puerto Rico.

Taylor, C. M., S. Silander, R. B. Waide, and W. J. Pfeiffer. 1996. Recovery of a tropical forest after gamma irradiation: A 23-year chronicle. In *Tropical forests: Management and ecology,* ed. A. E. Lugo and C. Lowe. *Ecological Studies.* Volume 112. New York: Springer-Verlag.

Teal, J. M. 1962. Energy flow in the salt marsh ecosystem of Georgia. *Ecology* 39: 614–24.

Telford, H. H. 1973. The Syrphidae of Puerto Rico. *Journal of Agriculture of the University of Puerto Rico* 57:217–46.

Terborgh, J. 1973. Chance, habitat and dispersal in the distribution of birds in the West Indies. *Evolution* 27:338–49.

Terborgh, J., and J. Faaborg. 1973. Turnover and ecological release in the avifauna of Mona Island, Puerto Rico. *Auk* 90:759–79.

Terborgh, J., and J. Faaborg. 1980. Saturation of bird communities in the West Indies. *American Naturalist* 116:178–95.

Terborgh, J., J. Faaborg, and H. J. Brockmann. 1978. Island colonization by Lesser Antillean birds. *Auk* 95:58–72.

Teskey, H. J. 1976. Diptera larvae associated with trees in North America. *Memoirs of the Entomological Society of Canada* 100:1–53.

Thomas, J. O. M. 1979. An energy budget for a woodland population of oribatid mites. *Pedobiologia* 19:346–78.

Thomas, M. E., 1972. Preliminary study of the annual breeding pattern and population fluctuations of bats in three ecologically distinct habitats in southwestern Colombia. Ph.D. diss., Tulane University, New Orleans.

Thomas, R. 1965. The feeding habits of captive amphisbaenids. *Herpetologica* 21:238.

Thomas, R. 1966. Additional notes on the amphisbaenids of Puerto Rico. *Breviora* 249:1–23.

Thomas, R. 1975. The *argus* group of West Indian *Sphaerodactylus* (Sauria: Gekkonidae). *Herpetologica* 31:177–95.

Thomas, R. 1985. Prey and processing in snakes of the genus *Typhlops*. (Abstract). *American Zoologist* 24:14A.

Thomas, R., and M. Leal. 1993. Feeding envenomation by *Arrhyton exiguum* (Serpentes: Colubridae). *Journal of Herpetology* 27:107–9.

Thomas, R., and J. A. Prieto Hernandez. 1985. The use of venom by the Puerto Rican snake, *Alsophis portoricensis*. *Decimo Simposio Recursos Naturales, 1983,* 13–22. San Juan, P.R.: Estado Libre Asociado de Puerto Rico. Departmento de Recursos Naturales.

Thomazini, L. I. 1974. Mycorrhiza in plants of the Cerrado. *Plant and Soil* 41: 707–11.

Thompson, F. C. 1981. The flower flies of the West Indies (Diptera: Syrphidae). *Memoirs of the Entomological Society of Washington* 9:1–200.

Tinker, P. B. 1980. Root-soil interactions in crop plants: Rhizosphere microorganisms. In *Soils and agriculture: Critical reports on applied chemistry*. Vol. 2, ed. P. B. Tinker, 1–34. Oxford: Blackwell Scientific Publications.

Tinkle, D. W., D. McGregor, and S. Dana. 1962. Home range ecology of *Uta stansburiana stejnegeri*. *Ecology* 43:223–9.

Todd, E. L. 1959. The fruit-piercing moths of the genus *Gonodonta* Hubner. U.S. Department of Agriculture Technical Bulletin, 1201.

Toft, C. A. 1980a. Feeding ecology of thirteen syntopic species of anurans in a seasonal tropical environment. *Oecologia* 45:131–41.

Toft, C. A. 1980b. Seasonal variation in populations of Panamanian litter frogs and their prey: A comparison of wetter and drier sites. *Oecologia* 47:34–38.

Toft, C. A. 1981. Feeding ecology of Panamanian litter anurans: Patterns in diet and foraging mode. *Journal of Herpetology* 15:139–44.

Toft, C. A. 1982. Community structure of litter anurans in a tropical forest, Makokow, Gabon: A preliminary analysis in the minor dry season. *Revue d'Ecologie: la Terre et la Vie* 36:223–32.

Toft, C. A. 1985. Resource partitioning in amphibians and reptiles. *Copeia* 1985: 1–21.

Toft, C. A., and T. W. Schoener. 1983. Abundance and diversity of orb spiders on 106 Bahamian islands: Biogeography at an intermediate trophic level. *Oikos* 41:411–26.

Torres, J. A. 1984a. Niches and coexistence of ant communities in Puerto Rico: Repeated patterns. *Biotropica* 16:284–95.

Torres, J. A. 1984b. Diversity and distribution of ant communities in Puerto Rico. *Biotropica* 16:296–303.

Torres, J. A. 1992. Lepidoptera outbreaks in response to successional changes after the passage of Hurricane Hugo in Puerto Rico. *Journal of Tropical Ecology* 8:285–98.

Torres, J. A., and M. Canals. 1983. Components of ant diversity and other miscellaneous notes on ants. *Science-Ciencia* 10:38–43.

Torres, J. A. 1994. Wood decomposition on *Cyrilla racemiflora* in a tropical montane forest. *Biotropica* 26:124–40.

Townsend, D. S. 1979. The relation of color polymorphism to morphology, behavior and latitudinal variation of the red-backed salamander (*Plethodon cinereus*) in Michigan. M.S. thesis, Central Michigan University, Mount Pleasant.

Townsend, D. S. 1984. The adaptive significance of male parental care in a Neotropical frog. Ph.D. diss., State University of New York at Albany.

Townsend, D. S. 1986a. The costs of male parental care and its evolution in a Neotropical frog. *Behavioral Ecology and Sociobiology* 19:187–95.

Townsend, D. S. 1986b. The ecology, population biology and reproduction of *Eleutherodactylus* frogs in eastern Jamaica. Unpublished research grant report to the National Geographic Society, Washington, D.C.

Townsend, D. S. 1989. The consequences of microhabitat choice for male reproductive success in a tropical frog (*Eleutherodactylus coqui*). *Herpetologica* 45: 451–8.

Townsend, D. S., and M. M. Stewart. 1985. Direct development in *Eleutherodactylus coqui* (Anura: Leptodactylidae): A staging table. *Copeia* 1985:423–36.

Townsend, D. S., M. M. Stewart, and F. H. Pough. 1984. Male parental care and its adaptive significance in a Neotropical frog. *Animal Behaviour* 32:421–31.

Townsend, K. V. 1985. Ontogenetic shift in habitat use by *Eleutherodactylus coqui*. M.S. thesis, State University of New York at Albany.

Traniello, J. F. A. 1981. Enemy deterrence in the recruitment strategy of a termite: Soldier-organized foraging in *Nasutitermes costalis*. *Proceedings of the National Academy of Sciences of the United States of America* 78:1976–9.

Traniello, J. F. A. 1982. Recruitment and orientation components in a termite trail pheromone. *Naturwissenschaften* 69:343–5.

Travers, J. R. 1938. Mayflies of Puerto Rico. *Journal of Agriculture of the University of Puerto Rico* 33:5–40.

Trojanowski, J., K. Haider, and A. Huettermann. 1984. Decomposition of carbon-14-labeled lignin holocellulose and lignocellulose by mycorrhizal fungi. *Archiv für Microbiologie* 139:202–6.

Turnbull, A. L. 1960a. The prey of the spider *Linyphia triangularis* (Clerck) (Araneae, Linyphiidae). *Canadian Journal of Zoology* 38:859–73.

Turnbull, A. L. 1960b. The spider population of a stand of oak (*Quercus robur* L.) in Wytham Woods, Berks., England. *Canadian Entomologist* 92:110–24.

Turnbull, A. L. 1973. Ecology of the true spiders (Araneomorphae). *Annual Review of Entomology* 18:305–48.

Turner, F. B., and C. S. Gist. 1970. Observations of lizards and tree frogs in an irradiated Puerto Rican forest. In *A tropical rain forest: A study of irradiation and ecology at El Verde, Puerto Rico,* ed. H. T. Odum and R. F. Pigeon, E25–E49. Oak Ridge, Tenn.: U.S. Atomic Energy Commission.

Tuttle, M. D., and M. J. Ryan. 1981. Bat predation and the evolution of frog vocalizations in the Neotropics. *Science* 214:677–8.

U.S. Forest Service. 1973, 1976, 1981, 1982, and 1984. *Forest Pest Management.* (Puerto Rico) U.S. Department of Agriculture, Forest Service, State & Private, Southern Area, Asheville Office.

Ubelaker, J. E., R. D. Specian, and D. W. Duseynski. 1977. Endoparasites. In *Biology of bats of New World family Phyllostomatidae,* Part II, ed. R. J. Baker, J. K. Jones, Jr., and D. C. Carter, 7–56. Special Publications. The Museum, Texas Tech University 13. Lubbock: Texas Tech Press.

Uetz, G. W. 1975. Temporal and spatial variation in species diversity of wandering spiders (Araneae) in deciduous forest litter. *Environmental Entomology* 4:719–24.

Uetz, G. W. 1979. The influence of variation in litter habitats on spider communities. *Oecologia* (Berlin) 40:29–42.

Uetz, G. W., A. D. Johnson, and D. W. Schemske, 1978. Web placement, web structure and prey capture in orb-weaving spiders. *Bulletin of the British Arachnological Society* 4:141–8.

Underwood, G. 1951. Reptilian retinas. *Nature* 167:183–5.

Van Berkum, F. H., F. H. Pough, M. M. Stewart, and P. F. Brussard. 1982. Altitudinal and interspecific differences in the rehydration abilities of Puerto Rican frogs (*Eleutherodactylus*). *Physiological Zoology* 55:130–6.

Van der Schalie, H. 1948. The land and fresh-water molluscs of Puerto Rico. Miscellaneous Publications. *University of Michigan Museum of Zoology 70. Ann Arbor: University of Michigan Press.*

Van Hook, R. I., Jr. 1971. Energy and nutrient dynamics of spider and orthopteran populations in a grassland ecosystem. *Ecological Monographs* 41:1–26.

Van Hook, R. I., Jr., M. G. Nielsen, and. H. H. Shugart. 1980. Energy and nitrogen relations for a *Macrosiphum liriodendri* (Homoptera: Aphididae) population in an east Tennessee *Liriodendron tulipifera* stand. *Ecology* 61:960–75.

Van Veen, J. A., and E. A. Paul. 1979. Conversion of biovolume measurements of soil organisms, grown under various moisture tensions, to biomass and their nutrient content. *Applied Environmental Microbiology* 37:686–92.

Velez, M. J. 1967. Checklist of the terrestrial and freshwater Decapoda of Puerto Rico. *Caribbean Journal of Science* 7:41–44.

Velez, M. J. 1979. Bibliografia selecta de la fauna y las comunidades naturales de Puerto Rico. Parte Tercera. *Science-Ciencia* 6:189–220.

Vial, J. L. 1968. The ecology of the tropical salamander, *Bolitoglossa subpalmata,* in Costa Rica. *Revista de Biologia Tropical* 15:13–115.

Villa, J., and D. S. Townsend. 1983. Viable frog eggs eaten by phorid fly larvae. *Journal of Herpetology* 17:278–81.

Villalobos-Figueroa, A. 1982. Decapoda. In *Aquatic biota of Mexico, Central America, and the West Indies,* ed. S. H. Hurlbert and A. Villalobos-Figueroa, 215–39. San Diego, Calif.: S. H. Hurlbert, Dept. of Biology, San Diego State University.

Villamil, J., and R. G. Clements. 1976. Some aspects of the ecology of the freshwater shrimps in the Upper Espiritu Santo River at El Verde, Puerto Rico. PRNC-206. Rio Piedras: Puerto Rico Nuclear Center.

Vitt, L. J. 1983. Ecology of an anuran-eating guild of terrestrial tropical snakes. *Herpetologica* 39:52–66.

Vogt, K. A., D. J. Vogt, P. Boon, A. P. Covich, F. N. Scatena, H. Asbjornsen, T. Siccama, and J. Bloomfield. Short- and long-term dynamics of above- and belowground litter transfers along topographic gradients following Hurricane Hugo, Luquillo Experimental Forest, Puerto Rico. *Psyche* (in review).

Wadsworth, F. H. 1949. The development of the forest land resources of the Luquillo Mountains, Puerto Rico. Ph.D. diss., University of Michigan, Ann Arbor.

Wadsworth, F. H. 1951. Forest management in the Luquillo Mountains. I. The setting. *Caribbean Forester* 11:93–132.

Wadsworth, R. K. 1970. Point-quarter sampling of forest type-site relations at El Verde. In *A tropical rain forest: A study of irradiation and ecology at El Verde, Puerto Rico,* ed. H. T. Odum and R. F. Pigeon, B97–B104. Oak Ridge, Tenn.: U.S. Atomic Energy Commission.

Waide, R. B. 1981. Interactions between resident and migrant birds in southern Campeche, Mexico. *Tropical Ecology* 22:134–54.

Waide, R. B. 1987. The fauna of Caribbean island ecosystems: Community structure and conservation. *Acta Cientifica* 1:64–71.

Waide, R. B. 1991a. Summary of response of animal populations to hurricanes in the Caribbean. *Biotropica* 23:508–12.

Waide, R. B. 1991b. The effect of Hurricane Hugo on bird populations in the Luquillo Experimental Forest, Puerto Rico. *Biotropica* 23:475–80.

Waide, R. B., and A. E. Lugo. 1992. A research perspective on disturbance and recovery of a tropical montane forest. In *Tropical forests in transition: Ecology of natural and anthropogenic disturbance processes,* ed. J. G. Goldammer, 173–90. Basel: Birkhäuser Verlag.

Waide, R. B., and P. M. Narins. 1988. Tropical forest bird counts and the effect of sound attenuation. *Auk* 105:296–302.

Waide, R. B., and D. P. Reagan. 1983. Competition between West Indian anoles and birds. *American Naturalist* 121:133–8.

Walker, L. R. 1991. Tree damage and recovery from Hurricane Hugo in Luquillo Experimental Forest, Puerto Rico. *Biotropica* 23:379–85.

Walker, L. R. 1995. Timing of post-hurricane tree mortality in Puerto Rico. *Journal of Tropical Ecology* 11:315–20.

Walker, L. R. 1994. The effects of fern thickets on tree seedlings in landslides in Puerto Rico. *Journal of Vegetation Science* 5:525–32.

Walker, L. R., N. V. L. Brokaw, D. J. Lodge, and R. B. Waide. 1991. Ecosystem, plant and animal responses to hurricanes in the Caribbean. *Biotropica* 23: 313–521.

Wallace, A. R. 1858. On the tendency of varieties to depart indefinitely from the original type. *Proceedings of the Linnaean Society of London* 3:53–62.

Wallace, A. R. 1880. *Island life.* London: Macmillan and Co.

Wallwork, J. A. 1967. Acari. In *Soil biology,* ed. W. A. Burges and F. Raw, 363–95. New York: Academic Press.

Wallwork, J. A. 1983. Oribatids in forest ecosystems. *Annual Review of Entomology* 28:109–30.

Walter, D. E., D. T. Kaplan, and T. A. Permar. 1991. Missing links: A review of methods used to estimate trophic links in soil food webs. *Agriculture, Ecosystems and Environment* 34:399–405.

Walter, H. 1971. Ecology of tropical and subtropical vegetation. New York: Van Nostrand Reinhold Co.

Wanner, H. 1970. Soil respiration, litter fall and productivity of tropical rain forest. *Journal of Ecology* 58:543–7.

Warren, P. H. 1989. Spatial and temporal variation in the structure of a freshwater food web. *Oikos* 55:299–311.

Warren, P. H., and A. H. Lawton. 1987. Invertebrate predator-prey body size relationships: an explanation for upper triangular food webs and patterns in food web structure? *Oecologia* (Berlin) 74:231–5.

Watanabe, H. 1980. A study of the three species of isopods in an evergreen broad-leaved forest in southwestern Japan. *Revue d'Ecologie et de Biologie du Sol* 17: 229–39.

Watkinson, S. C. 1984. Morphogenesis of the *Serpula lacrimans* colony in relation to its functions in nature. In *The ecology and physiology of the fungal mycelium*, ed. D. H. Jennings and A. D. M. Rayner, 165–84. Cambridge: Cambridge University Press.

Weaver, P. L., and P. G. Murphy. 1990. Forest structure and productivity in Puerto Rico's Luquillo Mountains. *Biotropica* 22:69–82.

Webb, D. P. 1977. Regulation of deciduous forest litter decomposition by soil arthropod feces. In *The role of arthropods in forest ecosystems*, ed. W. J. Mattson, 57–69. New York: Springer-Verlag.

Webb, J. P., Jr., and R. B. Loomis. 1977. Ectoparasites. In *Biology of bats of the New World family Phyllostomatidae*. Part II, ed. R. J. Baker, J. K. Jones, Jr., and D. C. Carter, 57–120. Special Publications. *The Museum, Texas Tech University 13*. Lubbock: Texas Tech Press.

Webster, J. R., and E. F. Benfield. 1986. Vascular plant breakdown in freshwater ecosystems. *Annual Review of Ecology and Systematics* 17:567–94.

Weinbren, M. P., B. M. Weinbren, W. B. Jackson, and J. B. Villella. 1970. Studies on the roof rat (*Rattus rattus*) in the El Verde forest. In *A tropical rain forest: A study of irradiation and ecology at El Verde, Puerto Rico*, ed. H. T. Odum and R. F. Pigeon, E169–E181. Oak Ridge, Tenn.: U.S. Atomic Energy Commission.

Weiner, J., and Z. Glowacinski. 1975. Energy flow through a bird community in a deciduous forest in southern Poland. *Condor* 77:233–42.

Welgolaski, F., ed. 1975. *Fennoscandian tundra ecosystems: analysis and systems ecology*. Berlin: Springer-Verlag.

Went, F. W., and N. Stark. 1968. Mycorrhiza. *BioScience* 18:1035–9.

Werner, E. E., and J. F. Gilliam. 1984. The ontogenetic niche and species interactions in size-structured populations. *Annual Review of Ecology and Systematics* 15:393–425.

Werner, R. A. 1976. Role of aboveground invertebrate animals in nutrient cycling and decomposition. In *The structure and function of a black spruce* (Picea mariana *[Mill] B.S.P.*) *forest in relation to other fire-affected taiga ecosystems*, K. van Cleve and T. Dryness, principal investigators, 111–6. Unpublished progress report NSF BMS 75-13998.

Westercamp, D., P. Andreieff, P. Bouysse, A. Mascle, and J. C. Baubron. 1985a. Geologie de l'archipel des Grenadines (Petites Antilles meridionales). *Documents du Bureau de Recherches Geologiques et Minieres* 92.

Westercamp, D., P. Andreieff, P. Bouysse, A. Mascle, and J. C. Baubron. 1985b. The Grenadines, southern Lesser Antilles. Part I. Stratigraphy and volcano-structural evolution. In *Caribbean geodynamics,* ed. A. Mascle, 109–18. Paris: Editions Technip.

Wetmore, A. 1916. Birds of Puerto Rico. U.S. Department of Agriculture Bulletin 326. Washington, D.C.: U.S. Government Printing Office.

Wetmore, A. 1927. The birds of Porto Rico and the Virgin Islands. In *Scientific survey of Porto Rico and the Virgin Islands.* Vol. 9, Pts. 3–4, 245–571. New York: New York Academy of Sciences.

Wever, E. G., and C. Gans. 1973. The ear in Amphisbaenia (Reptilia); further anatomical observations. *Journal of Zoology* 171:189–206.

Weygoldt, P. 1969. Beobachtungen zur fortpflanzensbiologie und zum verhalten der geisselspinne *Tarantula marginemaculata* C. L. Koch (Chelicerata, Amblypygi). *Zeitschrift für Morphologie der Tiere* 64:338–60.

Weygoldt, P. 1970. Lebenszyklus und postembryonale entwicklung der geisselspinne *Tarantula marginemaculata* C. L. Koch (Chelicerata, Amblypygi) im laboratorium. *Zeitschrift für Morphologie der Tiere* 67:58–85.

Weygoldt, P. 1975. Untersuchungen zur embryologie und morphologie der geisselspinne *Tarantula marginemaculata* C. L. Koch (Arachnida, Amblypygi, Tarantulidae). *Zoomorphologie* 82:137–99.

Wheeler, M. R., and H. Takala. 1963. A revision of the American species of Hycodrosophila (Diptera: Drosophilidae). *Annals of the Entomological Society of America* 56:292–9.

Wheeler, Q., and M. Blackwell, eds. 1984. *Fungus-insect relationships: Perspectives in ecology and evolution.* New York: Columbia University Press.

Wheeler, W. M. 1908. The ants of Porto Rico and the Virgin Islands. *Bulletin of the American Museum of Natural History* 24:117–58.

Wheeler, W. M. 1936. Ecological relations of ponerine and other ants to termites. *Proceedings of the American Academy of Arts and Sciences* 71:159–243.

White, L. R., R. Powell, J. S. Parmerlee, Jr., A. Lathrop, and D. D. Smith. 1992. Food habits of three syntopic reptiles from the Barahona Peninsula of Hispaniola. *Journal of Herpetology* 26:518–20.

Whitford, W. G., Y. Steinberger, and G. Ettershank. 1982. Contributions of subterranean termites to the "economy" of Chihuahuan desert ecosystems. *Oecologia* 55:298–302.

Whitham, T. G., and S. Mopper. 1985. Chronic herbivory: Impacts on architecture and sex expression of pinyon pine. *Science* 228:1089–91.

Wickler, W. 1968. *Mimicry in plants and animals.* New York: McGraw-Hill.

Wiegert, R. G. 1965. Leaf fall, decomposition, and litter animals. In *The rain forest project annual report,* 162–65. Rio Piedras: Puerto Rico Nuclear Center.

Wiegert, R. G. 1970a. Effects of ionizing radiation on leaf fall, decomposition, and litter microarthropods of a montane rain forest. In *A tropical rain forest: A study of irradiation and ecology at El Verde, Puerto Rico,* ed. H. T. Odum and R. F. Pigeon, H89–H100. Oak Ridge, Tenn.: U.S. Atomic Energy Commission.

Wiegert, R. G. 1970b. Energetics of the nest building termite *Nasutitermes costalis* (Holm.) in a Puerto Rican forest. In *A tropical rain forest: A study of irradiation and ecology at El Verde, Puerto Rico,* ed. H. T. Odum and R. Pigeon, I57–I64. Oak Ridge, Tenn.: U.S. Atomic Energy Commission.

Wiegert, R. G., and D. C. Coleman. 1970. Ecological significance of low oxygen consumption and high fat accumulation by *Nasutitermes costalis* (Isoptera: Termitidae). *BioScience* 20:663–5.

Wiegert, R. G., and P. Murphy. 1970. Effects of season, species and location on the disappearance rate of leaf litter in a Puerto Rican rain forest. In *A tropical rain forest: A study of irradiation and ecology at El Verde, Puerto Rico,* ed. H. T. Odum and R. F. Pigeon, H101–H104. Oak Ridge, Tenn.: U.S. Atomic Energy Commission.

Wiens, J. A. 1977. Model estimation of energy flow in North American grassland bird communities. *Oecologia* 31:135–51.

Wieser, W. 1978. Consumer strategies of terrestrial gastropods and isopods. *Oecologia* (Berlin) 36:191–201.

Wiggins, G. B. 1977. Larvae of the North American caddisfly genera. Toronto: University of Toronto Press.

Wiggins, G. B. 1978. Trichoptera. In *An introduction to the aquatic insects,* ed. R. W. Merrit and K. W. Cummins, 147–85. Dubuque, Iowa: Kendall/Hunt Publishers.

Wignarajah, S., and J. Phillipson. 1977. Numbers and biomass of centipedes (Lithobiomorpha: Chilopoda) in a *Betula-Alnus* woodland in N. E. England. *Oecologia* (Berlin) 31:55–66.

Wiley, J. W. 1991. Ecology and behavior of the Zenaida Dove. *Ornitologica Neotropical* 2:49–75.

Wiley, J. W., and G. P. Bauer. 1985. Caribbean National Forest, Puerto Rico. *American Birds* 39:12–18.

Will, T. 1991. Birds of a severely hurricane-damaged Atlantic coast rain forest in Nicaragua. *Biotropica* 23:497–507.

Williams, E. C., Jr. 1941. An ecological study of the floor fauna of the Panamanian rain forest. *Bulletin of the Chicago Academy of Sciences* 6:63–124.

Williams, E. E. 1969. The ecology of colonization as seen in the zoogeography of anoline lizards on small islands. *Quarterly Review of Biology* 44:345–89.

Williams, E. E. 1972. The origin of faunas: Evolution of lizard congeners in a complex island fauna: A trial analysis. In *Evolutionary biology,* Vol. 6, ed. T. M. Dobzhansky, M. Hecht, and W. Steere, 47–89. New York: Appleton-Century-Crofts.

Williams, E. E. 1976. West Indian anoles: A taxonomic and evolutionary summary. 1. Introduction and a species list. *Breviora* 440:1–21.

Williams, E. E. 1983. Ecomorphs, faunas, island size, and diverse end points in island radiations of *Anolis*. In *Lizard ecology: Studies of a model organism,* ed. R. B. Huey, E. R. Pianka, and T. W. Schoener, 326–70. Cambridge, Mass.: Harvard University Press.

Williams, E. H. 1980. Disjunct distributions of two aquatic predators. *Limnology and Oceanography* 25:99–1006.

Williams, G. 1962. Seasonal and diurnal activity of harvestman (Phalangida) and spiders (Araneida) in contrasted habitats. *Journal of Animal Ecology* 31:23–42.

Williamson, M. 1981. *Island populations.* Oxford: Oxford University Press.

Willig, M. R. 1982. A comparative study of Caatingas and Cerrado chiropteran communities: Composition, structure, morphometrics, and reproduction. Ph.D. diss. University of Pittsburgh, Pittsburgh.

Willig, M. R. 1983. Composition, microgeographic variation, and sexual dimorphism in Caatingas and Cerrado bat communities from northeast Brazil. *Bulletin of the Carnegie Museum of Natural History* 23:1–131.

Willig, M. R. 1985a. Reproductive activity in female bats from northeast Brazil. *Bat Research News* 26:17–20.

Willig, M. R. 1985b. Reproductive pattern in bats from Caatingas and Cerrado biomes of northeast Brazil. *Journal of Mammalogy* 66:668–81.

Willig, M. R. and A. Bauman. 1984. Notes on bats from the Luquillo Mountains of Puerto Rico. CEER-T-194:1–12

Willig, M. R., and G. R. Camilo. 1991. The effect of Hurricane Hugo on six invertebrate species in the Liquillo Experimental Forest of Puerto Rico. *Biotropica* 23: 455–61.

Willig, M. R. and M. A. Mares. 1989. Mammals from the Caatinga: an updated list and summary of recent research. *Revista Brasileira de Biologia* 49:361–7.

Willig, M. R., R. W. Garrison, and A. J. Bauman. 1986. Population dynamics and natural history of a Neotropical walking stick, *Lamponius portoricensis* Rehn (Phasmatodea: Phasmatidae). *Texas Journal of Science* 38:121–37.

Willig, M. R., G. R. Camilo, and S. J. Noble. 1993. Dietary overlap in frugivorous and insectivorous bats from edaphic Cerrado habitats of Brazil. *Journal of Mammalogy* 74:117–28.

Willig, M. R., E. A. Sandlin, and M. R. Gannon. 1993. Structural and taxonomic components of habitat selection in the Neotropical folivore, *Lamponius portoricensis* (Phasmatodea: Phasmatidae). *Environmental Entomology* 22:634–41.

Willis, E. O. 1966. The role of migrant birds at swarms of army ants. *Living Bird* 5:187–231.

Willis, E. O. 1976. Seasonal changes in the invertebrate litter fauna on Barro Colorado Island, Panama. *Revista Brasileira de Biologia* 36:643–57.

Willis, E. O. 1980. Ecological roles of migratory and resident birds on Barro Colorado Island, Panama. In *Migrant birds in the Neotropics: Ecology, behavior, distribution and conservation,* ed. A. Keast and E. S. Morton, 205–26. Washington, D.C.: Smithsonian Institution Press.

Willson, M. F. 1973. Tropical plant production and animal species diversity. *Tropical Ecology* 14:62–65.

Wilson, D. E. 1979. Reproductive patterns. In *Biology of bats of the New World family Phyllostomatidae,* Part III, ed. R. J. Baker, J. K. Jones, Jr., and D. C. Carter, 317–78. Special Publications. *The Museum, Texas Tech University* 16. Lubbock: Texas Tech Press.

Wilson, D. E. 1983. Checklist of mammals. In *Costa Rican natural history,* ed. D. H. Janzen, 443–7. Chicago: University of Chicago Press.

Wilson, D. E. 1990. Mammals of La Selva, Costa Rica. In *Four Neotropical rainforests,* ed. A. H. Gentry, 273–86. New Haven, Conn.: Yale University Press.

Wilson, D. S. 1976. Deducing the energy available in the environment: An application of optimal foraging theory. *Biotropica* 8:96–103.

Wilson, E. O. 1959. Some ecological characteristics of ants in New Guinea rain forests. *Ecology* 40:437–47.

Wilson, E. O. 1987. The arboreal ant fauna of Peruvian Amazon forests: A first assessment. *Biotropica* 19:245–51.

Wilson, E. O. 1988. The current state of biological diversity. In *Biodiversity,* ed. E. O. Wilson and F. M. Peter, 3–18. Washington, D.C.: National Academy Press.

Winemiller, K. O. 1983. An introduction to the freshwater fish communities of Corcovado National Park, Costa Rica. *Brenesia* 21:47–66.

Winemiller, K. O. 1989a. Must connectance decrease with species richness? *American Naturalist* 134:960–8.

Winemiller, K. O. 1989b. Ontogenetic diet shifts and resource partitioning among piscivorous fishes in the Venezuelan llanos. *Environmental Biology of Fishes* 26:177–99.

Winemiller, K. O. 1990. Spatial and temporal variation in tropical fish trophic networks. *Ecological Monographs* 60:331–67.

Wise, D. H. 1984. The role of competition in spider communities: Insights from the field experiments with a model organism. In *Ecological communities: Conceptual issues and the evidence,* ed. D. R. Strong, Jr., D. Simberloff, L. G. Abele, and A. B. Thistle, 42–53. Princeton, N.J.: Princeton University Press.

Witkamp, M. 1970. Aspects of soil microflora in a gamma-irradiated rain forest. In *A tropical rain forest: A study of irradiation and ecology at El Verde, Puerto Rico,* ed. H. T. Odum and R. F. Pigeon, F29–F33. Oak Ridge, Tenn.: U.S. Atomic Energy Commission.

Witkamp, M. 1974. Direct and indirect counts of fungi and bacteria as indexes of microbial mass and productivity. *Soil Science* 118:150–5.

Wolcott, G. N. 1924. The food of Porto Rican lizards. *Journal of the Department of Agriculture of Porto Rico* 7:5–37.

Wolcott, G. N. 1948. The insects of Puerto Rico. *Journal of Agriculture of the University of Puerto Rico* 32:1–975.

Wolcott, G. N. 1953. The food of the mongoose (*Herpestes javanicus auropunctatus* Hodgson) in St. Croix and Puerto Rico. *Journal of Agriculture of the University of Puerto Rico* 37:241–7.

Wolda, H. 1978a. Fluctuations in abundance of tropical insects. *American Naturalist* 112:1017–45.

Wolda, H. 1978b. Seasonal fluctuations in rainfall, food and abundance of tropical insects. *Journal of Animal Ecology* 47:369–81.

Wolda, H. 1979. Abundance and diversity of Homoptera in the canopy of a tropical forest. *Ecological Entomology* 4:181–90.

Wolda, H. 1980a. Fluctuationes estacionales de insectos en el tropico: Sphingidae. In *Memorias del VI Congreso de la Sociedad Colombiana de Entomologia "SOCOLEN"* (July 1979), 10–58. Cali, Colombia: SOCOLEN.

Wolda, H. 1980b. Seasonality of tropical insects. I. Leafhoppers (Homoptera) in Las Cumbres, Panama. *Journal of Animal Ecology* 49–277–90.

Wolda, H. 1982. Seasonality of leafhoppers (Homoptera) on Barro Colorado Island, Panama. In *Ecology of a tropical forest,* ed. E. G. Leigh, Jr., A. S. Rand, and D. M. Windsor, 319–30. Washington, D.C.: Smithsonian Institution Press.

Wolda, H. 1983a. Diversity, diversity indices and tropical cockroaches. *Oecologia* (Berlin) 58:290–98.

Wolda, H. 1983b. Spatial and temporal variation in abundance in tropical animals. In *Tropical rain forest: Ecology and management,* ed. S. L. Sutton, T. C. Whitmore, and A. C. Chadwick, 93–105. Oxford: Blackwell Scientific Publications.

Wolda, H., and R. W. Flowers. 1985. Seasonality and diversity of mayfly adults (Ephemeroptera) in a "nonseasonal" tropical environment. *Biotropica* 17:330–5.

Wolda, H., and R. Foster. 1978. *Zunacetha annulata* (Lepidoptera: Dioptidae), an outbreak insect in a Neotropical forest. *Geo-Eco-Trop* 2:443–54.

Wood, P. A., and P. D. Gabbutt. 1978. Seasonal vertical distribution of pseudoscorpions in beech litter. *Bulletin of the British Arachnological Society* 4:176–83.

Wood, S. L. 1982. The bark and ambrosia beetles of North and Central America (Coleoptera, Scolytidae): A taxonomic monograph. *Great Basin Naturalist Memoirs* 6. Provo, Utah: Brigham Young University.

Wood, S. L. 1983. *Scolytodes atratus panamensis* (Escarabajito de Guarumo, Cecropia Petiole Borer). In *Costa Rican natural history,* ed. D. H. Janzen, 768–9. Chicago: University of Chicago Press.

Wood, T. G. 1978. Food and feeding habits of termites. In *Production ecology of ants and termites,* ed. M. V. Brian, 55–80. International Biological Programme 13. New York: Cambridge University Press.

Wood, T. G., and W. A. Sands. 1978. The role of termites in ecosystems. In *Production ecology of ants and termites,* ed. M. V. Brian, 245–92. International Biological Programme 13. New York: Cambridge University Press.

Wood, T. G., R. A. Johnson, S. Bacchus, M. O. Shittu, and J. M. Anderson. 1982. Abundance and distribution of termites (Isoptera) in a riparian forest in the southern Guinea savanna vegetation zone of Nigeria. *Biotropica* 14:25–39.

Woodring, J. P., and M. S. Blum. 1965. The anatomy, physiology, and comparative aspects of the repugnatorial glands of *Orthcricus arboreus* (Diplopoda: Spirobolida). *Journal of Morphology* 116:99–108.

Woods, C. A. 1989a. The biogeography of West Indian rodents. In *Biogeography of the West Indies: Past, present and future,* ed. C. A. Woods, 741–98. Gainesville, Fla.: Sandhill Crane Press.

Woods, C. A. 1989b. The land mammals of Madagascar and the Greater Antilles: Comparison and analysis. In *Biogeography of the West Indians: Past, present and future,* ed. C. A. Woods, 799–826. Gainesville, Fla.: Sandhill Crane Press.

Woods, F. W., and C. M. Gallegos. 1970. Litter accumulation in selected forests of the Republic of Panama. *Biotropica* 2:46–60.

Woods, L. E., C. V. Cole, E. T. Elliot, R. V. Anderson, and D. C. Coleman. 1982. Nitrogen transformations in soil as affected by bacterial-microfaunal interactions. *Soil Biology and Biochemistry* 14:93–98.

Woodwell, G. M., and R. H. Whittaker. 1968. Primary production in terrestrial ecosystems. *American Zoologist* 8:19–30.

Woolbright, L. L. 1985a. Patterns of nocturnal movement and calling by the tropical frog *Eleutherodactylus coqui. Herpetologica* 41:1–9.

Woolbright, L. L. 1985b. Sexual dimorphism in body size of the subtropical frog, *Eleutherodactylus coqui.* Ph.D. diss., State University of New York at Albany.

Woolbright, L. L. 1989. Sexual dimorphism in *Eleutherodactylus coqui:* Selection pressures and growth rates. *Herpetologica* 45:68–74.

Woolbright, L. L. 1991. The impact of Hurricane Hugo on forest frogs in Puerto Rico. *Biotropica* 23:462–7.

Woolbright, L. L. The effect of artificial gaps on the herpetofauna of a Puerto Rican forest. *Conservation Biology* (in review).

Woolbright, L. L., and M. M. Stewart. 1987. Foraging success of the tropical frog, *Eleutherodactylus coqui:* The cost of calling. *Copeia* 1987:69–75.

Wray, D. L. 1953. New Collembola from Puerto Rico. *Journal of Agriculture of the University of Puerto Rico* 37:140–50.

Wright, S. 1921. Correlation and causation. *Journal of Agricultural Research* 20: 557–85.

Wright, S. J. 1981. Extinction-mediated competition: The *Anolis* lizards and insectivorous birds of the West Indies. *American Naturalist* 117:181–92.

Wunderle, J. M., D. J. Lodge, and R. B. Waide. 1992. Short-term effects of Hurricane Gilbert on terrestrial bird populations in Jamaica. *Auk* 109:148–66.

Wunderle, J. M., Jr. 1992. Sexual habitat segregation in wintering black-throated blue warblers in Puerto Rico. In *Ecology and conservation of Neotropical migrant land birds,* ed. J. M. Hagan III and D. W. Johnston, 299–307. Washington, D.C.: Smithsonian Institution Press.

Wunderle, J. M., Jr., A. Díaz, I. Velázquez, and R. Scharron. 1987. Forest openings and the distribution of understory birds in a Puerto Rican rainforest. *Wilson Bulletin* 99:22–37.

Yodzis, P. 1984. How rare is omnivory? *Ecology* 65:321–3.

You, C. 1991. Population dynamics of *Manilkara bidentata* (A.DC.) Cher. in the Luquillo Experimental Forest, Puerto Rico. Ph.D. diss., University of Tennessee, Knoxville.

You, C., and W. H. Petty. 1991. Effects of Hurricane Hugo on *Manikara bidentata,* a primary three species in the Luquillo Experimental Forest. *Biotropica* 23: 400–406.

Young, A. M. 1980. Observations on feeding aggregations of *Orthemis ferruginea* (Fabricus) in Costa Rica (Anisoptera: Liebellulidae). *Odonatologica* 9:325–8.

Young, D. A. 1953. Empoascan leafhoppers of the *Solana* group with descriptions of new species. *Journal of Agriculture of the University of Puerto Rico* 37:151–60.

Zaret, T. M., and A. S. Rand. 1971. Competition in tropical stream fishes: Support for the competitive exclusion principle. *Ecology* 52:336–42.

Zimka, J. R. 1974. Predation of frogs, *Rana arvalis* Nilss., in different forest site conditions. *Ekologia Polska* 22:31–63.

Zimmerman, B. L., and M. T. Rodrigues. 1990. Frogs, snakes, and lizards of the INPA-WWF reserves near Manaus, Brazil. In *Four Neotropical rainforests,* ed. A. H. Gentry, 426–54. New Haven, Conn.: Yale University Press.

Zimmerman, J. K., E. M. Everham, III, R. B. Waide, D. J. Lodge, C. M. Taylor, and N. V. L. Brokaw. 1994. Responses of tree species to hurricane winds in subtropical wet forest in Puerto Rico: Implications of tropical tree life histories. *Journal of Ecology* 82:911–22.

Zimmerman, J. K., W. M. Pulliam, D. J. Lodge, V. Quiñones-Orfila, N. Fetcher, R. B. Waide, S. Guzmán-Grajales, J. A., Parrotta, C. E. Asbury, and L. R. Walker. 1995. Nitrogen immobilization by decomposing woody debris and the recovery of tropical wet forest from hurricane damage. *Oikos* 72:316–22.

Zou, X., C. P. Zucca, R. B. Waide, W. H. McDowell. 1995. Long-term influence of deforestation on tree species composition and litter dynamics of a tropical rain forest in Puerto Rico. *Forest Ecology and Management* 78:147–57.

Index

Figures and tables are denoted by "f" or "t" following page references.